An Introduction to Physical Biochemistry

An Introduction to Physical Biochemistry

Henry B. Bull, Ph.D.

Research Professor of Biochemistry
University of Iowa
Iowa City, Iowa

Edition 2

F. A. Davis Company, Philadelphia, Pa.

 Manufactured in the United States of America. Library of Congress Catalog Card Number 70-123784. ISBN 0-8036-1400-4

Preface

This book is a guide to that broad area of physical chemistry related to biochemistry; it could as well be titled "Outlines of Physical Biochemistry." The purpose is to bring to the student of the biological sciences a picture of physical chemistry which is meaningful to him. An experimental rather than a theoretical approach has been adopted.

It has been six years since the appearance of the first edition of *An Introduction to Physical Biochemistry,* and during this time there have been significant developments in many of the topics presented. Whereas all of the chapters have received attention and have been modified to some extent, important changes have been made in the chapters Electrolytes and Water, Acids and Buffers, Biopolymers, Osmotic Pressure and Related Topics, and Solution Optics. The appearance of a new edition has also given me the opportunity of simplifying and clarifying much of the older material.

Some years ago solution theory was in the mainstream of physical chemistry. Examination of the current issues of the *Journal of Physical Chemistry* and *Journal of Chemical Physics* reveals that solution theory has been relegated to a minor role, and other matters are of concern to the professional physical chemist. On the other hand, solution theory and physical experiments done in solution remain of great interest to the biochemist. I have attempted to respond to the needs of the biochemist in the treatment of physical chemistry and, accordingly, have focused on events occurring in aqueous solutions.

Many people have been involved in the writing of this book. First, there are the many students who have contributed through the years and whose names are too numerous to list. Keith Breese did the figures and was otherwise of great help. A number of individuals have called my attention to typographical and other kinds of errors in the first edition, and I am most grateful for their help. Especially useful have been the suggestions of Dr. Lewis B. Barnett.

I also wish to thank Drs. Dipti K. Chattoraj and Santibrata Ghosh for their help.

HENRY B. BULL

Table of Contents

PAGE

CHAPTER 1. MATHEMATICAL REVIEW. 1

Numbers and Exponents 1
Solution of Equations 3
Differential Calculus and Rates of Change 8
Integration 10
Differential Equations 13
Expansion Series 14
Trigonometric Functions 16
Probability and Error 21
Method of Least Squares 26
Dimensional Analysis 27
Comments on Computers 29

CHAPTER 2. ENERGETICS 32

Heat 32
Enthalpy 37
Work 38
Survey of Thermodynamics. 40
Free Energy 43
Free Energy and Temperature 46
Evaluation of Free Energies 47
Solutions 49
Partial Molar Quantities. 50
Activity and Choice of Standard States 51
ΔF of Chemical Reactions 52
The Fuel Supply 56

Coupled Reactions 58
Organic Phosphates 59
ΔF for Hydrolyses of ATP 60
Experimental ATP–ΔF Values 61
Energetics in Biology 63

CHAPTER 3. ELECTROLYTES AND WATER 67
Ions 67
Ionic Strength 67
Activity of Electrolytes 69
Elementary Electrostatics 69
Ionic Interaction 71
Ionic Activity 74
Water 76
Ion Hydration 80
Hofmeister (Lyotropic) Ion Series 83
Ions in Biology 84
Ionic Complexes 86

CHAPTER 4. OXIDATION-REDUCTION POTENTIALS 89
Equation for a Chemical Cell 91
Membrane Electrodes 93
Concentration Cells 94
Amperometric Methods 94
Soluble Oxidation-Reduction Systems 96
Influence of Hydrogen Ions 97
Stepwise Oxidations 98
Biological Oxidations 100

CHAPTER 5. ACIDS AND BUFFERS 104
Ionization of Water 105
Titration Curves 106
Ionization of Weak Acids 108
Temperature and Ionization 110
Buffer Capacity 111
Practical Buffers 114
Blood Buffers and Respiration 115
Multiple Binding 117
Electrostatic Effects 120
Amino Acids 122
Protein Titration 126
Titration of Nucleic Acids 130
Ion Binding in General 130

CHAPTER 6. BIOPOLYMERS 134
Hydrogen Bonds 136
Hydrophobic Interactions 137
Van der Waal Forces 139
Electrical Effects 139
Peptide Structure 139
Molecular Model Building 141
Native Protein Structure 144

Hemoglobin. 145
Protein Denaturation 148
Solubility of Proteins 150
Nucleic Acids 152

CHAPTER 7. OSMOTIC PRESSURE AND RELATED TOPICS 158
Vapor Pressure. 158
Measurement of Vapor Pressure 161
Isopiestic Method 161
Heat of Vaporization 165
Freezing Point Depression 166
Osmotic Pressure 167
Osmotic Pressure Measurements 169
Departure from Ideality 171
Associating Systems 173
Donnan Equilibrium 174
Donnan Potential 178
Osmosis in Biology 179
Osmotic Behavior of Small Ions 180
Osmotic Coefficients. 180

CHAPTER 8. SOLUTION OPTICS 184
Elementary Theory 184
Refraction of Light 186
Double Refraction. 187
Optical Rotation 189
Rotatory Dispersion 192
Light Scatter 195
Concentration Fluctuations 199
Larger Particles 201
Still Larger Particles 204
Spectroscopy. 206
Radiofrequency Spectroscopy 207
Electron Spin Resonance 207
Nuclear Magnetic Resonance 208
Infrared 210
Electronic Configuration 212
Raman Spectra 214
Fluorescence. 214
Absorption of Light 215
Measurement of Absorption 217
Electronic Absorption Spectra 218
Solvent Perturbation Spectra. 222
Photochemical Changes 223
Photosynthesis 224
Vision 225

CHAPTER 9. SURFACES AND INTERFACES. 229
Surface Tension 229
Capillary Rise 230
Drop Weight Method 230
Pendant Drop Method 231

du Nouy Ring Method 232
Wilhelmy Slide 232
Energy of a Surface 232
Cohesion and Adhesion 234
Contact Angles 234
Wetting Balance 236
Surface Tension of a Solution 236
Surface Potentials 239
Surface Viscosity 240
Adsorption from Solution 242
Energy of Adsorption 244
Binding and Adsorption 245
Chromatography 246
Spread Monolayers 246
Lecithin Films 248
Protein Films 249
Gaseous Films 251
Film Penetration 253
Bifacial Films. 255
Deposited Monolayers 256

CHAPTER 10. VISCOSITY AND THE FLOW OF LIQUIDS 261
Flow in a Capillary 262
Methods. 265
Suspension of Spherical Particles 268
Expression for the Viscosity of Suspensions 269
Viscosity and Particle Asymmetry 270
Viscosity and Molecular Weights 271
Hydrodynamic Hydration 273
Non-Newtonian Viscosity 273

CHAPTER 11. DIFFUSION 276
Porous Disk Method 277
Fick's Second Law 279
Concentration Gradients 281
Optical Methods 282
Schlieren Optics 282
Rayleigh Interference Optics 284
Concentration and Heterogeneity 286
Diffusion in Gels 286
Molecular Sieves 288
Diffusion to a Flat Surface 289
Diffusion through Membranes 290
Penetration through Cellular Membranes 292
Thermodynamics of Diffusion 295
Diffusion and Molecule Size 296
Diffusion and Shape of Molecules 297
Rotary Diffusion Coefficient 299
Measurement of Rotary Diffusion Coefficients 299

CHAPTER 12. ION TRANSPORT 303
Conductance in Solution 303

Conductance Measurements 304
Independent Migration of Ions 305
Ionization Constants from Conductance 307
Conductance of Suspensions 308
Diffusion of Ions 309
Liquid Junction Potentials 311
Membrane Potentials 313
Anomalous Osmosis 315
Cellular Membranes 317
Conductance and Frequency 317
Accumulation of Ions by Cells 319
Ussing's Short-Circuit 321
Nerve Conduction 324

CHAPTER 13. ELECTROPHORESIS AND ELECTROKINETIC POTENTIALS 327
Electrophoretic Mobility 327
Electrophoretic Measurements 330
Moving Boundary 334
Boundary Conditions 336
Boundary Anomalies 338
Diffuse Boundaries 338
Heterogeneity 339
Interacting Systems 340
Electrophoresis of Complex Mixtures 341
Zone Electrophoresis 341
Electroosmosis and Streaming Potential. 342
Electrophoresis and Particle Size and Shape 344
The Zeta Potential 346
Interfacial Charge 347
Charge on Spherical Particles ($Kr < 300$) 348
Protein Charge and Electrophoresis 349
Electrophoresis of Adsorbed Proteins 351
Surface Potential and Ionization 353
Electroviscous Effect 355
Electrostatic Forces 357

CHAPTER 14. SEDIMENTATION 360
Movement of Particles 360
Microscopic Particle Sizes 361
The Ultracentrifuge 362
Hydrodynamic Hydration 365
Equilibrium Method 366
Archibald Method 368
Partial Specific Volume 369
Molecular Weight Methods 370

CHAPTER 15. KINETICS AND ENZYME ACTIVATION 374
Zero Order 375
First Order Reactions 375
Second Order Reactions 377
Change in a Given Time 379
Time for a Given Change 380

Reversible Reactions 380
Simultaneous Reactions 381
Consecutive Reactions 383
Synthesis and Degradation of Polymers 384
Metabolic Reactions 385
Molecular Collisions and Reaction. 388
Influence of Temperature 390
Transition State Theory 392
Protein Denaturation 394
Fast Reactions 396
Enzyme Activation 400
Enzyme Kinetics 401
Steady State Kinetics 402
Two Substrates. 405
Inhibitors 407
Activators. 409
Enzyme-Substrate Affinity 413
Inhibition by Excess Substrate 414
Allosteric Enzymes 416
Reversible Reactions 417
Depolymerizing Enzymes 418
Enzyme Concentration 420
Mechanism of Enzyme Action 420

CHAPTER 16. ELASTICITY AND STRUCTURE 425
Phase Separations 425
Gels 426
Thixotropy 427
Swelling of Gels 428
Rigidity of Gels 429
Elasticity of Gels 432
Molecular Basis of Elasticity 434
Elasticity of Protein Fibers 436
Silk Fibroin 438
Keratin 439
Collagen. 440
Elastin 443
Resilin 443
Muscle Fibers 445
Microscopy 448
X-ray Diffraction. 450

TABLE OF FREQUENTLY USED CONSTANTS 455
GREEK ALPHABET 456
INDEX 457

CHAPTER 1

Mathematical Review

It is certainly not possible to teach mathematics in a chapter or in a few lectures, but it has been our experience that it is very helpful to students to recall that which has been learned previously and to set this material in a new context. It is not necessary to be a skilled mathematician to understand and to use a large part of physical chemistry, but it is necessary to have some knowledge and facility with simpler mathematical skills and ideas.

Numbers and Exponents. The number system in everyday use is based on ten because most people have ten fingers. Other systems are possible and indeed some would be more useful; for example, a twelve system would have more even divisors than the ten system although two new numbers would have to be invented. The binary system has assumed great importance in connection with modern computers. The central nervous system also uses the binary system; a neuron either fires in response to a stimulus or it does not.

An arithmetic series of numbers increases by a constant amount as the series is ascended from number to number. The integers 1, 2, 3, 4, 5, 6, · · · comprise such a series. In a geometric series the ratio of successive numbers is constant and 2, 4, 8, 16 · · · is an example. It is informative to compare the above arithmetic series of numbers with the corresponding geometric series (see Table 1).

TABLE 1. ARITHMETIC, GEOMETRIC AND EXPONENTIAL SERIES

Arithmetic	*Geometric*	*Exponential*
1	2	2
2	4	2^2
3	8	2^3
4	16	2^4
5	32	2^5
6	64	2^6
7	128	2^7
8	256	2^8

Inspection of Table 1 shows that addition in the arithmetic series is equivalent to multiplication in the geometric series. For example, to multiply 4 by 64, 2 is added to 6, and 8, the sum, corresponds to 256 in the geometric series which is the product of 4 times 64. This very simple operation illustrates the use of logarithms to the base 2. For example the statement

$$2^n = N \tag{1}$$

can be written

$$\log_2 N = n \tag{2}$$

The relation between these two equalities can be generalized in the forms

$$B^n = N \text{ or } \log_B N = n \tag{3}$$

where B is the base of the logarithm. Further comparison of logarithms with exponents reveals three relations of great utility and these are

$$\log x \cdot y = \log x + \log y \tag{4}$$
$$\log x/y = \log x - \log y \tag{5}$$
$$\log x^n = n \log x \tag{6}$$

Any number could be selected for the base of the logarithms and the relation between the logarithms to two different bases can be obtained as follows: Select some number N and write

$$N = a^x \text{ and } N = b^y \tag{7}$$

Comparing these two statements it is evident that

$$a^x = b^y \tag{8}$$

and as logarithms

$$\log_a a^x = \log_a b^y \tag{9}$$

But by definition x is equal to $\log_a N$ and y is equal to $\log_b N$. Substituting these values in eq. 9 it is found that

$$\log_a N \log_a a = \log_b N \log_a b \tag{10}$$

It is evident that $\log_a a$ must equal unity and eq. 10 becomes

$$\log_a N = \log_b N \log_a b \tag{11}$$

and permits the transformation of a logarithm from one base (a) to another base (b).

The two bases in more frequent use are 10 and e. The base e is significant because natural processes progress as exponents of e and logarithms to the base e are called natural logarithms. In a chemical change such as that of the inversion of sucrose to fructose and glucose in the presence of acid, the concentration of sucrose (C) at any time (t) is

$$C = C_o e^{-kt} \tag{12}$$

or

$$\log_e \frac{C}{C_o} = -kt \tag{13}$$

where C_o is the original concentration of sucrose at the beginning of the experiment and k is the first order reaction constant; the numerical value of e is 2.71828.

The relation between logarithms to the base e and to the base 10 can be computed by the use of eq. 11; substituting 10 for b and e for a, there results

$$\log_e N = \log_{10} N \log_e 10 \tag{14}$$

or

$$\log_e N = 2.303 \log_{10} N \qquad 15$$

In what is to follow the natural logarithm will be denoted by ln and logarithms to the base 10 by log. Logarithms to the base 10 are more often used in numerical calculations than those to the base *e*; pH is expressed as

$$\text{pH} = -\log C_{H+} = \log \frac{1}{C_{H+}} \qquad 16$$

where C_{H+} denotes the hydrogen ion concentration. In Chapter 5 it will be shown that the above expression for pH is only true in the limit as the hydrogen ion concentration approaches zero.

Figure 1 shows plots of logarithms to the bases *e* and 10 against the corresponding numbers.

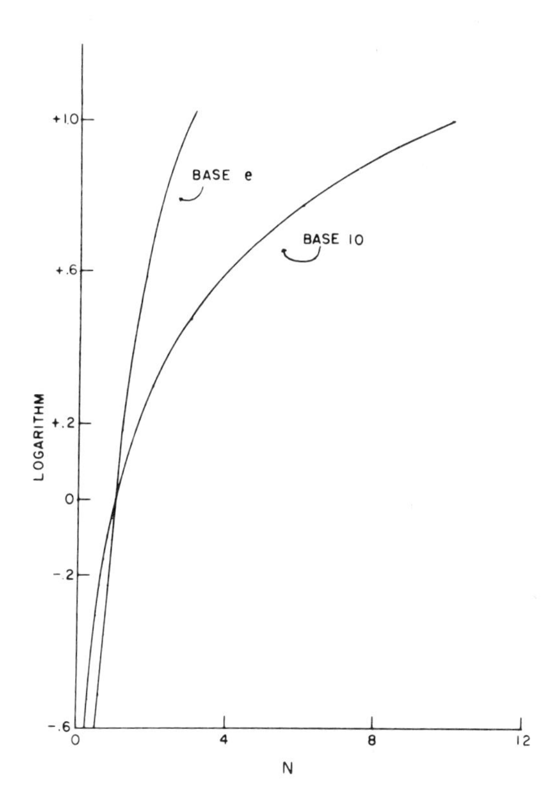

Fig. 1. Logarithms to the bases *e* and 10 plotted against the corresponding numbers.

Solution of Equations. Given the equation

$$ax + b = 0 \qquad 17$$

obviously the solution is

$$x = -\frac{b}{a} \qquad 18$$

Usually, an equation has in it an independent and a dependent variable such that

$$y = a + bx \qquad 19$$

By experiment, the values of y corresponding to those of x are known and the constants a and b are to be determined. Selecting two sets of x and y values as widely separated in magnitude as possible the two simultaneous equations are

$$y' = a + bx' \tag{20}$$

and

$$y'' = a + bx'' \tag{21}$$

where y' and y'' are the experimental values corresponding to x' and x'' respectively. The constants can now be eliminated between eqs. 20 and 21 to yield

$$b = \frac{y'' - y'}{x'' - x'} \tag{22}$$

and

$$a = \frac{y''x' - y'x''}{x' - x''} \tag{23}$$

With a large number of experimental data the solution of simultaneous equations to take advantage of all of the available data becomes tedious. A practical approach is to graph y as a function of x and draw by inspection the best straight line through the points. The slope of the line gives the value of b (eq. 19) and the intercept of the straight line on the y-axis yields the a-constant.

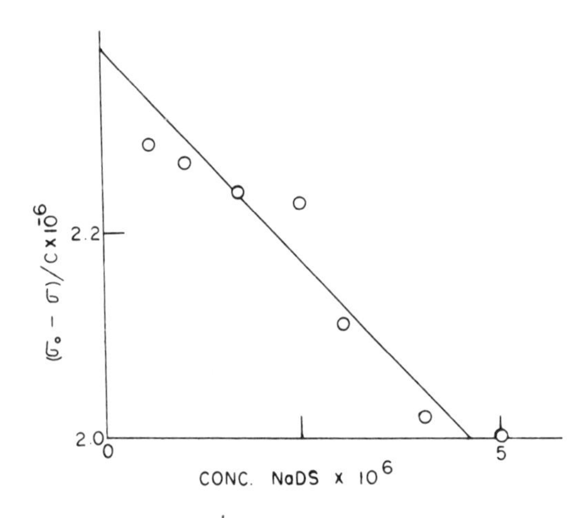

Fig. 2. Interfacial tension lowering in dynes per centimeter divided by molar concentration of sodium dodecyl sulfate plotted against molar concentration of sodium dodecyl sulfate, water-paraffin oil interface, 25°C.

For example, in Figure 2 is plotted the interfacial tension lowering at a water-paraffin oil interface produced by the addition of sodium dodecyl sulfate to the aqueous phase divided by the molar concentration of sodium dodecyl sulfate and plotted against the molar concentration of sodium dodecyl sulfate, the aqueous phase consisting of mixtures of sodium chloride and of hydrochloric acid to produce an ionic strength of 0.05 and a pH of 1.65. Over the concentration range considered (Fig. 2), the relation is linear and the slope of the line is b and the intercept on the y-axis is a in the equation of the line which is

$$\frac{\sigma_0 - \sigma}{C} = a + bC \tag{24}$$

Presently, it will be shown that the best straight line through a series of experimental points can be drawn by the method of least squares.

In addition to linear equations whose plots give straight lines, there are equations of higher orders such as quadratic, cubic, etc. The roots of a quadratic equation such as

$$ax^2 + bx + c = o \qquad 25$$

are

$$x = \frac{-b \pm (b^2 - 4ac)^{1/2}}{2a} \qquad 26$$

The meaningful root is to be selected from the physical situation. Suppose, however, $4ac$ turns out to be larger than b^2; then we would have to take the square root of a negative number and, obviously, this is impossible. It becomes necessary to define a number i such that i^2 is equal to -1. And the quadratic formula becomes

$$x = \frac{-b \pm i(4ac - b^2)^{1/2}}{2a} \qquad 27$$

which can be solved in terms of i.

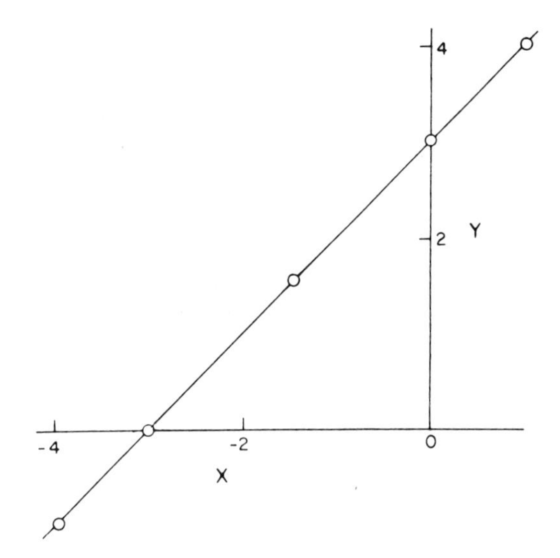

Fig. 3. Plot of eq. 29.

Provided the roots are real, an equation can always be solved graphically; every intersection with the x-axis represents a numerical solution of the equation. For example, in the expression

$$x + 3 = 0 \qquad 28$$

x is obviously equal to -3. This solution can be achieved graphically by writing

$$x + 3 = y \qquad 29$$

and plotting y against x for assumed values of x (see Fig. 3); the root of the equation is -3. The equation

$$x^2 - 8x + 9 = 0 \qquad 30$$

could be solved with the help of the quadratic formula (eq. 26) and would yield the roots 1.354 and 6.646. Eq. 30 can also be set equal to y and the y-values plotted against assumed values of x. Such a plot is shown in Figure 4 (curve 1).

However, if eq. 30 is changed to read

$$x^2 - 8x + 18 = 0 \qquad 31$$

the roots are imaginary and are $4 \pm 1.41i$; the plot of y against x does not cross the x-axis (see curve 2 Fig. 4).

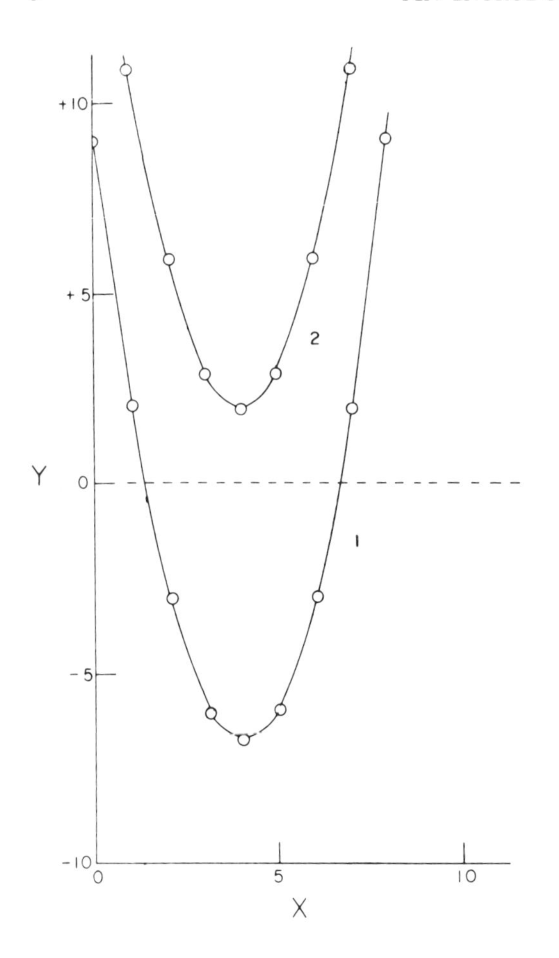

Fig. 4. Plot of eq. 30; the roots are approximately 1.4 and 6.6 (curve 1). Curve 2 has no real roots (eq. 31).

Often, it is possible to linearize an equation of higher order by the appropriate plot and for the equation

$$y = ax^2 + b \qquad 32$$

y can be plotted against x^2 for given values of x. Such a plot will, of course, yield a straight line from which the values of the constants a and b can be obtained from the slope of the line and the intercept on the y-axis respectively.

The well known Michaelis-Menten equation for the velocity of an enzymatically catalyzed reaction is

$$V = \frac{V_m C}{K_m + C} \qquad 33$$

in which V is the initial velocity of the reaction at substrate concentration C, V_m is the maximum velocity of the reaction which occurs when the enzyme is saturated by the substrate and K_m is the dissociation constant of the enzyme-substrate complex. A plot of V against C will yield a curve of the form shown in Figure 5. Equation 33 can be linearized by inversion and

$$\frac{1}{V} = \frac{K_m + C}{V_m C} = \frac{K_m}{V_m C} + \frac{1}{V_m} \qquad 34$$

or both sides of eq. 34 can be multiplied by C to give

$$\frac{C}{V} = \frac{K_m}{V_m} + \frac{C}{V_m} \tag{35}$$

or eq. 34 can be multiplied by $\frac{VV_m}{K_m}$ and rearranged to give

$$\frac{V}{C} = \frac{V_m}{K_m} - \frac{V}{K_m} \tag{36}$$

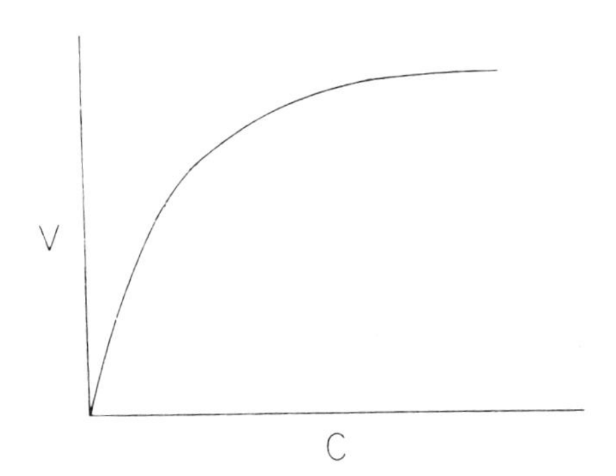

Fig. 5. Hypothetical plot of the velocity of enzymatic reaction against substrate concentration (eq. 33).

Straight lines should thus be obtained if $\frac{1}{V}$ is plotted against $\frac{1}{C}$, or $\frac{C}{V}$ against C, or $\frac{V}{C}$ against V and the constants K_m and V_m can be evaluated from each type of plot (see Fig. 6).

It is desirable to linearize a relation because: (1) Such an operation brings out the simplest possible connection between variables. (2) It is possible to detect experimental errors with greater ease. (3) The constant terms can be evaluated by inspection. (4) The method of least squares can be applied and the most probable values assigned to the constant terms.

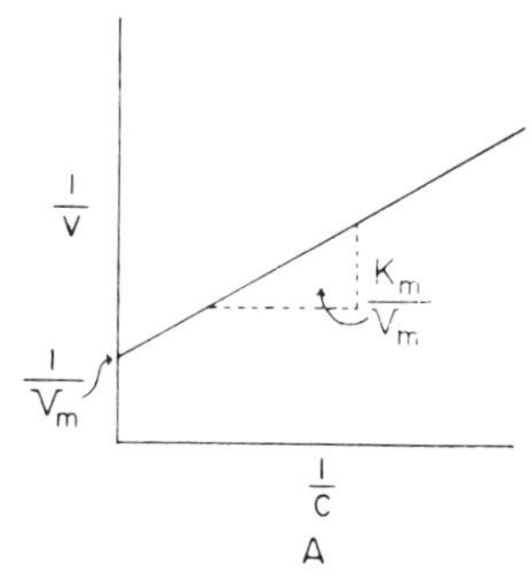

FIG. 6. Hypothetical plots of eqs. 34 (Fig. A), 35 (Fig. B) and 36 (Fig. C).

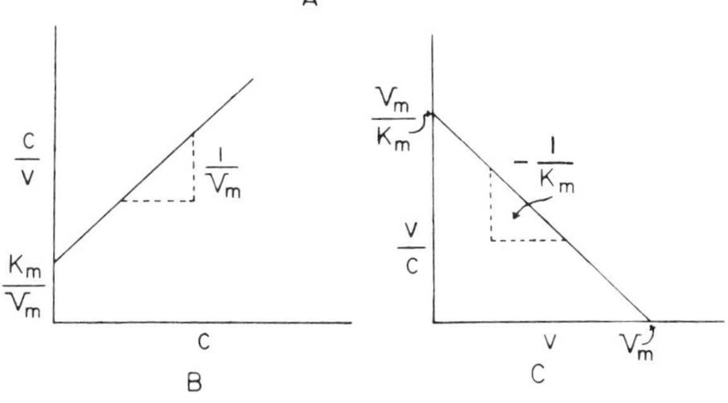

Sometimes it is required to plot very small or very large numbers on a graph and use is often made of an exponent in the legend of the figure to obtain a condensed scale and to exhibit the data in a clear manner. The same device is frequently used in the construction of tables and space is conserved with no loss of meaning by writing an exponent at the head of a column in the table. Since, in both the figures and tables, a function multiplied by an exponent is equal to a given value, then the function itself is equal to the value divided by the exponent and this is how such tables and figures are to be interpreted; the sign of the exponent in the legend is opposite to that of the interpolated value.

Differential Calculus and Rates of Change. Differential calculus is probably the easiest branch of mathematics to understand and to use; in modern times the rates of change are so great in so many areas that the concept of rates of change has become intuitive.

If the distance (S) an object travels is plotted against time (t) and the speed is constant, a straight line is obtained, the equation for which is

$$S = kt \qquad 37$$

The slope of the straight line is k and is the velocity of travel and is expressed by the statement

$$\text{Velocity} = \frac{dS}{dt} = k \qquad 38$$

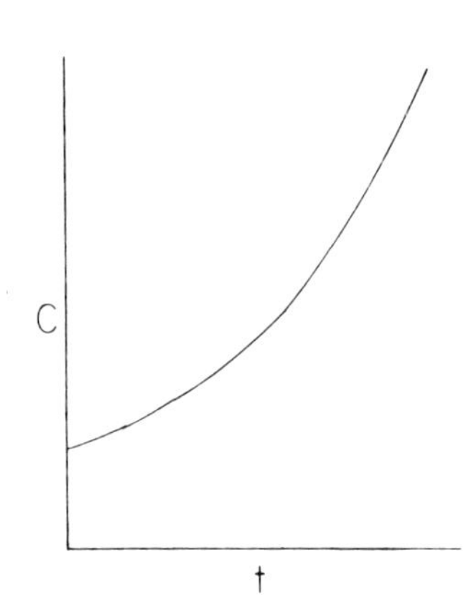

Fig. 7. Plot of eq. 39.

Suppose the concentration of a substance in a chemical reaction is given by the relation

$$C = at^2 + b \qquad 39$$

where t is the elapsed time and a and b are constants. The plot of concentration against time would give a graph such as is shown in Figure 7, the slope of which at any time, t, is given by the differentiation of eq. 39 and

$$\frac{dC}{dt} = 2at \qquad 40$$

The velocity, dC/dt, is that of an autocatalytic reaction. The rate of change of the velocity with time is the acceleration and from eq. 40

$$\frac{d\frac{dC}{dt}}{dt} = \frac{d^2C}{dt^2} = 2a \qquad 41$$

Maxima and minima can be located by use of differential calculus. Returning to eq. 30 and Figure 4, the differential of y relative to x is

$$\frac{dy}{dx} = 2x - 8 \qquad 42$$

and, at the minimum of curves 1 and 2, the rate of change of y in respect to x is zero or x is equal to 4 which locates the minimum.

It is usually a simple matter to obtain the differential for a given expression. The following expressions are some of the standard differential forms where a is a constant and x and y are the variables:

$$da = O \qquad dx^a = ax^{a-1}dx$$

$$dax = adx \qquad d \ln x = \frac{dx}{x}$$

$$dxy = xdy + ydx \qquad d \log x = 0.43d \ln x = \frac{0.43dx}{x}$$

$$d\left(\frac{x}{y}\right) = \frac{ydx - xdy}{y^2} \qquad \frac{d\frac{dx}{dy}}{dy} = \frac{d^2x}{dy^2}$$

Fig. 8. Illustrating the derivation of a differential expression.

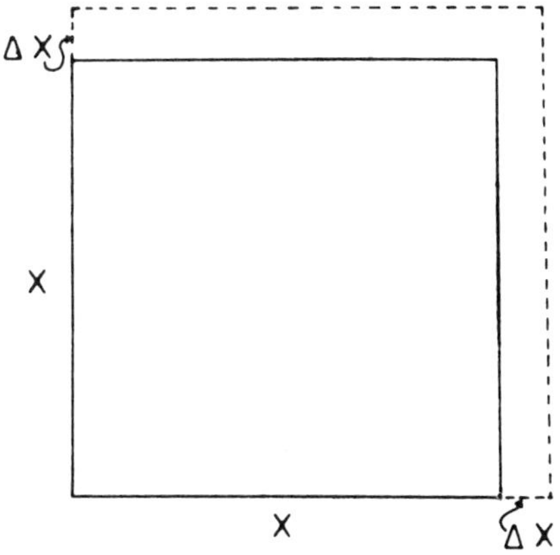

Proof for the above expressions and for others is the subject of courses in calculus; however, to illustrate the method, the expression dx^a equals $ax^{a-1}dx$ can be shown to be true for a particular case. Imagine a square whose sides are length x and whose area is y. Thus x^2 equals y. The relation is diagramed in Figure 8. Suppose the value of x is increased by Δx; the new area of the square is now

$$(x + \Delta x)^2 = x^2 + 2x\Delta x + \Delta x^2 \qquad 43$$

and the increment in area is

$$y' - y = \Delta y = (x + \Delta x)^2 - x^2 = 2x\Delta x + \Delta x^2 \qquad 44$$

The smaller the increment, Δx, becomes, Δx^2 becomes that much smaller and in the limit as Δx approaches zero it can be neglected and eq. 44 becomes

$$dy = 2xdx \qquad 45$$

The most general method of differentiation is graphical; however complex the relation between the variables, the slope of the line at a given point on the curve relating the variables is equal to the rate of change of the dependent variable in respect to the independent variable; a tangent to the curve is drawn and the slope of the tangent estimated.

The drawing of a tangent to a curve, due to the subjective element, can suffer from large numerical errors and it is useful to employ some optical device to aid in the measurement of the slopes. The accuracy of such estimates can be improved considerably by the use of a simple device consisting of a plane mirror fixed at right angles to a straight edge. When the edge of the mirror rests on the curve, the reflection appears to be a continuation of the curve and a line parallel to the tangent of the curve can be drawn along the straight edge. A still more effective device is the so-called tangentmeter which consists of a triangular glass prism mounted in the center of a circular piece of plate glass which can be rotated. The tangentmeter is placed over the curve with the coordinates of the graph coinciding with the rectangular mountings of the circular glass plate and the curved line in the figure observed through the prism. The prism is rotated until the curve appears smooth and continuous when viewed through the prism. The tangent of the slope of the curve at the chosen point is then read on the scale attached to the circular piece of glass.

Integration. Integration obtains the sum of an infinite number of infinitely small quantities; it is the reverse process of differentiation. Equation 40 is the differential form of eq. 39 and eq. 40 can be integrated by writing

$$C = 2a \int t dt \qquad 46$$

or

$$C = at^2 + \text{constant} \qquad 47$$

Compare eqs. 39 and 47.

There are standard integral forms just as there are standard differential expressions and

$$\int a dx = ax + \text{constant}$$

$$\int x^a dx = \frac{x^{a+1}}{a+1} + \text{constant}$$

$$\int e^x dx = e^x + \text{constant}$$

$$\int \frac{dx}{x} = \ln x + \text{constant}$$

An expression can be integrated as shown in the standard forms above which include a constant of integration or it can be integrated between certain specified limits. For example, between the limits t_1 and t_2 eq. 46 becomes

$$2a \int_{t_1}^{t_2} t dt = at^2 \Big]_{t_1}^{t_2} = a [t_2^2 - t_1^2] \qquad 48$$

Usually more mathematical skill is required to integrate than to differentiate. The easiest and most general method of integration is the graphical method and this method also illustrates the real meaning of integration. Suppose the velocity of a process is directly proportional to time as expressed in eq. 40. The amount of substance (m) produced, if the volume of the reaction mixture remains constant, is evidently equal to the velocity of transformations multiplied by time and for an infinitesimally small time, dt, the amount transformed (dm) is

$$dm = \frac{VdC}{dt} dt \qquad 49$$

where V is the volume of the reaction mixture or

$$m = V \int \frac{dC}{dt}\, dt \tag{50}$$

It is a necessary consequence of eq. 40 that a linear relation will be obtained if the velocity of the reaction is plotted against time, and such a plot is shown in Figure 9.

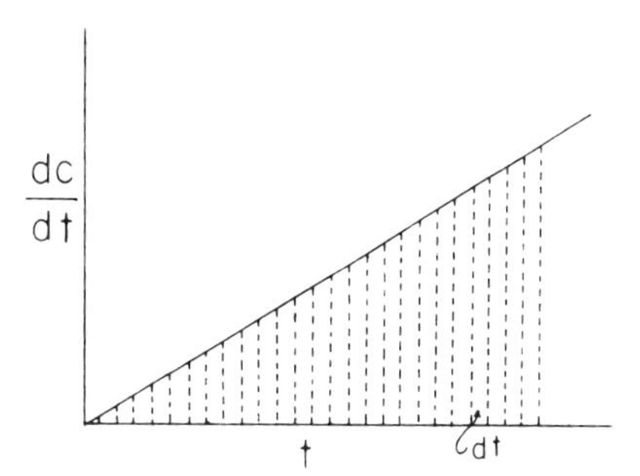

Fig. 9. Plot of velocity of reaction against time (eq. 40).

The area of each of the vertical elements is $(dC/dt) \cdot dt$ and, accordingly, if the area under the straight line from the origin to t_1 is measured, the total concentration change of the substance will be obtained; if this concentration increase is multiplied by the volume of the reaction mixture, the amount of the substance (m) will be obtained (see eq. 50). The amount of substance (m) formed in this time interval can also be calculated since there exists an explicit relation between the velocity of the reaction and the time (eq. 40). Integrating eq. 40, eq. 46 results and putting in the limits t equal zero to t_1

$$C = 2a \int_0^{t_1} t dt \tag{51}$$

or

$$m = 2aV \int_0^{t_1} t dt \tag{52}$$

or

$$m = aVt^2 \Big]_0^{t_1} = aVt_1^2 \tag{53}$$

The area under a curve can be measured in a simple way by graphing the curve and cutting out the area under the curve and weighing the piece of paper with a good analytical balance. The area can also be measured with a planimeter which is an instrument used to trace the outline of an area and to sum the infinitesimal area elements in polar coordinates.

The relation between the dependent and independent variables can be very complex. Consider, for example, an electrophoretic diagram of the proteins of blood serum. By means of an ingenious optical arrangement, it is possible to measure the change of the protein concentration with distance along the electrophoretic cell (dC/dx) as a function of the distance x. A plot of a typical electrophoretic diagram of blood serum is shown in Figure 10. The total amount of the albumin component (A) is clearly

$$A = a \int_{x_1}^{x_2} \left(\frac{dC}{dx}\right) dx \tag{54}$$

where $\int_{x_1}^{x_2} (dC/dx)\, dx$ is the area under the curve between the limits x_1 and x_2 and a is a constant whose value must be determined. In general, if values of dy/dx corresponding to given values of x are known from experiment, graphical integration can be employed and no explicit analytical relation between x and y is needed.

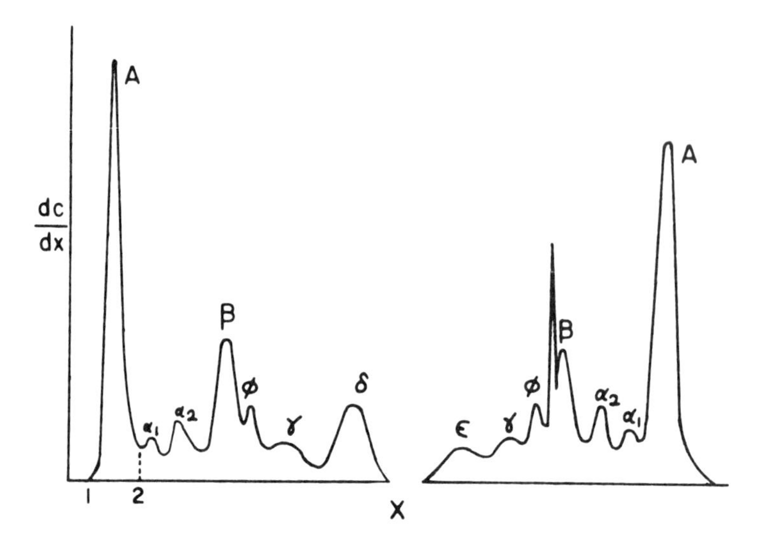

Fig. 10. Electrophoretic diagram of human blood serum at pH 8.6 veronal buffer ionic strength 0.10.

As noted, a differential can be integrated analytically providing an explicit relation between the variables exists. An example of analytical integration was given at the beginning of the discussion on integration in which an exact relation between concentration and time was assumed for a chemical reaction. To illustrate this procedure again, consider the work done in the compression of a gas. The gas is to be enclosed in a cylinder equipped with a movable piston and work done in moving the piston against frictional forces is to be neglected. The situation is diagramed in Figure 11.

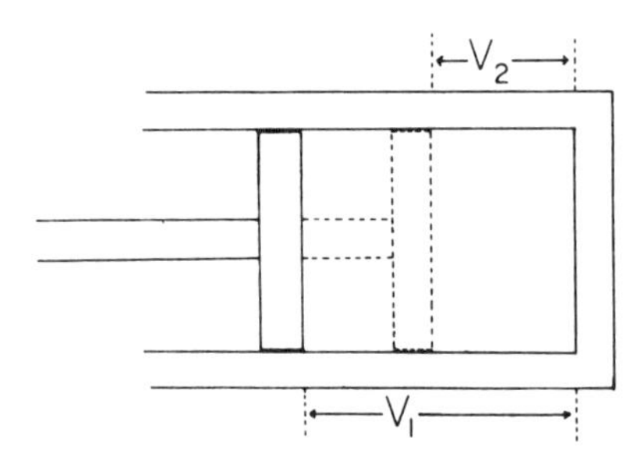

Fig. 11. Compression of a gas at constant temperature from volume V_1 to volume V_2.

The work done is equal to the force exerted multiplied by the distance. As the gas is compressed the force which must be exerted increases continuously and the work is

$$\text{work} = \int_{S_1}^{S_2} F dS \qquad 55$$

where F is the force exerted on the gas and S is the distance through which the piston has been moved.

Now the volume change is equal to the area of the piston head multiplied by the distance through which the piston has been moved and

$$dS = \frac{dV}{A} \tag{56}$$

and the force (F) is equal to the gas pressure (P) multiplied by the area (A) of the piston head. Substituting this information along with eq. 56 in eq. 55 gives

$$\text{work} = \int_{V_1}^{V_2} \frac{PAdV}{A} = \int_{V_1}^{V_2} PdV \tag{57}$$

Per mole of ideal gas there exists the explicit relation between the pressure and the volume, i.e.

$$PV = RT \tag{58}$$

where R is the gas constant and T is the absolute temperature. Substituting the value of P from eq. 58 into eq. 57, there results

$$\text{work} = RT \int_{V_1}^{V_2} \frac{dV}{V} \tag{59}$$

and integrating between the limits V_1 and V_2

$$\text{work} = RT \ln V \Big]_{V_1}^{V_2} = RT \ln \frac{V_2}{V_1} \tag{60}$$

The work of compression of the gas could also be calculated by graphical integration by plotting P against V as measured by experiment and determining the area under the curve between V_1 and V_2. The numerical values for the work of compression of the gas by these two methods would not, in general, agree because no real gas will follow exactly the explicit relation between the pressure and the volume given by the ideal gas law (eq. 58). Comparison of the compression of a real and an ideal gas is shown in Figure 12.

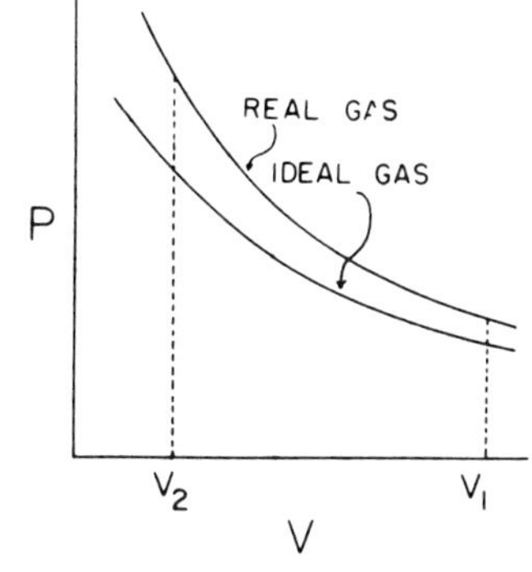

Fig. 12. Diagrammatic representation of the pressure as a function of the volume of real and of ideal gas.

Differential Equations. A differential equation is a relation in which the derivative of a function involves the function itself, thus

$$\frac{dy}{dx} = ay \tag{61}$$

is a differential equation the solution of which is

$$\ln y = ax + c \tag{62}$$

and can be written in the form

$$y = Ae^{ax} \tag{63}$$

where the integration constant has been set equal to ln A.

Differential equations can come in all degrees of difficulty and there is hardly any branch of mathematics in which skill acquired through practice is so important; unfortunately, this skill is rapidly lost unless used.

Suppose a small, microscopic, charged spherical particle of mass m and radius r suspended in water with a viscosity η is placed in an electrical field such that electrophoresis will occur. How long will it take for the particle to reach 0.9 of its terminal velocity? The basic equation is

$$F = F_e - F_r \qquad 64$$

where F is the net force acting on the particle to produce acceleration, F_e is the electrical force and F_r is the resisting force due to the viscous drag exerted by the water. The resisting force is given by Stokes factor $6\pi r\eta V$ where V is the velocity of the particle. The electrical force (F_e) is evidently equal to the resisting force when the terminal velocity has been attained and is equal to $6\pi r\eta U$ where U is the terminal velocity. Equation 64 can be written

$$\frac{md^2x}{dt^2} = 6\pi r\eta U - 6\pi r\eta \frac{dx}{dt} \qquad 65$$

Equation 65 can also be written

$$\frac{mdV}{dt} = 6\pi r\eta U - 6\pi r\eta V = 6\pi r\eta (U - V) \qquad 66$$

and

$$\frac{dV}{U - V} = \frac{6\pi r\eta}{m} dt \qquad 67$$

On integration, eq. 67 becomes

$$\ln (U - V) = \frac{-6\pi r\eta t}{m} + C \qquad 68$$

when t is zero, C is equal to ln U and

$$\ln \frac{U}{U - V} = \frac{6\pi r\eta t}{m} \qquad 69$$

A particle whose density is one and whose radius is 1×10^{-4} cm will attain 0.9 of its terminal electrophoretic mobility in water at 25°C in 0.57×10^{-6} second.

Expansion Series. Provided a variable is a continuous function of another variable and however complex the relation may be, the two variables may be expressed as a power series such as

$$y = a + bx + cx^2 + dx^3 + \qquad 70$$

where a, b, c, d, etc. are constants whose values are to be determined. There are many types of series; the arithmetic series and the geometric series have been discussed in connection with exponents and logarithms.

It is obvious that $(a + x)^2$ is equal to $a^2 + 2ax + x^2$ and $(a + x)^3$ is equal to $a^3 + 3a^2x + 3ax^2 + x^3$. Such results as these can be generalized in the form which permits the expansion of $(a + x)^n$; this is a statement of the binomial theorem.

The Maclaurin theorem for the expansion of an expression can be illustrated by again considering the expansion of $(a + x)^n$. First, it is assumed that $(a + x)^n$ can be set equal to a power series such that

$$(a + x)^n = a_0 + a_1x + a_2x^2 + \qquad 72$$

It remains to determine the values of the coefficients a_0, a_1, a_2, etc. To this purpose eq. 72 is differentiated in respect to x and

$$d(a+x)^n = n(a+x)^{n-1}\,dx = a_1dx + 2a_2xdx + \qquad 73$$

Dividing eq. 73 through by dx gives

$$n(a+x)^{n-1} = a_1 + 2a_2x + \qquad 74$$

Differentiating eq. 74 and again dividing through by dx yields

$$n(n-1)(a+x)^{n-2} = 2a_2 + \qquad 75$$

This process could be continued to include as many terms of the series as wished.

If eq. 72 is true for any value of x, it is true for all values of x including x equal zero; the same can be said of eqs. 73 and 75 and as many other steps as have been included. Setting x equal to zero in each of these equations, there is obtained

From eq. 72

$$a^n = a_0 \qquad 76$$

From eq. 74

$$na^{n-1} = a_1 \qquad 77$$

From eq. 75

$$n(n-1)a^{n-2} = 2a_2 \text{ or } a_2 = \frac{n(n-1)a^{n-2}}{2} \qquad 78$$

Inserting values of a_0, a_1 and a_2 into eq. 72 gives

$$(a+x)^n = a^n + na^{n-1}x + \frac{n(n-1)a^{n-2}x^2}{2!} + \qquad 79$$

Other functions can be expanded by the use of the Maclaurin theorem. For example, Szyszkowski showed that the surface tension of surface active compounds is related to the concentration of the added compound by the equation

$$\frac{\sigma_0 - \sigma}{\sigma_0} = B \log\left(\frac{C}{A} + 1\right) \qquad 80$$

where σ_0 is the surface tension of the pure solvent in dynes per centimeter, σ is the surface tension at the concentration C of the solute and A and B are constants to be determined by experiment. As written, eq. 80 is awkward to use and the evaluation of the constants is not easy. A more manageable expression can be obtained by expanding the $\log\left(\frac{C}{A} + 1\right)$ term by means of the Maclaurin theorem. Thus

$$\log\left(\frac{C}{A} + 1\right) = a_0 + a_1C + a_2C^2 + \qquad 81$$

Differentiating eq. 81 and dividing by dC in successive steps, the following equations are obtained

$$\frac{0.4344}{C+A} = a_1 + 2a_2C \qquad 82$$

$$-\frac{0.4344}{(C+A)^2} = 2a_2 \qquad 83$$

Setting C equal to zero in eqs. 81, 82 and 83, there results

$$a_0 = O \tag{84}$$

$$a_1 = \frac{0.4344}{A} \tag{85}$$

$$a_2 = -\frac{0.4344}{2A^2} \tag{86}$$

and

$$\log\left(\frac{C}{A} + 1\right) = \frac{0.4344C}{A} - \frac{0.4344C^2}{2A^2} \tag{87}$$

and the expanded Szyszkowski equation becomes

$$\frac{\sigma_0 - \sigma}{\sigma_0} = \frac{0.4344BC}{A} - \frac{0.4344BC^2}{2A^2} \tag{88}$$

Rearranging eq. 88 gives

$$\frac{\sigma_0 - \sigma}{C} = \frac{0.4344B\sigma_0}{A} - \frac{0.4344B\sigma_0 C}{2A^2} \tag{89}$$

Compare eq. 89 with eq. 24; for low concentrations of surface active solute, the terms beyond the last term on the right in eq. 89 are too small to affect the results. Evidentlv, a in eq. 24 is equal to $0.4344B\sigma_0/A$ and b in eq. 24 is equal to $-0.4344B\sigma_0/2A^2$.

The sum of a series of terms is customarily indicated by a large Greek sigma and

$$f(x) = \sum_{n=0}^{n=n'} a_n x^n \tag{90}$$

means that all the terms of the expansion series from n equal zero to n equal n' have been added together. If a series approaches a limit as the number of terms increases, the series is said to be convergent whereas if no such limit exists, the series is divergent; a series whose terms become progressively smaller with each successive term is convergent. Convergent series are usually of more interest in physical operations and, furthermore, to be of use as a close approximation, a series must converge very rapidly; otherwise the number of terms which must be considered becomes too large for convenience. There are a number of tests for convergence which can be found in appropriate mathematical handbooks.

Trigonometric Functions. It will be recalled that sides of a right triangle are related by definition to the angle α as follows:

$$\sin \alpha = \frac{a}{h}; \cos \alpha = \frac{b}{h}; \tan \alpha = \frac{a}{b}$$

where a, b, h, and α are shown in Figure 13; the reciprocal relations are known as the cosecant, secant and the cotangent respectively.

It is evident from an inspection of Figure 13 that as α increases from zero to 90° the sine increases from zero to unity and the cosine decreases from unity to zero. The tangent on the other hand increases from zero to infinity. As the angle α becomes greater than 90°, the sine decreases and reaches zero at 180°, the cosine becomes negative and is minus one at 180°. The sign of the various functions in each of the four quadrants can be predicted from a consideration of Figure 13. Figure 14 shows plots of the sine and the tangent as functions of the corresponding cosine. The degrees are indicated on the quarter circle plot of the sine.

For many purposes, it is more convenient to use radians instead of degrees. A radian is the angle subtended by an arc equal to the radius of the circle. Since the

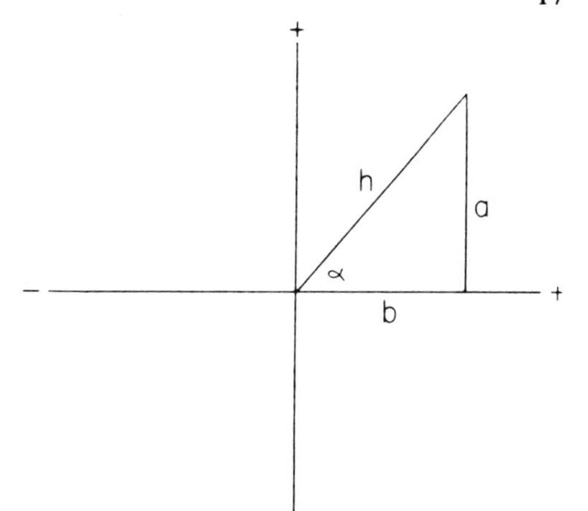

Fig. 13. Right triangle showing sides and angle.

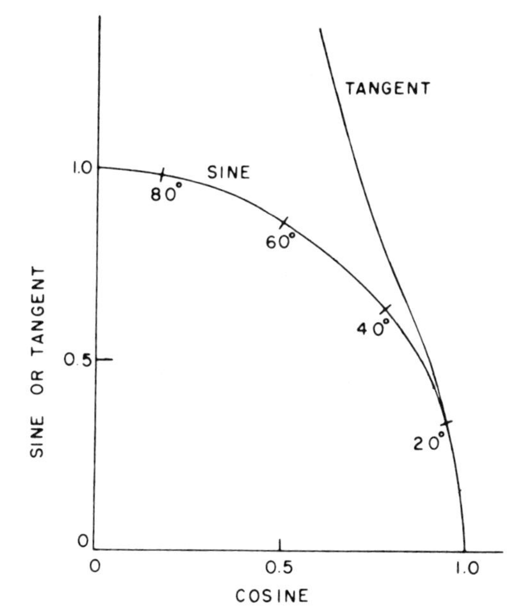

Fig. 14. Sine and tangent as functions of the cosine. Angles are indicated as quarter circle plot of sine.

circumference of a circle is $2\pi r$, the circumference corresponds to 2π radians and since there are 360° in a circle, one radian is equivalent to 57° 17′ 42″. Suppose a circle rotates on its axis as shown in Figure 15 and y' represents the distance normal to the x-axis. It is evident that the value of y' oscillates between 1 and -1. By definition

$$y' = r \sin \alpha \qquad 91$$

and r is the amplitude of the curve. If ω is the angular velocity of rotation of the circle and t the elapsed time, then

$$\alpha = \omega t \qquad 92$$

and the time required for a complete revolution of the circle is $2\pi/\omega$ and the reciprocal of this function, which is the number of revolutions per unit time, is the frequency and 2π radians is the wavelength. The amplitude as a function of time is

$$y' = r \sin \omega t \qquad 93$$

The velocity of propagation of the wave is equal to the frequency multiplied by the wavelength or is ω.

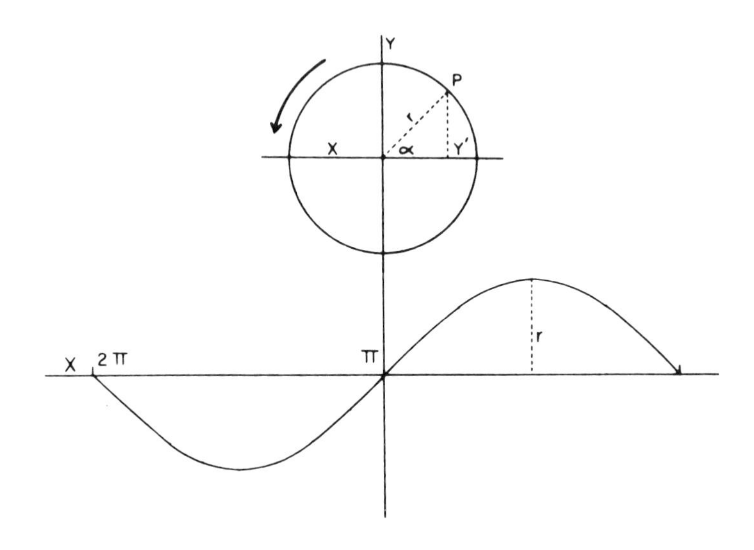

Fig. 15. Rotating circle of radius r and resulting sine wave described by point P.

Sine waves can be out of phase with each other and lead to reinforcement and cancellation at various parts of the cycle. The amplitude of a sine series which is composed of several waves superimposed on each other is

$$y = \sin x - \tfrac{1}{2} \sin 2x + \tfrac{1}{3} \sin 3x + \qquad 94$$

Plotting each of the right-hand terms separately, there are obtained the curves shown in Figure 16. Curve 1 is the plot of $\sin x$, curve 2 that of $\frac{1}{2} \sin 2x$ and curve 3 that of $\frac{1}{3} \sin 3x$ and represent overtones or harmonics. Curve 4 is the algebraic sum of the ordinates of curves 1, 2 and 3. Any repetitive, oscillating function can be represented by the proper combination of terms of a sine series such, for example, as the waves obtained from an electrocardiogram. Such a series is an example of the Fourier series.

A sine wave is an example of harmonic motion such as described by a vibrating spring with a particle of mass, m, attached to the end of the spring. Assuming there is no viscous drag on the particle, the basic equation to describe the forces acting is

$$F = -F_s \qquad 95$$

where F is the force due to the acceleration of the particle and is equal to ma, where a is the acceleration. F_s is the force exerted by the elastic spring and acts in an opposite sense from the acceleration; hence the negative sign. Equation 95 can be expressed in the form of a differential equation as

$$m \frac{d^2x}{dt^2} = -kx \qquad 96$$

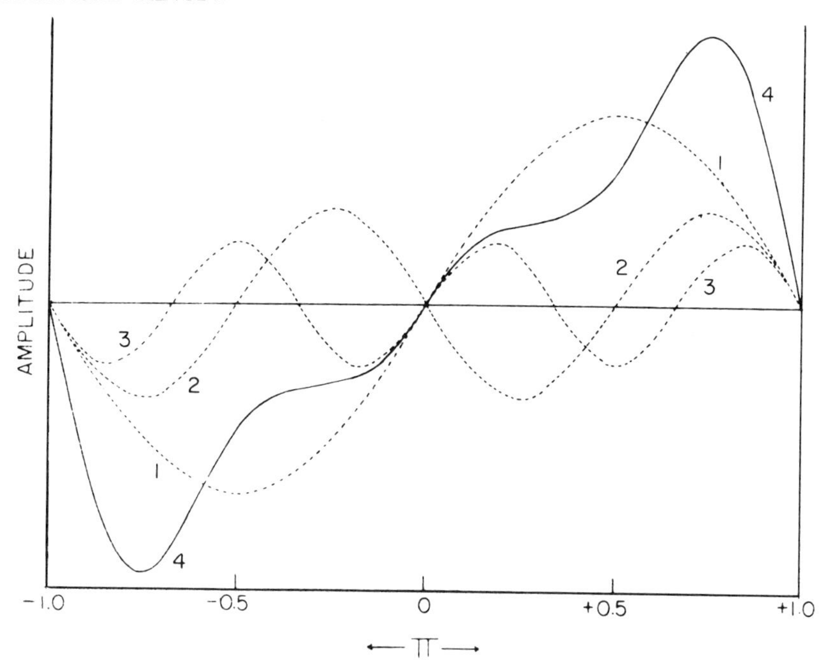

Fig. 16. Harmonics of the sine curve.

where x is the distance of displacement and k is an elastic coefficient. It appears likely that the solution of eq. 96 will be in the form of a sine or cosine function and the simplest way of proceeding is to propose a trial solution and see if this solution satisfies the original differential equation (eq. 96). First, eq. 96 is divided by m and the resulting coefficient in front of x which is k/m is set equal to ω^2. Equation 96 then becomes

$$\frac{d^2x}{dt^2} + \omega^2 x = 0 \qquad 97$$

The trial solution of eq. 97 is written as

$$x = A \cos(\omega t + \theta) \qquad 98$$

The first differentiation of eq. 98 results in

$$\frac{dx}{dt} = -A\omega \sin(\omega t + \theta) \qquad 99$$

and the differentiation of eq. 99 in

$$\frac{d^2x}{dt^2} = -A\omega^2 \cos(\omega t + \theta) \qquad 100$$

Substituting eq. 98 into eq. 100, it is found that

$$\frac{d^2x}{dt^2} = -\omega^2 x \qquad 101$$

which is identical with eq. 97 and, accordingly, it is concluded that eq. 98 is the solution of eq. 97. The constants A and θ in eq. 98 are the two integration constants. A is evidently the amplitude of the simple harmonic motion and θ is the phase angle and x is equal to $A \cos \theta$ when t is zero. The natural frequency of oscillation of the spring is equal to $\omega/2\pi$ or to $\frac{1}{2\pi}(k/m)^{1/2}$ and depends on the ratio of the elastic coefficient to the mass of the particle. If instead of a cosine function, a sine function had been used as a trial solution of eq. 97, an equally satisfactory outcome would have been achieved; the phase angle would have been different.

There are many important problems associated with harmonic motion and are apt to be encountered where oscillation occurs such as in electrical circuits, the pulsing of blood through an artery, the passage of light waves through a medium, etc. There are many variations on the simple theme given above. For example, the particle can be subject to a viscous drag, in which case the harmonic motion is dampened and the oscillations die out. The system can be subject to forced oscillations where a periodic external force is applied. If the frequency of the external force is the same as the natural frequency, resonance occurs and the amplitude of the oscillations increases greatly. In a mechanical system, resonance is achieved when the externally applied force has the frequency $\frac{1}{2\pi}(k/m)^{1/2}$ where k is the elastic coefficient and m is the mass. In an electrical circuit, resonance occurs when the frequency of the alternating current is equal to $\frac{1}{2\pi}(1/LC)^{1/2}$ where C is the capacitance and L is the inductance; the current and voltages are in phase.

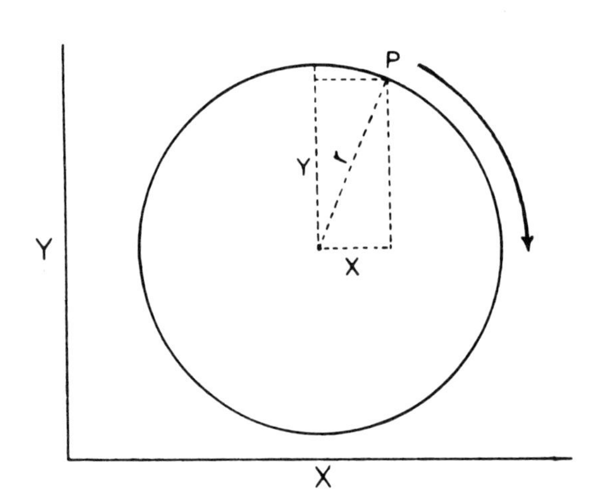

Fig. 17. Resolution of x and y components of the motion of the head of a centrifuge.

Biochemists make frequent use of the centrifuge; it is one of their bread and butter instruments. Consider the forces acting in a centrifuge and to do this place the centrifuge in x and y coordinates (see Fig. 17). A point, P, on the perimeter of the centrifuge executes simple harmonic motion relative to the x-axis as well as towards the y-axis as the centrifuge rotates. The motion is thus analogous to the oscillations of a particle at the end of a spring and the acceleration of the particle relative to the center of the centrifuge can be resolved as follows: The acceleration in the x-direction (a_x) and that in the y-direction (a_y) respectively are given by

$$a_x = -\omega^2 x \qquad 102$$

and

$$a_y = -\omega^2 y \qquad 103$$

The acceleration normal to the tangent to the perimeter of the centrifuge, and thus acting toward the center of the centrifuge, is

$$a = (a_x^2 + a_y^2)^{1/2} \qquad 104$$

Substituting in eq. 104 the values of a_x and a_y gives

$$a = [(-\omega^2 x)^2 + (-\omega^2 y)^2]^{1/2} \qquad 105$$

$$= (\omega^4 r^2)^{1/2} = \omega^2 r \qquad 106$$

The centrifugal force exerted by the centrifuge is then

$$F = am = \omega^2 rm \qquad 107$$

where m is the mass at point P. Since ω is to be expressed in radians per second and a radian is a ratio and not a length, it is clear that eq. 107 has the proper dimensions.

Since one revolution is equal to 2π radians, the revolutions per second (R.P.S.) is equal to 2π radians per second and eq. 107 becomes

$$F = 4\pi^2 r\,(\text{R.P.S.})^2 m \qquad 108$$

and the centrifugal force expressed as multiples of the force of gravity is

$$N = \frac{4\pi^2 r\,(\text{R.P.S.})^2 m}{gm} = \frac{4\pi^2 r\,(\text{R.P.S.})^2}{g} \qquad 109$$

$$= 0.0402r\,(\text{R.P.S.})^2 \qquad 110$$

The relation between e, i and the trigonometric functions is a simple one. If the exponential of e is expanded into a series by methods already described, it is found that

$$e^x = 1 + x + \frac{x^2}{2} + \frac{x^3}{6} + \frac{x^4}{24} + \frac{x^5}{120} + \qquad 111$$

x in eq. 111 can be replaced by $i\theta$ where i is the square root of minus one and θ is an angle expressed in radians. Equation 111 can then be written as

$$e^{i\theta} = 1 + i\theta - \frac{\theta^2}{2} - \frac{i\theta^3}{6} + \frac{\theta^4}{24} + \frac{i\theta^5}{120} + \qquad 112$$

The expansion of $\sin\theta$ results in

$$\sin\theta = \theta - \frac{\theta^3}{2} + \frac{\theta^5}{120} + \qquad 113$$

and the corresponding series for $\cos\theta$ is

$$\cos\theta = 1 - \frac{\theta^2}{2} + \frac{\theta^4}{24} + \qquad 114$$

Comparing eqs. 112, 113, and 114, it is seen that

$$e^{i\theta} = \cos\theta + i\sin\theta \qquad 115$$

Equation 115 describes two waves 90° out of phase. The real part of $e^{i\theta}$ is $\cos\theta$ and the imaginary part is $i\sin\theta$. If θ is equal to π, $\sin\theta$ is zero and $\cos\theta$ is equal to -1 and, accordingly, $e^{i\theta}$ is equal to -1.

Probability and Error. It has become clear that many of the laws of science are statistical in nature. If, for example, a penny is flipped 10,000 times, it would be expected that "heads" would show nearly 5,000 times and "tails" nearly 5,000 times; the equation for the osmotic pressure of a solution is no less and no more compelling than the simple proposition of a flipped penny.

Suppose a penny is flipped eight times; the chance that all eight flips will show heads is $(\frac{1}{2})^8$ or is one chance in 256. The chance that 7 heads will show is $8 \times (\frac{1}{2})^8$. The chance that 6 heads will show is $28 \times (\frac{1}{2})^8$ and the chance that 5 heads will show is $56 \times (\frac{1}{2})^8$ and the chance that 4 heads will show is $70 \times (\frac{1}{2})^8$ or is 70/256 or is 0.273. If the chances of heads showing is plotted against the number of heads showing for each of a series of 8 flips, the result shown in Figure 18 is obtained.

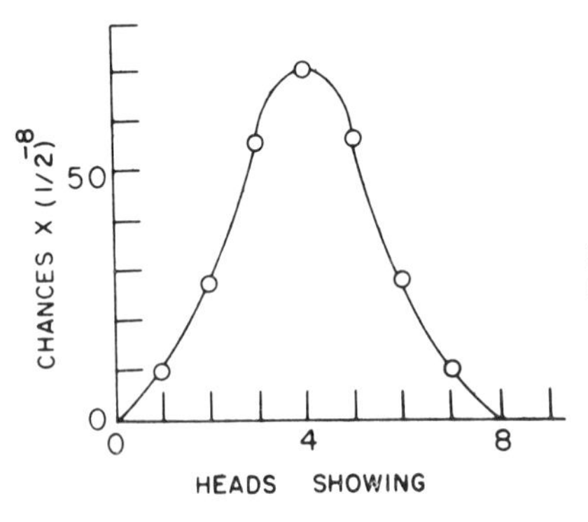

FIG. 18. Chances of "heads" showing out of eight flips of a coin.

The plot shown in Figure 18 is that of a probability curve; it is also to be noted that the terms $(\frac{1}{2})^8$, $8(\frac{1}{2})^8$, $28(\frac{1}{2})^8$, $56(\frac{1}{2})^8$, etc., are terms in the binomial expansion of $(x + y)^n$ where x and y are equal to $\frac{1}{2}$ and n is 8.

It is well known that if a quantitative determination of a physical quantity is made, the values measured will vary about the average value. If the deviations from the average are divided into equal intervals, the number of determinations in each interval counted and these numbers plotted against the deviation from the average, a typical bell-shaped curve will be obtained provided the number of determinations is large enough. The curve will resemble that shown in Figure 18 and is a probability or Gaussian distribution curve and in a more generalized form is shown in Figure 19.

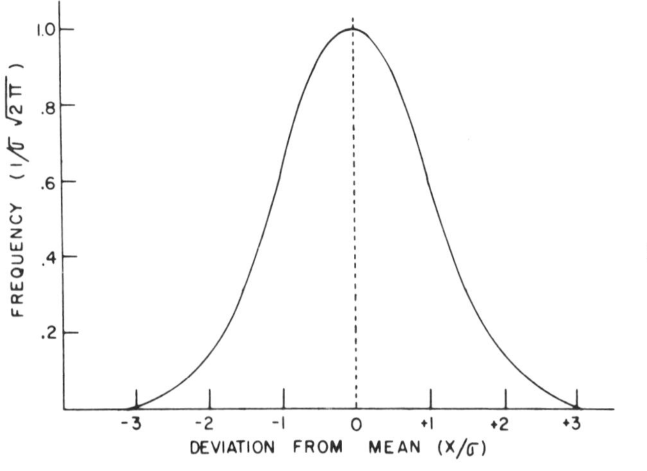

Fig. 19. Probability curve.

The equation for the probability curve must satisfy certain conditions. It must yield a maximum ordinate when the error is zero; it must be symmetrical with respect to the y-axis and, as the error (x) increases numerically, the frequency (y) must become vanishingly small. Such a curve is represented by the equation

$$y = y_0 e^{-a^2x^2} \qquad 116$$

where y_0 and a are constants into whose nature we shall inquire. Consideration of eq. 116 shows that y_0 will control the height of the curve and a the spread of errors.

The probability associated with a change of error, dx, is

$$dP = y_0 e^{-a^2x^2} dx \qquad 117$$

or

$$P = y_0 \int_{x_1}^{x_2} e^{-a^2x^2} dx \qquad 118$$

and between the limits x equals $\pm\infty$ the probability must obviously be one and

$$1 = y_0 \int_{-\infty}^{\infty} e^{-a^2x^2} dx = y_0 \frac{(\pi)^{1/2}}{a} \qquad 119$$

then y_0 is equal to $a/(\pi)^{1/2}$.

The meaning of the constant a can now be defined. Substituting value of y_0 into eq. 116, there results

$$y = \frac{a}{(\pi)^{1/2}} e^{-a^2x^2} \qquad 120$$

and the probability in the interval dx is

$$P = \frac{adx}{(\pi)^{1/2}} e^{-a^2x^2} \qquad 121$$

The probability of the occurrence of a set of independent errors x_1, x_2, ------x_n is the product of their separate probabilities or

$$P = \frac{a}{(\pi)^{1/2}} e^{-a^2x_1^2} \frac{a}{(\pi)^{1/2}} e^{-a^2x_2^2} \qquad 122$$

$$= \frac{a^n}{(\pi^n)^{1/2}} e^{-a^2\Sigma(x^2)} \qquad 123$$

Differentiating P in respect to a in eq. 123, there results

$$\frac{dP}{da} = \frac{na^{n-1}}{(\pi^n)^{1/2}} e^{-a^2\Sigma(x)^2} - \frac{2a\Sigma(x^2)a^n}{(\pi^n)^{1/2}} e^{-a^2\Sigma(x)^2} \qquad 124$$

For any set of observations, a has a maximum value and, to find this maximum, dP/da is set equal to zero and from eq. 124

$$a = \left(\frac{n}{2\Sigma(x^2)}\right)^{1/2} \qquad 125$$

The term $\left(\frac{\Sigma(x^2)}{n}\right)^{1/2}$ is known as the standard deviation and is usually indicated by σ or

$$a = \frac{1}{2^{1/2}\sigma} \qquad 126$$

Substituting the value of a into eq. 120 gives

$$y = \frac{1}{(2\pi)^{1/2}\sigma} e^{-\frac{x^2}{2\sigma^2}} \qquad 127$$

Actually, the standard deviation is best defined as

$$\sigma = \left(\frac{\Sigma(x^2)}{n-1}\right)^{1/2} \qquad 128$$

Since σ represents the standard deviation of a single observation, the comparison is made with $n-1$ observations and not the total number of observations. As the number of observations becomes very large the two definitions of σ become identical.

The standard deviation of the mean of a set of observations is

$$\sigma_m = \left[\frac{\Sigma(x^2)}{n(n-1)}\right]^{1/2} = \frac{\sigma}{n^{1/2}} \qquad 129$$

and the standard deviation (or standard error as it is sometimes called) of the difference between two means is

$$\sigma_d = \left(\frac{\sigma_1^2}{N_1} + \frac{\sigma_2^2}{N_2}\right)^{1/2} \qquad 130$$

where σ_1 is the standard deviation of the mean of the first sample and σ_2 that of the second sample. N_1 is the number of items in the first sample and N_2 that in the second. The standard error of the sum of two means is

$$\sigma_s = (\sigma_1)^2 + (\sigma_2)^2 \qquad 131$$

For the product of two means, it is

$$\sigma_P = \left(\frac{\sigma_1}{\bar{x}_1}\right)^2 + \left(\frac{\sigma_2}{\bar{x}_2}\right)^2 \qquad 132$$

where $\bar{x}_1$ and $\bar{x}_2$ are the arithmetic means of the first and second sample respectively.

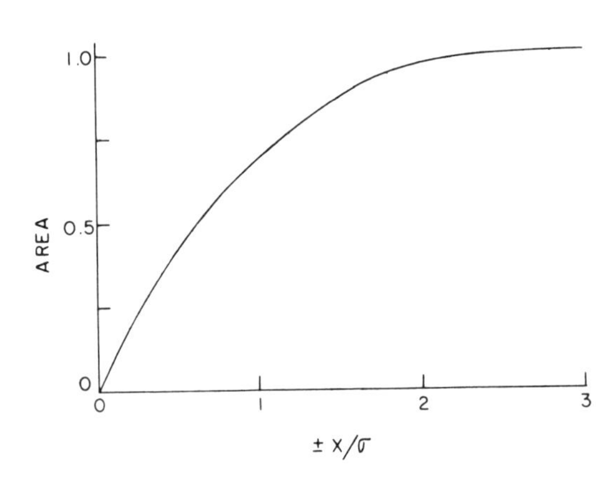

Fig. 20. Relative area under a probability curve as a function of the interval $\pm x/\sigma$ from the arithmetic mean.

The standard deviation is evidently a measure of the dispersion and of the goodness of observations. The meaning of the standard deviation becomes clearer if the area under the probability curve is plotted against the interval $\pm x/\sigma$ where x is the deviation from the arithmetic mean and σ is the standard deviation (see Fig. 20). It is to be noticed from Figure 20 that a value which deviates $\pm$ one σ from the mean covers 68.26 percent of the area under the distribution curve and 31.74 percent of the area lies outside of $\pm\sigma$. We, therefore, say that the chances of a value of the measurement deviating from the mean by $\pm\sigma$ is 0.3174 out of one; the deviation is evidently not significant. If the deviation is $\pm 2\sigma$ the chances are 0.0454 out of one and if 3σ, the chances are 0.0027 out of one that the particular value could have arisen by chance alone. Ordinarily, if a value differs by $\pm 2\sigma$ from the mean, it is said to be significantly different from the mean.

Sometimes the term probable error is used instead of the standard deviation. A deviation from the mean of plus or minus one probable error embraces just one half of the total area under the probability curve and

$$PE = 0.6745\sigma \qquad 133$$

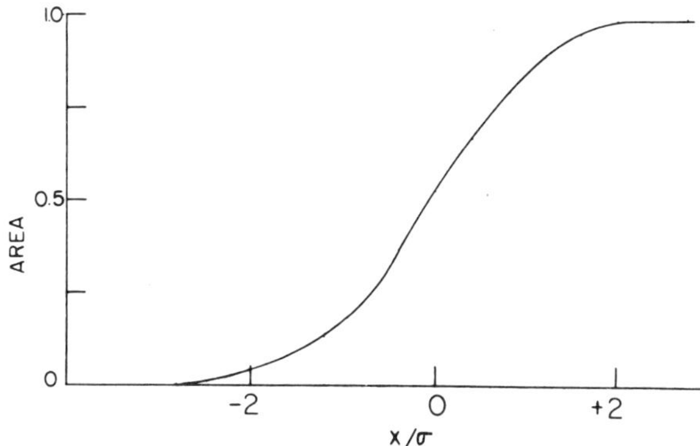

Fig. 21. Plot of the area under a probability curve as a function of x/σ.

It is also instructive to plot the area under a probability curve in a somewhat different manner from that shown in Figure 20. In Figure 21 is plotted the area against the scale of x/σ extending from -3 to $+3$ and as can be seen a sigmoid type of curve is obtained.

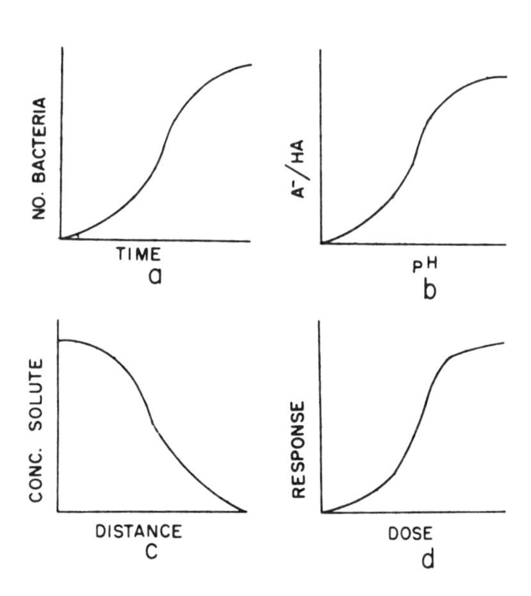

Fig. 22. Sigmoid curves: a, Growth of bacteria; b, titration of a weak acid (HA) with a strong base; c, diffusion at the boundary between a solution and the solvent; d, dose response curve of a drug.

There are many natural phenomena which give sigmoid curves when the proper plots are made; for example, the growth of bacteria with time, the pH of a solution of a weak acid plotted as a function of the added base, the dose response curve of a number of drugs, the concentration of a solute as a function of the distance the solute diffuses into the pure solvent. Shown in Figure 22 are hypothetical plots of these functions. The general equation for sigmoid curves is

$$x = x_0 + \log \frac{A}{1-A} \qquad 134$$

where A can have any value from zero to one; the sigmoid curve can evidently be linearized by plotting x against $\log \frac{A}{1-A}$.

The differential of eq. 134 is

$$dx = d \log A - d \log (1-A) = \frac{0.43\, dA}{A(1-A)} \qquad 135$$

or

$$\frac{dA}{dx} = 2.3A\,(1-A) \qquad 136$$

and the plot of dA/dx against A yields a bell-shaped (parabolic) curve.

The Gaussian or probability distribution is only one of the many kinds of distributions possible. Distribution curves which depart asymmetrically from the normal curve are said to be skewed.

Method of Least Squares. The most satisfactory equation to represent a linear relation of the type

$$y = a + bx \qquad 137$$

is obtained when the differences between the observed and calculated values are as small as possible and the constants a and b have been assigned their most probable values. If the data could be expressed in exact equations, for each pair of values for x and y

$$y_1 = a + bx_1 \qquad 138$$
$$y_2 = a + bx_2 \qquad 139$$
$$y_n = a + bx_n \qquad 140$$

Adding eqs. 138, 139 and 140 there results

$$\Sigma y = Na + b\Sigma x \qquad 141$$

where N is the number of sets of data.
Multiplying eq. 137 by the corresponding x-value gives

$$y_1x_1 = ax_1 + bx_1^2 \qquad 142$$
$$y_2x_2 = ax_2 + bx_2^2 \qquad 143$$
$$y_nx_n = ax_n + bx_n^2 \qquad 144$$

The addition of eqs. 142, 143 and 144 gives

$$\Sigma yx = a\Sigma x + b\Sigma x^2 \qquad 145$$

Comparing the simultaneous equations 141 and 145 permits the evaluation of the constants a and b. The substitution of the determined values of a and b into eq. 137 gives the least square equation, a plot of which is known as the line of regression.

It is important to realize that the magnitude of the constants a and b depends on the selection of the dependent and independent variable. It is assumed that there is no error in the independent variable and all of the experimental variations occur in the dependent variable. In eq. 137, x is always considered to be the independent variable and y the dependent variable. It can be shown that the standard deviations of a and b are

$$\sigma_a = \left(\frac{\Sigma d^2}{N-2}\right)^{1/2} \left(\frac{\Sigma x^2}{N\Sigma(x^2) - [\Sigma(x)]^2}\right)^{1/2} \qquad 146$$

and

$$\sigma_b = \left(\frac{\Sigma d^2}{N-2}\right)^{1/2}\left(\frac{N}{N\Sigma(x^2) - [\Sigma(x)]^2}\right)^{1/2} \qquad 147$$

where d represents the difference between the calculated and observed values of y.

The significance of the least square equation can be stated in still another way, namely, in terms of the correlation coefficient between the independent and dependent variable. It is evident that a comparison of deviations of the experimental values of y from the values of y calculated by the least square equation with the standard deviation of y should yield a measure of the scatter of the data. The standard error of estimate is the average (quadratic mean) of the deviations about the line of regression and is given by

$$S_y = \left(\frac{\Sigma d^2}{N}\right)^{1/2} \qquad 148$$

where d is again the difference between the calculated and observed values of y. The correlation coefficient is then given by the relation

$$r = \left(1 - \frac{S^2{}_y}{\sigma^2{}_y}\right)^{1/2} \qquad 149$$

where σ_y is the standard deviation of the values of y from the arithmetic mean of y. Evidently, if the relation between y-observed and y-calculated was exact, S_y would be zero and the correlation coefficient would be unity. If there were no relation between y and x, S_y and σ_y would be equal and the correlation coefficient would be zero. It is possible to have a negative correlation between the variables and the sign of the slope of the line of regression is attached to r.

If the number of data are small, eq. 149 has to be adjusted in order to yield a fair estimate and the corrected value of r becomes

$$r\ \text{corr} = \left[1 - (1 - r^2)\left(\frac{N-1}{N-2}\right)\right]^{1/2} \qquad 150$$

where N is number of experimental observations. The method of least squares is general and can deal with nonlinear equations as well as with multiple independent variables. For example, in a quadratic equation of the form

$$y = a + bx^2 \qquad 151$$

we can write

$$\Sigma y = Na + b\Sigma x^2 \qquad 152$$

Equation 151 can be multiplied by x and

$$\Sigma xy = a\Sigma x + b\Sigma x^3 \qquad 153$$

Comparison of eqs. 152 and 153 enables the best values of a and b to be calculated. If the dependent variable is a linear function of two independent variables x_1 and x_2, the appropriate equations can be written as follows:

$$\Sigma y = Na + b\Sigma x_1 + c\Sigma x_2 \qquad 154$$

$$\Sigma x_1 y = a\Sigma x_1 + b\Sigma x_1{}^2 + c\Sigma x_1 x_2 \qquad 155$$

$$\Sigma x_2 y = a\Sigma x_2 + b\Sigma x_1 x_2 + c\Sigma x_2{}^2 \qquad 156$$

The above simultaneous equations can be solved for the constant a and the coefficients b and c using available experimental data.

Dimensional Analysis. Of considerable interest and utility are the dimensions of physical quantities. The units of the C.G.S. system are centimeters, grams and seconds and the dimensions corresponding to these units are length (L), mass (M) and time (t). Temperature (T) and the dielectric constant (D) must be added to this list for completion.

An equation representing physical quantities must be homogeneous in respect to dimensions. That is, each term of an equation must have the same dimensions as every other term and the validity of even the most complex equation can be tested without delay by a dimensional analysis. The dimensions of various physical quantities such as viscosity, diffusion, etc., can be assigned by inspection and Table 2 lists the dimensions and customary units for several physical quantities. Note that when concentration is expressed in moles per unit volume, its dimensions are L^{-3} since the number of moles is a simple numerical ratio of the weight of the substance divided by its molecular weight. Such a dimensional quantity is called a numeric.

TABLE 2. DIMENSIONS AND UNITS OF SOME PHYSICAL QUANTITIES

Quantity	*Dimensions*	*Units*
Weight concentration	0	Grams per 100 grams solution
Molar concentration	L^{-3}	Moles per liter
Velocity	Lt^{-1}	cm. per second
Diffusion	L^2t^{-1}	sq. cm. per second
Flow	L^3t^{-1}	cc. per second
Acceleration	Lt^{-2}	cm. per second per second
Force	MLt^{-2}	Dynes
Flux	Mt^{-1}	Grams per second
Tension	Mt^{-2}	Dynes per cm.
Pressure	$ML^{-1}t^{-2}$	Dynes per sq. cm.
Action	ML^2t^{-1}	Calories per second
Work or energy	ML^2t^{-2}	Ergs, calories or Joules
Power	ML^2t^{-3}	Ergs per sec.

Some rules of dimensional analysis can be listed as follows:

1. Pure numbers such as π, $\frac{1}{2}$ and 50 are dimensionless (numeric).
2. Complete exponents are dimensionless.
3. Ratios of dimensionally identical quantities are dimensionless.
4. Sines, cosines, etc., are dimensionless.
5. Differentials retain the dimensions of the variable.
6. If quantities are additive, they must have the same dimensions.

By the use of dimensions and some intuition, it is possible to formulate equations relating various physical quantities. Consider, for example, the volume of a viscous liquid flowing per unit time through a capillary. Intuitively, one feels that the volume of liquid flowing should be related to pressure exerted on the liquid, the viscosity of the liquid and on the radius and length of the capillary. That is

$$V = CP^a\eta^b r^c l^d \qquad 157$$

where V is the volume of the liquid flowing per unit time and whose dimension is L^3t^{-1}. C is a numeric constant. The dimension of pressure are $L^{-1}Mt^{-2}$, of viscosity $L^{-1}Mt^{-1}$ and the dimension of r and of l is L. Substituting the above dimensions in eq. 157

$$L^3t^{-1} = (ML^{-1}t^{-2})^a(ML^{-1}t^{-1})^b L^c L^d \qquad 158$$

The following equations must be true since eq. 158 must be homogeneous in respect to each dimension

$$\text{Length } (L) = -a - b + c + d = 3 \qquad 159$$
$$\text{Mass } (M) = a + b = 0 \qquad 160$$
$$\text{Time } (t) = -2a - b = -1 \qquad 161$$

Solving for a, b, c and d by the use of eqs. 159, 160 and 161, we find

$$a = 1$$
$$b = 1$$
$$c + d = 3$$

d represents the exponent of the length of the capillary and, intuitively, the volume of the liquid flowing should be inversely proportional to the length of the capillary. This means that d is equal to -1 and c must, therefore, equal 4. It is now possibie to write the equation for the flow of a viscous liquid through a capillary and

$$V = \frac{CPr^4}{\eta l} \qquad 162$$

Compare eq. 162 with the Poiseuille equation for the flow of liquids in capillaries.

Comments on Computers. People have been using computers of one kind or another for a very long time. An abacus along with the human operator constitutes a digital computer; the human supplies the memory, retrieval, and programming. A slide rule represents numbers by lengths and is, accordingly, an analogue computer.

Modern electronic computers come in two kinds, analogue and digital; the analogue computer is essentially an example of model building. The elements of an electrical circuit such as condensers, resistances, amplifiers, etc., represent stresses, strains, times, etc., of the physical situation. The analogue computer can carry out various mathematical operations including addition, multiplication and integration and the results can be displayed on an oscilloscope screen or otherwise recorded. An analogue computer is not very good at differential calculus; the integral of the white noise of the equipment is zero but the differential is not. It is, however, effective in dealing with differential equations and in curve fitting; the coefficients can be varied until agreement with experiment is achieved. Since the operation depends on the magnitudes of the capacitances, resistances, and amplifiers, and these elements are not always exact, the analogue computer suffers from errors. Analogue computers come in various sizes and complexities and at correspondingly variable prices and they can be constructed to meet individual model-building needs. The simpler computers are relatively inexpensive and the space requirements are not large and, accordingly, a single laboratory can purchase and operate an analogue computer.

In principle, a digital computer is simpler than is the analogue computer, but in detail it is a far more elaborate instrument. All mathematical operations have been reduced to addition and subtraction and it usually but not necessarily employs the binary number system; unlike the analogue computer, it is an exact instrument. The accessory units are what make the digital computer expensive, complex, and space demanding. The cost is considerable and, in order to be useful, it needs a highly skilled staff; the cost is such that only the more prosperous institutions can afford ownership.

There are certainly large areas of scientific endeavor which do not need computer help. On the other hand, some modern developments would have been impossible without computer facilities. For example, calculations related to X-ray diffraction studies on crystals of proteins could not possibly have been done without digital computers and interpretations of these diffraction pictures in terms of detailed molecular structure could not have been made.

About the only decision that the average biochemist has to make is whether or not to ask for the fairly expensive aid of the university computer center. The rule of thumb usually given is that if he is able to do the job with a desk calculator in three weeks, he is better off on his own. In principle, anything that a computer can do a human can do, but not nearly so fast, that is, after the programming has been arranged.

GENERAL REFERENCES

J. G. Defares and I. N. Sneddon: An Introduction to the Mathematics of Medicine and Biology. The Year Book Publishers, Inc., Chicago, 1960.
H. E. Huntley: Dimensional Analysis. Rinehart and Co., Inc., New York, 1951.
J. W. Mellor: Higher Mathematics for Students of Chemistry and Physics. Reprint Dover Publications Inc., New York, 1955.

PROBLEMS

1. Given $\log_{10} 4 = 0.6021$ and $\log_{10} 8 = 0.9031$
   (a) Find $\log_{10} (25.6)^{1/2}$
   (b) Find $\log_4 8$
   (c) Solve for x if $4^x = 8$

   Ans: (a) 0.70415, (b) 1.50, (c) 1.50

2. Show how you would obtain a linear plot of the equation

$$y^2 = a + \frac{b}{x^3}$$

   where x and y are variables and a and b are constants. Indicate how you would obtain a graphical evaluation of the constants.

   Ans: plot y^2 vs $1/x^3$

3. Given the equation $y^2 = a + bx$
   (a) Find the slope of a plot of y against x at x equal 2 assuming a is equal to 5 and b equals 10. Obtain a graphical as well as an analytical solution.
   (b) Find the area under the curve from x equal zero to x equal 2 both analytically and graphically (assume the values of a and of b to be the same as in 3a).

   Ans: (a) 1.0, (b) 7.605

4. A solution containing m_2 moles of protein per liter with Z_2 positive charges per mole is dissolved in a solution of sodium chloride and separated from a solution of sodium chloride by means of a semi-permeable membrane. The osmotic pressure exerted by this protein solution is

$$\Pi = \frac{RTC_2}{M_2}\left[1 + \frac{y}{m_2}\left(1 - \left[1 - \frac{Z_2 m_2}{y}\right]^{1/2}\right)^2\right]$$

   where R is the gas constant, T the absolute temperature, M_2 the molecular weight of the protein, C_2 is the concentration of the protein in grams per cc. of solution and y is the total number of moles of chloride ions on both sides of the membrane. Simplify the above expression by means of expansion and rearrange to the simplest form.

   Ans:
$$\Pi = \frac{RTC_2}{M_2}\left[1 + \frac{Z_2^2 m_2}{4y} + \frac{Z_2^3 m_2^2}{8y^2}\right]$$

5. A human hair 10 cm. long and 71 microns in diameter was stretched by progressive increase in weights attached to the vertical fiber with the following results:

Length of hair in cm.	*Weight applied in grams*
10.000	0
10.091	2.06
10.136	5.15
10.244	8.24
10.282	10.30
10.324	13.39
10.405	17.51
10.413	18.54
10.545	19.57
11.230	20.60
11.810	21.63
12.240	22.66
12.570	23.69

   Calculate the work in ergs required to stretch the hair to 12.5 centimeters.

   Ans: 46,500 ergs

6. (a) The optical density (O.D.) of a series of solutions of egg albumin was determined at a wavelength of 280 mμ with the following results:

Protein concentration in grams per 100 ml.	*Optical density*
0.00083	0.005
0.00165	0.010
0.00413	0.027
0.00578	0.039
0.00826	0.055
0.01239	0.085
0.01652	0.117
0.02477	0.172
0.03303	0.236

Plot the optical density against the protein concentration in grams per 100 ml. and draw the best straight line through these points by the method of least squares. Calculate the standard deviations in the slope and in the intercept of the straight line.

(b) The optical density of a single solution of egg albumin whose concentration was 0.005 grams per 100 ml. was determined and found to be 0.039. What is the chance that this result is due to experimental error?

Ans: (a) $a = -0.0024 \pm 0.0009$; $b = 7.1525 \pm 0.0594$
(b) $\Delta y/\sigma_y \doteq 3.46$

7. A chromatographic column is being eluted with a 0.01 M phosphate buffer pH 7.0 at a rate of 10 ml. per hour from a 100 ml. container kept filled with buffer. After a given time, the influx of 0.01 M buffer into the container is stopped and a switch is made to 0.10 M buffer at the same previous rate. The contents of the 100 ml. container are well stirred. The chromatogram is 50 cm. long and contains 15 ml. of liquid. Plot the concentration of the buffer emerging from the chromatogram as a function of time in minutes.

Ans: After 90 minutes, $-\log(0.10 - C) = 1.046 + 0.725 \times 10^{-3}t$

CHAPTER 2

Energetics

Energy associated with living systems has great interest and much of modern biochemistry deals with sources of energy derived from intermediary metabolism and the use of this energy in tissue. Energy is defined as the ability to do work and can assume various forms such as mechanical, heat, electrical, osmotic, surface, chemical, radiant and atomic energy.

Heat. As with all forms of energy, heat is the product of an intensity factor and a capacity factor; the intensity factor is temperature and the heat capacity, the capacity factor. It is to be recalled that the calorie is the quantity of heat required to raise one gram of water from 15° to 16°C. Heat is a physiological sensation; as Mark Twain once remarked, a cat who has sat upon a hot stove will not sit upon a cold one. Heat may be defined as that form of energy which passes from one body to another solely as the result of a difference of temperature. Heat manifests itself as disordered motion of molecules and this motion can be resolved into (1) translation of molecules from one position to another, (2) rotation of the molecules and parts of molecules, and (3) vibration of parts of molecules in relation to each other. Each type of molecular motion contributes to the heat capacity of a body, i.e., the amount of heat absorbed by a body for a given rise in temperature (see Fig. 1).

The pressure exerted by a gas is due to the change in momentum of the gas molecules as they collide with the walls of the container. For one mole of ideal gas (all gases obey the ideal gas laws at very low pressures)

$$P = \frac{RT}{V} \qquad 1$$

where P is the gas pressure, V is the volume of the gas and T is the absolute temperature. Substituting numerical values of P, V and T at standard conditions and solving for R, it turns out that R is equal to 8.315×10^7 ergs per degree per mole; and, since one calorie is equal to 4.185×10^7 ergs, R is also equal to 1.987 calories per degree per mole; R is a constant because it has been defined as such. The Boltzmann con-

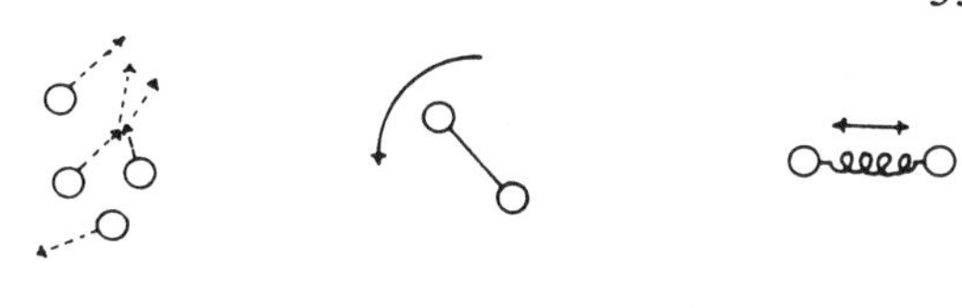

Fig. 1. Three types of molecular motion which contribute to the heat capacity of a body.

stant is the gas constant per molecule instead of per mole and is equal to R divided by Avogadros number or is 1.38×10^{-16} erg per degree.

It can be shown by simple means that the average translational energy per mole of molecules is equal to $\frac{3}{2}\ RT$ and the translational energy in any one of the three directions perpendicular to each other is $\frac{1}{2}RT$. (At 25°C. RT is equal to 592 calories.) It is inherently improbable that all the molecules of a given kind in a system would have identically the same velocities, energies, etc. The distribution of velocities and energies of molecules at a given temperature can be calculated by Maxwell's distribution law. Figure 2 shows the relative number of molecules having a particular energy of translation, the energies being expressed in calories per mole at 25°C. The average energy (889 cal) is indicated by a broken vertical line. Note that the distribution of energies is skewed towards the lower energies. This skewness is more pronounced at lower temperatures and with increasing temperature the distribution curve becomes flatter and approaches a Gaussian distribution.

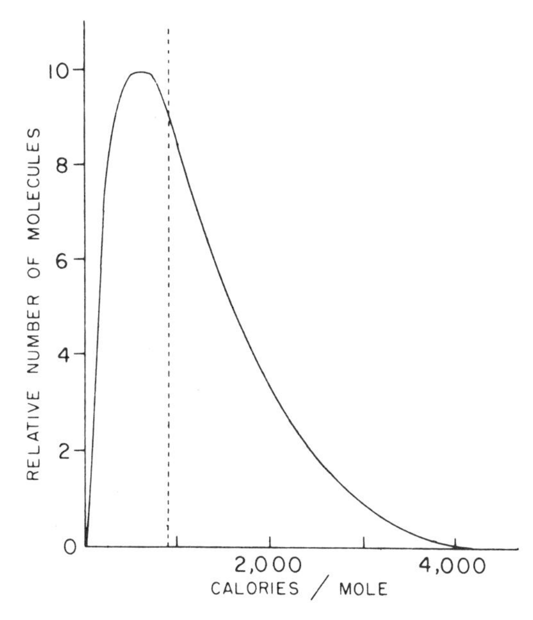

Fig. 2. Distribution of molar translational energies at 25°C.

The kinetic energy of a molecule depends only on the temperature and not at all on the nature of the medium, i.e., at a given temperature the kinetic energy is the same whether the molecules exist as a gas or in solution; a molecule can be defined as a particle which has a total translational energy of $\frac{3}{2}\ RT$ calories per mole.

In addition to the translational energies considered above, molecules also have rotational energies. Rotation in each of three directions perpendicular to each other contributes $\frac{1}{2}\ RT$ calories per mole to the kinetic energy so that the total rotational kinetic energy is $\frac{3}{2}\ RT$ calories per mole.

An outstanding characteristic of heat is to induce expansion in a homogeneous object and this property is taken advantage of in the measurement of the intensity of heat, because, in general, the amount of expansion over a limited temperature range is almost directly proportional to the intensity of the heat (temperature), and the usual laboratory thermometer depends on the thermal expansion of mercury and is calibrated against the freezing and boiling points of water and graduated in the centigrade scale. Actually, the triple point of water at which the water vapor, ice and liquid water exist at the vapor pressure of water (4.58 mm. Hg) is 0.0100°C.

The expansion of the volume of an ideal gas at constant pressure and at zero degrees centigrade is $1/273.15^{th}$ per degree centigrade and, accordingly, absolute zero temperature is at −273.15°C and the temperature on the absolute scale (Kelvin scale) is

$$T = T_c + 273.15 \quad (2)$$

where T_c is the temperature in degrees centigrade.

Whereas, under many experimental conditions a mercury thermometer is entirely suitable as a temperature indicator, there is sometimes a demand for more flexible and sensitive temperature probes. Thermocouples and more recently thermistors have proven extremely useful in thermometry.

If two unlike metals are placed in contact with each other, a potential difference develops between them and this potential varies with the temperature. A thermocouple consists of two such junctions in series, one of these junctions being maintained at a given temperature (usually the cold junction) and the other junction acting as the thermometer. There are a number of combinations of metals which can be used although junctions made of copper and constantan (copper-nickel alloy) and of iron and constantan are popular. Diagramed in Figure 3 is an apparatus which illustrates the basic arrangement for a thermocouple.

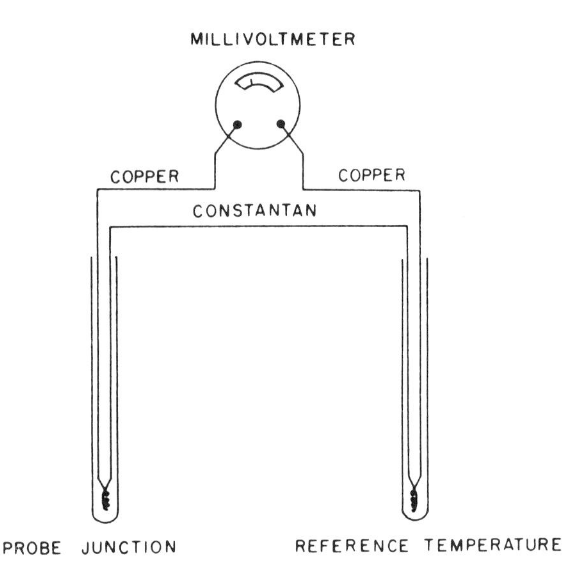

Fig. 3. Thermocouple with two junctions.

Shown in Figure 4 is a graph of the potential difference in millivolts between two copper-constantan junctions with the reference junction at zero degrees centigrade. Note that the relation between the potential difference and the temperature is not exactly linear. By increasing the number of junctions and connecting the junctions in series, the response to temperature can be made very sensitive; such a combination of thermocouples is known as a thermopile.

In general, the electrical resistance of metals increases with increasing temperature and it would be possible to measure the temperature by determining the resistance of a coil of metallic wire. The temperature coefficient of the resistivity of conductors is, however, small and semi-conductors such as crystals of silicon and germanium are far more suitable. Semi-conductor resistance elements are known as

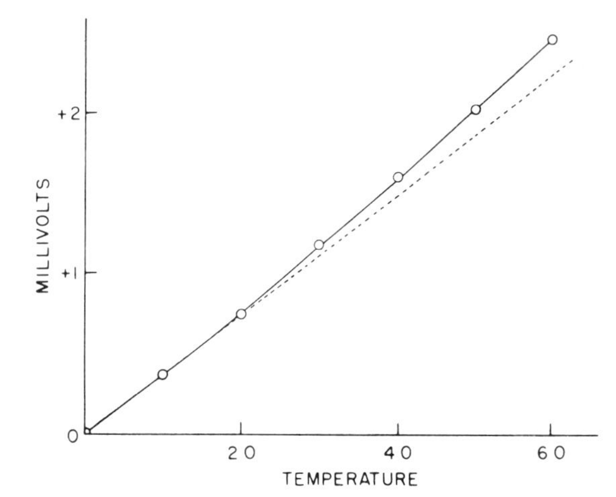

Fig. 4. Potential of a copper-constantan junction as function of temperature.

thermistors and are characterized by both a significant resistance at ordinary temperatures and a temperature coefficient of resistance of about 10 times that of commonly used conductors; the temperature coefficient is negative. Thermistors have a small volume, low heat capacity and a fast response to changes in temperatures. Shown in Figure 5 is a typical resistance-temperature plot of a thermistor. For temperature control and measurement with a thermistor, a semi-conductor is ordinarily used as one arm of a wheatstone bridge, the balance of the wheatstone bridge being disturbed by the change of resistance of the semi-conductor in response to a change in temperature.

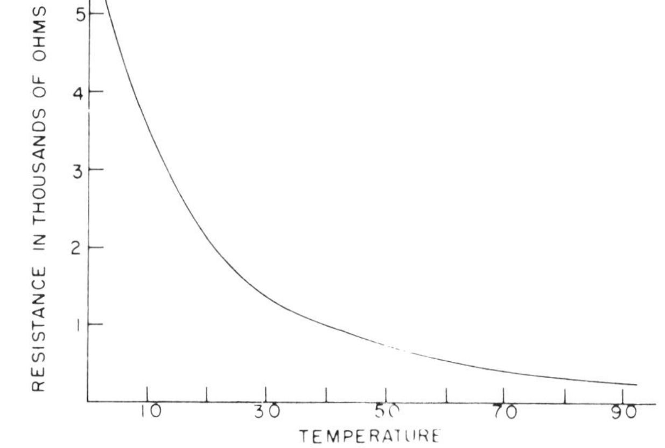

Fig. 5. Resistance of germanium thermistor as a function of the temperature (centigrade).

The heat capacity of a body is the property which when multiplied by the temperature change gives the quantity of heat in calories which has entered or departed when a body is brought into contact with another body having a different temperature and

$$C = \frac{Q}{\Delta T(T_2 \rightarrow T_1)} \qquad 3$$

It is possible to express heat capacities as a function of temperature over a considerable range of temperatures in the form of an empirical power series and

$$C = a + bT + cT^2 + dT^3 + \qquad 4$$

With few exceptions, heat capacities increase with increasing temperature. Shown in Figure 6 is the variation of the specific heat of water as a function of temperature; interesting is the broad minimum at about 35°C.

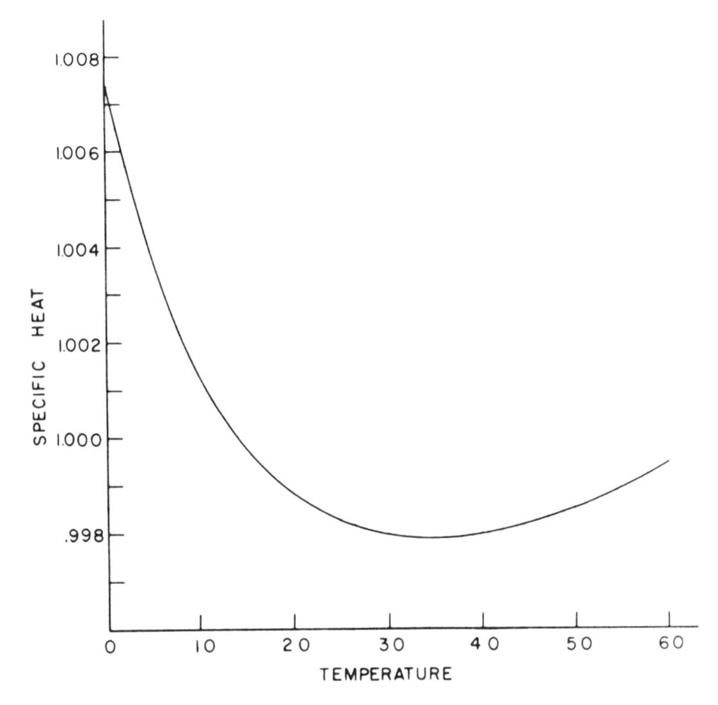

Fig. 6. Specific heat of water as a function of temperature; the specific heat of ice at 0°C is 0.502.

The heat evolved in chemical reactions and other processes is measured in a calorimeter and requires a determination of the change in temperature resulting from the reaction. It is necessary to know the volumes, densities, concentrations and temperatures of the initial solutions, the specific heats of the solutions along with the heat capacity of the apparatus. Corrections also have to be applied for the loss or gain of heat from the surroundings. The measurement of heats to about 2 percent accuracy is easily accomplished with a Dewar flask in which the mixing of the solutions takes place; thus the apparatus is simple and is diagramed in Figure 7.

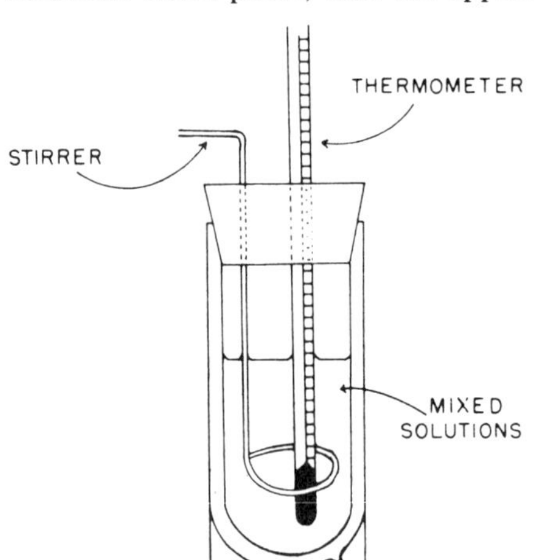

Fig. 7. Simple calorimeter.

A calorimeter, however simple or complicated it may be, has to be calibrated with the release of a known amount of heat. This is accomplished in either of two ways: (1) A reaction is allowed to proceed whose heat is exactly known; a favorite reaction is simple neutralization of a strong acid with a strong base. Thus the neutralization of one mole of HCl with one mole of NaOH both in one molal solutions at 25°C is 13,360 calories. (2) The heat in calories developed in a circuit by an electric current of i amperes flowing through a resistance of R ohms under a potential difference E in volts for t seconds releases q calories.

$$q = \frac{Eit}{4.185} = \frac{Ri^2t}{4.185} \qquad 5$$

There are several different kinds of calorimeters. In an adiabatic calorimeter, every effort is made to isolate the reaction mixture from the environment and to insulate against heat flow into or out of the calorimeter. Unfortunately, no substance is opaque to heat and such calorimeters tend to become complex and overburdened with instrumentation. More recently heat-burst calorimeters have been developed which use small volumes of reactants and permit the measurement of quite small heat releases with surprising accuracy. Such calorimeters can be very useful in the study of enzymic reactions. Heat-burst calorimeters provide for heat flow through a thermopile and into a metal block which acts as a heat sink. The potential derived from the pile is amplified and continuously recorded, and the total heat is represented by the area under the heat-flow curve. A variation on the heat-burst calorimeter is the flow calorimeter. Reactants are mixed and flow through a platinum tube of small diameter. The heat produced crosses a thermopile and is drained off into a heat sink. This equipment is commercially available.

In calorimeters suitable for the measurement of the heat capacity of a substance, heat is supplied to the system by means of a heating coil and the increase in temperature is noted. A variation on the ordinary heat-capacity calorimeter is the differential calorimeter, in which heat is supplied at a constant rate and the difference in temperature between a standard and the sample is noted as a function of temperature. The technique is useful in the study of the thermal denaturation of proteins and of nucleic acids. Differential thermal analyzers, as differential calorimeters are called, are commercially available.

Of great interest is the heat evolved from a muscle during contraction and from a nerve during conduction. Such measurements require careful attention to the design of the calorimeter. Practically all work along these lines is patterned after the simple, elegant and pioneer studies of A. V. Hill. The rise of temperature in a muscle twitch is small, being of the order of $3 \times 10^{-3\circ}$ and the time of heat release is short and complete after about 0.4 second. The method used to measure the heat of muscle contraction is to place the small muscle fiber in contact with a thermopile and to stimulate the muscle to contract; the rise in temperature is recorded with a fast galvanometer connected to the thermopile. The apparatus is calibrated by passing known currents through the killed muscle and noting the galvanometer reading. The thermopile described by Hill consisted of thirty couples of palladium-gold and iron; its resistance was 17.1 ohms and gave 1.363 millivolts per degree. It was about 40 microns thick and had an exceedingly fast response.

Enthalpy. The burning of one mole of solid D-glucose to carbon dioxide and liquid water in the presence of oxygen at 25°C and at 760 mm. mercury pressure releases 673,000 calories of heat and the reaction is written

$$C_6H_{12}O_6 + 6O_2 \rightarrow 6CO_2 + 6H_2O; \Delta H = -673{,}000 \qquad 6$$

Note the negative sign of the heat; since heat was evolved, the system is poorer by 673,000 calories and, accordingly, the heat has a negative sign. The heat change indicated by ΔH is known as the enthalpy and is the heat of reaction at constant temperature and pressure provided no work other than the work of expansion is done on or by the system. The enthalpy of a process is independent of the number and nature of the steps involved in proceeding from the initial to the final state. Thus, in the burning of glucose in the animal body through a series of complex transformations to carbon dioxide and water, the heat change or enthalpy is identical with that obtained by the combustion of glucose in a bomb calorimeter, providing the initial and final temperatures of the bomb are both at 37°C and also that a small correction be applied to the physiological combustion since the oxygen and carbon dioxide in the body are not in their standard states.

The biochemist is usually interested in processes conducted at constant pressure instead of at constant volume, but it is easier to determine the heats of reaction at constant volume (in a bomb calorimeter). The relation between the heat at constant volume and constant pressure is per mole

$$\Delta H = Q_v + P\Delta V \tag{7}$$

where Q_v is the heat at constant volume, and ΔV is the volume change at pressure P. If n_1 is the number of moles of gaseous reactants of the reaction and n_2 is the number of moles of gaseous products of the reaction, the process is accompanied by an increase of $n_2 - n_1$ or Δn moles of gas, the work done by the expansion of the gas is ΔnPV and, since PV per mole is, as has been noted, equal to RT, eq. 7 can be written

$$\Delta H = Q_v + RT\Delta n \tag{8}$$

The burning of glucose to liquid water and gaseous carbon dioxide involves no change in the volume so that Δn is zero and ΔH is equal to Q_v.

The enthalpy of a reaction is equal to the difference between the heat content of the products of the reaction and the heat content of the reactants. The heat of formation of a compound is the increase of the heat content when one mole of the substance is formed from its elements at a given temperature and pressure; the heat content of all elements in their standard states is considered to be zero, and, since the heat of formation of the compound is the difference between the heat content of the compound and that of the elements, it follows that the heat content of a compound is equal to its heat of formation.

Work. Work is defined as the product of the force acting (intensity factor) times the distance through which the force acts (capacity factor) and the unit of work is the erg. An erg is the work done in exerting a force of one dyne through a distance of one centimeter and a dyne is the force which will produce an acceleration of one centimeter per second per second in a mass of one gram. Thus a mass of one gram exerts a force downward equal to the acceleration of gravity (about 981 dynes).

It has been known for many years that there is an equivalency between work and heat. For example, a paddle wheel is caused to rotate in water by a falling weight and the rise in temperature of the water is measured and the equivalency between work and heat obtained. When this type of experiment is performed with all of its modern refinements, it is found that one calorie is equal to 4.185×10^7 ergs. Heat

TABLE 1. COMPARISON OF THE VARIOUS FORMS OF ENERGY

Energy Forms	*Mechanical*	*Heat*	*Electrical*	*Radiant*	*Chemical*
Dimensions	force × distance	temperature × heat capacity	voltage × current	frequency × h	chemical potential × moles
Usual units	erg	calorie	joule	quanta	calorie
Equivalency of units	1 erg	$= 2.389 \times 10^{-8}$ calorie	$= 1 \times 10^{-7}$ joule	$= 1/6.62 \times 10^{-27}\nu$*	

* ν is the frequency of the radiation.

and work can also be expressed in terms of joules, a joule being the work expended by an electric current of one ampere flowing through a resistance of one ohm. One joule is equal to 1×10^7 ergs.

Anticipating the discussion of the various forms of energy, a comparison of some of these forms is shown in Table 1.

It has also been observed that whereas it is possible to convert work energy completely into heat it is not possible to convert heat completely into work; there always remains a quantity of heat energy which is unavailable to do useful work. It turns out that this unavailable energy is related to the disorder of the system.

In Chapter 1, the work done in compressing a gas with a snugly fitting and frictionless piston at constant temperature was given as

$$\text{work} = \int_{v_1}^{v_2} PdV \qquad 9$$

Instead of compressing a gas, consider the work of concentrating a solution. An aqueous solution of a solute is placed in a cylinder and the head of the piston is equipped with a semipermeable membrane through which water molecules but not the solute can pass. The arrangement is diagramed in Figure 8.

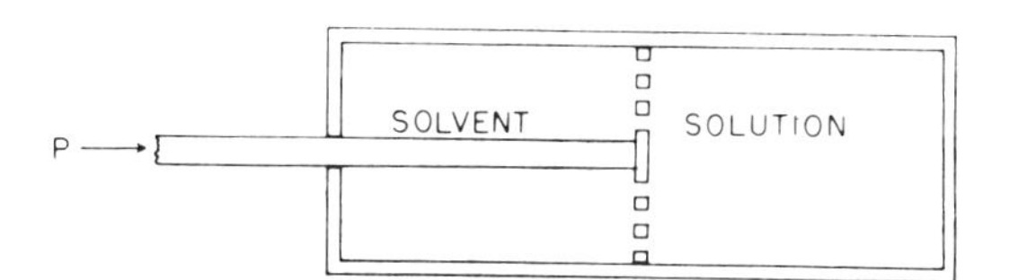

Fig. 8. Hypothetical apparatus for concentrating a solution. The piston head is permeable to solvent but not to solute.

The solution is concentrated slowly by exerting a pressure on the piston. The pressure which will neither concentrate nor permit dilution of the solution is the osmotic pressure of the solution. To concentrate the solution, the external pressure is permitted to exceed the osmotic pressure very slightly. It will be found that, if the osmotic pressure exerted by the solution is divided by the concentration and plotted against the concentration, over a limited concentration range straight lines will result; such a plot is shown in Figure 9.

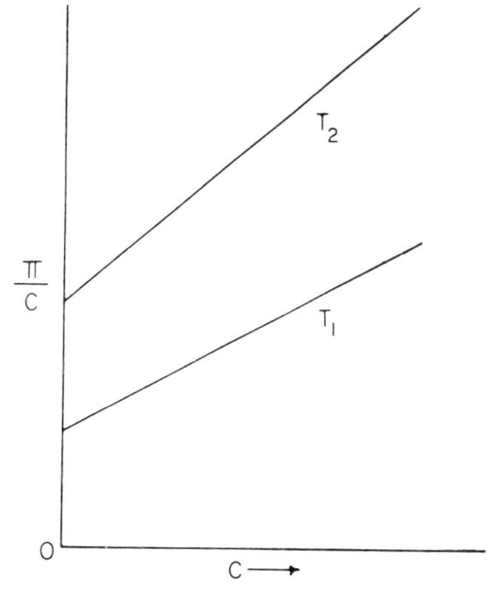

Fig. 9. Plot of the osmotic pressure (Π) divided by the molar concentration of the solute (C) against C at temperature T_1 and at a higher temperature T_2.

The slopes of the lines in Figure 9 depend on the nature of the solute and of the solvent used, but per mole and at a given temperature Π/C always extrapolates at zero concentration to the same value of Π/C for all solutes and solvents. At another and higher temperature, T_2, the extrapolation of Π/C to zero concentration yields a larger value of Π/C and, in fact, Π/C is directly proportional to the absolute temperature and the y-axis in Figure 9 is an osmotic thermometer. Without further delay, the osmotic pressure of an ideal solution (as C approaches zero) can be written

$$\Pi = CRT \tag{10}$$

where R is the proportionality factor between Π/C and T, and which is equal to the molar gas constant. The osmotic pressure exerted by a solution is analogous to that exerted by a gas, but, as will be noted in Chapter 7, the molecular basis for the existence of an osmotic pressure is quite different from the pressure exerted by a gas.

Returning to the work of concentrating a solution (Fig. 8) and substituting for V in eq. 9 its equal, $1/C$, there results

$$\text{work} = -\int_{C_1}^{C_2} \frac{\Pi dC}{C^2} \tag{11}$$

Substituting eq. 10 in eq. 11 it is found that

$$\text{work} = -RT \int_{C_1}^{C_2} \frac{dC}{C} \tag{12}$$

Integrating eq. 12 between the limits C_1 and C_2

$$\text{work} = -RT \ln \frac{C_2}{C_1} \tag{13}$$

Equation 13 gives the osmotic work performed in concentrating an ideal solution from C_1 to C_2. If the solution had been diluted from C_2 to C_1, the osmotic work used in concentrating the solution could have been regained provided the osmotic machine was coupled with the piston.

Earlier, the distribution of velocities and energies in a uniform field of force was considered (Maxwell distribution). The Boltzmann distribution describes the molecular concentrations in a non-uniform field of force, for example, in the gravitational field of the earth. The Boltzmann equation is derived by equating the rate of transfer of molecules from a region of higher potential to one of less. The equation is

$$\frac{n_1}{n_2} = e^{-\psi/RT} \tag{14}$$

where n_1 is the number of molecules per unit volume at one location and n_2 is that at a second location and ψ is the difference in potential energy of the molecules in the two locations. The difference in potential energy may arise as the result of any form of force acting on the molecules, such, for example, as the centrifugal force in an ultracentrifuge or an electrical potential acting on ions.

Note that eq. 14 can be written

$$\psi = -RT \ln \frac{n_1}{n_2} \tag{15}$$

Compare eqs. 13 and 15.

Survey of Thermodynamics. During the last century and the first part of the present century there developed a system of natural logic based upon three postulates or laws. The first and second laws were generalizations derived mostly from measurements on heat engines of various kinds. The third law has a somewhat more complex origin. These three laws form the basis of thermodynamics. The emphasis and indeed the great value of thermodynamics is in consideration of systems in equilibrium and perhaps a better name would be thermostatics. It will be impossible to give anything like a complete exposition of the science of thermodynamics, but there are certain basic ideas which can be presented and the notion and use of chemical affinity can be illustrated along with some applications to biological systems.

To use thermodynamics, the systems to which it applies must be defined. An open system permits the outflow and inflow of some or all constituents and of all kinds of energy; a living system is typically an open system although it is possible to restrict the flow of certain constituents and energy forms for shorter or longer periods of time. A closed system permits exchange of energy but not of the constituents with the surroundings. An isothermal process in which the temperature is maintained constant and heat transfers into or out of the system is typical of a closed system. A reaction vessel in a constant temperature water bath could be an example provided no substances are added and none allowed to escape during the course of the reaction. An isolated system permits neither energy nor constituent exchange with the surroundings. This condition can be closely approached in practice by insulating the reaction vessel and many calorimeters are systems of this kind. A reaction conducted in an isolated system is known as an adiabatic process. The open, closed and isolated systems are symbolized in Figure 10.

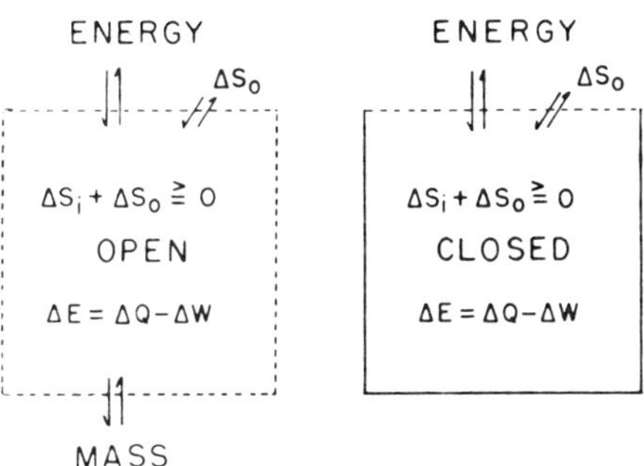

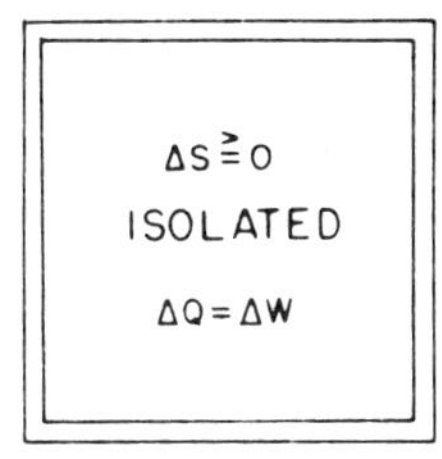

Fig. 10. Defined systems of thermodynamics.

The first law of thermodynamics is a statement of the conservation of energy; energy can change in form but cannot be created or destroyed. This idea can be formalized as

$$\Delta E = \Delta Q - \Delta W \qquad 16$$

where ΔE is the change in internal energy of a system, ΔQ is the heat absorbed by the system and ΔW is the work done by the system. For an isolated system, ΔE must be zero and the work done is always equal to the heat change; this is a statement of the equivalency of heat and work.

For a closed system, the change of internal energy must always equal $\Delta Q - \Delta W$ and not only is $\Delta Q - \Delta W$ independent of the arrangement for a given mechanical process but it is also independent of the pathway of a chemical process. In the complete combustion of a given quantity of glucose a certain amount of heat is evolved irrespective of the pathway of this combustion. This, however, is only true if no external work such as lifting a weight had been accomplished by the combustion of the glucose. The heat would be less by the amount of work done. It should be clear from what has been said that if a system is returned to its initial conditions ΔE for the completed cycle is zero. The first law of thermodynamics has nothing to say about the direction of a process, whereas the second law of thermodynamics concerns itself primarily with the spontaneity and feasibility of a process and direction is a central idea.

As noted above, in the isothermal dilution of an ideal solution from C_1 to C_2 the maximum osmotic work which can be obtained is

$$\text{Osmotic work} = -RT \ln \frac{C_2}{C_1} \tag{17}$$

Since the energies of the solute and solvent molecules in the concentrated and diluted solution must be identical (the solution is ideal, the temperature is constant; the translational, rotational and vibrational energies of the solute molecules are unchanged by simple dilution), the total energy change must, therefore, be zero and from the first law of thermodynamics

$$\Delta E = \Delta Q - \Delta W = O \tag{18}$$

and

$$\Delta Q = \Delta W = -RT \ln \frac{C_2}{C_1} \tag{19}$$

where ΔQ is the amount of heat which has flowed into the solution from the outside to maintain a constant temperature (isothermal process).

Evidently, here is a measure of spontaneity of a process; obviously, it is impossible for the solution to concentrate itself. Equation 19 can be rearranged to read

$$\frac{\Delta Q}{T} = -R \ln \frac{C_2}{C_1} = \Delta S \tag{20}$$

where $\Delta Q/T$ is equal to the entropy change (ΔS) of the process of dilution and is always positive in a spontaneous process; ΔQ as used above is called the reversible heat of the reaction. The above argument constitutes a statement of the second law of thermodynamics.

Suppose the concentrated ideal solution is diluted simply by pouring in some solvent and no osmotic work is done. Since the entropy of the system depends only upon the initial and final states and not at all on the pathway of the process, the entropy change involved in the reversible dilution of the solution while doing the maximum osmotic work is exactly equal to that of the simple and irreversible dilution of the solution in which no work was done. However, the entropy change of the workless dilution is no longer equal to the heat change divided by the temperature. In fact, the heat of the workless dilution must have been zero since no work was done. It is impossible to dilute a real solution without some change in internal energy; forces of attraction and of repulsion exist between the solute and solvent molecules, and when the solution is diluted the solute molecules are moved apart and heat is either absorbed or released. The entropy change in the reversible dilution of a real solution in which the maximum work is performed is equal to $\Delta Q/T$, but ΔQ is not equal to the external work done because the internal energy of the real, diluted solution is not equal to that of the concentrated solution.

In an isolated system (adiabatic process) the entropy change must be positive for the process to occur spontaneously, and in a closed system (isothermal process), in which heat has flowed into or out of the reacting system, the entropy change of the reacting system plus that obtained from the surroundings must be positive. What we are really saying is that for any system or systems in which the total energy change is zero the entropy change must be positive for a process to occur.

For an irreversible reaction (a reaction conducted not at its maximum work efficiency), the total entropy change is always greater than $\Delta Q/T$. There are many processes occurring in nature which are essentially irreversible such, for example, as the diffusion of a solute into a solvent, the flow of a liquid through a capillary, the passage of an electric current through a conductor. Indeed, life itself is inherently irreversible.

The entropy change is a measure of the disorder introduced into a system. Thus if P_1 is the probability or the number of ways the molecules can be arranged associated with one situation and P_2 that of the second situation, the entropy change in proceeding from the first to the second situation, providing there has been no net energy change, is per mole

$$\Delta S = R \ln \frac{P_2}{P_1} \qquad 21$$

What is meant, therefore, by the statement that the entropy must increase in a spontaneous reaction is that the reaction mixture must proceed towards a more probable state; entropy is a measure of the number of possible configurations of a system having a given energy. Probability used in the above sense is not the same as ordinary probability. Ordinary probability extends over the interval from zero to unity; probability associated with entropy changes refers to the total number of possible configurations, which is usually a very large number.

The relation between probability and entropy constitutes the connecting link between classical thermodynamics and statistical thermodynamics. The application of statistical thermodynamics to the elastic properties of fibers and to the behavior of solutions of high molecular weight polymers has been especially fruitful and some of these problems will be considered in more detail later in this book.

The concepts of probability and of temperature have little meaning when applied to single molecules or even to a small group of molecules, and thermodynamics is not appropriate to their study.

Free Energy. Whereas the entropy change provides a test for the spontaneity and extent of a reaction providing there has been no change in the internal energy, this restriction imposes a severe limitation on its utility. The change in internal energy also provides such a test if there has been no change in entropy. A sensible approach to the problem of feasibility of a reaction is to consider a combination of the changes in internal energy as well as in the entropy.

At constant pressure, the change in internal energy is

$$\Delta E = \Delta Q - \Delta W = \Delta H - P\Delta V \qquad 22$$

and for a reversible reaction

$$\Delta Q = T\Delta S \qquad 23$$

Substituting eq. 22 into eq. 23 gives

$$\Delta E = T\Delta S - \Delta W = \Delta H - P\Delta V \qquad 24$$

Rearranging eq. 24 yields

$$-\Delta W + P\Delta V = \Delta H - T\Delta S \qquad 25$$

The free energy change, ΔF, at constant pressure and temperature is defined as equal to $-\Delta W + P\Delta V$, and evidently represents the maximum net work which can be obtained from a system; by definition

$$\Delta F = \Delta H - T\Delta S \qquad 26$$

The free energy change as defined is known as Gibbs free energy. The Helmholtz free energy is defined as

$$\Delta A = \Delta E - T\Delta S \qquad 27$$

It is evident that if the ΔE, the change in internal energy, is equal to ΔH, the enthalpy, the change in the Gibbs free energy is equal to the change in the Helmholtz free energy. This is true when $P\Delta V$ is either zero or trivial. In most biochemical systems $P\Delta V$ is, in fact, trivial.

For a reversible reaction at constant temperature and pressure, the decrease in free energy is equal to the maximum work done in excess of the pressure-volume work. It should be noted that the sign of the work done is opposite to that of the free energy change. It is clear from the above discussion that work can be accomplished either by a decrease in the enthalpy, ΔH, or by an increase in the entropy, ΔS, or by both. Any process which can occur can be classed as being primarily enthalpic or primarily entropic. For example, the work done by the contraction of a rubber band is primarily entropic, whereas the work done by the combustion of glucose in the body is primarily enthalpic.

The burning of one mole of D-glucose to gaseous carbon dioxide and liquid water in a machine which is completely efficient and reversible in its operation liberates 688,160 calories of work. This is the free energy change for the reaction and

$$C_6H_{12}O_6 + 6O_2 \rightarrow 6CO_2 + 6H_2O : \Delta F = -688{,}160 \qquad 28$$

The negative sign indicates that work has been done by the reaction: the system contains 688,160 calories less potential work. The burning of glucose releases free energy and is a spontaneous process; such reactions are exergonic. If energy had had to have been added to make the reaction proceed, the reaction is obviously not spontaneous; such reactions are endergonic.

It will be noticed that the maximum work obtainable from the combustion of glucose is 688,160 calories, whereas the enthalpy is 673,000 calories; there is a difference of 15,160 calories, that is, during the reversible isothermal burning of a mole of glucose 15,160 calories of heat flowed into the reaction from the outside and were converted into work. The entropy change is, therefore, 15,160/298 or is 50.8 e.u. for the process.

The free energy function has some useful and important properties. Not only is the free energy change capable of predicting the direction of a chemical change, it is also capable of predicting the extent of the reaction at equilibrium; a simple and explicit relation exists between the free energy change and the equilibrium constant of the reaction. This relationship can be derived in the following manner.

Consider the following reaction at constant temperature and pressure

$$A + B \rightleftharpoons C + D \qquad 29$$

The initial concentrations are A_1, B_1, C_1 and D_1. The reaction is allowed to proceed to equilibrium and the concentrations become A_2, B_2, C_2 and D_2. The individual work terms for the reversible dilution or concentration of the components are

$$W_a = -RT \ln \frac{A_2}{A_1} \qquad 30$$

$$W_b = -RT \ln \frac{B_2}{B_1} \qquad 31$$

$$W_c = -RT \ln \frac{C_2}{C_1} \qquad 32$$

$$W_d = -RT \ln \frac{D_2}{D_1} \qquad 33$$

The maximum net work for the reaction is then

$$W = W_a + W_b - W_c - W_d = -\Delta F \qquad 34$$

Substituting the values for the individual work terms into eq. 34 gives

$$-\Delta F = -RT \ln \frac{A_2}{A_1} - RT \ln \frac{B_2}{B_1} + RT \ln \frac{C_2}{C_1} + RT \ln \frac{D_2}{D_1} \quad 35$$

and collecting terms in eq. 35 yields

$$-\Delta F = RT \ln \frac{A_1 \times B_1}{A_2 \times B_2} \times \frac{C_2 \times D_2}{C_1 \times D_1} \quad 36$$

If the initial concentrations of the reactants and products are all unity, then at equilibrium

$$\Delta F^\circ = -RT \ln \frac{C_2 \times D_2}{A_2 \times B_2} = -RT \ln K \quad 37$$

and at 25° and converting the logarithm to base 10

$$\Delta F^\circ = -1324.25 \log K \quad 38$$

Equation 38 is completely general for a system at constant temperature and pressure and is valid no matter how many reacting substances are present, provided that the reactants are initially at unit concentration and provided the substances obey the ideal solution laws. Later it will be shown that to apply eq. 38 to real solutions, activities of the substances must be employed instead of molar concentrations.

A plot of the free energy change for the conversion of substance A to substance B as a function of the percent of A present is shown in Figure 11. The ΔF° for this reaction is assumed to be zero and the temperature is 25°C.

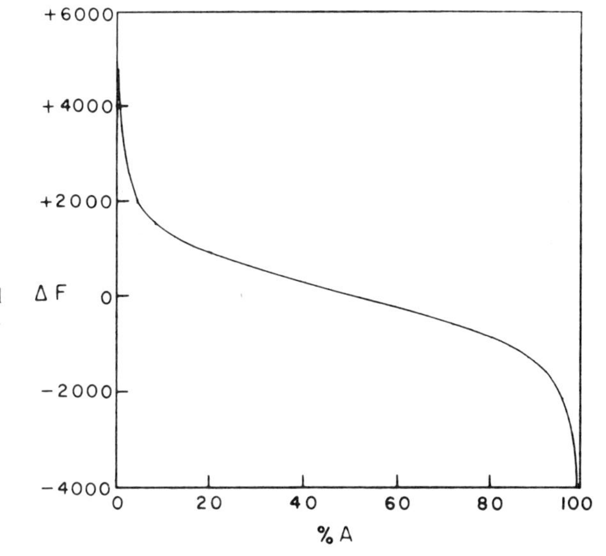

Fig. 11. ΔF as a function of the initial percent composition of the reaction A ⇌ B in respect to A at 25°, $\Delta F^\circ = 0$.

The free energy change involves the concept of equilibrium. At equilibrium a system's capacity for doing work is zero; the free energy of the system is at a minimum. Its entropy is at a maximum and it is in its most probable state. The standard free energy change ΔF° represents the free energy change in proceeding from unit concentration (activity) of the reactants to equilibrium. If the initial concentrations of the reactants are different from unity, the free energy change for the reaction will not be equal to ΔF°. For example, if the concentrations of the reactants are chosen equal to those at equilibrium, the free energy change for the reaction would be zero; this constitutes a definition of a state of equilibrium, i.e., the free energy change is always zero at equilibrium.

Free Energy and Temperature. Starting with the definition of the free energy change (eq. 26)

$$\Delta F = \Delta H - T\Delta S \qquad 39$$

and if ΔH and ΔS do not vary with temperature, the differential of eq. 39 is

$$d\Delta F = -\Delta S dT \qquad 40$$

or

$$\left(\frac{d\Delta F}{dT}\right)_P = -\Delta S \qquad 41$$

Substituting eq. 41 into eq. 39

$$\Delta F = \Delta H + \left(\frac{d\Delta F}{dT}\right)_P T \qquad 42$$

which is the Gibbs-Helmholtz equation. It is only when the free energy change is independent of temperature that ΔF is equal to ΔH and ΔH is a measure of chemical affinity or the feasibility of a reaction.

Substituting the relation $\Delta F° = -RT \ln K$ in the Gibbs-Helmholtz equation (eq. 42), there results

$$-RT \ln K = \Delta H° + T\frac{d(-RT \ln K)}{dT} \qquad 43$$

Differentiating and rearranging eq. 43, it is found that

$$\Delta H° = \frac{T^2 R d \ln K}{dT} \qquad 44$$

or

$$\frac{d \ln K}{dT} = \frac{\Delta H°}{RT^2} \qquad 45$$

The integration of eq. 45, conversion to logarithm to the base 10 and the substitution of the numerical value of R, gives

$$\log K = -\frac{\Delta H°}{4.574T} + C \qquad 46$$

or between the limits T_1 and T_2

$$\log \frac{K_2}{K_1} = \frac{\Delta H°}{4.574}\left(\frac{T_2 - T_1}{T_1 T_2}\right) \qquad 47$$

The above expressions are variations of the Van't Hoff equation.

Eisenberg and Schwert investigated the equilibrium between native and denatured chymotrypsinogen, an inactive form of a pancreatic proteolytic enzyme, as a function of the absolute temperature. They represented the equilibrium constant of the reaction by

$$K = \frac{\text{Denatured Protein}}{\text{Native Protein}} \qquad 48$$

Figure 12 shows a plot of their data both at pH 2.0 and at 3.0. From the slopes of the lines in Figure 12, the enthalpy per mole for the denaturation of chymotrypsinogen is 99,600 calories at pH 2.0 and 143,000 calories at pH 3.0. At pH 2.0 and at 43.6° the free energy change per mole is 300 calories and the entropy change is 316 e.u.

The variation of the enthalpy with temperature is a function of the specific heats of the reactants and of the products of the reaction and since these, in general, increase with increasing temperature, the net effect is a difference which tends to remain constant. Over a very wide temperature interval, there is sufficient change in the enthalpy to warrant correction, but over the narrow temperature range which the biochemist is apt to work the change with temperature is usually trivial, and it is customary to neglect the change in ΔH with temperature.

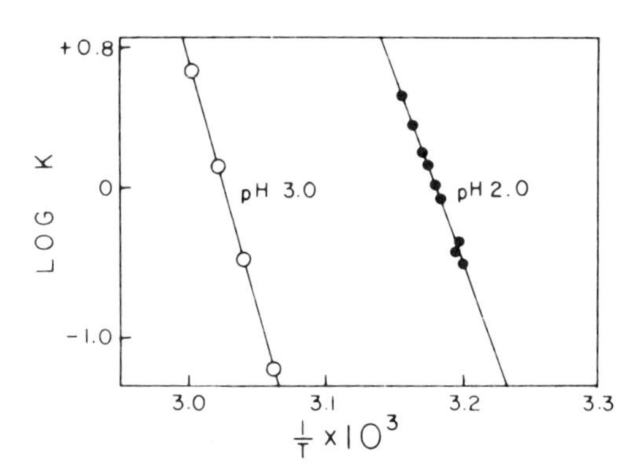

Fig. 12. Plot of log K against $1/T$ for the denaturation of chymotrypsinogen at pH 2.0 and at pH 3.0. (M. A. Eisenberg and G. W. Schwert: J. Gen. Physiol. *34*; 583, 1951.)

There should be complete agreement between the heat of reaction as obtained in a calorimeter and that calculated from the variation of the equilibrium constant with temperature. There can be, however, a pitfall in such calculations. Consider two successive first order reversible reactions

$$A \overset{K_1}{\rightleftharpoons} B \text{ and } B \overset{K_2}{\rightleftharpoons} C \qquad 49$$

A calorimetric heat would yield the net heat for the two reactions whereas, if the equilibrium constant for the first reaction only were determined, a false result would be obtained and the two heats would obviously not agree. It would be necessary to study the variation of the product K_1K_2 with temperature.

Evaluation of Free Energies. It is evident from eq. 38 that the free energy change for a reaction can be calculated from the equilibrium constant of the reaction. The reactants are added at any convenient concentrations to a reaction flask at constant temperature; the reaction is permitted to come to equilibrium and the concentrations of the reactants and products are measured by ordinary analytical procedure.

Provided the reaction involves a reversible oxidation-reduction step, the free energy change can be also obtained from the oxidation-reduction potential. The standard free energy, $\Delta F°$, is related to the oxidation-reduction potential by the relation (see Chapter 4 for more detail)

$$\Delta F° = -nFE° \qquad 50$$

where n is the number of moles of electrons transferred, F is the Faraday constant and is 96,487 coulombs per equivalent. Thus, if ferrocytochrome C is mixed with ferricytochrome C in a ratio of 1 to 1, the voltage at pH 7 is 0.11. Then

$$\Delta F° = -\frac{96{,}487}{4.185} \times 0.11 = -2{,}540 \text{ cals/mole} \qquad 51$$

Accordingly, cytochrome C will tend to oxidize spontaneously to the ferristate.

It is clear from the definition of the free energy change at constant pressure and temperature (eq. 26)

$$\Delta F = \Delta H - T\Delta S \qquad 52$$

that if there existed a method by which the entropy of the reactants and products could be obtained, the free energy change for a reaction could be calculated from the entropy change and the heat of the reaction.

The evaluation of the entropies of chemical compounds is, in fact, possible through the use of the third law of thermodynamics which can be stated as follows: If the entropy of each element in some crystalline state be taken as zero at the absolute zero of temperature, every substance has a finite positive entropy; but at absolute zero of temperature the entropy may become zero, and does so become in the case of perfect crystalline substances.

The above means that as the temperature of a substance is lowered, the free energy and the enthalpy must approach each other more closely and finally become equal at absolute zero (see Fig. 13).

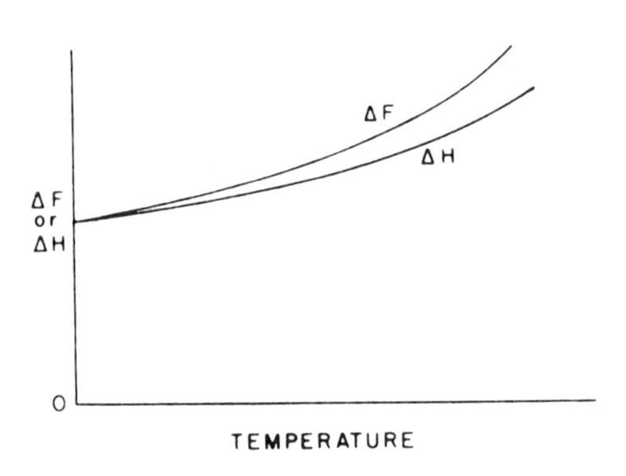

Fig. 13. Diagrammatic change of the enthalpy (ΔH) and of the free energy (ΔF) as functions of the temperature.

For a reversible process

$$\Delta S = \frac{Q}{T} \tag{53}$$

where Q is the reversible heat of reaction at constant pressure and

$$d\Delta S = \frac{dQ}{T} = \frac{C_p dT}{T} \tag{54}$$

where C_p is the heat capacity at constant pressure. The integration of eq. 54 gives

$$S = \int_0^T \frac{C_p dT}{T} + S_0 \tag{55}$$

Since from the third law of thermodynamics S_0 for a perfect crystalline solid is zero, eq. 55 can be integrated by determining C_p as a function of temperature down to very low temperatures. The enthalpy of formation of chemical compounds is then measured by the appropriate combination of heats of combustion. From the entropy of formation and the heats of formation the free energy of formation is then calculated with the use of eq. 26.

Figure 14 shows the specific heat of anhydrous zinc insulin as a function of the absolute temperature (Kelvin).

The free energies of formation can be used to calculate the free energy change for a chemical reaction. Consider the synthesis of the peptide DL-leucylglycine from glycine and DL-leucine

$$\text{gly} + \text{leu} \rightarrow \text{leu-gly} + H_2O \tag{56}$$

The free energies of formation in kilocalories are: gly − 89.26, leu − 81.76, leu-gly − 110.90 and H_2O − 56.69. Addition of the free energies of formation for gly and leu gives −171.02 Kcals and the sum for leu-gly and water is −167.59 Kcals. Then the free energy change for the reaction as written is −167.59 − (−171.02) or is +3.43 Kcals; evidently, the synthesis of leu-gly is endergonic. There are complications to the calculation of free energy changes for reactions in solution and, accordingly, it is best to review some of the properties of solutions.

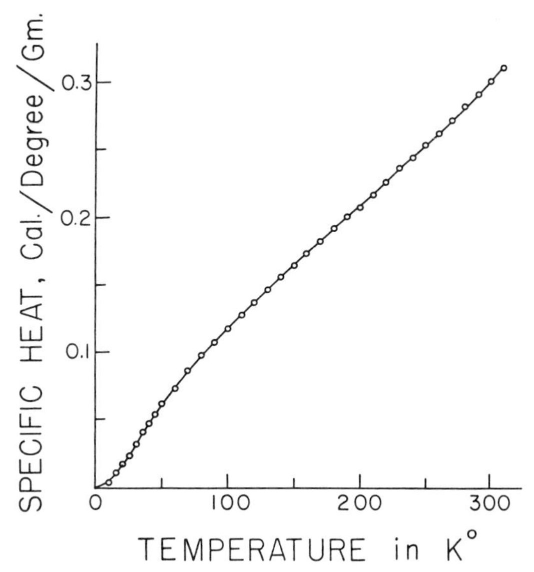

Fig. 14. Specific heat of anhydrous bovine zinc insulin as a function of the absolute temperature. (Data of J. O. Hutchens, A. G. Cole and J. W. Stout: J. Biol. Chem. *244:* 26, 1969.)

Solutions. A more elaborate discussion of solutions will be presented in Chapters 3 and 7 but in order to deal with free energy changes of reactions in solutions it is necessary to anticipate with consideration of activity, chemical potential and related topics.

Concentrations may be expressed in a variety of ways. A molar solution contains one gram molecular weight of the substance in one liter of solution; due to the dependence of density on temperature this expression for concentration is also temperature dependent. One mole of a substance dissolved in 1,000 grams of solvent is a molal solution and the concentration is independent of temperature. The mole fraction is the ratio of the number of moles of the substance in solution to the total number of moles of all substances present in solution including that of the solvent. In a solution containing n_1 moles of the first substance and n_2 moles of the second, the mole fractions are

$$N_1 = \frac{n_1}{n_1 + n_2} \text{ and } N_2 = \frac{n_2}{n_1 + n_2} \qquad 57$$

and

$$N_1 + N_2 = 1 \qquad 58$$

Irrespective of the number of components, the sum of the mole fractions is always unity. The mole fraction of any solute in a molal solution in water is

$$N_2 = \frac{1}{\frac{1{,}000}{18.016} + 1} = 0.01770 \qquad 59$$

The idea of concentration can be expressed in terms of activity; activity is a thermodynamic concentration and takes into consideration the departure from ideal behavior. The ratio between activity and concentration is the activity coefficient and

$$\text{Activity coefficient } (\gamma) = \frac{\text{activity } (a)}{\text{molal concentration } (m)}$$

A more complex expression than the above is needed for the mean activity coefficient of an electrolyte; this is given in Chapter 3.

An adequate discussion of activity requires a definition of the standard state of reference and this is a somewhat slippery concept which is best postponed until partial molar quantities have been presented.

Partial Molar Quantities. If a mole of a solute is dissolved in a very large quantity of the solution—so large a volume that the concentration is not significantly changed—the resulting increase in volume is known as the partial molar volume of the added material. A more feasible experimental approach to this problem is to maintain the number of moles of the solvent constant and measure the increase in volume as the solute is added. The volume of the solution is plotted against the number of moles of solute added and the slope of the resulting plot at any concentration is equal to the partial molar volume of the solute or

$$\frac{dV}{dn_2} = \bar{V}_2 \qquad 60$$

where V is the volume of the solution, n_2 is the moles of solute added and $\bar{V}_2$ is the partial molar volume of the solute. If the amount of solute is expressed in grams instead of in moles, the resulting volume increase is the partial specific volume of the solute. The plot of the volume of the solution against the added weight of the solute is, usually, very nearly linear and the slope of the line, at least for non-electrolytes, is also nearly equal to the specific volume of the solute in its pure state.

If the number of moles of solute is fixed and the volume measured as a function of the change in concentration of the solvent, the partial volume of the solvent can be obtained and, in general, the relation

$$dV = \bar{V}_1 dn_1 + \bar{V}_2 dn_2 \qquad 61$$

holds where V is the total volume of the system, $\bar{V}_1$ and $\bar{V}_2$ are the partial volumes of the solvent and solute respectively and n_1 and n_2 are the moles (or grams) of the solvent and solute respectively. A more complete discussion of partial volumes is found in Chapter 14.

Any extensive property of a solution can be treated in the same manner as has been done for the volume as shown above and the resulting plots give partial molar free energies, partial molar heats, partial molar entropies, etc.

The partial molar free energy is of special interest and is known as the chemical potential and is expressed as

$$\left(\frac{dF}{dn_1}\right)_{T,\,P,\,n_2,\,n_3} = \bar{F}_1 = \mu_1 \qquad 62$$

There is a chemical potential for each component of the solution and for an infinitesimal change in the free energy of a system

$$dF = \left(\frac{dF}{dT}\right)_{P,\,n_1,\,n_2} dT + \left(\frac{dF}{dP}\right)_{T,\,n_1,\,n_2} dP + \mu_1 dn_1 + \mu_2 dn_2 \qquad 63$$

For a process at constant temperature and pressure eq. 63 becomes

$$(dF)_{T,\,P} = \mu_1 dn_1 + \mu_2 dn_2 \qquad 64$$

At equilibrium, the change in free energy is zero, therefore, for a two component system

$$\mu_1 dn_1 + \mu_2 dn_2 = O \qquad 65$$

In principle, the measurement of the chemical potential of one of the components of a two component system along with the concentrations of both components permits the calculation of the chemical potential of the other component.

The chemical potential is an intensity factor in the same sense as are pressure and temperature, and, at equilibrium, the chemical potential of a given component must be uniform throughout the system irrespective of the number of components or phases present. Suppose, for example, iodine is dissolved in carbon tetrachloride and the solution is shaken with water. At equilibrium the iodine will distribute itself between the two liquid phases, but the concentration of the iodine in the carbon tetrachloride will be about 85 times what it is in the water phase. Evidently, at equilibrium, the vapor pressure of the solute (iodine) is the same in both phases in contact, otherwise the solute would distill from one phase to the other; the chemical potential of the solute in the two phases is identical. However, for a given molar concentration of solute, the vapor pressure of the solute in one of the phases may be and frequently is much larger than it is in the other phase; the two immiscible solutions can depart greatly from ideality.

Activity and the Choice of Standard States. Intuitively, it can be seen that there should exist a relation between the activity and the chemical potential of the solute. It has been shown that the free energy of dilution of an ideal solution from C_1 to C_2 is

$$\Delta F = n_2 RT \ln \frac{C_2}{C_1} \qquad 66$$

where n_2 is the number of moles of solute being diluted. For a real solution, the concentrations have to be replaced by activities and

$$\Delta F = n_2 RT \ln \frac{a_2}{a_1} \qquad 67$$

The differentiation of eq. 67 keeping the ratio a_2/a_1 constant gives

$$\frac{dF}{dn_2} = RT \ln a_2 - RT \ln a_1 = \mu_2 \qquad 68$$

and when a_2 is set equal to unity

$$\mu_0 = -RT \ln a_1 \qquad 69$$

Then substituting eq. 69 into eq. 68, there results

$$\mu = \mu_0 + RT \ln a_2 \qquad 70$$

Equation 70 is, in fact, an exact definition of the activity and from which it is seen that whereas the log activity is proportional to a chemical potential, it is not equal to the chemical potential.

The discussion of activity leads immediately to the problem of the choice of the standard reference state. The selection is really a matter of convenience. Convention, however, dictates that the activity of pure liquids and of pure solids is set equal to unity. Solutions are treated somewhat differently. It is considered that the activity of the pure solvent is unity but a different standard state is usually assigned to the solute.

It is found experimentally that, as the solution becomes more and more dilute, it approaches more nearly ideality—i.e., the activity coefficient approaches unity as the concentration approaches zero. It would, however, be awkward to select zero concentration as our standard state because the activity of the solute would be zero and the chemical potential would be minus infinity. To escape from this difficulty, the standard state of the solute selected is such that its activity is unity for an ideal solution. Such a solution is known as a hypothetical one molal solution (on the molal scale). Thus, the standard state of the solute is defined in terms of no real solution, just as an ideal gas is defined in terms of no real gas. For example, at 25° 1.734m KCl has a mean molal activity coefficient of 0.577 and, accordingly, the activity is $(1.734 \times 0.577)^2$ or is unity. This solution, however, is not the hypothetical one molal; it is a physical accident.

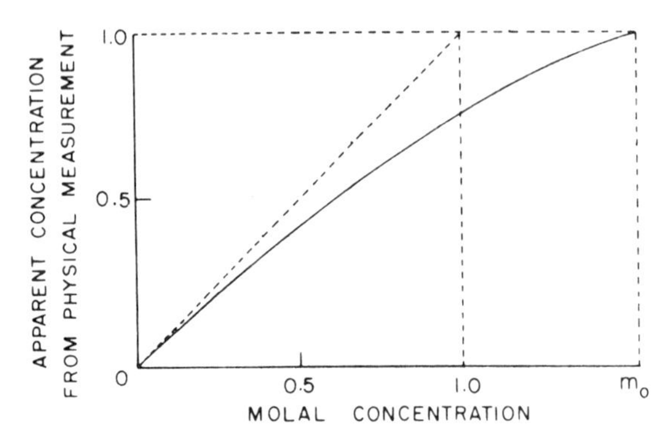

Fig. 15. Plot of the apparent concentration of a solute as determined from physical measurements against the actual molal concentration. m_o is the solute concentration at which the solute activity is unity.

Figure 15 shows the apparent concentration of a solution plotted against the molal concentration. The apparent concentration has been determined by measuring some appropriate property of the solution such as the freezing point depression, the boiling point lowering or osmotic pressure as a function of the solute concentration. The apparent concentration of the solute as revealed by the physical measurement is a reflection of the activity of the solute, and the dotted line in Figure 15 is the activity of an ideal solution and m_0 is the molal concentration of the real solution which has an activity of unity. In principle, the choice of the standard reference state requires an extrapolation of the physical measurement to an infinitely dilute solution in which the activity coefficient is unity in order to predict the behavior of the ideal solution.

The chemical potential of a solute in a saturated solution in the presence of the solid solute is identical with the chemical potential of the solid, but the activity of the solute in solution would depend on the choice of the standard reference state for the solute in this condition. If it were desired, the activity of the solute in equilibrium with the solid could be set equal to unity, in which event the activities of the solute in solution and in the solid phase would be equal, by definition. On the other hand the activity of the solute in solution could refer to a standard state of the hypothetical one molal or to some other standard state, in which event the activities of the solute in solution and in the solid state would no longer be the same but their chemical potentials would be identical at equilibrium.

This problem is somewhat analogous to reading the temperature with centigrade and Fahrenheit thermometers; the temperature is the same for both thermometers but they read differently.

ΔF of Chemical Reactions. The free energy change under standard condition in solution (hypothetical one molal solutions) for the reaction

$$A + B \rightleftharpoons C + D \tag{71}$$

is simply the free energy of formation of A plus that of B at unit activities (hypothetical one molal solution) subtracted algebraically from the sum of the free energies of C plus that of D. If the free energies of formation in aqueous solution in the standard state are unavailable, then the free energy change for dilution from a mole fraction of unity (pure liquid or solid) to a mole fraction corresponding to that of a one molal solution in water is calculated. The free energy of dilution is always negative; the free energy of formation in the liquid or solid state is increased in a negative sense by the free energy of dilution to obtain the free energy of formation in solution. To calculate the free energies of dilution exactly, the activity coefficient of the solute is needed. Frequently, this information is not known and the next best thing is done and concentrations instead of activities are used; the free energy change for the transfer of one mole of pure solute to 1,000 grams of water is (for an ideal solution at 25°C)

$$\Delta F = RT \ln \frac{1}{55.5 + 1} = -2,410 \text{ calories} \qquad 72$$

If the solute has a limited solubility, the free energy of formation of the solute in a saturated solution is identical with that of the pure solute and, accordingly, the free energy of dilution is then calculated from the change in concentration from the saturated solution to a molal solution. The calculation of the free energies of dilution of non-electrolytes from concentrations rather than activities is apt to be far more successful than for electrolytes. The reason for this will be discussed in Chapter 3.

Table 2 is a collection of free energies of formation per mole and, with few exceptions which are indicated, the energies refer to aqueous solutions at 25°C at one molal activity.

In practice, two complications are apt to be encountered in free energy calculations of reactions in solution. One of these complications arises when the solvent itself participates in the reaction as is true of all hydrolysis reactions conducted in water as the solvent. The other complication has to do with the presence of various ionized forms of the reactants and products.

Water ionizes as

$$H_2O \rightleftharpoons H^+ + OH^- \qquad 73$$

and the equilibrium constant of this reaction is

$$K = \frac{H^+ \times OH^-}{H_2O} \qquad 74$$

It has been found by experiment that at 25°C the product $H^+ \times OH^-$ is equal to 0.9978×10^{-14} and the concentration of water in pure water is 1,000/18.016 or is 55.506 molal and, accordingly, K is equal to 1.80×10^{-16} and ΔF for the ionization of water is calculated to be 21,471 calories per mole. This calculation is based upon initial concentrations of one molal for water, for hydrogen ions and for hydroxyl ions. A more meaningful calculation is based on an initial concentration of water of 55.5 molal and with the hydrogen and hydroxyl ions each one molal. Under these conditions the ionization constant (K_w) is equal to 55.5 K or is 0.9978×10^{-14} at 25°C and the free energy change is 19,098.7 calories. From Table 2, the free energy of formation of liquid water is −56,690 calories, that of hydroxide ions is −37,600 calories and zero for hydrogen ions, and the free energy change for the ionization of water is 19,090 calories per mole.

A simple reaction which was much studied in the early development of physical chemistry was the hydrolysis of ethyl acetate. The reaction is

$$CH_3\overset{\overset{\displaystyle O}{\|}}{C}\text{—}O\text{—}CH_2CH_3 + H_2O \rightleftharpoons CH_3\overset{\overset{\displaystyle O}{\|}}{C}\text{—}OH + CH_3CH_2OH \qquad 75$$

the equilibrium constant of which is

$$K = \frac{\text{Ethanol} \times \text{Acetic Acid}}{\text{Ethyl acetate} \times \text{Water}} \qquad 76$$

Expressing each of the reactants in terms of mole fractions, the experimentally determined value for the equilibrium constant is about 0.35 and the calculated free energy change at 25°C is about 620 calories. The heat of the reaction is about 2,000 calories per mole and, accordingly, the equilibrium constant changes only slightly with temperature. The evaluation of the equilibrium constant was accomplished by mixing a given number of moles of each of the reactants and allowing the mixture to come to equilibrium using HCl as a catalyst and with subsequent analysis of the equilibrium mixture. The free energy of formation of liquid ethyl acetate is

TABLE 2. FREE ENERGIES OF FORMATION IN KCALS AT 25°C AND AT ATMOSPHERIC PRESSURE IN AQUEOUS SOLUTION AT ONE MOLAL (HYPOTHETICAL) UNLESS OTHERWISE STATED.

Compound	kcal	Compound	kcal
Acetaldehyde	−33.4	Glyoxylate$^-$	−112.0
Acetic acid	−95.5	Hydrogen ion	0.0
Anion$^-$	−89.0	Hydrogen chloride	−31.4
Acetoacetate$^-$	−118.0	Hydrogen peroxide	−32.7
Acetone	−38.5	Hydrogen sulfide	−6.5
cis Aconitate^{3-}	−220.5	Anion$^-$	+3.0
L-Alanine	−88.8	Hydroxide ion	−37.6
DL-Alanine	−89.1	β Hydroxybutyric acid	−127.0
L-Alanylglycine	−114.6	Anion$^-$	−121.0
Ammonia (gas)	−4.0	Isocitrate^{3-}	−277.7
Ammonia (NH_3)	−6.4	L-Isoleucine	−82.2
Ammonia (NH_4^+)	−19.0	Isopropanol	−44.4
L-Asparagine · H_2O	−182.6	α-Ketoglutaric acid	−190.6
L-Aspartic acid	−172.4	Lactate$^-$	−123.8
Anion$^-$	−167.0	α-Lactose	−362.2
Anion^{2-}	−155.0	β-Lactose	−375.8
Butyric acid	−90.9	L-Leucine	−82.0
Anion^{-1}	−84.3	DL-Leucine	−81.8
n-Butanol	−41.1	DL-Leucylglycine	−110.9
Calcium ion^{2+}	−132.2	Lithium ion$^+$	−70.2
Carbon dioxide (gas)	−94.3	L-Malate^{2-}	−202.0
Carbon dioxide	−92.3	β-Maltose	−357.8
Carbonic acid	−149.0	Mannitol	−225.3
Anion$^-$	−140.3	Methanol	−41.9
Anion^{2-}	−126.2	L-Methionine	−120.2
Carbon monoxide (gas)	−32.8	Nitrate$^-$	−26.4
Chloride ion$^-$	−40.0	Nitrite$^-$	−8.3
Citrate^{3-}	−279.2	Oxalate^{2-}	−161.3
Creatine	−63.2	Oxalacetic^{2-}	−190.5
Creatinine	−6.9	Palmitic acid (solid)	−80.0
L-Cysteine	−81.2	L-Phenylalanine	−49.5
L-Cystine	−159.4	Potassium ion$^+$	−67.5
Ethanol	−43.4	Potassium chloride	−97.6
Formaldehyde	−31.2	n-Propanol	−42.0
Fructose	−218.8	Pyruvate$^-$	−113.4
Formic acid	−85.1	L-Serine	−122.1
Anion$^-$	−80.0	Sodium ion$^+$	−62.6
Fumaric acid	−154.7	Sodium chloride	−93.9
Anion^{2-}	−144.4	Sorbitol	−225.3
α-D-Galactose	−220.7	Succinic acid	−178.4
α-D-Glucose	−219.2	Anion^{2-}	−165.0
L-Glutamic acid	−173.0	Sucrose	−370.9
Anion^{2-}	−165.9	Sulfate^{2-}	−177.3
L-Glutamine	−126.6	L-Threonine	−123.0
Glycerol	−116.8	L-Tryptophane	−26.9
Glycine	−89.3	L-Tyrosine	−87.3
Glycogen		Urea	−48.7
(per glucose unit)	−158.3	L-Valine	−85.3
Glycolate$^-$	−126.9	Water (gas)	−54.6
Glycylglycine (solid)	−116.6	Water (liquid)	−56.7

somewhat dubious, but −77,600 calories is the best value and this along with the appropriate values from Table 2 gives a free energy change for the reaction of −1,230 calories. The agreement with the free energy change from the equilibrium constant is only fair and the discrepancy probably lies in the uncertain value for the free energy of formation of liquid ethyl acetate; it is also possible that there is substantial departure from ideality of the reaction mixture.

It is more meaningful to conduct this reaction by composing a mixture which is one molal in respect to ethyl acetate, to ethanol and to acetic acid in pure water and to permit the mixture to attain equilibrium. One mole each of the components has been mixed with 55.5 moles of water and the mole fraction of each with the exception of water is 1/58.551 or is 0.01708 and the calculated net free energy of dilution is −2,410 calories. Combining this with the free energies of formation of the reactants and products, the free energy change for the hydrolysis of ethyl acetate is −2,170 calories per mole. By selecting pure water as the reference state, the equilibrium point of the reaction has been shifted significantly towards hydrolysis; there is more water present.

The hydrolysis of ethyl acetate can be used to illustrate the influence of ionization on the free energy change of a reaction. Ethyl acetate, acetic acid and ethanol are again dissolved in pure water such that the concentration of each of these components with the exception of water is one molal. Now, however, the standard reference state is changed by permitting acetic acid to ionize, and the concentration of hydrogen ions and, consequently, the concentration of acetic acid are regulated by the addition of alkali. What in effect is done is that the hydrolysis of ethyl acetate is coupled to the neutralization reaction

$$HAc + OH^- \rightleftharpoons H_2O + Ac^- \qquad 77$$

the free energy change of which per mole is about $-10{,}500$ calories. The problem is to calculate the concentration of acetic acid as a function of the hydrogen ion concentration; by neutralization, the acetic acid will be diluted from one molal to the concentration which exists at a given hydrogen ion concentration. Over a fairly wide range of hydrogen ion concentration, the ionization of acetic acid can be expressed as

$$K = \frac{H^+ \times Ac^-}{HAc} \qquad 78$$

The concentration of the acetate ion (Ac^-) present is evidently equal to the original concentration of acetic acid, which is one molal minus the concentration of the unionized acetic acid (HAc) and

$$K = \frac{H^+ \times (1 - HAc)}{HAc} \qquad 79$$

or

$$HAc = \frac{H^+}{K + H^+} \qquad 80$$

The free energy change for the ionization of acetic acid which had an initial concentration of one molal is per mole

$$\Delta F = RT \ln \frac{H^+}{K + H^+} \qquad 81$$

The free energy change for the hydrolysis of ethyl acetate at any hydrogen ion concentration, all components except water and hydrogen ions being initially one molal, is

$$\Delta F = -2{,}170 + RT \ln \frac{H^+}{K + H^+} \qquad 82$$

Figure 16 shows a plot of the free energy change for the hydrolysis of ethyl acetate as a function of pH as predicted by eq. 82. The free energy yield from the hydrolysis of ethyl acetate has been very greatly increased by coupling the hydrolysis reaction to the neutralization reaction; coupled reactions are very common in biochemistry.

In light of the above and anticipating further discussion, it appears best to define free energy changes under several conditions as Burton and Krebs have done. The symbols to be employed and their definitions are as follows:

1. ΔF means the increment of free energy for a given reaction under arbitrarily specified conditions.

2. ΔF° is the free energy change for standard conditions, i.e., the reactants in their standard states.

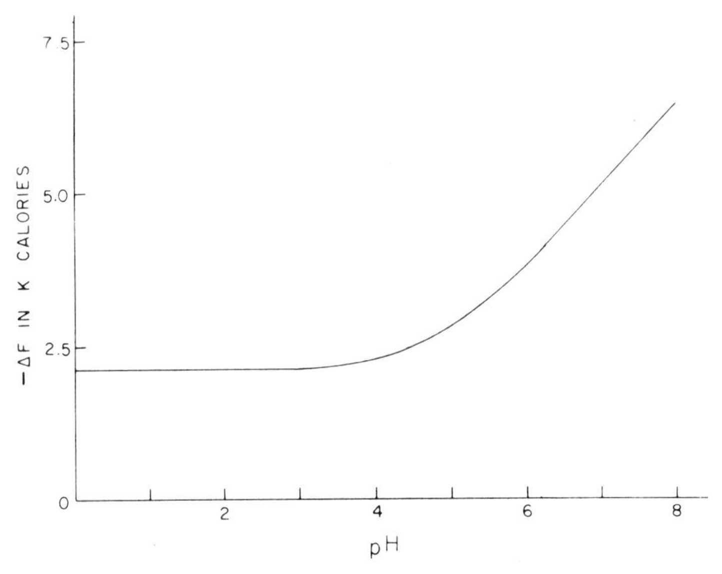

Fig. 16. Free energy change at 25°C for the hydrolysis of ethyl acetate as a function of pH.

3. $\Delta F'$ refers to the free energy change in solution with all reactants except hydrogen ion in their standard states.

4. $\Delta F°f$ is the free energy of formation of a compound from its elements under standard conditions (see Table 2).

The Fuel Supply. A living system has to do work of various kinds to survive. For example, animals do muscle work and transmit nerve impulses. They also actively accumulate solutes against concentration gradients across cellular membranes. They synthesize various molecules and some of these syntheses require a large expenditure of free energy. Electric eels can do impressive electrical work and still other animals produce light. In all these systems it is necessary to have a fuel supply (metabolism) coupled to a molecular engine of some kind. Much is known about the fuel supply, but the nature of the molecular engines to convert the chemical energy into work is still largely obscure. The relation between the fuel supply and the use of the energy derived from metabolism is illustrated in Figure 17 in a highly diagrammatic fashion.

A very substantial part of modern biochemistry concerns itself with the supply of fuel to cells, and during the last 30 years a fairly complete if to a great extent qualitative picture of intermediary metabolism has emerged. This picture in all of its details is exceedingly complex.

The energy released by digestion is unavailable for useful work because the hydrolytic reactions are not coupled with other reactions; the energy is lost as heat. The hydrolyzed foodstuffs enter the body cells and there they undergo various degradations; this is the second phase of energy release. In this phase is included such reactions as glycolysis in which glucose is degraded to pyruvate. The energy yield for these changes is substantial and much of it can be converted into work. Finally, the partially degraded products enter the Krebs tricarboxylic acid cycle and are further oxidized. This last and third phase includes oxidative phosphorylation and accounts for about two thirds or more of the total energy which can be derived from the ingested foodstuffs.

There are a large number of steps in metabolism and, accordingly, the overall free energy change has been resolved into many parts so that each step is more

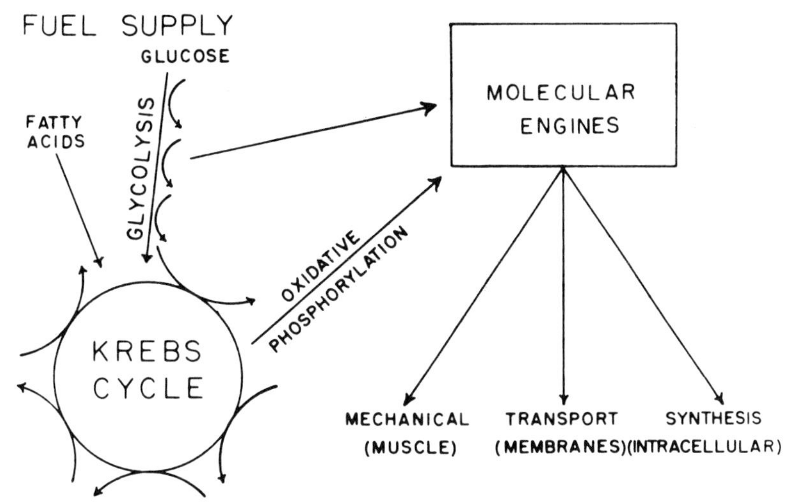

Fig. 17. Diagrammatic sketch of relation between fuel supply and transducers in tissue.

nearly an equilibrium reaction; this makes for a more efficient utilization of the energy released. The free energy changes of many metabolic reactions are now known with fair accuracy.

It would be impossible in the space available to review the metabolic reactions; much intensive research has gone into the elucidation of each of them. To gain some impression of the quantitative methods used, the comparatively simple lactate-pyruvate equilibrium can be looked at. This reaction is not on the main metabolic stream; under reducing conditions, lactate accumulates in tissue and during the early days of biochemistry lactate was thought to play a fundamental role in muscle contraction—a view no longer held.

The reaction for the oxidation of lactate can be written

$$\begin{array}{c} CH_3 \\ | \\ H\text{—}C\text{—}OH \\ | \\ COO^- \end{array} \rightleftharpoons \begin{array}{c} CH_3 \\ | \\ C{=}O \\ | \\ COO^- \end{array} + 2H^+ + 2e \qquad 83$$

where e indicates the electron. From Table 2, the $\Delta F(f)$ of lactate is $-123{,}760$ calories per mole and for pyruvate it is $-113{,}440$ calories per mole, which gives 10,320 calories for the free energy change per mole for the oxidation of lactate, all reactants being at unit activity.

Racker investigated this reaction experimentally, using a lactic acid dehydrogenase prepared from rabbit muscle, and composed mixtures of lactate, pyruvate, nicotinamide adenine dinucleotide (NAD) and the enzyme. The NAD participates in hydrogen transport and has the ability to accept two hydrogens from the lactate ion. The reaction is

$$\text{Lactate}^- + \text{NAD}^+ \rightleftharpoons \text{Pyruvate}^- + \text{NAD} + \text{H}^+ \qquad 84$$

and the equilibrium constant is

$$K = \frac{\text{Pyruvate}^- \times \text{NAD} \times \text{H}^+}{\text{Lactate}^{-1} \times \text{NAD}^+} \qquad 85$$

The nucleotide has the merit that the reduced form NAD has strong absorbancy of light at a wavelength of 340 mμ, whereas the oxidized form NAD$^+$ does not absorb. Accordingly, knowing the initial concentrations of pyruvate, lactate, NAD, NAD^+ and the hydrogen ions, the value of K can be calculated by studying the light absorbancy at 340 mμ. Figure 18 shows the value of K as a function of the pH of the reaction mixture at 25°C. At unit activity of hydrogen ions, the equilibrium constant of the reaction would be 4.4×10^{-12} and the free energy change for the reaction would be 15,550 calories per mole. The free energy change for the oxidation of NAD at unit activity of hydrogen ions is −5,210 calories.

Combining the free energy changes for the oxidation of lactate with the reduction of NAD^+, the free energy change for the conversion of lactate to pyruvate becomes 10,290 calories per mole. Since hydrogen ions are not directly involved in the hypothetical conversion of lactate to pyruvate, this value would also be the free energy change at pH 7.0. It will be recalled that 10,320 calories per mole were obtained for the reaction using values for the free energies of formation listed in Table 2.

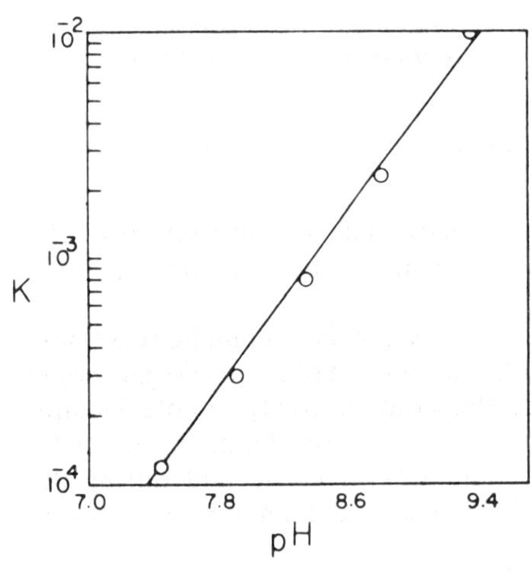

Fig. 18. Equilibrium constant for lactic acid dehydrogenase as a function of pH at 25°C. (E. Racker: J. Biol. Chem. *184:* 313, 1950.)

Coupled Reactions. The transference of chemical energy can be accomplished with coupled reactions. Consider the following hypothetical reactions

$$A + B \rightarrow C + D \qquad \Delta F_1 \qquad 86$$

$$C + E \rightarrow F + G \qquad \Delta F_2 \qquad 87$$

Adding the two reactions gives

$$A + B + E \rightarrow D + F + G \qquad \Delta F_3 = \Delta F_1 + \Delta F_2 \qquad 88$$

The second reaction removes C from the reaction mixture and drives the reaction forward with the accumulation of D, F and G. E is the coupling agent and C is the common intermediate.

Coupled reactions can be illustrated by the production of glucose -6-phosphate from glucose and adenosine triphosphate (ATP) as follows:

$$\text{Glucose} + HPO_4^{-2} \rightarrow \text{Glucose -6-}PO_4 + H_2O \qquad \Delta F \cong 4.5 \text{ Kcals} \qquad 89$$

$$ATP + H_2O \rightarrow ADP + HPO_4^{-2} \qquad \Delta F \cong -7.7 \text{ Kcals} \qquad 90$$

The sum of these two reactions is

$$\text{Glucose} + ATP \rightarrow \text{Glucose -6-}PO_4 + ADP \qquad \Delta F \cong -3.2 \text{ Kcals} \qquad 91$$

In the above ADP stands for adenosine diphosphate, H_2O is the common intermediate, and ATP is the coupling agent. Coupled metabolic reactions are of frequent occurrence; however, there are other means by which energy can be transferred. For example, radiant energy can be absorbed and converted into chemical energy (photosynthesis). Also osmotic or cell membrane (mitochondrial membranes) coupling appears to be a possible means by which energy can be transferred without the use of a common intermediate.

Organic Phosphates. The importance of phosphate in biochemical reactions was early recognized, particularly in relation to fermentation reactions and carbohydrate metabolism in general.

Lipmann emphasized what he called high energy phosphate bonds and their role in coupled reactions. The term high energy phosphate bond is an unfortunate one; it is not in accord with the customary usage of this term and the physical chemist means something entirely different when he speaks of the energy of chemical bonds. In the Lipmann sense, the term high energy bond means that the hydrolysis of such bonds proceeds nearly to completion before equilibrium is reached and the free energy change for hydrolysis is large and negative. Perhaps a more suitable term would have been hypergonic bonds.

Briefly, what Lipmann noticed was that certain phosphate esters such as that of glucose-1-phosphate underwent hydrolysis with only a moderate change in free energy whereas other phosphates such as adenosine triphosphate, 1,3 diphosphoglyceric acid, acetyl phosphate, phosphoenol pyruvic acid, creatine phosphate and arginine phosphate upon hydrolysis suffered a much higher free energy change. Initially, the estimates of the free energy changes were little more than informed guesses. The situation is much better now but still needs careful attention.

The approximate free energies of hydrolysis of various bonds including some phosphate bonds expressed in Kcal per mole at pH 7 and at 25°C are shown in Figure 19. The free energies listed could be called group transfer potentials, H_2O being the acceptor.

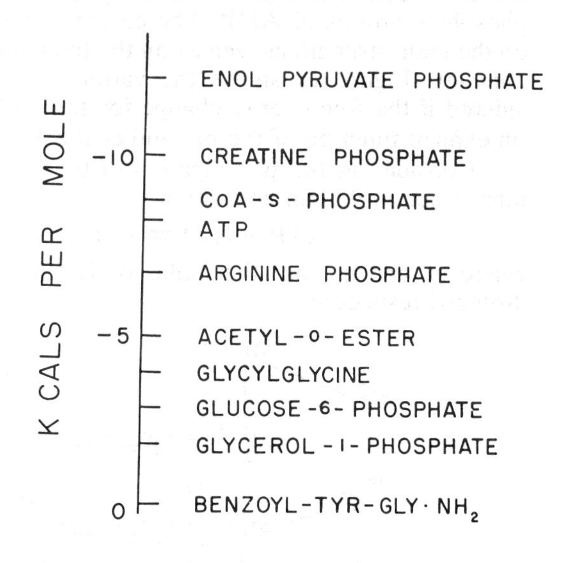

Fig. 19. Group transfer potentials in Kcals per mole pH 7.0, 25° water being the acceptor.

Of the high energy phosphate bonds, the most versatile as a coupling agent is that in adenosine triphosphate (ATP). In fact, it appears that to provide energy for useful work, the other high energy compounds must first transfer their free energy to ATP. It is established that ATP plays a key role as an energy transmitter in muscle contraction. It furnishes energy for the production of light by the firefly and is the fuel source for the current produced by the electric eel. It is believed by some to participate in the active transport of solutes across cellular membranes. ATP is also used in the synthesis of many molecules of biological importance.

The projection of a scaled molecular model of ATP is shown in Figure 20.

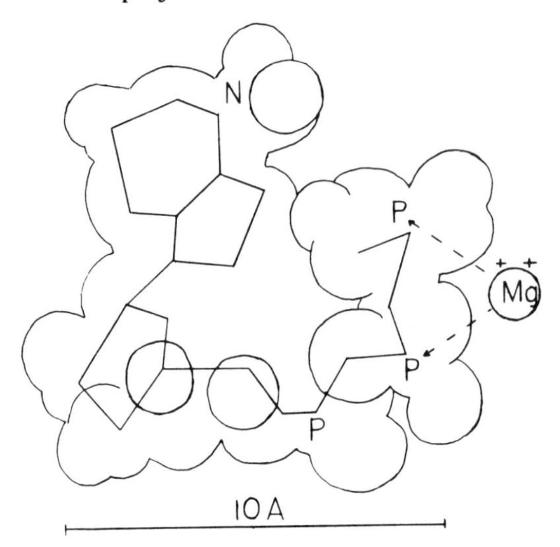

Fig. 20. Projection of a scaled molecular model of ATP. Mg^{++} complex also shown.

ΔF for Hydrolyses of ATP. The simple hydrolysis of ATP to adenosine diphosphate (ADP) and inorganic phosphate (Pi) can be written

$$ATP + H_2O \rightleftharpoons ADP + Pi \qquad 92$$

However, ATP, ADP and Pi are all capable of the ionization of protons and, furthermore, the phosphates have the ability of complexing with various cations especially Mg^{++}. Nuclear magnetic resonance studies indicate that Mg^{++}, Ca^{++} and Zn^{++} complex with the β and γ phosphate groups of ATP and with the α and β phosphate groups of ADP. The extent of the hydrolysis of ATP will also depend on the ionic strength as well as on the temperature, so that the situation is inherently complex. Figure 21 shows the various ionization reactions which must be considered if the free energy change for the hydrolysis of ATP is to be expressed as an explicit function of the pH and of the Mg^{++} concentration.

Considering the participation of the hydrogen ions as well as the magnesium ions, reaction 92 can be written

$$ATP + H_2O \rightleftharpoons ADP + P_i + n_H\, H^+ + n_{Mg}\, Mg^{+2} \qquad 93$$

where n_H and n_{Mg} are the moles of H^+ and Mg^{+2} liberated per mole of ATP hydrolyzed respectively.

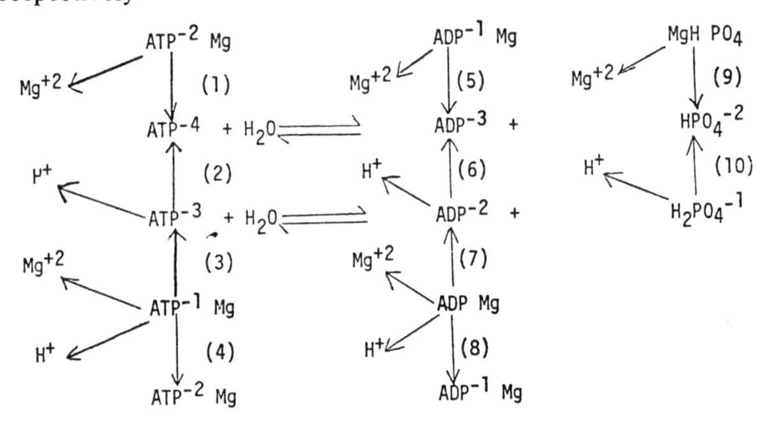

Fig. 21. Acid and magnesium dissociation reactions involved in the hydrolysis of ATP. The values for the dissociation constants corresponding to the circled numbers are shown in Table 3.

The equilibrium constant for reaction 93 is

$$K = \frac{ADP \times Pi \times (H^{+})^{nH} \times (Mg^{+2})^{nmg}}{ATP} \quad 94$$

the activity of the water being set at unity.

It will be realized that the free energy change calculated by the use of the equilibrium constant from eq. 94 refers all reactants and products including the hydrogen and magnesium concentrations to one molal solutions. Providing the various dissociation constants are known, the equilibrium constant for the hydrolysis of ATP at any pH and magnesium concentration can be calculated from a single experimental determination of the equilibrium concentrations of ATP, ADP, Pi, H^+ and Mg^{+2}. Clearly, however, such calculations are best done with a computer. Table 3 gives the dissociation constants corresponding to the reactions shown in Figure 21.

TABLE 3. ACID AND MAGNESIUM DISSOCIATION CONSTANTS OF ATP, ADP AND INORGANIC PHOSPHATE, 25°. ZERO IONIC STRENGTH. REACTION NUMBER REFERS TO THE NUMBERS SHOWN IN FIGURE 21. (Data of R. C. Phillips, P. George and R. J. Rutman: J. Biol. Chem. *244*:3330, 1969.)

Reaction Number	*K*	*Reaction Number*	*K*
1	1.5×10^{-6}	6	6.3×10^{-8}
2	2.1×10^{-8}	7	2.6×10^{-3}
3	2.5×10^{-4}	8	4.2×10^{-6}
4	3.6×10^{-6}	9	1.8×10^{-3}
5	5.3×10^{-5}	10	6.6×10^{-8}

It will be noticed from Table 3 that the formation of the magnesium complexes of the phosphates is accompanied by an appreciable negative free energy. Since the stability constants of the ATP magnesium complexes are significantly larger than the ion complexes of the other phosphates, the presence of magnesium tends to reduce the overall free energy available from the hydrolysis of ATP; however, this reduction is pH dependent.

It is difficult to specify the conditions prevailing in the living cell because the actual concentrations of ATP, ADP, Pi, Mg^{+2} and H^+ at a given location in the cell are unknown. The hydrogen ion concentration is, however, in the neighborhood of neutrality and would remain practically constant due to the tissue buffers. Thus, in tissue, the hydrolysis of ATP is coupled to the neutralization of the hydrogen ions by hydroxyl ions with the formation of water. The free energy contributed by the neutralization reactions is considerable; it is largely because the concentration of the hydrogen ions has been reduced to such low levels in the tissue that the hydrolysis of ATP is strongly exergonic.

Whereas the hydrolyses of ATP to ADP and ADP to adenylic acid (AMP) are highly exergonic, the hydrolysis of AMP to adenosine gives only a moderate free energy yield; it is a low energy phosphate bond. This is understandable because the hydrolysis of inorganic phosphate from AMP exposes the third hydroxyl of phosphoric acid whose pK is 12.4; the proton does not dissociate at physiological pH values and the hydrolysis reaction is not coupled with a neutralization.

Experimental ATP – ΔF Values. Having defined the free energy change for the hydrolysis of ATP which is meaningful for our needs, it is useful to describe the experimental approaches which have been devised. The difficulty of evaluating the free energies of hydrolysis of ATP for conditions which approximate those inside the living cell arises principally from the extreme shift of the reaction in the direction of hydrolysis. Thus a free energy change of 10,000 calories means that the equilibrium constant is about 5.6×10^{-8}. Such a value of K would require extraordinarily accurate and sensitive analytical methods. It is necessary to separate the hydrolysis of ATP into two reactions each of which contributes more moderate free energies to the overall reaction.

The first attempt along this line was made by Levintow and Meister who studied the reaction

$$\text{Glutamate} + \text{ATP} + \text{Ammonia} \rightleftharpoons \text{Glutamine} + \text{ADP} + \text{Pi} \qquad 95$$

They found that a glutamyltransferase from green peas was able to catalyze the above reaction, and, by a study of the concentration of inorganic phosphate along with the initial concentrations of the other materials, found K for this reaction to be 1.2×10^3 at pH 7.0 and at 37°C. This value of K yields a calculated $-4,300$ calories for the free energy change of the reaction. Reaction 95 is really made up of two reactions

$$\text{Glutamate} + \text{Ammonia} \rightleftharpoons \text{Glutamine} + H_2O \qquad 96$$

and

$$\text{ATP} + H_2O \rightleftharpoons \text{ADP} + \text{Pi} \qquad 97$$

Thus, if the free energy change for the hydrolysis of glutamine were available, the free energy change for the hydrolysis of ATP could be calculated by combining reactions 95 and 96. Benzinger and Hems determined the equilibrium constant for the hydrolysis of glutamine by an ingenious technique in which they added small quantities of glutamine to a fixed concentration of ammonium glutamate in the presence of a glutaminase prepared from Clostridium welchii and measured the rate of heat release or uptake as a function of the amount of glutamine added. Evidently, at equilibrium no heat would be absorbed or released by the reaction. By these means they calculated a free energy change of $-3,420$ calories per mole at 25° and pH 7. Figure 22 shows the plot of the experimental heat data of Benzinger and Hems for the glutaminase reaction. Based upon these results and on the Levintow-Meister data along with the ionization constants for H^+ and Mg^{+2} (see Table 3), Phillips, George and Rutman have made extensive calculations of the free energy change for the hydrolysis of ATP as a function of the pH and of the magnesium ion concentrations. Some of the results of their calculations are shown in Figure 23.

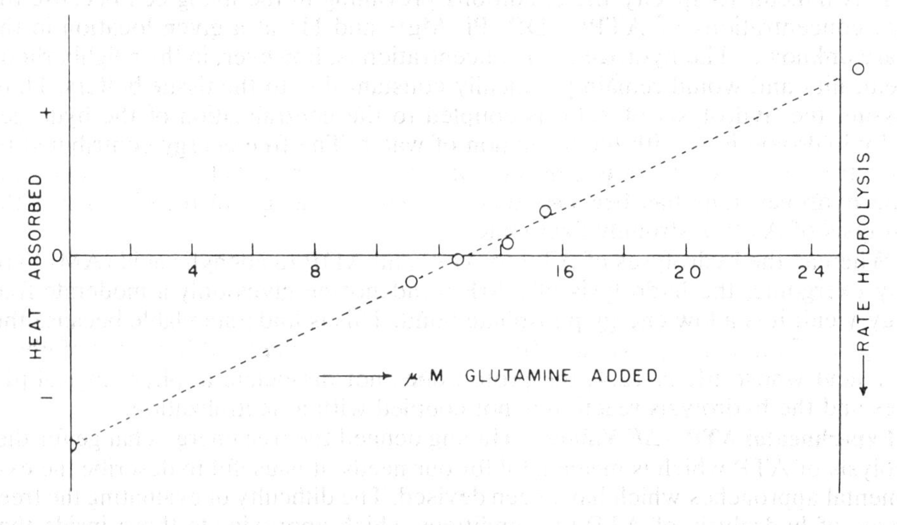

Fig. 22. Rates of glutamine hydrolysis (heat evolved) and synthesis (heat absorbed) as a function of the amount of glutamine added to a 0.884 M solution of ammonium glutamate. (T. H. Benzinger and R. Hems: Res. Report. Naval Med. Res. Inst. 14:949, 1956.)

The enthalpy of hydrolysis of ATP can be measured with much greater ease than can the free energy change. It is only necessary to subject a solution of ATP at a given pH to the phosphatase activity of myosin and measure the heat evolved in a

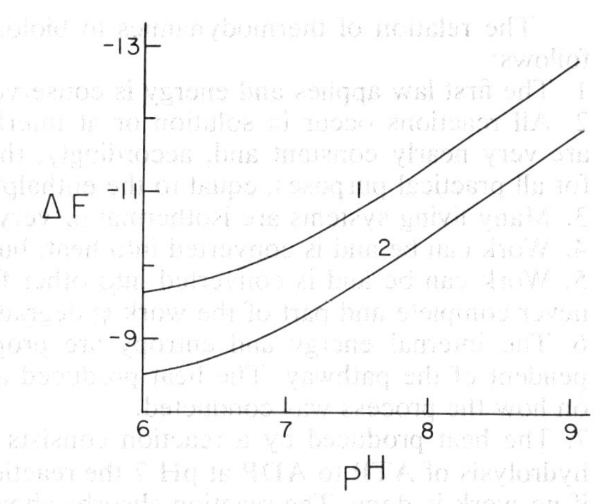

Fig. 23. ΔF, the free energy change in Kcals per mole for the hydrolysis of ATP. Curve 1, no Mg^{+2} present; curve 2, 25 mM Mg^{+2}. Ionic strength 0.15. 25°. (Data of R. C. Phillips, P. George and R. T. Rutman: J. Biol. Chem. *244:* 3330, 1969.)

suitable calorimeter. The only complication is that during the hydrolysis of ATP a proton is released and the heat of neutralization of this proton must be considered. The enthalpy for the hydrolysis of ATP to ADP is about −5,000 calories per mole.

A very considerable amount of work remains to be done on the free energy changes of biological reactions. Far too many of the reactions of importance in biochemistry have been described only in a qualitative manner and little precise knowledge regarding the free energies is available. The discussion in this chapter attempts to outline some of the difficulties associated with research of this character.

Energetics in Biology. Thermodynamics recognizes no special role for the biological. Biological systems are, however, typically open systems and are continually interacting with the environment, and the entropy of the biological system can and does decrease as a result of this interaction. A living system is never in complete equilibrium although equilibrium can be closely approached and perhaps actually achieved in respect to certain constituents. Certainly, however, most of the multitudinous reactions which are continually in process do not proceed under equilibrium conditions and cannot yield the maximum free energy; there is a large element of irreversibility with a consequent increase of entropy and the production of heat. In reference to the living, the conception of a steady state rather than of a state of equilibrium has meaning; in the steady state the system exhibits no net change of energy or mass.

The question of whether or not an isolated system can depart from equilibrium has been discussed. To put this problem in a different context, suppose that a living cell membrane could show one-way permeability to some constituent, perhaps glucose. The glucose could enter the cell by simple diffusion and the cell, by doing no work, block the exit of glucose molecules. This situation is exactly analogous to that of the famous Maxwell Demon who could operate a trap door to a container to admit gas molecules from the outside but slam the door shut to prevent the exit of gas molecules. In due time the cell would build up a considerable concentration of glucose in its interior and it could use this excess concentration to do osmotic work although the cell itself had done no work in the process of accumulating the glucose. Such a situation would be completely contrary to thermodynamics and it is clear that a membrane, however living it may be, cannot show one-way permeability without doing the required osmotic work.

Free energy changes are central in a consideration of energetics in biology. It is, however, important to remember that free energy does not need to be conserved and for a given reaction the actual amount of energy converted to useful work will depend upon how the process is conducted and the work efficiency can vary from zero to 100 percent.

The relation of thermodynamics to biological systems can be summarized as follows:

1. The first law applies and energy is conserved.
2. All reactions occur in solution or at interfaces and both pressure and volume are very nearly constant and, accordingly, the change in internal energy, ΔE, is, for all practical purposes, equal to the enthalpy, ΔH.
3. Many living systems are isothermal or very nearly so.
4. Work can be and is converted into heat, but this conversion is irreversible.
5. Work can be and is converted into other forms of work but this conversion is never complete and part of the work is degraded into heat at each step.
6. The internal energy and entropy are properties of the system and are independent of the pathway. The heat produced and the work done, however, depend on how the process was conducted.
7. The heat produced by a reaction consists of two parts: $T\Delta S$ and ΔH. In the hydrolysis of ATP to ADP at pH 7 the reaction evolves about 4,800 cals per mole if no work is done. The reaction absorbs about 2,900 cals per mole if the reaction is coupled to reversible work. The actual heat change will be somewhere between these values. If the efficiency of the reaction in respect to work done is 38 percent, no heat will be produced or absorbed.

The standard free energy of a reaction has been defined as the maximum work which can be obtained from the reaction when all reactants and products are initially in their standard state (frequently in the hypothetical one molal solution) and allowed to proceed reversibly to equilibrium. This, however, is not the usual way in which reactions are conducted either in vitro or in vivo.

Metabolic reactions, for example, flow continuously from left to right, the reactants being supplied and products removed. It is likely that many or even most of these reactions operate close to equilibrium. The question naturally arises as to the amount of work required to make the reaction proceed or the amount of work obtainable from the reaction. We have seen that the free energy of the hydrolysis of ATP to inorganic phosphate and ADP at pH 7 is about $-8{,}000$ cals. This means that the equilibrium constant for the reverse reaction (the synthesis of ATP) is about 10^{-6}. If there exists an active mechanism for the transport of ATP across the mitochondrial membrane, it is evident that the synthesis of ATP would proceed and, furthermore, the work done to synthesize a mole of ATP need not be large; in fact, if the reaction is very close to equilibrium, the work could be infinitely small. The substantial work term would involve concentrating the mole of synthesized ATP to a usable concentration level. Living systems ordinarily concentrate by a specific and active transport of the solute and not by an active transport of water.

GENERAL REFERENCES

Alberty, R. A.: Effect of pH and metal ion concentration on the equilibrium hydrolysis of ATP to ADP. J. Biol. Chem. *243*:1337, 1968.

Alberty, R. A.: Standard Gibbs free energy, enthalpy and entropy changes as a function of pH and pMg for several reactions involving adenosine phosphate. J. Biol. Chem. *244*:3290, 1969.

Benzinger, T. H. and C. Kitzinger: Microcalorimetry, new methods and objectives. *In* Temperature—Its Measurement and Control in Science and Industry, Vol. 3. Reinhold Publishing Co., New York, 1963.

Dawes, E. A.: Quantiative Problems in Biochemistry, 4th ed. Williams and Wilkins, Baltimore, 1967.

Edsall, J. T. and J. Wyman: Biophysical Chemistry, Vol. 1. Academic Press, New York, 1958.

Klotz, I. M.: Chemical Thermodynamics. Benjamin, New York, 1964.

Klotz, I. M.: Energy Changes in Biochemical Reactions. Academic Press, New York, 1967.

Montgomery, R. and C. A. Swenson: Quantitative Problems in Biochemical Sciences. W. H. Freeman, San Francisco, 1969.

Phillips, R. C., P. George, and R. J. Rutman: Potentiometric studies of the secondary phosphate ionization of AMP, ADP and ATP and calculations of thermodynamic data for the hydrolysis reaction.

Biochem. *2*:501, 1963.

Phillips, R. C., P. George, and R. J. Rutman: Thermodynamic data for the hydrolysis of adenosine triphosphate as a function of pH, Mg^{2+} ion concentration and ionic strength. J. Biol. Chem. *244*:3330, 1969.

Wadso, I.: Design and testing of a microreaction calorimeter. Acta Chem. Scand. *22*:927, 1968.

PROBLEMS

1. Drop a 25 gram brass weight with a density of 8.40 in air, 25 centimeters through water at 25°C. Calculate the rise in temperature of the brass weight and 100 grams of water in the column, neglecting the heat absorbed by the container. The specific heat of water at 25°C is .998 cals. per gram per degree and that of brass is 0.087 cals. per gram per degree.

   Ans: 1.265×10^{-4} degrees

2.* Calculate the equilibrium constants at 38°C for the following reactions:
   (a) $\text{Oxaloacetate}^{=} + H_2O \rightleftharpoons \text{Pyruvate}^{-} + HCO_3^{-}$
   (b) $\text{Fumarate}^{=} + 2H_2O \rightleftharpoons \text{Lactate}^{-} + HCO_3^{-}$
   (c) $\text{Fumarate}^{=} + H_2O \rightleftharpoons \text{Malate}^{=}$
   The values for the standard free energies of formation of these compounds at 38°C in their standard states in aqueous solution are:
   Pyruvate⁻, −106,406 cals; lactate⁻, −117,960 cals; HCO_3^{-}, −139,200 cals; malate⁼, −199,430 cals; H_2O, −56,200 cals; fumarate⁼, −142,525 cals; oxaloacetate⁼, −184,210 cals.

   Ans: (a) 4.887×10^3, (b) 37.15, (c) 3.126

3.* The malic dehydrogenase enzyme from horse heart catalyses the reversible reaction

$$L - \text{malate}^{=} + NAD^{+} \rightleftharpoons \text{oxaloacetate}^{=} + NAD + H^{+}$$

   The equilibrium constant of this reaction was measured by following the change in optical density at 340 mμ; this enabled the concentration of reduced NAD to be obtained and that of oxaloacetate was assumed to be the same. (Burton and Wilson: Biochem. J., *54:* 86, 1953.) The concentrations of L-malate and NAD^+ were obtained from the initial concentrations and those of NAD formed. These equilibrium figures are presented below for 25°C.

	pH	NAD M × 10⁵	NAD⁺ M × 10⁵	L-Malate M × 10³
a)	8.81	2.82	32.4	5.27
b)	8.83	3.27	41.3	5.14
c)	7.55	0.79	43.7	5.19

   Calculate the equilibrium constant for each experiment and comment on the influence of pH on its value. The activity coefficients may be assumed to be unity in every case. Calculate the standard free energy change of the reaction.

   Ans: (a) 7.21×10^{-13}, (b) 7.45×10^{-13}, (c) 7.75×10^{-13}; $\Delta F° = 16{,}638$ calories.

4. Derive the relation between the free energy change and the equilibrium constant.

   Ans: See text.

5. The composition of the parietal secretion produced by the stomach has the following composition: H^+ − ions 0.159 molar, Cl^- − ions 0.166 molar, K^+ − ions 0.0074 molar. The concentration of blood plasma in respect to these ions is: H^+ − ions 4.0×10^{-8}, Cl^- − ions 0.105 molar, K^+ − ions 0.0047 molar. Calculate the minimum work in calories which the body must do to secrete one liter of parietal juice.

   Ans: 1542.5 calories.

6. The following table shows the ionization constant of water as a function of the temperature:

T°C	$K_w \times 10^{14}$
10	0.293
20	0.681
30	1.471

   Calculate the molar change in free energy, in the heat and in the entropy involved in the ionization of water at 25°C.

   Ans: $\Delta F° = 19{,}104$; $\Delta H° = 13{,}760$; $\Delta S° = -17.9$

*These problems are from Quantitative Problems in Biochemistry by Edwin A. Dawes, E. and S. Livingstone Ltd., Edinburgh and London, 1963, and are being used with the permission of the author and publishers.

7. Calculate the free energy change for the oxidation of one gram of D-glucose in a 0.001 molar solution to liquid water and dissolved carbon dioxide at a carbon dioxide pressure of 50 mm. Hg and an oxygen pressure of 10 mm. Hg at pH 7.0 and at 25°.

   Ans: −3,768 calories

8. A reaction mixture consisting of the following was incubated 20 minutes at 30°:

Glucose-6-phosphate, 2.40×10^4 C.P.M./μM	42.9μM
ADP	47.3μM
$MgCl_2$	23.8μM
Serum albumin	1.2 Mgs
Solution of crystalline hexokinase	0.1 ml
Mixture made to 6.0 mls and at pH 6.0	

   One ml. of the reaction mixture was added to 0.8 ml. of 1:1 perchloric acid and one ml. of water on ice bath, and 25 mg. of glucose were added as a carrier. The mixture was centrifuged and the supernatant neutralized to pH 6 with standard base, chilled and the potassium perchlorate removed by centrifugation. The supernatant solution was decanted on to a resin column. Eluates collected and analyzed for glucose and radioactivity. To the eluate were added 10 μM of glucose-6-phosphate and the eluate passed through another resin column and assayed. Results:

Effluent	*1st column*	*2nd column*
mg. glucose/ml.	1.33	0.612
plated mg.	1.064	0.979
net C.P.M.	450	382

   Calculate the equilibrium constant and the free energy changes for the hexokinase reaction and discuss these results in relation to the free energy change for the hydrolysis of ATP. Note: Read Robbins and Boyer: J. Biol. Chem. *224:* 121, 1957. This problem was submitted on request by Dr. Paul Boyer.

   Ans: $K = 304$, $\Delta F - 3,440$

9. (a) Calculate the equilibrium constant for the hydrolysis of DL-alanyl glycine at 37.5°. The free energies for the compounds in their standard state in aqueous solutions at 37.5° are: alanyl glycine, −114,680 cals; alanine, −87,300 cals; glycine, −87,710 cals; water, −56,260 cals.
   (b) Calculate the contribution of the ionization of the reactants and the products to the free energy of hydrolysis; obtain an expression which will permit the estimation at any pH and which involves the ionization constants of the various groups.

   Ans: (a) K = 805

10. The aldehyde mutase reaction consists of two coupled reactions:

$$CH_3CHO + NADH_2 \rightarrow CH_3CH_2OH + NAD$$
$$CH_3CHO + H_2O + NAD \rightarrow CH_3COOH + NADH_2$$

   (a) Calculate the equilibrium of the overall reaction at 25°C. The standard free energies of formation at 25°C in water are: acetaldehyde, −33,400 cals; acetic acid, −95,500 cals; ethanol, −43,400; and liquid water, −56,690 cals.
   (b) What concentration of acetaldehyde will be in equilibrium with 0.1 M acetic acid and 0.1 M ethanol?

   The assumption has been made above that the acid is not appreciably dissociated; the reaction is usually carried out in a bicarbonate $-CO_2$ buffer and may be represented as

$$2CH_3CHO + HCO_3^- \rightleftharpoons CH_3COO^- + CH_3CH_2OH + CO_2\ (g).$$

   The free energy of formation of HCO_3^- is −140,300 cals; CH_3COO^-, 89,000 cals; and gaseous CO_2 −92,300 cals. Calculate (c) the equilibrium constant of this reaction and (d) the concentration of acetaldehyde in equilibrium with 1 M acetate and ethanol if HCO_3^- is 0.03 M and the partial pressure of CO_2 is 0.05 atmospheres.

   Ans: (a) $K = 1.98 \times 10^{11}$; (b) 2.25×10^{-7}; (c) $K = 7.95 \times 10^{12}$; (d) 4.58×10^{-7}

CHAPTER 3

Electrolytes and Water

The roles of water and electrolytes in biology have long been the subject of intensive investigation but much still remains to be done; many of the problems associated with water and ions are far from clear.

Ions. Ions in aqueous solution can arise from: (1) oxidation of a metal or the reduction of a non-metal; (2) solution of an ionic crystal; (3) ionization of a neutral molecule.

Ions are produced when a metallic element yields electrons to an electronegative element and both acquire the electronic structure of a rare gas; such structures have completed outer electronic shells and tend to be stable. It should be noted, however, that when Cr, Mn, Fe, Co, Ni, Cu, Ag, Hg and Zn form ions, a rare gas electronic structure does not result and such ions have residual valences which permit them to complex with various molecules more easily.

In crystals of sodium chloride, sodium and chlorine do not exist as atoms but as ions. Other salts such as KCl, NaBr, etc., also form ionic crystals, and when such crystals are dissolved in water an ionic solution results.

The presence of water tends to promote ionization of a neutral molecule in two ways: (1) because the dielectric constant of water at room temperature is about 78, the force of attraction between the positive and negative ions is reduced to about 1/78th of what it would be in air and (2) the water molecules having a relatively large dipole moment are oriented towards and bound to the ions; the ions become hydrated and the extent of ionization is increased as a result of the release of the energy of hydration.

The radii of ions are customarily assigned from crystallographic studies on ionic crystals. Figure 1 shows the ionic sizes of several inorganic ions to scale. Note that the ionic sizes are all roughly comparable to that of the water molecule.

Ionic Strength. Ionic concentrations may be expressed in a variety of ways, any one of which may be appropriate to the needs at hand. Ionic strength has proven to be especially meaningful and Lewis and Randall enumerated the important rule: "In dilute solutions, the activity coefficient of a given strong electrolyte is the same

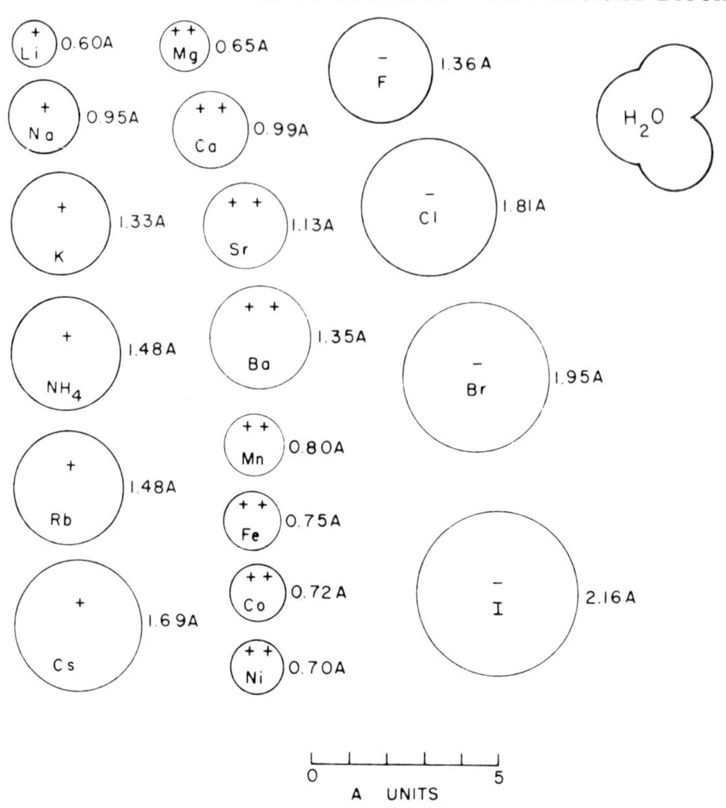

Fig. 1. Sizes of ions drawn to scale; note size of water molecule. Radii given in A-units.

in all solutions of the same ionic strength." They defined the ionic strength, Γ/2, as equal to the sum of the ionic concentrations, each ionic concentration being multiplied by the square of the ionic valence. The whole sum is divided by 2

$$\Gamma/2 = \frac{C_1Z_1^2 + C_2Z_2^2 + C_3Z_3^2}{2} \tag{1}$$

where C_1, C_2, etc., are the individual ionic concentrations in mole ions per liter and Z_1, Z_2, etc., are the respective valences of the ions. Thus ionic strengths of one tenth molar solutions of NaCl, of Na_2HPO_4 and of $MgSO_4$ are 0.1, 0.3, and 0.4 respectively. The rule of Lewis and Randall given above, that solutions of all strong electrolytes at the same ionic strength exhibit the same ionic effects, was found to be valid up to an ionic strength of about 0.1. There are a number of symbols used to represent ionic strength; Γ/2 has been adopted because there is less chance of confusion with other symbols.

The principle of ionic strength is an important one for the biochemist. Frequently, for example, a study is made of the variation of some factor as a function of the pH of a solution, buffers being employed to yield the proper pH. In order, however, for the effect observed to be due only to the variation in pH and not obscured by secondary salt effects, it is necessary to make up the series of buffers to the same ionic strength. This does not always suffice, as many physiological responses have specific ion effects. Nevertheless, a constant ionic strength does represent the least that the biochemist can do to make his experiments as unambiguous as possible.

Activity of Electrolytes. It was noted in Chapter 2 that the activity coefficient of an undissociated molecule is simply the ratio of the activity to the concentration. The relation for an electrolyte is more complex.

The chemical potential of an electrolyte must equal the sum of the chemical potentials of the individual ions and

$$\mu_{MA} = \mu_M^+ + \mu_A^- \quad 2$$

and

$$\mu_{MAo} + RT \ln a_{MA} = \mu_{Mo}^+ + RT \ln a_M^+ + \mu_{Ao}^- + RT \ln a_A^- \quad 3$$

Since

$$\mu_{MAo} = \mu_{Mo}^+ + \mu_{Ao}^- \quad 4$$

eq. 3 can be written

$$a_{MA} = a_M^+ \times a_A^- \quad 5$$

The activities of the individual ions can be expressed as the products of their molal concentrations m and their activity coefficients γ_+ and γ_- respectively; then

$$a_M^+ = m\gamma_+ \text{ and } a_A^- = m\gamma^- \quad 6$$

Substitution of eq. 6 into eq. 5 yields

$$a_{MA} = (m\gamma_+)(m\gamma_-) = m^2\gamma_\pm^2 \quad 7$$

where γ_+ is the mean activity coefficient for a univalent electrolyte (such as NaCl). It is evident from eq. 7 that

$$\gamma_\pm = (\gamma_+\gamma_-)^{1/2} \quad 8$$

The mean activity coefficient $\gamma_\pm$ can be measured experimentally; the individual ion activity coefficients cannot. The general expression for the mean activity of an electrolyte of any valence type is

$$a_\pm = m^v(v_+\gamma_+)^{v+}(v_-\gamma_-)^{v-} \quad 9$$

where m is the molal concentration of the electrolyte which upon ionization gives rise to v_+ moles of cations and v_- moles of anions. v is equal to $v_+ + v_-$. The corresponding expression for the mean activity coefficient is

$$\gamma_\pm = (\gamma_+^{v+}\gamma_-^{v-})^{1/(v_+ + V_-)} \quad 10$$

Elementary Electrostatics. Since ions are charged particles, a brief survey of electrostatic theory is necessary for an understanding of ionic behavior. The charge on the electron is 4.802×10^{-10} electrostatic charge units (e.s.u.); one mole of monovalent ions, therefore, carries $4.802 \times 10^{-10} \times 6.0238 \times 10^{23}$ or 2.8926×10^{14} e.s.u. Since one e.s.u. charge is equal to 3.336×10^{-10} coulombs, the number of coulombs carried by one mole of univalent ions is 96,493; this quantity of charge is known as a Faraday.

There are three systems of electrical units in use and confusion sometimes arises as to which system is appropriate. The systems are the practical, the electromagnetic and the electrostatic. Table 1 shows numerical comparisons of the units.

Unit electrostatic charge is defined as that quantity of electricity which will repel, in a vacuum, a like quantity of electricity of the same sign at a distance of one centimeter with the force of one dyne, the two charges being point charges. The force between two point charges at a distance, r, apart is given by (Coulomb's law)

$$F = \frac{Q_1 \times Q_2}{Dr^2} \quad 11$$

where Q_1 and Q_2 are the charges and, if they have the same sign, the force (F) is repulsive and, if unlike, attractive. D is the dielectric constant of the medium and is the ratio of the electrostatic force in a vacuum to that in a given medium.

TABLE 1. NUMERICAL COMPARISON OF THE UNITS OF THE PRACTICAL, ELECTROMAGNETIC (E.M.U.) AND THE ELECTROSTATIC (E.S.U.) SYSTEMS

Quantity	*Name of Practical Unit*	*Number of practical units in the corresponding e.m.u. and e.s.u. units*	
		e.m.u.	*e.s.u.*
Charge	Coulomb	10	3.3356×10^{-10}
Potential	Volt	1×10^{-8}	*299.796
Resistance	Ohm	1×10^{-9}	8.98776×10^{11}
Capacitance	Farad	1×10^{9}	1.11263×10^{-12}
Current	Ampere	10	3.3356×10^{-10}
Inductance	Henry	1×10^{-9}	8.98776×10^{11}

*The potential ratio e.s.u./e.m.u. is equal to the speed of light in a vacuum.

The electrostatic potential at any point is equal to the work in ergs necessary to transport a unit charge from an infinite distance to the point in question. The difference in potential between two points A and B is equal to the work in ergs required to move a unit charge from infinity to B and is independent of the pathway employed. The work done is also equal to the free energy change of the process providing only electrical work has been accomplished. Figure 2 represents the work (potential) of moving a point charge from position A through an irregular path towards another point charge of like sign located at B.

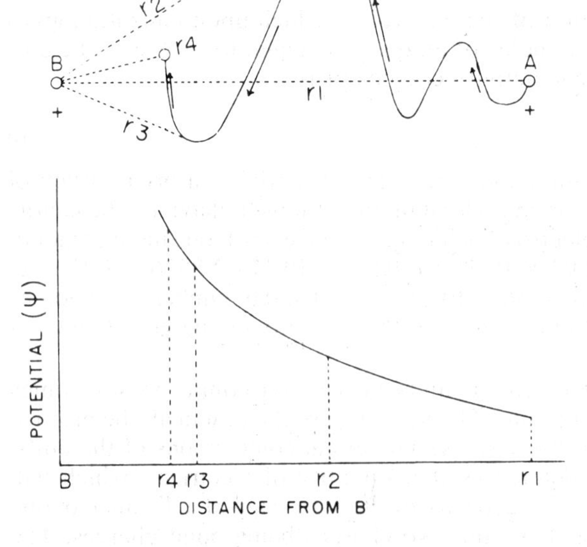

Fig. 2. Diagrammatic representation of the potential, ψ, as a function of the distance (r) between charges A and B.

The field strength or electrical intensity is the rate of change of the potential with distance ($d\psi/dr$) and the field strength multiplied by the charge gives the force acting on the charge. The work of charging a conducting sphere of radius a in a medium of dielectric constant D can be calculated as follows: Let α be the fraction of the total charge Q; as the charging process proceeds, α goes from zero to unity. The potential at the surface of the sphere is $\alpha Q/Da$ and the work increment, dw, of adding additional charge $Qd\alpha$ is

$$dw = \psi Q d\alpha = \frac{Q^2 \alpha d\alpha}{Da} \tag{12}$$

and the total work of charging is

$$w = \int_{\alpha=0}^{\alpha=1} \frac{Q^2 \alpha d\alpha}{Da} = \frac{Q^2}{2Da} \tag{13}$$

The ratio of the charge (Q) to the potential (ψ) is the capacity (C) of the system and

$$C = \frac{Q}{\psi} \tag{14}$$

The relation between the charge and the potential depends on the geometry of the conductors. In any particular case the relation is obtained from the integration of Poisson's equation (see eq. 18). Table 2 shows the capacities of several simple arrangements of conductors.

TABLE 2. CAPACITIES OF VARIOUS ARRANGEMENTS OF CONDUCTORS

System	*Capacity*	*Potential*
Sphere of radius a	Da	$\frac{Q}{Da}$
Concentric spheres of radii a_1 and a_2	$\frac{Da_1a_2}{a_2 - a_1}$	$\frac{Q(a_2 - a_1)}{Da_1a_2}$
Concentric cylinders of radii a_1 and a_2 and length L	$\frac{DL}{2\ln^{a_2}/_{a_1}}$	$\frac{2Q\ln^{a_2}/_{a_1}}{DL}$
Parallel plates of area A and distance apart of d	$\frac{DA}{4\pi d}$	$\frac{4\pi Qd}{DA}$

The dielectric constant of a medium has an important bearing on its electrostatic behavior. The dielectric constant in a vacuum is defined as unity and that of air is very nearly unity. Shown in Table 3 are values of the dielectric constant of several common substances at 25°C.

TABLE 3. DIELECTRIC CONSTANTS OF SOME COMMON LIQUIDS AT 25°C

Substance	*Dielectric Constant*
Water	78.30
Nitrobenzene	34.89
Methanol	33.1
Ethanol	25.0
n-Propanol	20.8
n-Butanol	17.0
Benzene	2.28
Petroleum oil	2.12

The dielectric constant of water can be expressed as a function of temperature by the empirical equation

$$\log D = 1.94409 - 0.001991T \tag{15}$$

where T is given in degrees centigrade.

Ionic Interaction. The electrostatic interactions of ions in solution was the basis of the well known Debye-Hückel theory of electrolytes which has been so fruitful in investigations on ionic solutions. This theory and ideas related to it are basic to discussions of electrophoresis, titrations of proteins and nucleic acids and a variety of other topics.

Due to electrostatic attractive forces, there will be an excess of positive ions in the immediate neighborhood of a negative ion and the difference in concentration between the cations and anions determines the charge density. The concentration of ions in the vicinity of an ion is given by the Boltzmann distribution equation which is

$$n = n_0 e^{\frac{Z\epsilon\psi}{kT}} \qquad 16$$

where n_0 is the number of ions per unit volume in the absence of an electrostatic potential (ψ) and n is the number of ions in the presence of the electrostatic field. Z is the valence of the ion and ϵ is unit electric charge. k is the Boltzmann constant and T is the absolute temperature. The sign of the exponent is positive for cations and negative for anions provided the central ion is an anion. $Z\epsilon\psi$ represents the electrostatic energy of the ions and kT, the kinetic energy, tends to dissipate the charge density.

The charge density per unit volume is evidently equal to the difference in concentration between that of the cations and anions

$$\rho = n^+ - n^- = Z_1\epsilon n_0 e^{\frac{Z_1\epsilon\psi}{kT}} - Z_2\epsilon n_0 e^{\frac{-Z_2\epsilon\psi}{kT}} \qquad 17$$

where ρ is the charge density and Z_1 and Z_2 represent the valences of the cation and anion respectively.

The potential distribution in the electrical double layer around the central ion is related to the space charge by Poisson's equation which states that $4\pi/D$ lines of force radiate from a unit electrostatic charge and in Cartesian coordinates is

$$\frac{d^2\psi}{dx^2} + \frac{d^2\psi}{dy^2} + \frac{d^2\psi}{dz^2} = -\frac{4\pi\rho}{D} \qquad 18$$

At a plane charged surface exposed to an ionic solution, the potential normal to the surface is evidently

$$\frac{d^2\psi}{dx^2} = -\frac{4\pi\rho}{D} \qquad 19$$

The combination of eqs. 17 and 19 leads to an equation which can be integrated and this will be discussed in Chapter 13, Electrophoresis.

For spherical symmetry such as exists around an ion, the Poisson equation in spherical coordinates is

$$\frac{1}{r^2}\frac{d}{dr}\left(\frac{r^2 d\psi}{dr}\right) = -\frac{4\pi\rho}{D} \qquad 20$$

where r is the radial distance from the central ion.

Unfortunately, when the value of ρ from the Boltzmann equation is substituted directly into the Poisson equation, the resulting non-linear differential equation cannot be integrated and use has to be made of the Debye-Hückel approximation which consists in the expansion of the Boltzmann equation and the neglect of the higher terms. This expansion gives

$$\rho = \Sigma n_i Z_i \epsilon - \Sigma n_i Z_i \epsilon \left(\frac{Z_i\epsilon\psi}{kT}\right) + \frac{\Sigma n_i Z_i \epsilon}{2}\left(\frac{Z_i\epsilon\psi}{kT}\right)^2 + \qquad 21$$

The term $\Sigma n_i Z_i \epsilon$ vanishes because of the condition of neutrality of the solution as a whole and if $kT >> Z_i\epsilon\psi$, only the term which is linear in ψ is appreciable and

$$\rho = -\frac{\Sigma n_i Z_i^2 \epsilon^2 \psi}{kT} \qquad 22$$

Substituting eq. 22 into eq. 20, there results

$$\frac{1}{r^2}\frac{d}{dr}\left(\frac{r^2 d\psi}{dr}\right) = \frac{4\pi\epsilon^2}{DkT}\Sigma n_i Z_i^2 \psi = \kappa^2 \psi \qquad 23$$

where κ has the dimensions of reciprocal distance and whose meaning will be discussed in more detail.

The solution of eq. 23 is

$$\psi = \frac{Ae^{-\kappa r}}{r} + \frac{Be^{\kappa r}}{r} \qquad 24$$

where A and B are integration constants. For very large values of r, the second term on the right of eq. 24 approaches infinity for any finite value of B. Since at infinity the potential must be zero, B must, therefore, be zero. Equation 24 then reduces to

$$\psi = \frac{Ae^{-\kappa r}}{r} \qquad 25$$

Combination of eqs. 22 and 25 gives

$$\rho = -\frac{Ae^{-\kappa r}}{r}\frac{\Sigma n_i Z_i^2 \epsilon^2}{kT} = -\frac{A\kappa^2 D}{4\pi}\frac{e^{-\kappa r}}{r} \qquad 26$$

The charge on the central ion must be equal and opposite to the net charge of the ion atmosphere surrounding the central ion and

$$\int_a^\infty 4\pi r^2 \rho dr = -Z\epsilon \qquad 27$$

where a is the distance of closest approach of the ions. The substitution of eq. 26 into eq. 27 yields

$$A\kappa^2 D \int_a^\infty re^{-\kappa r} dr = Z\epsilon \qquad 28$$

which on integration gives

$$A = \frac{Z\epsilon}{D}\frac{e^{\kappa a}}{1+\kappa a} \qquad 29$$

and substituting the value of A into eq. 25 gives

$$\psi = \frac{Z\epsilon e^{\kappa(a-r)}}{D(1+\kappa a)r} \qquad 30$$

At the surface of the ion where r is equal to a

$$\psi = \frac{Z\epsilon}{D(1+\kappa a)a} = \frac{Z\epsilon}{Da} - \frac{Z\epsilon\kappa}{D(1+\kappa a)} \qquad 31$$

The first term on the right side of eq. 31, $\frac{Z\epsilon}{Da}$, is the potential of an isolated charged sphere (see Table 2). The second term on the right side of eq. 31 represents the potential due to the ion atmosphere.

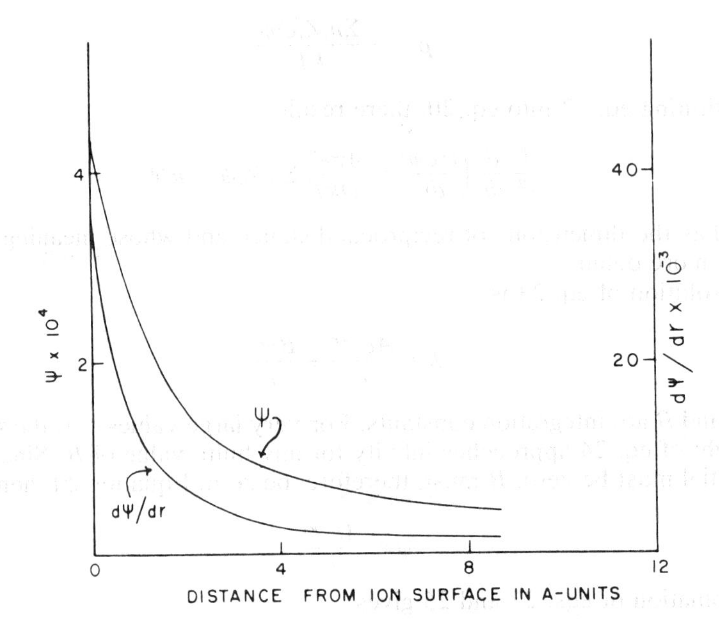

Fig. 3. Plot of the potential, ψ, in electrostatic volts as a function of the distance from the surface of a potassium ion. Also shown is the field strength ($d\psi/dr$). Ionic strength 0.01 and 25°C. ($1/\kappa$ is 30.5 *A*.)

Figure 3 shows a plot of the potential ψ as a function of the distance from the surface of a potassium ion at an ionic strength of 0.01, eq. 30 being used. Also shown is the field strength ($d\psi/dr$) as a function of the distance from the ion surface. Note the very large potential gradients in the neighborhood of an ion, amounting to about three million ordinary volts per centimeter at a distance of about one *A*-unit.

Comparing eq. 31 with the expression for the capacity of two concentric spheres (Table 2), it is seen that $1/\kappa$ corresponds to the distance between two concentric spheres, i.e., between the central ion and the center of gravity of the ion atmosphere (see Fig. 4) and $1/\kappa$ is equal to $a_2 - a_1$ (Table 2).

According to eq. 23, κ can be written

$$\kappa = \left(\frac{8\pi N^2\epsilon^2\Gamma/2}{1{,}000\ DRT}\right)^{1/2} \qquad 32$$

where $\Gamma/2$ is the ionic strength. Substituting numerical values for the constants at 25°C in water

$$\kappa = 3.282 \times 10^7\ \Gamma/2^{1/2}\ \text{cm}^{-1} \qquad 33$$

Ionic Activity. The work of charging the ionic atmosphere around the central ion can now be obtained just as was done previously for an isolated conducting sphere (eq. 13) where the charge is increased reversibly in small increments from zero to unit fraction of the total charge. [As noted, the potential ψ due to the ion atmosphere is given by $-Z\epsilon\kappa/D\ (1 + \kappa a)$ (see eq. 31)]. When this is done, the work or free energy change for the charging process is

$$\Delta F_e = -\frac{Z^2\epsilon^2}{2D}\left(\frac{\kappa}{1 + \kappa a}\right) \qquad 34$$

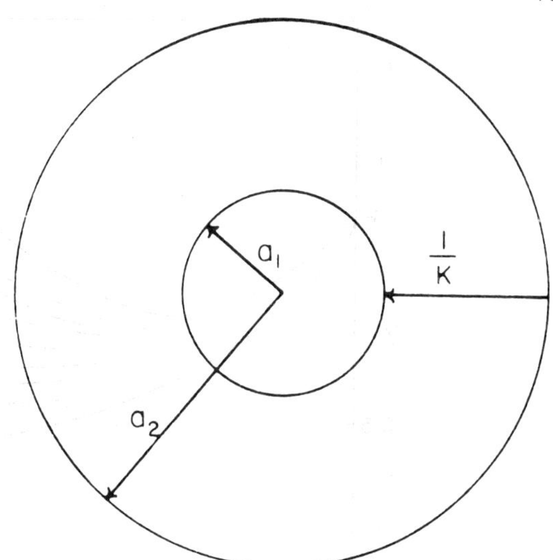

Fig. 4. The central ion and its ion atmosphere acting at a distance $1/\kappa$.

It is now necessary to relate the electrical work to the activity coefficient of the electrolyte. The free energy change in diluting a mole of ions from concentration C_1 to C_2 ideally is

$$\Delta F = RT \ln \frac{C_2}{C_1} \tag{35}$$

If the ions form an ideal solution at concentration C_1 but not at concentration C_2, the free energy change is

$$\Delta F = RT \ln \frac{\gamma C_2}{C_1} \tag{36}$$

where γ is the activity coefficient of the ions at concentration C_2. In the particular case being considered C_1 is equal to C_2. Substituting eq. 36 into eq. 34 and rearranging and per mole of ions

$$-RT \ln \gamma = \frac{NZ_1Z_2\epsilon^2}{2D} \frac{\kappa}{(1+\kappa a)} \tag{37}$$

When values of the constants are substituted into eq. 37 at 25° and converted to log base 10, there results

$$\log \gamma = \frac{-0.509Z_1Z_2\Gamma/2^{1/2}}{1 + 3.3 \times 10^7 a\Gamma/2^{1/2}} \tag{38}$$

For values of $\Gamma/2$ less than about 0.1 eq. 38 reduces to

$$\log \gamma = -0.509Z_1Z_2\Gamma/2^{1/2} \tag{39}$$

Equation 39 is a statement of the Debye-Hückel limiting law according to which log γ approaches linearity in the square root of the concentration at high dilutions.

The Debye-Hückel equations have been subject to exhaustive test and found to yield accurate values for the activity coefficients of dilute solutions of electrolytes. Difficulties have, however, been experienced in more concentrated solutions and there are several probable causes for these difficulties at higher concentrations. A number of approximations were made in the derivation, and the dielectric constant was assumed to be independent of the electrolyte concentration. Most important,

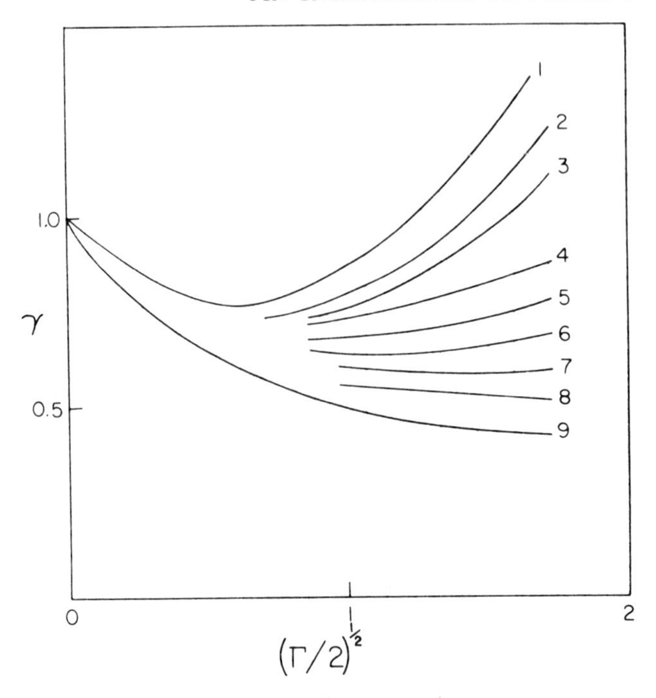

Fig. 5. Mean activity coefficients γ of 1-1 halides at 25° curves 1, LiI; 2, LiBr; 3, LiCl; 4, NaI; 5, NaBr; 6, NaCl; 7, KBr; 8, RbI; 9, CsI.

however, is the effect of the intense electrostatic fields of the ions as they approach more closely at higher concentrations on the structure of the water.

Figure 5 shows the mean activity coefficients of 1-1 halides at 25°C. It will be noted from Figure 5 that there is a strong tendency for a minimum to occur in the activity-concentration relation. The Debye-Hückel equation cannot account for such a minimum; an additional term is needed, a term which increases with concentration; a so-called salting-out term. Including the salting-out term the Debye-Hückel equation becomes

$$-\log \gamma = \frac{0.509 Z_1 Z_2 \Gamma/2^{1/2}}{1 + 3.3 \times 10^7 a \Gamma/2^{1/2}} - B\Gamma/2 \qquad 40$$

where B is an empirically determined salting-out constant. Figure 6 shows a plot of the limiting and extended Debye-Hückel equations.

For the purposes of plotting and approximate calculations, eq. 40 can be put into a somewhat more convenient form by assuming that a, the radii of the ions, is 3.0×10^{-8} cm. and that B, the salting-out constant, is 0.1. Then eq. 40 becomes

$$-\log \gamma = \frac{0.509 Z_1 Z_2 \Gamma/2^{1/2}}{1 + \Gamma/2^{1/2}} - 0.10\Gamma/2 \qquad 41$$

The discussion of the constant B will be extended in connection with ion hydration and particularly in relation to the lyotropic or Hofmeister series.

Although the Debye-Hückel equations were originally derived for strong electrolytes, they are equally applicable to weak and intermediate electrolytes, and the ionic strength is obtained from the degree of dissociation of the weak electrolyte.

Water. Water is the only biological solvent; biological reactions are conducted in an aqueous medium and in point of quantity; water is the principal component of tissue; about 60 percent of the human body is water. In the mammalian body the

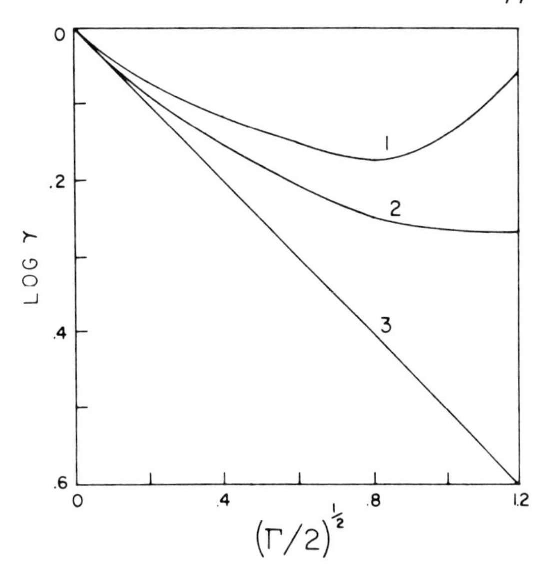

Fig. 6. Plot of the Debye-Hückel equations: (1) eq. 40, (2) eq. 38, (3) eq. 39.

water is conveniently considered to be located in three compartments: (a) the blood plasma which accounts for about 7 percent of the body water, (b) the interstitial fluid which lies outside of the blood system and which bathes the cells and constitutes about 21 percent of the total water, and finally (c) the water inside the cells (the intracellular water) which makes up the largest quantity of water—about 71 percent of the total. The water distribution in the body is calculated from dilution studies on substances which penetrate into the various areas of the body. The water balance of the body and of the three water compartments bear on important and complex problems in biochemistry, in physiology and in clinical medicine, and as yet are poorly understood. Figure 7 illustrates the water distribution in the mammalian body in a simple and highly diagrammatic manner.

A water molecule is in the form of an isosceles triangle. The O—H bond is 0.99A and the H—O—H angle is 104°35′; due to the attraction of the oxygen nucleus for electrons, electrons are drawn away from the hydrogens leaving them positive. This separation of charge leads to a high dipole moment for the molecule of 1.87×10^{-18} electrostatic units. The dipole acts along the bisector of the H—O—H angle with the negative end towards the oxygen. Figure 8 shows a scaled drawing of an isolated water molecule.

The structure of ice indicates that the water molecule contains, in addition to the two positive charges on the hydrogen atoms, two negatively charged regions, the centers of the negative charges being such that the four charges occupy the corners of a tetrahedron. In ice the tetrahedra are associated to form a structure diagramed in Figure 9. In ice the H—O—H bond angle has been increased to 109°28′ due to the interaction of the water molecules through hydrogen bonding. Actually, nine modifications of the ice crystal exist, depending upon pressure and temperature.

The ordinary ice crystal is a very open structure and the individual water molecules retain a considerable degree of freedom and are capable of rotation in the crystal which is reflected in the relatively high dielectric constant of ice. The density of ice at 0°C is 0.917, whereas that of water at this temperature is 0.999. If ice were a close packed structure, its density would be about 1.7. When ice melts to form liquid water the number of nearest neighbors of a given water molecule actually increases. For example, X-ray diffraction studies indicate the number of nearest neighbors of a water molecule at 1.5°C and also at 13°C to be 4.4 which rises to 4.6 at 30°C. The average distance from the central molecule, however, increases with

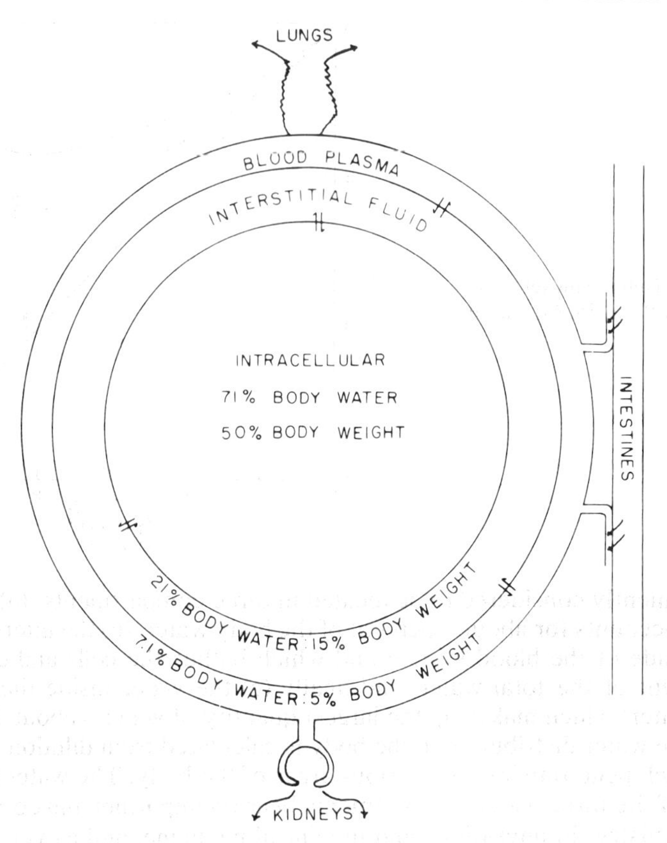

Fig. 7. The three water compartments of the mammalian body: (a) blood plasma, (b) interstitial fluid, and (c) intracellular water. The relative areas are approximately proportional to the relative amounts of water.

increasing temperature and at 30°C it is 2.9A as compared with 2.76A in ice. About 15 percent of the hydrogen bonds linking the molecules of water in ice are broken when ice melts.

Water molecules not only form hydrogen bonds with other water molecules, they also form such bonds with dissolved molecules which contain appropriate groups. The water molecule having a large dipole moment reacts vigorously with ions; water is a very sticky substance.

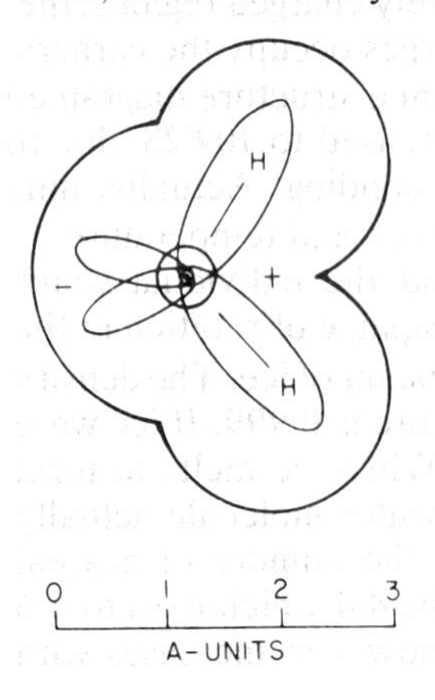

Fig. 8. Scaled drawing of a gaseous water molecule (projected area of the water molecule is $10.9A^2$).

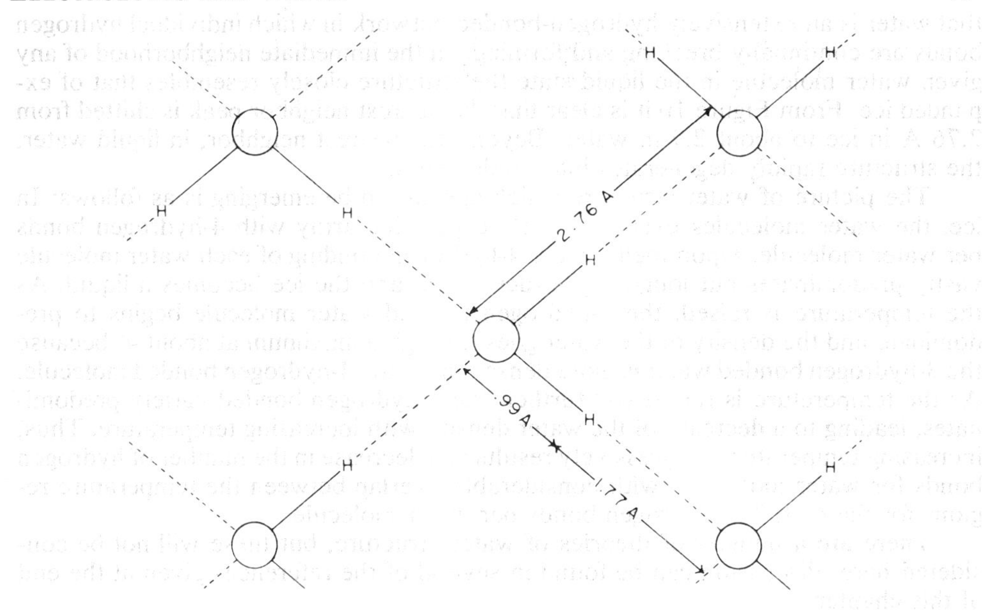

Fig. 9. Diagrammatic arrangement of the water molecule in an ice crystal showing tetrahedral structure.

There is considerable evidence from the physical properties of water that this substance is highly associated. For example, if the boiling and melting points of water are compared with those of closely related hydrides—e.g., NH_3, H_2S, HCl—it is found that the transition temperatures of water are significantly higher.

There has been and is much speculation and even controversy regarding the structure of liquid water. To a considerable degree this controversy is a matter of semantics; two molecules of water associated together through hydrogen bonds constitute a structure. The question arises as to how many molecules a structure needs to have and what its lifetime needs to be before it can be recognized as a structure. Elementary fluctuation theory indicates that clustering of molecules in any liquid is to be expected. Light scatter experiments on water suggest that molecular clustering in water is about 10 percent larger than would be predicted from fluctuation theory. Other physical measurements suggest that the cluster size must be small but might lie in the range of 10 to 100 molecules. There is a serious question about whether or not it is meaningful or useful to talk about water structure. It was Roentgen, the discoverer of X-rays, who first proposed that water could be represented as being composed of two kinds of substances—an ice-like, bulky species and a denser component. The notion that water consists of a mixture of two or more components persists today. If quasi-crystalline structures do exist, it is well to remember that there are no less than nine crystalline forms of water and that these do not differ greatly in energy content. Many of these forms could exist in water clusters in the liquid. Incidentally, optical evidence indicates that the concentration of the non-hydrogen bonded monomer in water is vanishingly small.

It is not possible to reach a definitive judgment regarding water structure from X-ray diffraction studies. X-ray diffraction pictures of water give diffuse concentric rings characteristic of liquids and detail is missing.

Figure 10 shows a plot of the radial distribution function, $4\pi r^2 p(r)$ (A^{-1}) for water at 4° contrasted with the same plot for ordinary ice as derived from X-ray diffraction studies. The radial distribution is an expression of the average distribution of oxygen-oxygen distances. X-ray diffraction results are consistent with the notion that water is an extensively hydrogen-bonded network in which individual hydrogen bonds are continually breaking and forming. In the immediate neighborhood of any given water molecule in the liquid state the structure closely resembles that of expanded ice. From Figure 10 it is clear that the nearest neighbor peak is shifted from 2.76 A in ice to about 2.9 in water. Beyond the nearest neighbor, in liquid water, the structure rapidly degenerates into randomness.

The picture of water structure which appears to be emerging is as follows: In ice, the water molecules exist in orderly crystalline array with 4-hydrogen bonds per water molecule. Upon melting, the 4-hydrogen bonding of each water molecule vastly predominates but long-range order is lost and the ice becomes a liquid. As the temperature is raised, the 3-hydrogen bonded water molecule begins to predominate and the density of the water goes through a maximum at about 4° because the 3-hydrogen bonded water is more dense than is the 4-hydrogen bonded molecule. As the temperature is raised still further, the 2-hydrogen bonded variety predominates, leading to a decrease of the water density with increasing temperature. Thus, increasing temperature progressively results in a decrease in the number of hydrogen bonds for water molecules with considerable overlap between the temperature regions for the 4, 3, 2, 1 hydrogen bonds per water molecule.

There are a number of theories of water structure, but these will not be considered here; discussion can be found in several of the references given at the end of this chapter.

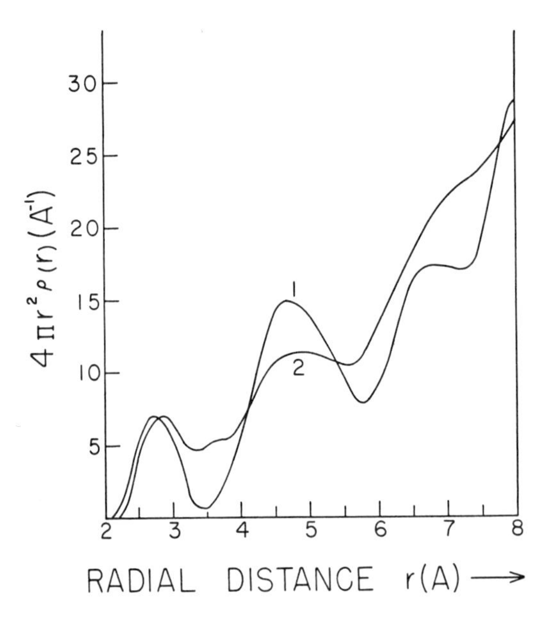

Fig. 10. Radial distribution function ($4\pi r^2 \rho$ (r) (A^{-1}) plotted against the radial distance form a water molecule. Curve 1, ice 1; curve 2, water at 4°.

Ion Hydration. There has long been a firm belief in ion hydration but, unfortunately, the extent of hydration of ions appeared to depend upon the methods used in its determination. Modern thinking on this subject dates from the paper by Bernal and Fowler.

It is evident that any method which will permit the measurement of the activity of an electrolyte, such as solubility, e.m.f. cells, vapor pressure lowering, etc., will allow the calculation of the partial molar free energy of the electrolyte in solution, and if the partial molar free energy is obtained as a function of temperature the partial molar heat and entropy can be calculated. If the partial molar energy of the hydrogen ion is assigned arbitrarily the value of zero in a hypothetical standard state of one gram ion per kilogram of water, the energies of all the other ions can be stipulated. Such measurements and calculations as these do indeed yield valuable information but are not directly what is needed. The ions are already hydrated in solution and, whereas the energies calculated by the above procedure are related to the energies of hydration, they are not equal to the energies of hydration.

To calculate the energies and entropies of hydration the following procedure has been adopted. The standard reference state is taken as a hypothetical one mole per liter of gaseous ions and the transfer is made to a hypothetical molal solution of aqueous ions all at constant temperature and pressure. The entropy of an ideal gas is calculated by the methods of statistical mechanics and the standard entropy of hydration of a given electrolyte S_h^0 is then equal to the difference between the entropy of the gaseous ions ($\bar{S}_g^0$) and the entropy in solution ($\bar{S}_{soln}^0$) as obtained by methods mentioned in the last paragraph and

$$\bar{S}_h^0 = \bar{S}_{soln}^0 - \bar{S}_g^0 \qquad 42$$

The heat of hydration is arrived at by considering the crystal lattice energy $\Delta\bar{H}_{crys}$ in conjunction with the heat of solution. The crystal lattice energy $\Delta\bar{H}_{crys}$ is defined as the energy required to separate ions in the crystal and move them an infinite distance apart, and the heat of hydration $\Delta\bar{H}_h$ is given by the relation

$$\Delta\bar{H}_h = \Delta\bar{H}_{soln} - \Delta\bar{H}_{cryst+} \qquad 43$$

Knowing the entropies and heats of hydration, the free energy change is then calculated from the relation

$$\Delta\bar{F}_h^0 = \Delta\bar{H}_h^0 - T\Delta\bar{S}_h^0 \qquad 44$$

The energies and entropies calculated refer to sums for the cations and accompanying anions. To obtain values for the individual ions a decision has to be made concerning the values for some individual cation or anion relative to which the energies of hydration of the other ions may be expressed. It is the convention to fix the entropy of the aqueous hydrogen ion at zero.

To obtain the free energy change $\Delta\bar{F}_h^0$ of a single ion, the energy of a sphere of unit charge and of a given radius immersed in a medium of the dielectric constant of water is calculated. Thus the free energy change for the transfer of chloride ions from a gas at standard conditions to water is calculated to be 84.2 Kcals per mole at 25°C. All other ions are then compared with this value for the chloride ion. The energies and entropies of hydration of some ions are shown in Table 4.

TABLE 4. FREE ENERGIES AND ENTROPIES OF IONS AT 25°

Ion	ΔF_h *Kcals/mole*	ΔH_h *Kcals/mole*	ΔS_h *E.U. per mole*	*radii* *A*
Li^+	−114.6	−121.2	−22	0.60
Na^+	−89.7	−94.6	−17	0.95
K^+	−73.5	−75.8	−8	1.33
Rb^+	−67.5	−69.2	−6	1.48
Cs^+	−60.8	−69.0	−4	1.69
F^-	−113.9	−122.6	−29	1.36
Cl^-	−84.2	−88.7	−15	1.81
Br^-	−78.0	−81.4	−12	1.95
I^-	−70.0	−72.1	−7	2.16

It is to be noted from Table 4 that there exists a close relation between the energies and entropies of hydration and the ionic radii; for a given valence type the charge density of an ion increases with decreasing ionic radii. In general, it is not particularly meaningful to express ion hydration in terms of a specific number of water molecules because water molecules are acted upon by decreasing electrostatic force as the distance from the ion increases. A proton, however, is very small and the energy of hydration is so great that this ion does not exist as such but only in the hydrated form, the hydronium ion (H_30^+).

Figure 11, in which the molar entropies of hydration of some of the ions are plotted against the ionic radii, serves to emphasize the importance of the ionic radii as well as to call attention to the differences between the behavior of the cations and anions.

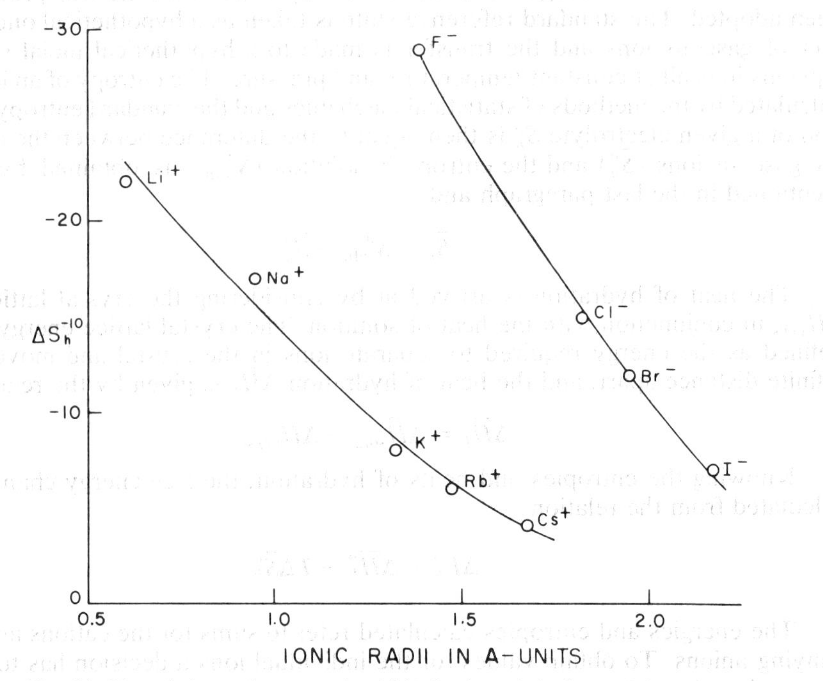

Fig. 11. Plot of the molar entropies of hydration of the ions against the ionic radii.

In the immediate vicinity of an ion there is a very intense electrostatic field amounting to millions of ordinary volts per centimeter (see Fig. 3) and it is generally agreed that water molecules very close to such small ions are essentially immobilized due to the interaction of the electrostatic forces with the dipoles of the water molecules. It is this type of hydration which has been invoked to explain the decrease of the compressibility of water with increasing salt concentration; the immobilized water has already been compressed as much as possible. Likewise the heat capacities of salt solutions decrease with increasing concentrations. The ionic mobilities are smaller, the smaller the ionic radii; the small ions immobilize more water and, accordingly, the effective radius of the small ion, including the hydration shell, is larger than that of the ions with a larger ionic radius.

Apparently, ions not only have the ability to orient water molecules and to create an hydration shell about themselves but also the ability to decrease the "iceness" of water. For example, a solution of 0.1 M CsCl is more fluid than is pure water at the same temperature; the ions appear to be breaking down the water structure. Also, when the entropies of ionic hydration are corrected for the immobilization of the first water layer around an ion, it appears that all of the alkali metal cations except Li^+ and Na^+ and all of the halide anions, except F^-, lose too little entropy when dissolved in water from the gaseous state. This conception of the action of ions on water is diagramed in Figure 12.

The innermost region A (Fig. 12) is one of immobilization and great order, the second region B is one in which the water is less ice-like, and region C has water polarized by relatively weak electrostatic field and leaves the water structure essentially unchanged. According to this conception, ions then are structure formers as well as structure breakers; for larger ions the structure breaking function predom-

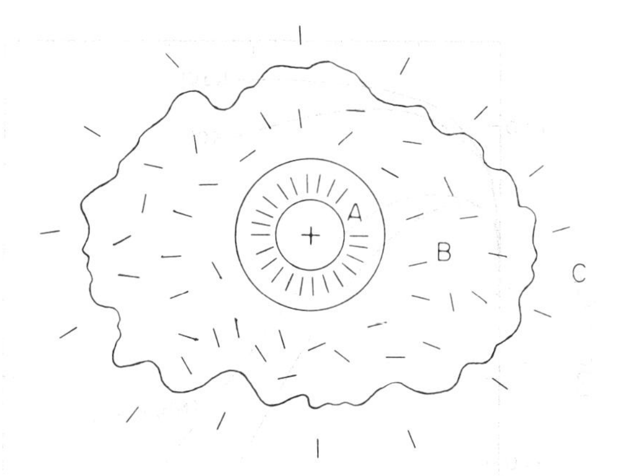

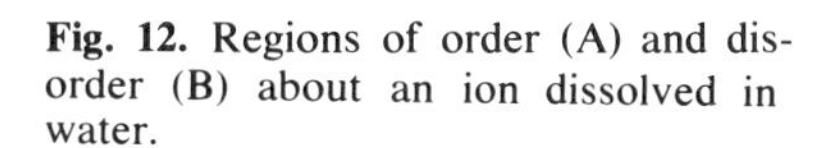
Fig. 12. Regions of order (A) and disorder (B) about an ion dissolved in water.

inates. Cations smaller or more highly charged than K^+ are net structure formers. K^+ is slightly structure breaking and the structure breaking tendency increases through Rb^+ and Cs^+. F^- is a structure former and the other halide ions are structure breakers and increasingly so as the size of the ions increases. NO_3^- and ClO_4^- are strongly structure breaking and $SO_4^=$ less so. The hydroxide ion appears to be a structure former.

The presence of a non-polar group (paraffin-like) in water tends to increase the ice-likeness of water. The non-polar molecule or group decreases the forces acting on single water molecules in the direction of the non-polar structure (there is less attraction between the water and the paraffin than there is between water and water) and, therefore, increases the tendency to iceness. The increased ordering of the water molecules in the presence of a hydrocarbon is shown by the fact that the solution of a paraffin in water is accompanied by a large negative excess partial molar entropy amounting to about -20 E.U., but only a small enthalpy change.

Surprising is the ability of many non-polar substances such as chloroform, methane and xenon to form crystalline hydrates called clathrate crystals. The water structure has been opened up sufficiently to provide chambers into which the non-polar molecules can be fitted. The number of water molecules associated with one molecule of the apolar compound is usually large. For example, chloroform forms the hydrate $CHCl_3 \cdot 17H_2O$. As a rule the stability of such crystals extends only a few degrees above the freezing point of water.

Hofmeister (Lyotropic) Ion Series. It is found by the German biochemist Hofmeister, in his early studies on the solubility of proteins, that the cations and anions could be arranged in series in relation to their effect upon the solubility of proteins. The ability of certain salts at sufficient concentration to precipitate proteins has been extensively employed in protein purifications. Shown in Figure 13 is the solubility of horse carboxyhemoglobin as a function of ionic strength.

At higher salt concentrations, the log of the protein solubility is a linear function of the ionic strength and the empirical relation

$$\log S = \log S_0 - K\Gamma/2 \qquad 45$$

appears to be valid where K is the so-called salting-out constant. The chemical potential of a component in a system in equilibrium is uniform throughout the system. A solute in equilibrium with the solid phase has the same chemical potential as the solid phase and a pure solid phase has a constant chemical potential at constant pressure and temperature. Since the solute (protein) in a saturated solution has a constant chemical potential, its activity must remain constant. Variation of solubility as the ionic environment is changed, therefore, means that the activity coefficient of the solute changes and, in fact, the activity coefficient is inversely propor-

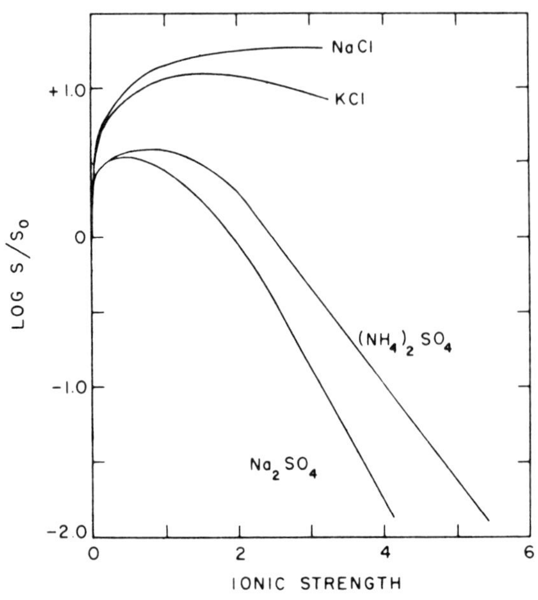

Fig. 13. Solubility of carboxyhemoglobin in aqueous salt solutions. S_o is the solubility in pure water, S that in salt solutions. (Data A. A. Green.)

tional to the solubility of the solute. What is, therefore, measured when the solubility of a protein is determined as a function of the salt concentration is the variation of the activity coefficient of the protein (or other solute).

It will be recognized immediately that there is a close relation between the Hofmeister series (the salting-out effect) and the constant B of the extended Debye-Hückel theory, and considerable effort has been expended to obtain a rational treatment of this effect; none has been entirely successful. What is being said is that there does not as yet exist a wholly adequate theory of concentrated electrolyte solutions.

A Hofmeister series is found in a wide variety of phenomena in addition to solubility of proteins. Sparingly soluble solutes of all kinds, as well as organic solutes such as ethyl acetate in water, exhibit an ionic salting-out effect. Such an ionic series also appears in viscosity of solutions, electrophoretic mobilities, enzymatic reactions and the surface tension of molten salts. The series usually given for anions is citrate^{-3} > tartrate^{-2} > SO_4^{-2} > acetate$^-$ > Cl^- > NO_3^- > Br^- > I^- > $CNS^=$. The series for the cations, though usually somewhat less marked than for the anions, is generally given as Th^{+4} > Al^{+3} > H^+ > Ba^{+2} > Sr^{+2} > Ca^{+2} > K^+ > Na^+ > Li^+.

Undoubtedly, the series has its origin in the intensity of the electrostatic field around the ions, the small ions of the same valence having a more intense field than the larger ions. The intense field of the smaller ions leads to greater hydration, the difference in hydration being the immediate cause of the series in aqueous systems. There are, however, some serious complications; for example, a highly hydrated anion such as $SO_4^=$ is a precipitating ion whereas the highly hydrated cation such as Li^+ is a salting-in ion. In fact, a strong solution of lithium bromide or thiocyanate can put a pair of silk stockings into solution.

Ions in Biology. The ion balance of any living system is of great importance and the subject of ions in relation to various biochemical and biological situations will be of recurring concern to us; a general and short statement is given here.

Ions play diverse roles in biology. Osmotically, they help maintain water balance in the various water compartments of the body; in addition, they have specific effects. There is probably no reaction in a living system that is not sensitive to the ionic environment; possibly the most dramatic effect of ions is on enzyme systems. Intermediates in biosynthetic pathways are all essentially ionic. The most frequent ionic forms contain carboxyl or phosphate groups.

There are also many fascinating but in most cases obscure problems associated with the transport of ions across living membranes (Chapter 12). The ionic concentrations of body fluids must be maintained within fairly narrow limits to sustain life. The blood plasma of different species varies considerably in absolute concentrations, but the ratios of the concentrations of the various ions to each other are approximately the same as found in sea water.

Knowledge about ion concentrations in biological systems has been greatly augmented through the use of the flame photometer. This is a fast, accurate method based on the intensity of the emission spectra of the inorganic metal ions in a flame resulting from the combustion and vaporization of a sample.

Figure 14 serves to give an idea of the ionic concentrations of human blood plasma, of the interstitial fluid and of the intracellular fluid; the concentrations are given in milliequivalents per liter of water. The ionic composition of blood plasma resembles rather closely that of the interstitial fluid, the principal difference being the lower content of protein (anion) in the interstitial fluid as contrasted with that of the plasma. Neglecting the contribution due to the protein, which is relatively small, the total ionic strength of blood plasma is about 0.165 which is just about the upper limit of the validity of the Debye-Hückel theory, i.e., without the salting-out correction. In the plasma and in the interstitial fluid the concentrations of the sodium and chloride ions far outweigh those of all the other ions put together. It is to be noted that the composition of the intracellular fluid is very much different from that of the plasma or of the interstitial fluid; the intracellular fluid is characterized by high concentrations of potassium, magnesium and phosphate. Even within the cells, however, the ionic concentrations are far from uniform. The ratios of the equivalents of potassium to sodium in the blood plasma are about 0.035, in the intracellular fluid about 13, in intact mitochondria from liver about 6.65 and in the membrane fraction of the liver mitochondria about 0.42.

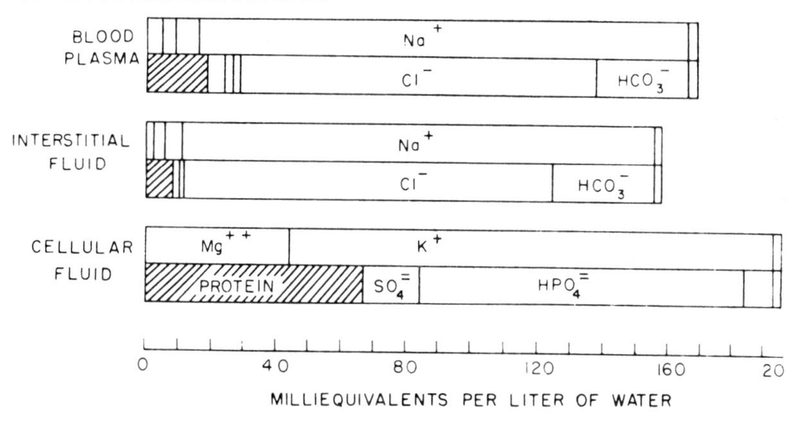

Fig. 14. Ionic composition of human body fluids in milliequivalents per liter of water.

The absolute salt concentration in a living system is not so important as the relative concentration of ions. Thus sodium in high concentrations is toxic, but the toxicity largely disappears if the increased concentration is matched by a corresponding increase in the concentration of the other cations. The most startling ionic antagonisms exist between magnesium and calcium ions. Magnesium ions have pronounced anesthetic effects upon an animal and calcium ions tend to prevent magnesium anesthesia.

Sodium ions are stored in the chondroitin sulfate of connective tissue and bone stores calcium and phosphate; both of these tissues act as glorified ion exchange beds for these ions.

Ionic Complexes. Ions have the ability to complex with many organic molecules and the specific biological effect of an ion is almost certainly due to the formation of specific ion complex with a cellular constituent. Fifteen metal cations have been found to activate one or more enzymes, namely, Na^+, K^+, Rb^+, Cs^+, Mg^{++}, Ca^{++}, Zn^{++}, Cd^{++}, Cr^{+++}, Cu^{++}, Mn^{++}, Fe^{++}, Co^{++}, Ni^{++}, Al^{+++} as well as NH_4^+. Mg^{++} and Mn^{++} are especially frequent activators. The metal ions appear to act in two ways: with some enzyme-metal ion associations the stability of the enzyme is increased and in others the catalytic site of the enzyme is activated. Not only are metal ion activators necessary for many enzymes but also metal ions can act as inhibitors of some enzymes.

There are all degrees of specificity and strength of ion binding to be expected; in a vague sense an ion in the electrostatic field of another ion of opposite charge can be said to be complexed. If the ionic radii are sufficiently small and ion charge large enough, ion-pair formation can occur in which the ions are held together by electrostatic bonding. Even ammonium acetate exhibits ion association in aqueous solutions, the association constant being about 0.5.

In general, there are two kinds of complexes: a simple complex, in which the ion attaches itself to a suitably charged site having the correct dimensions to accommodate it, and, secondly, a chelate complex. In a simple complex the ion is apt to retain its essential ionic character and the complex can be dissociated by dilution although some simple complexes can form insoluble aggregates and are more difficult to dissociate. Other simple complexes such as the metallic sulfides of organic compounds exhibit a high degree of stability. For example, there exists a powerful affinity between the sulfhydryl group (—SH) in proteins and heavy metal ions, and this affinity parallels rather closely the relative insolubility of the corresponding inorganic metal sulfides.

A weak acid such as acetic acid in its un-ionized form can be considered to be an ion complex of a proton and the acid anion; in such a complex the quantum forces predominate and the ions have lost their ionic character.

If the substance which combines with the metal ion contains two or more donor groups so that one or more rings are formed, the resulting structure is said to be a chelate compound or metal chelate. The electron-pair bond formed between the electron-accepting metal and the electron-donating complexing agent may be essentially ionic or essentially covalent depending on the metal ion and the complexing agents. The coordination number or covalence of the metal ion varies. The numbers 4 and 6 are most common, but a few metals have coordination numbers of 2 and 8. The covalence of the metal ion chelate is the result of reactive sites on the chelating agent and in aqueous solution these sites are occupied by water molecules; in the presence of an ion the water molecules are displaced from the ion and the agent and the complex forms. Four membered chelate rings are relatively rare and evidence indicates that practically all chelates have 5 or 6 membered rings.

The stability of ion complexes is expressed as an association constant, that is

$$K = \frac{MA}{M^+ \times A^-} \qquad 46$$

where M^+ is the metal ion, A^- the complexing agent and MA the complex. Since the values of K are frequently very large, it is usual to use $\log K$ instead of K itself. In many cases, there is a one-to-one molar ratio between the complexed ion and the complexing agent in which event eq. 46 is appropriate. In other situations, and particularly in relation to high molecular weight complexing agents such as proteins, there is often multiple binding of ions; multiple ion binding will be considered in Chapter 5.

The determination of the stability constants of a metal complex must involve some method for the estimation of the concentration of the metal ion, of the com-

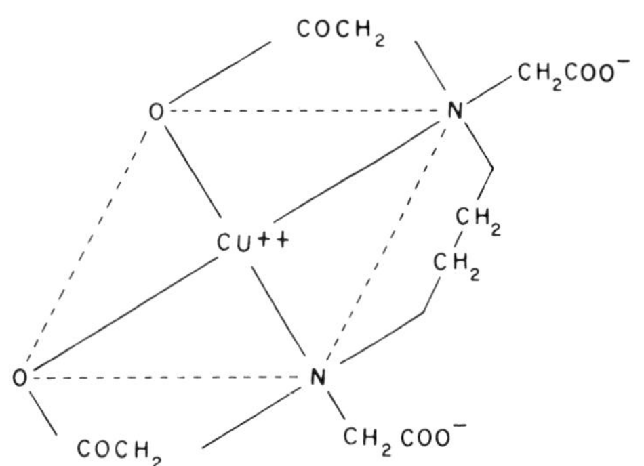

Fig. 15. Chelate complex of cupric ion with "versene."

plexing agent or of the complex in solution; a number of physical methods have been used. One of the most frequently used procedures is to determine the shift of the pH of the solution as a result of complex formation. The hydrogen ion is displaced from the complexing agent by the metal ion and the hydrogen ion is released into solution. For example, the interaction of imidazole with zinc and cupric ions has been studied with such a method. In both cases the coordination number of the ion for imidazole was 4. Cupricimidazole showed marked absorption in the visible spectrum between 590 and 690 mμ and in the ultraviolet between 230 and 290 mμ, and indicates that the imidazole residues of histidine are involved in chelation with the copper.

A mercury electrode has been used as an indicator electrode for a metal in the presence of a chelating agent. The agent must, however, form a 1:1 complex with mercury as well as with the metal in question; this appears to be a simple and rapid method.

Other methods for determining stability constants are: (1) ion exchange, (2) optical absorption methods, (3) oxidation potentials, (4) polarographic measurements, (5) solubility measurements, (6) reaction rates, (7) biological assay, (8) displacement reactions, (9) electrical conductance.

It is not always easy to discriminate between ordinary ion complex formation (simple ionic complexes) and chelation. The most useful criteria are: (1) Chelation frequently involves a pronounced color change. (2) The metal suffers changes in properties. (3) Chelation frequently leads to two optical forms which can be resolved.

A powerful chelating agent which has enjoyed wide popularity for the removal of metal ions from enzymatic and other biochemical systems is ethylenediaminetetraacetic acid or "versene" whose coordination with a cupric ion is shown in Figure 15.

Although there is considerable individuality in the strength of binding of metal ions by chelating agents, nevertheless there is a marked tendency for the relative binding strengths to fall into an ion order. The alkali metals Li^+, Na^+, K^+ have small binding constants. Other ions can be arranged in the order of increasing binding constants of $Ca^{++} < Mg^{++} < Mn^{++} < Zn^{++} < Co^{++} < Ni^{++} < Hg^{++} < Cu^{++}$.

Amino acids form complexes with many metals through the carboxyl and amino groups. Table 5 gives the log$K$ values for several metal ions with glycine, alanine and glycylglycine.

TABLE 5. LOG K VALUES OF SEVERAL METAL IONS*

	Glycine	*Alanine*	*Glycylglycine*
Cu^{++}	8.62	8.51	6.05
Ni^{++}	6.18	5.96	4.49
Zn^{++}	5.52	5.21	3.80
Pb^{++}	5.47	5.00	3.23
Co^{++}	5.23	4.82	3.49
Mn^{++}	3.44	3.02	2.15
Mg^{++}	3.44	1.96	1.06

* Data of C. B. Monk: Trans. Faraday Soc. *47*: 297, 1951.

It has been mentioned that Mn^{++} and Mg^{++} are the outstanding enzyme activators, but, as noted in Table 5, these two ions form the weakest chelate complexes with amino acids.

GENERAL REFERENCES

Dorsey, N. E.: The Properties of Ordinary Water Substance in All its Phases. ACS Monograph 81, Reinhold Publishing Co., New York, 1940; reprinted by Hafner Publishing Co., New York, 1969.

Eisenberg, D. and W. Kauzmann: The Structure and Properties of Water. Oxford University Press, New York, 1969.

Kavanau, J. L.: Water and Solute-Water Interactions. Holden Day, San Francisco, 1964.

Narten, A. H. and H. A. Levy: Observed diffraction pattern and proposed models of liquid water. Science *165*:447, 1969.

Rich, A. and N. Davidson: Structural Chemistry and Molecular Biology. W. H. Freeman, San Francisco, 1968.

Robinson, R. A. and R. H. Stokes: Electrolyte Solutions, 2nd ed. Butterworth, London, 1968.

PROBLEMS

1. (a) Calculate the activity coefficients of 0.01M $CaCl_2$ and of 0.10 M $CaCl_2$ in solution in water at 25°C. Compare your calculated values with those experimentally determined.
   (b) What would be the free energy of dilution per mole of 0.10 M $CaCl_2$ to 0.01M $CaCl_2$?
   (c) What is the value of kappa for these two solutions?

   Ans: (a) 0.71, 0.46 (b) −3,500 calories (c) 5.62×10^6, 1.78×10^7

2. Calculate the ionic strengths of the following solutions: (a) 0.1 N HCl; (b) 0.1 M H_2SO_4; (c) 0.05 M Na_2HPO_4; (d) a solution which is 0.1 M in respect to NaH_2PO_4 and 0.1 M in respect to Na_2HPO_4.

   Ans: (a) 0.1, (b) 0.3, (c) 0.15, (d) 0.40

3. Relate the influence of ions on the iceberg structure of water to the existence of the lyotropic series of ions in respect to the swelling of a gelatin gel.

   Ans: Speculative

4. (a) What is a chelate structure? (b) How can you detect the presence of a chelate structure? (c) Design an experiment by which you would measure in a quantitative manner the stability of a metal chelate.

   Ans: See text.

CHAPTER 4

Oxidation-Reduction Potentials

Valence changes are accompanied by electron transfers and

$$\text{Oxidant}^{z} + \text{ne} \rightleftharpoons \text{Reductant}^{z-n} \qquad 1$$

where z is the valence, n is the number of electrons (e) transferred per mole. Oxidation-reduction systems can assume a variety of forms which embrace calomel half cells as well as the cytochrome C system. The common property of these systems is the ability to convert chemical energy into electrical energy.

Diagramed in Figure 1 is an arrangement of a calomel half cell and a hydrogen electrode (platinum black and H_2) capable of producing an electric current; such an assembly is known as an electromotive cell.

Upon connecting the external leads (Fig. 1), current will flow; the mercurous ions are reduced to metallic mercury and the hydrogen is oxidized to hydrogen ions. The overall reaction is

$$\tfrac{1}{2}H_2(g) + Hg^{+} \rightleftharpoons Hg(l) + H^{+} \qquad 2$$

The cell (Fig. 1) in which this reaction takes place is written

$$\bar{P}t,\ H_2(g) \mid \text{soln } x \parallel \text{KCl (sat)},\ Hg_2Cl_2(S) \mid \overset{+}{Hg} \qquad 3$$

The calomel electrode will be positive (mercurous ions have deposited on the mercury) and the hydrogen electrode negative (protons have been released at this electrode). Upon delivery of an external potential opposite in sign and slightly larger than that of the cell, the chemical reaction will be reversed and hydrogen gas will be given off from the hydrogen electrode and mercurous ions produced at the mercury electrode; the cell is reversible.

Precautions must be taken in the measurement of the voltage of electromotive force cells such that little or no current will be drawn from the cell; otherwise the voltage measured will be too small and will not represent the voltage of the cell while performing reversibly. The measurement of the maximum voltage is best accom-

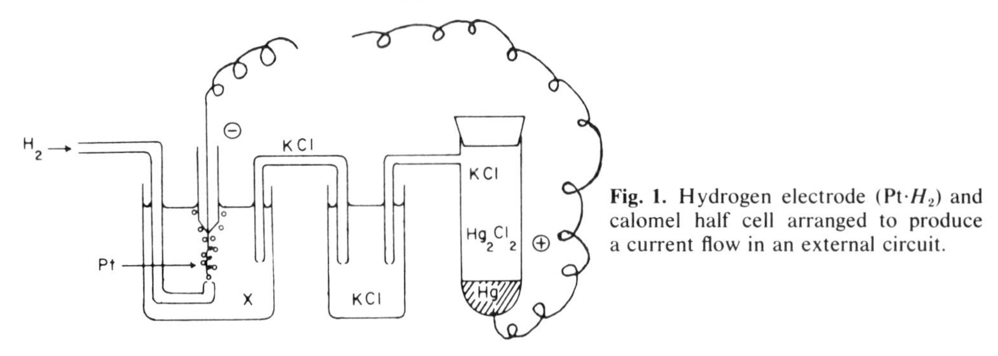

Fig. 1. Hydrogen electrode (Pt·H_2) and calomel half cell arranged to produce a current flow in an external circuit.

plished with a potentiometer which permits the balancing of a known voltage against the unknown and the current flow is exceedingly small. A simple potentiometer circuit is diagramed in Figure 2.

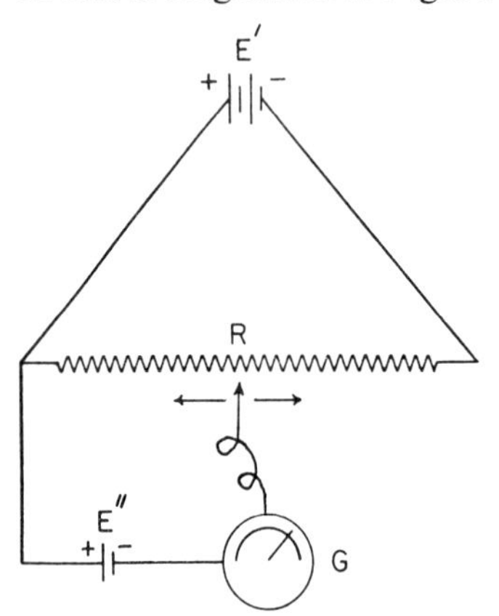

Fig. 2. Simple potentiometer circuit. E' is source of known voltage, E'' is the unknown voltage, G is the galvanometer, R is the resistance wire.

The work done by a cell is equal to the product of the current flowing and the voltage; for a reversible process the electrical work is equal to the free energy change. The amount of current per equivalent of ions transferred is equal to 96,494 coulombs (one Faraday). The free energy change ΔF for the reaction is then

$$\Delta F = -nFE = -96{,}494nE \quad 4$$

where E is the reversible voltage of the cell. Equation 4 gives the free energy change in joules; to convert to calories it is only necessary to recall that one calorie is equal to 4.185 joules and ΔF in calories is

$$\Delta F = -23{,}062nE \quad 5$$

By convention, E, for a spontaneously discharging cell, is positive and, accordingly, since the free energy change for such a process must be negative, the signs in eqs. 4 and 5 are negative.

Any element in the presence of its ions is called a half cell, and, obviously, it is always necessary to have two half cells connected together to realize a voltage which is capable of being measured. The hydrogen half cell has been selected as the standard against which all other half cells are compared. The normal hydrogen half cell is defined as having zero voltage at unit activity of hydrogen ions under 760 mm. Hg pressure of hydrogen gas and at any temperature. Each element has its own normal half cell which is defined as a cell consisting of the element in its standard state in a solution of its ions at unit activity at 25°C (see Table 1).

TABLE 1. STANDARD ELECTRODE POTENTIALS AT 25°.

Electrode	E^0
Li^+; Li	−3.045
K^+; K	−2.925
Na^+; Na	−2.714
Zn^{++}; Zn	−0.763
Fe^{++}; Fe	−0.440
H^+; H_2, Pt	0.000
Cu^{++}; Cu	+0.337
I^-; I_2(S), Pt	+0.536
Fe^{+++}, Fe^{++}, Pt	+0.771
Br^-; Br_2(1), Pt	+1.056
Cl^-; Cl_2(g), Pt	+1.360

Equation for a Chemical Cell. The free energy change for the reaction

$$\mathrm{aA + bB \rightleftharpoons cC + dD} \qquad 6$$

can be written

$$\Delta F = \Delta F^0 + RT \ln \frac{a^c \times a^d}{a^a \times a^b} \qquad 7$$

where a denotes the respective activities of the reactants. Conversion of the free energy changes to voltages (see eq. 4) gives

$$E = E^0 - \frac{RT}{nF} \ln \frac{a^c \times a^d}{a^a \times a^b} \qquad 8$$

At 25°C, substituting numerical values of the constants and converting to log to base 10, eq. 8 becomes

$$E = E^0 - \frac{8.314 \times 298.15 \times 2.3026}{96{,}494n} \log \frac{a^c \times a^d}{a^a \times a^b} \qquad 9$$

$$= E^0 - \frac{0.0591}{n} \log \frac{a^c \times a^d}{a^a \times a^b} \qquad 10$$

It is thus possible to measure ionic activities by the use of electromotive cells although it must be noted that it is impossible to ascertain the activities of individual ions within the framework of thermodynamics. What is measured in an electromotive cell is the mean ion activity. The mean activity coefficient of an electrolyte is given by the expression

$$\gamma_\pm = (\gamma_+^{v+} \gamma_-^{v-})^{1/(v_+ + v_-)} \qquad 11$$

where v_+ and v_- are the number of moles of cations and anions respectively. For a salt such as sodium chloride v_+ and v_- are both unity and $\gamma_\pm$ is equal to $(\gamma_+\gamma_-)^{1/2}$.

In Figure 1, the hydrogen ion concentration in solution x can be changed at will leaving the condition of the calomel half cell intact (the activity of the mercurous ions remains unaltered). Then from eq. 10 the equation of the cell becomes

$$E = E' - \frac{RT}{F} \ln a_{H^+} \qquad 12$$

where E' is equal to the sum of the standard potential of the calomel half cell and the liquid junction potential; the standard potential of the hydrogen half cell is by definition zero. The liquid junction potential which is essentially due to the different ionic mobilities at the junction between the salt bridge and the solution is, in general, unknown, but is usually of the order of a few millivolts. Converting to logarithms to base 10 and rearranging, eq. 12 becomes

$$-\log a_{H^+} = \frac{(E - E')F}{2.3026RT} = \text{pH} \qquad 13$$

In eq. 13, the pH which is a measure of the hydrogen ion activity has been defined in terms of the electrode potential E'; E is the voltage which is actually read on the potentiometer. In the original and monumental publication of Sørensen, the pH was explicitly and formally defined as being equal to $-\log H^+$ concentration. This definition has turned out to be meaningless. Fortunately, Sørensen's substantial and operational definition was in effect the same as given in eq. 13 except that his value of E' differs about 2.3 millivolts from the E' value used today. This voltage difference corresponds to an addition of 0.04 pH units to the Sørensen pH scale. The problem of establishing a pH scale and of pH standards thus requires an assignment of a numerical value to E', the electrode potential of the reference cell.

The technique for the determination of E' and hence of the formulation of standard buffers is well described by Bates. In principle, the method consists in using the ionization constants of weak acids to calibrate the scale. The pH of a solution of a weak acid in the presence of its salt of a strong base can be calculated by the relation (see Chapter 5)

$$\text{pH} = \text{pK} + \log \frac{\text{salt}}{\text{acid}} \qquad 14$$

where pK is equal to $-\log K$. The ratio of salt to acid is fixed in a dilute buffer and the total electrolyte concentration varied by the addition of neutral chloride. The calculated pH is extrapolated to zero electrolyte concentration and the value of the standard potential of the reference electrode which yields the correct pH as calculated from eq. 13 is considered to be the correct value of electrode voltage. A table of standard buffer solutions is shown in Chapter 5, Table 3.

Whereas the hydrogen electrode is the standard against which other electrodes are compared, due to experimental inconvenience it has been almost entirely abandoned for the routine determination of pH. The glass electrode or pH-meter as it is frequently called has superseded all other methods for the evaluation of pH.

The glass electrode consists of a glass membrane of suitable composition on one side of which is a non-polarizable electrode, such as silver-silver chloride. On the other side is the unknown solution, which is connected by means of a salt bridge to another non-polarizable electrode. The observed voltage is a straight line function of pH; the slope of this relation, however, varies with temperature. At 25°C the voltage changes 59 millivolts per unit pH; whereas the slope of the pH-voltage line is dependent only on temperature, the position of the line has to be fixed by the use of a buffer of known pH, i.e., the glass electrode has to be calibrated.

It has been long known that the glass electrode is subject to sodium errors in alkaline solutions; the pH read on the meter is less than the actual pH. Glasses containing trivalent oxides such as those of aluminum and of boron are sensitive to the metal cations, whereas a high silica glass has little error even in strongly alkaline solutions. Lithium aluminum silica glass of the proper composition develops a high sodium specificity and to such an extent that such glasses can be used as excellent sodium electrodes with little interference from potassium ions and none from the alkaline earths. The electrodes are, however, responsive to hydrogen ions and, in fact, below pH 4 the hydrogen function becomes predominant even in high concentrations of sodium ions.

Potassium ion sensitive glass electrodes have also been developed. Both the sodium and potassium electrodes are commercially available.

The biochemist and physiologist employ electromotive force cells in their work so frequently that they are apt to forget how often. It must be emphasized that in measuring potentials in any solution, biological or otherwise, non-polarizable electrodes are required; there must exist a reversible reaction at the electrode which can transfer current from the electrode to the solution.

Two favorite non-polarizable reference half cells are the calomel electrode introduced above and the silver-silver chloride electrode. The amount of care needed for the preparation of these reference electrodes depends upon the use to which they are to be put. If they are to be used as secondary standards to yield the accepted voltages, many precautions must be taken in their preparation. If, however, they are to be employed simply as a source of constant potential relative to a changing potential their preparation is simple.

The calomel electrode consists of metallic mercury in contact with a paste of calomel and a solution of potassium chloride saturated with calomel. Electrical contact is made through a platinum wire in the liquid mercury. The voltage of the cell depends on the potassium chloride concentration and, at 25°C and using a saturated KCl salt bridge, the tenth normal calomel electrode has a potential of +0.3356 volts and the electrode with saturated potassium chloride gives a potential of +0.2444 volts relative to the hydrogen electrode.

The calomel electrode, as used, has a salt bridge (see Fig. 1) which can assume a variety of forms. One of these consists of a glass tube filled with agar saturated with potassium chloride. A favorite salt bridge in a glass electrode assembly is a fiber saturated with potassium chloride in a capillary hole connecting the calomel half cell to the solution.

A silver-silver chloride electrode may be prepared in various ways. One method is as follows: A platinum wire is covered with silver by electrolysis (4 volts, 1 milliampere) of a solution obtained by mixing equal volumes of 13 percent KCN solution and 18 percent $AgNO_3$ solution. After washing, the silver-plated platinum wire is electrolytically covered with a thin layer of AgCl, a normal solution of HCl and a current of 3.5 milliamperes being used for 20 minutes.

Usually, silver-silver chloride electrodes are used without a salt bridge and the silver-silver chloride electrode in the form of a wire is plunged directly into the test solution. The voltage of the electrode at 25°C is +0.22234 volts (molar chloride solution). In addition to serving as a non-polarizable electrode in the measurement of bioelectric potentials and other potentials, the silver-silver chloride electrode can be used as a chloride electrode and the activity of a chloride-containing electrolyte can be determined with it. Silver chloride is relatively insoluble; accordingly, the amount of silver ions in solution depends on the ion product constant of silver chloride and the electrode potential of the silver-silver chloride can be written

$$E = E_0 + \frac{RT}{F} \ln \frac{K_{AgCl}}{a_{Cl^-}} \qquad 15$$

where K_{AgCl} is the ion product constant of silver chloride and a_{Cl^-} is the activity of the chloride ions in solution. Unfortunately, the silver-silver chloride electrode will not behave as a reliable chloride electrode in the presence of proteins.

Membranes Electrodes. Oxidized collodion membranes carry negative charge sites (carboxyl groups) in their pores and, if sufficiently dense, are nearly impermeable to anions but permit cations to pass. The potential across such membranes as measured by two calomel half cells on the two sides of the membrane is responsive to the difference in cation concentration across the membrane. Thus the activity of the alkali metal cation electrolytes can be measured by such membrane electrodes. It has been found, however, that only membranes whose negative charge arises from sulfonic acid groups are suitable for measurement of the alkaline earth cations. Collodion membranes treated with protamine (a highly basic protein) permit anions to pass but not cations; such membranes act as anion electrodes. Unfortunately, collodion membranes show little individual ion selectivity and measurements have to be confined to a single electrolyte at a time. (There is too much interference from other ions.)

More recently, a number of liquid ion exchange membrane electrodes have been reported; several such electrodes are commercially available and certainly the most important of these is a calcium electrode. Previously, there had not been a satisfactory calcium electrode; calcium ion activity had been determined by monitoring the rate of the beat of an isolated frog heart bathed with the experimental solution.

The liquid exchange electrode for calcium consists of a glass tube one cm in diameter sealed at the lower end with dialysis tubing. The bottom of the tube, to about 2 cm depth, contains the ion exchange solution of 0.1 M calcium salt of didecylphosphoric acid dissolved in di-n-octylphenyl phosphonate. A narrow glass tube containing 0.1 M $CaCl_2$ in a 2 percent agar gel provides electrical contact between the organic phase and a silver-silver chloride electrode. When the electrode, with a reference silver-silver chloride, is inserted in a $CaCl_2$ solution, the observed potential is given by

$$E = E'_o + \frac{3}{2}\frac{RT}{F} \ln m\gamma_{\pm} \qquad 16$$

where m is the molal concentration and $\gamma_{\pm}$ the mean ionic molal activity coefficient of $CaCl_2$. The electrode is reported to be almost devoid of response to all ions except calcium.

Concentration Cells. Two electrode chambers containing solutions of potassium chloride at two different concentrations connected by a salt bridge and having silver-silver chloride electrodes in each chamber constitute a concentration cell. The free energy of dilution of one mole of chloride ions from one activity to another is

$$\Delta F = RT \ln \frac{a_2}{a_1} \qquad 17$$

and since ΔF is equal to $-nFE$, eq. 17 becomes, for potassium chloride (n is unity)

$$E = -\frac{RT}{F} \ln \frac{a_2}{a_1} \qquad 18$$

Knowing the activity of the potassium chloride in either chamber, it is then possible to calculate the activity of the electrolyte in the other. The above neglects the liquid junction potential between the electrode chambers and the mean activities rather than single ion activities are intended.

It is possible to construct electromotive force cells without a liquid junction. For example, if a hydrogen electrode and a silver-silver chloride electrode are both immersed in the same electrode vessel containing hydrochloric acid, the liquid junction disappears. If two such cells contain hydrochloric acid at different concentrations, their combined voltages will be the voltage of a true concentration cell and

$$E_2 - E_1 = E = \frac{2RT}{F} \ln \frac{a_2}{a_1} \qquad 19$$

where a_2 and a_1 are the activities of the hydrochloric acid in the two cells.

Amperometric Methods. The polarograph in which the current is measured as a function of an externally applied potential difference at a dropping mercury electrode was devised by Jaroslav Heyrovsky. The reduction or oxidation of material at the mercury electrode provides for current flow. The voltage is characteristic of the substance and the amount of current is proportional to the concentration of the substance. The method is useful for the identification and estimation of trace amounts of a substance in solution. There are several related methods in which current flow is measured. They are irreversible in their operation.

In principle, the reactions in an oxygen and hydrogen cell can be written

$$\tfrac{1}{2}\,O_2 + H_2 \rightleftharpoons O^{=} + 2H^+ \qquad 20$$

$$O^{=} + 2H^{+} \rightleftharpoons H_2O \qquad 21$$

Since the reduction of oxygen is a two electron transfer and the free energy of formation of water is 56,690 calories, the potential of the oxygen electrode should be 1.2292 volts. Oxygen electrodes, however, have not been successful although there are metallic electrodes whose potentials are responsive to the oxygen tension; this dependence is attributed to films of metallic oxide on the electrode. Such electrodes are not, however, useful because of their erratic behavior. To measure oxygen tension in physiological media and in other solutions a polarographic arrangement has been developed. This is not a true oxygen electrode; the oxygen is electrolytically reduced at an inert metallic electrode and the rate of reduction is measured by the current flowing. The electrodes consisting of a metallic silver anode and a platinum wire as cathode are covered with an oxygen-permeable thin membrane, such as polyethylene, thus insulating the electrodes from the test solution except for the oxygen which penetrates the membrane to yield the same oxygen tension to the electrode solution as that in the external solution; the electrodes are thereby protected from poisoning by the biological solution under investigation. The adjustable potential across the electrodes is gradually increased until the platinum electrode is several tenths of a volt negative to the silver electrode, at which point the oxygen dissolved in the medium is electrolytically reduced at the surface of the platinum and current begins to flow through the circuit. At still higher voltages, all of the oxygen arriving at the platinum electrode by diffusion is reduced and a plateau in the current-voltage plot results. Figure 3A shows a plot of the current against voltage in a 0.10 M NaCl solution saturated with air. Figure 3B gives a calibration curve using 0.10 M NaCl and 0.7 volt.

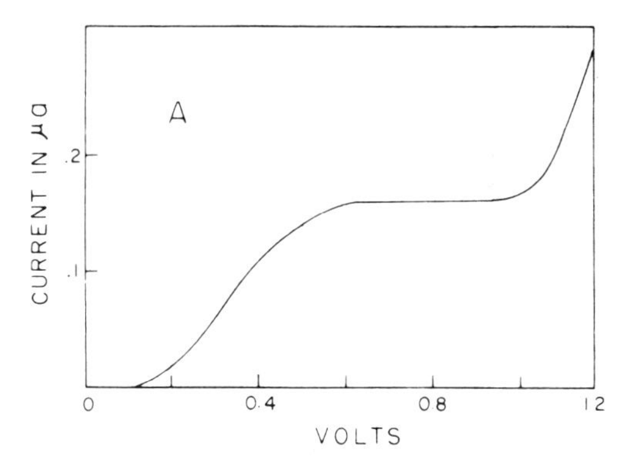

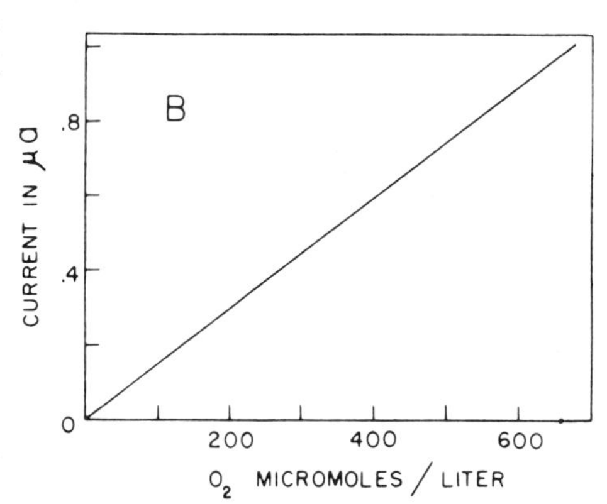

Fig. 3. A, Current-voltage plot for a bare platinum electode in 0.10 M NaCl. Solution unstirred but saturated with air, 25°C. B, Calibration curve for bare platinum wire in 0.10 M NaCl solution.

The two reactions which are believed to occur at the inert electrode are

$$O_2 + 2H^+ + 2e \rightarrow H_2O_2 \qquad 22$$

and

$$H_2O_2 + 2H^+ + 2e \rightarrow 2H_2O \qquad 23$$

This oxygen-polarographic electrode has found considerable favor in physiological work and is widely used.

An amperometric method is used for the estimation of sulfhydryl, $-SH$, groups and depends upon the removal of silver ions from solution by the $-SH$ groups. In this method the substance is dissolved in tris buffer. Tris (hydroxymethyl-amino-methane) complexes with the silver ions and prevents their precipitation by anions present. The tris also serves to maintain the pH in the desired range. A suitable reference electrode is $Hg \cdot HgO$ saturated with $Ba(OH)_2$ and gives a potential of -0.10 volt against the saturated calomel half cell. A rotating platinum electrode is used in connection with the reference electrode. The rotating platinum electrode is immersed in the tris buffer containing the dissolved thiol compound and the solution is connected by a salt bridge to the $Hg \cdot HgO$ electrode. The solution in the beaker is now titrated with a standard solution of $AgNO_3$. Up to the endpoint, the silver ions react strongly with the SH-groups preventing the accumulation of the silver-tris complex in solution and little or no current flows through the external circuit. After the SH-groups have been completely titrated, further addition of $AgNO_3$ solution leads to current flow which is registered on a sensitive galvanometer. Figure 4 shows the titration of the SH-groups of sheep hemoglobin (there are 8 moles of titratable SH-groups per mole of sheep oxyhemoglobin) in tris buffer pH 7.4 with a $AgNO_3$ solution.

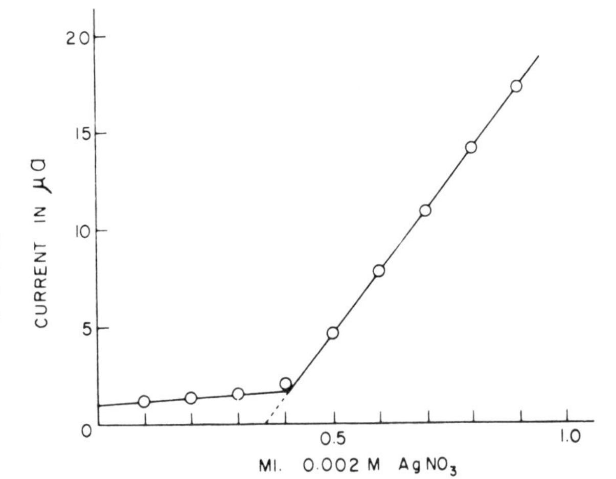

Fig. 4. Amperometric titration of sheep oxyhemoglobin in tris buffer at pH 7.4. (Figure modified from R. E. Benesch, H. A. Lardy and R. Benesch: J. Biol. Chem. *216:* 663, 1955.)

Soluble Oxidation-Reduction Systems. The activities of sodium and of potassium in electrodes employing mercury amalgams of these metals can be varied at will. Also the activity of hydrogen in the hydrogen electrode can be altered by changing the pressure of the hydrogen, but, in general, oxidation-reduction systems formed by metals in the presence of their ions are restricted to the activity of the pure metal whose activity is taken as unity. In soluble oxidation-reduction systems such, for example, as salts of ferro-ferri cyanide the concentrations of the oxidant and reductant can, of course, be made to any desired concentrations.

The free energy change for an oxidation-reduction reaction is

$$\Delta F = \Delta F^0 - RT \ln \frac{\text{ox}}{\text{red}} \qquad 24$$

and if the reaction can be conducted in such a manner as to produce a reversible potential, eq. 24 can be changed to read

$$E = E^0 + \frac{RT}{nF} \ln \frac{\text{ox}}{\text{red}} \qquad 25$$

where E^0 is the standard red-ox potential of the system and is equal to E, the observed potential when the ratio ox/red is unity.

Red-ox potentials are measured by inserting a bright platinum wire into the solution which in turn is connected by a salt bridge to an appropriate half cell such as calomel. The exclusion of oxygen is necessary if the reductant is spontaneously oxidized by atmospheric oxygen; the presence of oxygen is also apt to give trouble in any sluggish red-ox system.

The plot of the red-ox potential as a function of the percentage of oxidant yields a typical S-shaped curve (see Fig. 5). From Figure 5, although system A has a higher normal red-ox potential than does system B, if B contains 90 percent oxidant B will tend to oxidize system A if that system has 90 percent reductant. What is really being said is that if the measured red-ox potential against a given reference electrode is

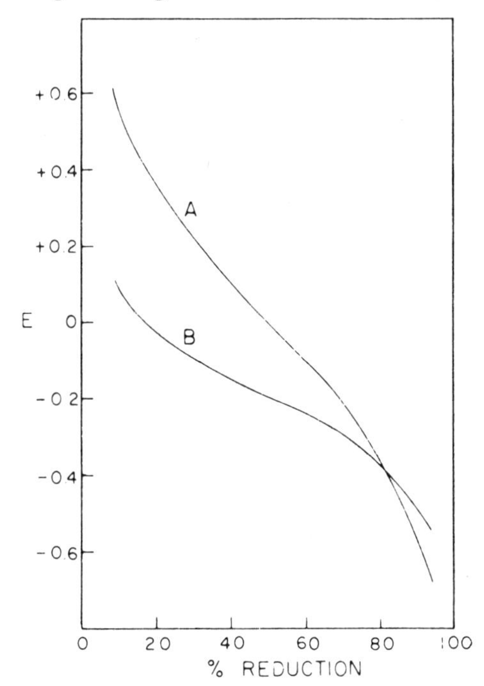

Fig. 5. Red-ox potentials as functions of the percentage of reductant present. *A* is for a single electron transfer and *B* is for a two electron transfer.

more positive for system B than for system A, system B will oxidize system A irrespective of the normal red-ox potentials.

The red-ox potential is a measure of the free energy change for the reaction and whereas a system of lower red-ox potential can never oxidize a system of higher potential it cannot be said that a system at a higher potential will definitely oxidize a system with a lower potential. The reaction may proceed but so slowly that it is of no practical interest.

Influence of Hydrogen Ions. The hydrogen ion concentration of most organic red-ox systems must usually be considered, because in these systems the reduced form can exist as an anion which can accept hydrogen ions and so become inoperative as far as contributing to the potential; naturally, the potential of such systems is greatly influenced by the hydrogen-ion concentration.

A two electron reduction involving the formation of an anion may be expressed as

$$ox + 2e \rightarrow red^{=} \qquad 26$$

and the electrode equation is

$$E = E^0 + \frac{RT}{2F} \ln \frac{ox}{red^{=}} \qquad 27$$

but the reduced form ionizes as

$$H_2\ red \rightleftharpoons H\ red^- + H^+ \qquad 28$$

and

$$H\ red^- \rightleftharpoons H^+ + red^{=} \qquad 29$$

then

$$K_1 = \frac{H^+ \times H\ red^-}{H_2\ red} \qquad 30$$

and

$$K_2 = \frac{H^+ \times red^{=}}{H\ red^-} \qquad 31$$

Also the total reduced form, red, is equal to the ionized and un-ionized forms and

$$red = H_2\ red + H\ red^- + red^{=} \qquad 32$$

Combining eqs. 30, 31, and 32 and substituting in eq. 27 there results

$$E = E^0 + \frac{RT}{2F} \ln \frac{ox}{red} + \frac{RT}{2F} \ln \frac{(H^+)^2 + K_1H^+ + K_1K_2}{K_1K_2} \qquad 33$$

If the ratio of ox to red is unity, eq. 33 can be written

$$E = E' + \frac{RT}{2F} \ln [(H^+)^2 + K_1H^+ + K_1K_2] \qquad 34$$

If K_1 and K_2 are small compared with the hydrogen ion concentration, eq. 34 becomes at 30°

$$E = E' + \frac{RT}{F} \ln H^+ = E' - 0.060\ pH \qquad 35$$

If on the other hand the concentration of hydrogen ions is small compared with K_1 but large as compared with K_2, eq. 34 becomes at 30°C

$$E = E'' + \frac{RT}{2F} \ln H^+ = E'' - 0.030\ pH \qquad 36$$

The dependence of the potentials of several oxidation systems on pH is shown in Figure 6.

The ionization constants of the red-ox systems are equal to the hydrogen ion concentration at the inflection points of the curves (see Fig. 6); that is, pH is equal to the pK at this point. When the sign of the change of the slope of the pH vs E_o curve is negative, the change is due to the ionization of the reductant; when positive, to the oxidant; systems 4, 5, 6 and 8 shown in Figure 6 all involve the ionization of the reductant.

Stepwise Oxidations. The reduction of an organic compound is a two electron addition and it was thought for many years that the electrons had to be transferred simultaneously. But, according to Michaelis, "All oxidations of organic molecules, although they are bivalent, proceed in two successive univalent steps, the intermediate state being a free radical."

Michaelis cited three criteria for deciding whether the electronic transfer occurs

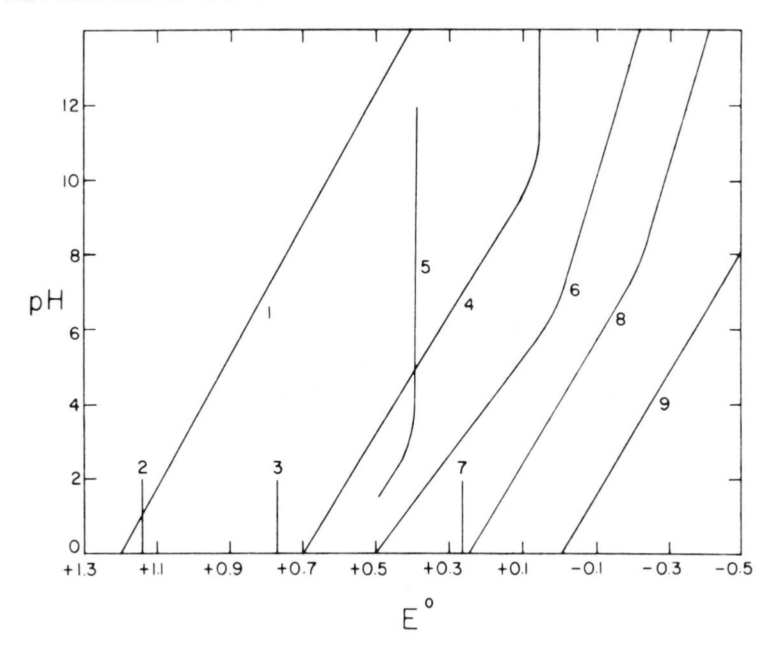

Fig. 6. Red-ox potentials as functions of pH: (1) hypothetical oxygen electrode, (2) O-phenanthroline, (3) ferrous-ferric, (4) quinhydrone, (5) ferro-ferri cyanide, (6) methylene blue, (7) calomel electrode, (8) indigo monosulfonate and (9) hydrogen electrode.

in one or two steps: (1) the color of the intermediate compound, (2) potentiometric oxidation titration curves, and (3) the study of the magnetic properties.

Provided the two oxidation steps are sufficiently separate, the formation of a free radical can be studied by potentiometric titration in which the completely reduced form is titrated with a convenient oxidizing agent and the oxidation-reduction potential measured. For a univalent oxidation at 30°C

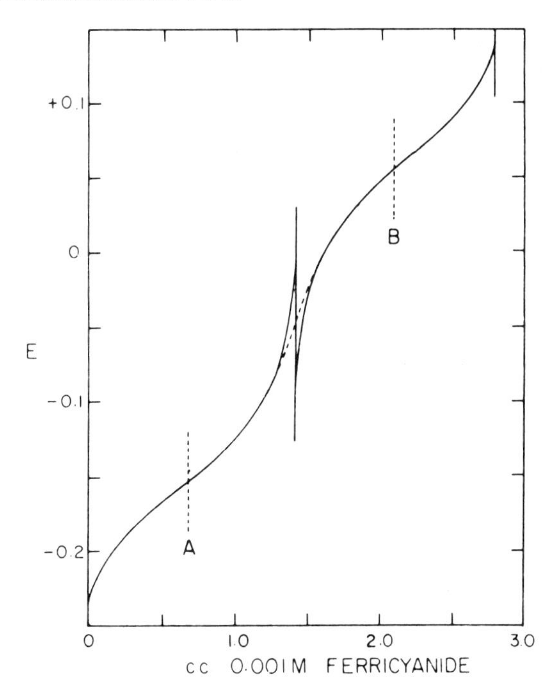

Fig. 7. Oxidation of pyocyanine with ferricyanide. *A* is the mid-point of the first electron transfer; *B* is that of the second. (E. A. H. Friedheim and L. Michaelis: J. Biol. Chem. *91:* 355, 1931.)

$$E = E^0 + 0.06 \log \frac{\text{ox}}{\text{red}} \tag{37}$$

whereas for a bivalent oxidation

$$E = E^0 + 0.03 \log \frac{\text{ox}}{\text{red}} \tag{38}$$

It is clear that if the oxidation occurs in two distinct univalent steps, the oxidation titration curve should reveal this. Figure 7 shows the titration of pyocyanine with ferricyanide; this is a clear example of a two-step oxidation.

If the amount of the free radical is large, the two steps of the titration are conspicuous and the amount of the free radical can be estimated from an oxidation titration curve. Provided the two electrons are released and bound completely independently of each other the electron dissociation constants for a two-electron transfer will bear a ratio of 1 to 4 to each other; this is purely a statistical effect and, under these circumstances, the red-ox titration curve will have only one inflection and the calculated constant for the electron dissociation will equal the square root of the product of the two individual dissociation constants.

Electrons have a spin and, in an ordinary covalent bond, the two electrons spin in an opposite sense so that they give rise to no net magnetic moment. In fact, when such substances with paired electrons are placed in a magnetic field a magnetic moment is induced which opposes the applied field and they tend to move out of the field. This is the general state of matter and these substances are said to be diamagnetic. On the other hand, a substance with unpaired electrons has a permanent magnetic moment and tends to become oriented in line with the applied magnetic field. These substances are paramagnetic. It is evident that magnetic studies on free radicals are important and are valuable in characterizing oxidation-reduction reactions.

Magnetic susceptibilities were formerly studied with some form of the Gouy balance in which the substance in question was suspended between the poles of an electromagnet from a sensitive analytical balance and magnetic force acting on the substance measured. Unfortunately, this method is not capable of detecting small quantities of a free radical, and very concentrated solutions had to be used. More recently the method of electron spin resonance has been employed with great success. Electron spin resonance will be discussed in Chapter 8, Solution Optics.

Biological Oxidations. In its simplest terms, physiological oxidation consists in the removal of hydrogen from the substrate and the combination of it with molecular oxygen to form water or hydrogen peroxide. The living cell does not carry out this process in such a simple, direct fashion. The hydrogen is taken from the substrate and passed along in successive steps until it is finally combined with oxygen. Four different types of substances are involved in cellular oxidation: dehydrogenases, hydrogen transports, oxidases and peroxidases.

Dehydrogenases are specific enzymes which activate the hydrogen of the substrate so that the hydrogen can be removed from the substrate. The hydrogen transport systems convey the hydrogen removed from the substrate to the oxidase enzymes. The function of the oxidase enzymes is to activate the oxygen in the tissue so that it will quickly oxidize the hydrogen that is supplied by the hydrogen transports. The function of the peroxidases is to destroy the peroxides that may be formed. Carbon dioxide production involves the decarboxylation of the substrate.

There are a very large number of red-ox systems in the body. Table 2 is a much abbreviated list of the electrode potentials at 30°C and at pH 7. This list becomes more meaningful if examined in connection with Figure 6.

TABLE 2. RED-OX POTENTIALS OF SOME SYSTEMS OF BIOLOGICAL IMPORTANCE; pH 7 AND 30°

System	*E′ in volts*
Hydrogen-H^+	−0.42
*Nicotinamide adenine dinucleotide (NAD)	−0.32
Riboflavin	−0.21
Lactate-pyruvate	−0.20
Yellow enzyme	−0.123
Flavin adenine dinucleotide [FAD]	−0.06
Succinate-fumarate	−0.02
Cytochrome C	+0.26
Cytochrome A	+0.29
Oxygen-water	+0.81

* Formerly known as diphosphopyridine nucleotide (DPN)

Respiration can be either aerobic, in which event the ultimate hydrogen acceptor is molecular oxygen, or anaerobic and the hydrogen acceptor is some organic molecule. Anaerobic respiration is similar to fermentation and is known as glycolysis when the substance undergoing respiration is glucose, the end product of which is pyruvate, and if the oxygen tension is sufficiently low, lactate; the energy yield from aerobic respiration is very much larger than it is from anaerobic respiration.

The oxidation chain for aerobic respiration is diagramed in Figure 8.

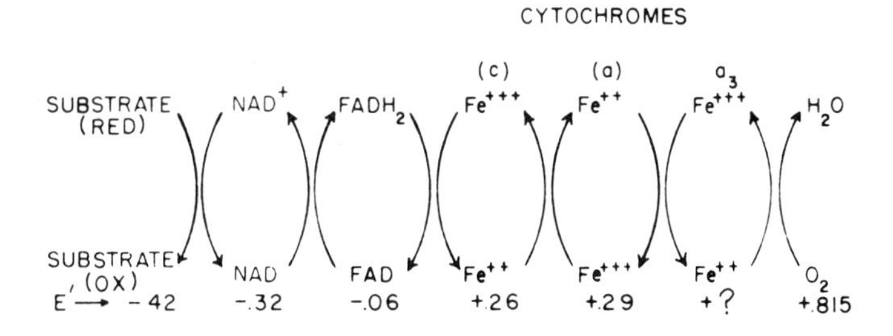

Fig. 8. Respiratory chain with approximate values of E' given for pH 7.

There need not be an actual transfer of hydrogen atoms. For example, hydrogen atoms yield electrons to the cytochrome system and are thereby converted to hydrogen ions. The iron of the cytochrome system is reduced by the electrons, and the oxygen activated by the cytochrome oxidase accepts the electrons from the cytochrome system. The reduced oxygen then combines with protons to yield water. The final result in both cases is water, but the mechanism differs as shown by the following reactions:

Transfer of hydrogens

$$\begin{matrix} R'\text{—}H \\ | \\ R'\text{—}H \end{matrix} + \begin{matrix} R'' \\ \| \\ R'' \end{matrix} \rightarrow \begin{matrix} R' \\ \| \\ R' \end{matrix} + \begin{matrix} R''\text{—}H \\ | \\ R''\text{—}H \end{matrix} \qquad (39)$$

$$\begin{matrix} R''\text{—}H \\ | \\ R''\text{—}H \end{matrix} + O_2 \rightarrow \begin{matrix} R'' \\ \| \\ R'' \end{matrix} + H_2O_2 \rightarrow H_2O + O_2 \qquad (40)$$

Transfer of electrons

$$\begin{matrix} R'\text{—}H \\ | \\ R'\text{—}H \end{matrix} + 2M^{3+} \rightarrow \begin{matrix} R' \\ \| \\ R' \end{matrix} + 2M^{2+} + 2H^+ \qquad (41)$$

$$4M^{2+} + O_2 + 4H^+ \rightarrow 2H_2O + 4M^{3+} \qquad (42)$$

The water produced by hydrogen transfer is derived from the decomposition of H_2O_2 and the hydrogens do not exchange with those of water, whereas the water produced by electron transfer will evidently exchange with that of the aqueous medium.

Figure 8 as drawn is evidently incomplete; no means are shown by which the body could derive benefit from the oxidation of the substrate. If the oxidation chain were not coupled to other reactions, such energy as might become available would be dissipated as heat. Somehow these oxidative reactions must be coupled with reactions which lead to the production of high energy phosphate bonds. The general process by which high energy phosphate bonds are produced as the result of oxidation is known as oxidative phosphorylation.

Whereas the oxidative enzymes and coenzymes associated with the oxidation of a substrate have all been isolated and purified and will function in this state, the ability to produce high energy phosphate bonds as a result of oxidation by oxygen is confined to intact subcellular particles called mitochondria, to chloroplasts during photosynthesis, and to bacteria. All efforts to isolate a coupling agent between the oxidative chain and the synthesis of ATP have been fruitless. The mechanism of oxidative phosphorylation is still unclear and is the only important link in the fuel supply which has not been elucidated. An alternative approach to direct chemical coupling through a common intermediate as described in Chapter 2 is coupling across the mitochondrial membrane.

It was suggested several years ago that if the oxidation-reduction chain were properly oriented across a cellular membrane, an ion pump would result. Figure 9 is a highly diagrammatic illustration of such an ion pump.

Electrical neutrality must be maintained both outside and inside the cell. This requires that an anion must move from the inside to the outside of the cell or that a cation must cross the cellular membrane from outside to inside. If the anion moving from inside to outside were a chloride ion, HCl would be produced; this idea was evoked to explain the secretion of HCl by the gastric mucosa. If the pump were reversed, sodium ions could exchange from the inside to the outside of the cell with the protons and a sodium pump would result.

Stoichiometry of the red-ox pump requires that the ratio of the moles of HCl secreted per mole of O_2 used be equal to 4; the experimental value is closer to 12. It is concluded for this reason and others that the simple picture of the ion-pump illustrated in Figure 9 needs elaboration. It is believed that the hydrolysis of ATP

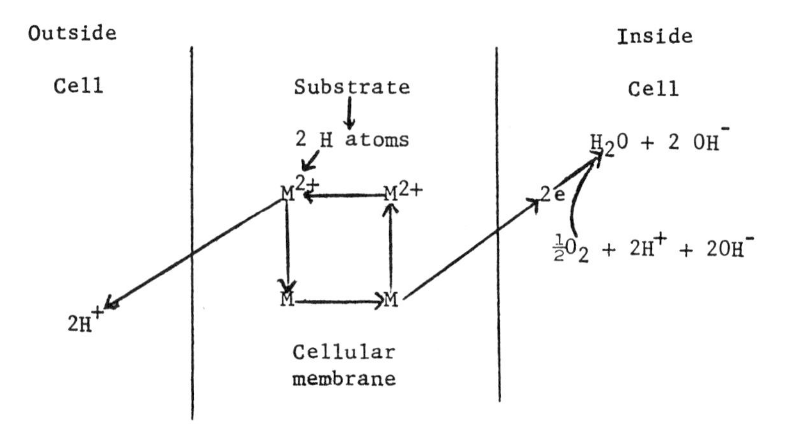

Fig. 9. Hypothetical red-ox ion pump. M^{2+} represents the oxidized red-ox system and M the reduced form; *e* is the electron.

provides the immediate energy for ion transport across cellular membranes. It is now considered possible, however, that a red-ox pump somewhat like that shown in Figure 9 operates in the mitochondrial membrane to provide for the synthesis of ATP; this is the basis of the chemiosmotic hypothesis of Mitchell.

The total spread of potential shown in Figure 8 is about 1.23 volts corresponding to the voltage of the oxygen-water electrode, which in turn corresponds to the free energy of formation of water with a two-electron transfer. If two-electron transfers are needed to produce one high energy phosphate bond and it is assumed that the energy of the phosphate bond is about 10,000 calories, the spread of 1.23 volts corresponds to about 5.6 phosphate bonds formed per molecule of water produced.

Finally, it should be emphasized that the oxidative phase of metabolism supplies most of the energy need of aerobic cells—estimated at about 90 percent of the total.

GENERAL REFERENCES

Bates, R. G.: Determination of pH. John Wiley and Sons, New York, 1964.
Ives, D. J. G. and C. J. Janz: Reference Electrodes. Academic Press, New York, 1961.
Mitchell, P.: Proton-translocation phosphorylation. Fed. Proc. *26*:1370, 1967.
Ross, J. W.: Calcium-selective electrode with liquid ion exchanger. Science *156*:1378, 1967.
Bioelectrodes. Ann. N.Y. Acad. Sci. *148*:1, 1968.

PROBLEMS

1. Suppose you have a 0.10M KCl—calomel electrode connected by a 0.10M KCl salt bridge to a silver-silver chloride half cell in 0.10M KCl at 25°C.
   (a) What would be the potential difference between the electrodes?
   (b) Which electrode would be negative?
   (c) What is the chemical reaction involved and what is the molar free energy change for this reaction?

   Ans: (a) 0.0474 volts, (b) silver-silver chloride, (c) −2,186 calories

2. Whereas the mechanism of oxidative phosphorylation is still unknown, there is strong evidence that the process is related to electron transfer. From your knowledge of the free energy of the hydrolysis of ATP to ADP at pH 7, what must be the minimum potential difference in the transfer of two electrons per phosphate bond?

   Ans: About 0.18 volts.

3. Explain why hydrogen ions influence the oxidation-reduction potential of many systems and not of others.

   Ans: See text.

4. 10 ml. of 0.1M $FeCl_3$ are mixed with 10 ml. of 0.1M potassium ferrocyanide solution at 25°C. The oxidation-reduction potential of ferro-ferricyanide is 0.36 volts and that of ferrous-ferric system is 0.771 volts. Calculate the concentration of the ferric ions after equilibrium is reached. What would be the oxidation-reduction potential of the mixture?

   Ans: 1.67×10^{-5} molar; $E' = 0.566$ volts.

CHAPTER 5

Acids and Buffers

There are various ways in which acids and bases can be defined and each is appropriate for the purpose at hand. The definition proposed independently by Bronsted and by Lowry appears adequate for our present needs. Acids are substances that are capable of yielding protons and a base is able to accept protons. This notion is expressed as

$$A \rightleftharpoons H^+ + B \qquad 1$$

where A is an acid and B is a base. Thus in the reaction

$$R \cdot COOH \rightleftharpoons R \cdot COO^- + H^+ \qquad 2$$

$R \cdot COO^-$ is a base and $R \cdot COOH$ is an acid. Also

$$R \cdot NH_3^+ \rightleftharpoons R \cdot NH_2 + H^+ \qquad 3$$

and RNH_3^+ is an acid and RNH_2 is a base.

The ionization of an amino group can also be represented as a basic dissociation and

$$R \cdot NH_2 + H_2O \rightleftharpoons R \cdot NH_3^+ + OH^- \qquad 4$$

the ionization constant of which is

$$K_\beta = \frac{R \cdot NH_3^+ \times OH^-}{R \cdot NH_2} \qquad 5$$

The acid ionization constant (from eq. 3) is

$$K_\alpha = \frac{RNH_2 \times H^+}{RNH_3^+} \qquad 6$$

It is to be noted that the product $K_\alpha K_\beta$ is equal to K_w and, further, the sum of reactions 3 and 4 is simply the ionization of water.

Ionization of Water. There are many weak acids but the most important of these is water; water ionizes as

$$H_2O \rightleftharpoons H^+ + OH^- \qquad 7$$

and

$$K = \frac{H^+ \times OH^- \times \gamma_{H^+} \times \gamma_{OH^-}}{H_2O \times \gamma_{H_2O}} \qquad 8$$

where γ_{H^+}, γ_{OH^-} and γ_{H_2O} are the activity coefficients of the hydrogen ion, the hydroxyl ion and the water respectively. Considering the activity of liquid water to be unity, eq. 8 becomes

$$K_w = H^+ \times OH^- \times \gamma_{H^+} \times \gamma_{OH^-} \qquad 9$$

In solutions near neutrality, γ_{H^+} and γ_{OH^-} are close to unity and

$$K_w = H^+ \times OH^- \qquad 10$$

The ionization constant of water can be calculated from conductance data along with a knowledge of the mobilities of H^+ and OH^-. However, more exact values can be obtained from electromotive force measurements on cells of the type

$$Pt, H_2(g) \mid KOH, KCl, Ag\ Cl(S) \mid \overset{+}{Ag} \qquad 11$$

The electrode equation for this cell can be written

$$E - E^0 + \frac{RT}{F} \ln \frac{C_{Cl^-}}{C_{OH^-}} = -RT \ln K_w \qquad 12$$

The plot of left terms of eq. 12 against the ionic strength yields a nearly linear relation which can be extrapolated to zero ionic strength and the intercept on the y-axis is $-RT \ln K_w$ from which K_w is obtained. Figure 1 shows pK_w of water as a function of temperature. The values of K_w at 25°C, 30° and 40°C are 1.008×10^{-14}, 1.469×10^{-14} and 2.919×10^{-14} respectively.

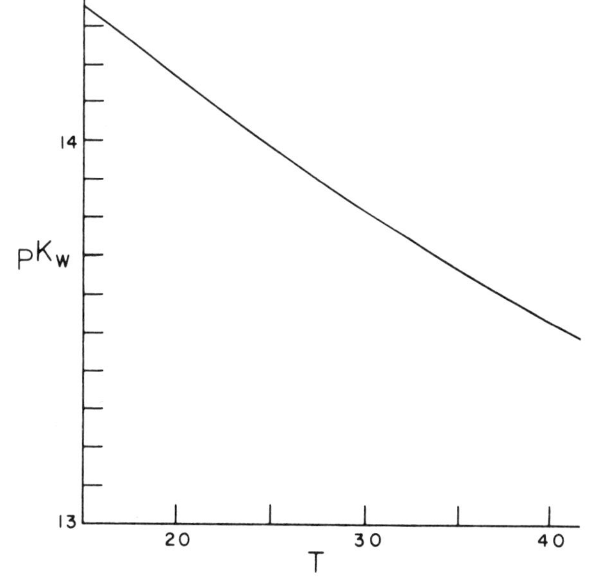

Fig. 1. pK_w as a function of temperature.

At 25°, pH plus pOH equals 14 and accordingly, in absolutely pure water, the pH should be 7. Due to absorption of CO_2 from the air, ordinary distilled water usually has a pH closer to 5.

In principle, water is a dibasic acid, but the second ionization constant is so small (approximately 2×10^{-28}) that the proton does not ionize from the hydroxyl ion.

It was mentioned in Chapter 3 that hydrogen ions are hydrated in solution to produce hydronium ions and naked protons do not exist in water. The reaction for the production of hydronium ions is

$$H_2O + H_2O \rightleftharpoons H_3^+O + OH^- \qquad 13$$

In the discussion to follow, the hydration of hydrogen ions will be ignored.

Titration Curves. With the aid of a pH meter it is easy to titrate an acid of known concentration with a strong base of known concentration. The plot of the measured pH against the amount of base added yields a titration curve. Figure 2 shows the titration of 50 ml. of 0.10N solutions of HCl and CH_3COOH and 100 ml. of 0.1N H_3PO_4 with 0.10N NaOH respectively.

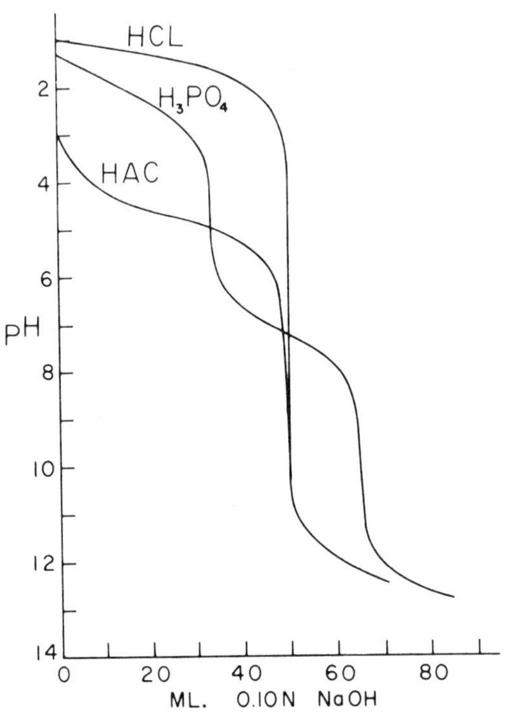

Fig. 2. Titration of 50 ml. 0.10N HCl and acetic acid (HAc) and 100 ml. 0.10 N H_3PO_4 with 0.10N NaOH.

During the titration of a weak acid by a strong base, the acid ionizes as

$$HA \rightleftharpoons H^+ + A^- \qquad 14$$

and

$$K = \frac{H^+ \times A^-}{HA} \qquad 15$$

Since the salt formed during the titration is completely ionized, the anions (A^-) present come almost entirely from the salt, and without making a large error eq. 15 can be written

$$K = \frac{H^+ \times \text{salt}}{\text{acid}} \qquad 16$$

Taking the logarithm of both sides of eq. 16 gives

$$\log K = \log H^+ + \log \frac{\text{salt}}{\text{acid}} \qquad 17$$

and from which results the well-known Henderson-Hasselbalch equation

$$pH = pK + \log \frac{\text{salt}}{\text{acid}} \qquad 18$$

It is thus easy to linearize a titration curve by plotting pH against the ratio of the salt to the acid concentration and its reciprocal as a semi-log plot. Such a plot is shown in Figure 3. As can be noted, the titration lines are now parallel and are displaced relative to each other, depending on the individual pK values.

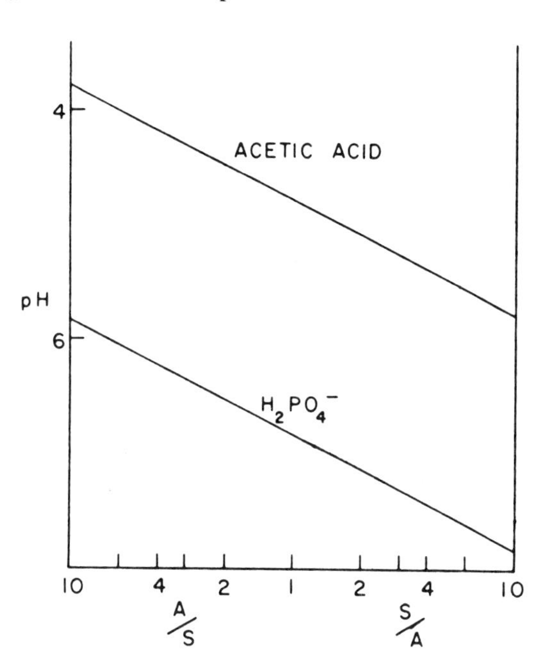

Fig. 3. Linearized titration curves on a semi-log plot. A is the concentration of the undissociated acid and S is the salt concentration.

It is a common experience that salts of strong acids and strong bases have neutral reactions. For example, NaCl, K_2SO_4, etc., dissolved in water are neutral. On the other hand, a solution of sodium acetate is alkaline, and one of ammonium chloride is acid.

Sodium acetate is practically completely ionized in solution

$$\text{Na Ac} \rightarrow \text{Na}^+ + \text{Ac}^- \qquad 19$$

The acetate ion so formed reacts with water

$$\text{Ac}^- + H_2O \rightleftharpoons \text{HAc} + \text{OH}^- \qquad 20$$

and acts in this instance as a base in the Bronsted sense, since it is able to accept a proton from water.

For the acetic acid formed from the hydrolysis of sodium acetate, the equlibrium equation is

$$\frac{\text{H}^+ \times \text{Ac}^-}{\text{HAc}} = K \qquad 21$$

and rearranging

$$\text{H}^+ = \frac{K \times \text{HAc}}{\text{Ac}^-} \qquad 22$$

The amount of acetic acid formed equals the amount of hydroxyl ions produced, and the amount of acetate ion very nearly equals the amount of sodium acetate (salt) in solution. When this information is incorporated in eq. 22 there results

$$H^+ = \frac{K \times OH^-}{\text{salt}} \qquad 23$$

Since

$$OH^- = \frac{K_w}{H^+} \qquad 24$$

and from 23 and 24

$$(H^+)^2 = \frac{K \times K_w}{\text{salt}} \qquad 25$$

Taking logarithms of both sides of eq. 25 and multiplying through by a minus one, there results

$$\text{pH} = 7 + \tfrac{1}{2}\,\text{pK} + \tfrac{1}{2}\log \text{salt} \qquad 26$$

Similarly, for a salt of a strong acid and a weak base, for example, ammonium chloride

$$\text{pH} = 7 - \tfrac{1}{2}\,\text{pK}_b - \tfrac{1}{2}\log \text{salt} \qquad 27$$

where K_b is the ionization constant of ammonium hydroxide.

Employing the same line of reasoning, it is found that the pH of a solution of a salt of a weak acid and a weak base, for example, ammonium acetate, is

$$\text{pH} = 7 + \tfrac{1}{2}\,\text{pK} - \tfrac{1}{2}\,\text{pK}_b \qquad 28$$

The pH of a solution of such a salt is independent of concentration.

Ionization of Weak Acids. The calculation of the pH of a pure solution of a weak acid can be made, provided the ionization constant of the acid is known. The hydrogen ion concentration is equal to the anion concentration, and

$$\frac{(H^+)^2}{\text{acid}} = K \qquad 29$$

or

$$(H^+)^2 = K \times \text{acid} \qquad 30$$

Taking the logarithms of both sides of eq. 30 and rearranging, it is found that

$$\text{pH} = \tfrac{1}{2}\,\text{pK} - \tfrac{1}{2}\log \text{acid} \qquad 31$$

where the acid concentration is really that of the undissociated acid, but for most purposes can be taken equal to the total acid concentration.

The ionization constant of a weak acid can be expressed either in terms of concentrations or of activities. These two constants are designated K_c and K_a respectively and for the ionization

$$HA \rightleftharpoons H^+ + A^- \qquad 32$$

when the concentrations are expressed in moles per liter

$$K_c = \frac{H^+ \times A^-}{HA} \qquad 33$$

And using activities

$$K_a = \frac{a_{H^+} \times a_{A^-}}{a_{HA}} \qquad 34$$

The relation between these constants is given by

$$K_a = \frac{\gamma_{H^+} \times \gamma_{A^-}}{\gamma_{HA}} \cdot K_c \qquad 35$$

where γ_{H^+}, γ_{A^-} and γ_{HA} are the activity coefficients of the hydrogen ions, anions and undissociated acid, respectively. The value of K_a is independent of electrolyte concentration; that of K_c will vary with electrolyte concentration. K_a is called the thermodynamic or true acid dissociation constant, and K_c is the stoichiometric constant; K_a is by far the more useful and fundamental.

There are two general methods used to evaluate acid ionization constants which are (1) from conductance measurements and (2) from electromotive force cells. The conductance method depends on the measurement of the conductance of a series of dilutions of a pure acid; the equivalent conductance is determined, and, since the conductance due to each ion is known, the ion product can be calculated (see Chapter 12, Ion Transport).

TABLE 1. pK_a VALUES, ENTHALPY AND ENTROPY CHANGE PER MOLE FOR THE IONIZATION OF SOME WEAK ACIDS IN AQUEOUS SOLUTION AT 25°C

Acid	pK_a	ΔH	ΔS
Acetic	4.756	−92	−22.1
Benzoic	4.201	104	−18.9
Boric	9.234	3,373	
n-Butyric	4.820		
Carbonic			
K_1	3.88		
K_1 (apparent)	6.352	2,240	−21.6
K_2	10.329	3,603	−35.2
Citric			
K_1	3.128		
K_2	4.761		
K_3	6.396	−802	
Ethanolammonium	9.498	12,080	−0.6
Formic	3.752	−23	−17.6
Fumaric			
K_1	3.019		
K_2	4.384		
Lactic	3.860	−99	−18.0
Maleic			
K_1	1.921		
K_2	6.225		
Malonic			
K_1	2.855		
K_2	5.696	−1,160	
Oxalic			
K_1	1.271		
K_2	4.166	−1,658	
Phosphoric			
K_1	2.148	−1,828	−16.0
K_2	7.198	987	−29.6
K_3	12.32	4,200	−42.7
Propionic	4.874	−163	−22.8
Pyruvic	2.5	2,900	−1.7
Tris	8.076	10,900	−0.3
n-Valeric	4.860		

Returning to the Henderson-Hasselbalch equation (eq. 18)

$$pH = pK + \log \frac{\text{salt}}{\text{acid}} \tag{36}$$

it is evident that if the ratio of salt to acid concentration be fixed at unity, the pH of the solution will equal pK of the weak acid. The value thus calculated will equal K_c and what is needed is to extrapolate the measured value of K_c to zero concentration of the mixture of salt and acid. The activity coefficients at zero concentration will be unity and K_c will equal K_a.

Such an extrapolation is accomplished with an electromotive force cell without a liquid junction consisting of the following

$$\overset{-}{Pt}, H_2(g) \mid HA, NaA, NaCl, AgCl(S) \mid \overset{+}{Ag} \tag{37}$$

where HA is the weak acid and NaA is the sodium salt of the weak acid. The equation for this cell is

$$E = E^0 - \frac{RT}{F} \ln \gamma_{H^+} \times \gamma_{Cl^-} \times m_{H^+} \times m_{Cl^-} \qquad 38$$

where m_{H^+} and m_{Cl^-} are the molalities of the hydrogen and of the chloride ions respectively. Substituting in eq. 38 the expression for the thermodynamic dissociation constant of a weak acid and rearranging, there results

$$E - E^0 + \frac{RT}{F} \ln \frac{m_{HA} \times m_{Cl^-}}{m_{A^-}} = -\frac{RT}{F} \ln \frac{\gamma_{H^+} \times \gamma_{Cl^-}}{\gamma_{H^+} \times \gamma_{A^-}} - \frac{RT}{F} \ln K_a \qquad 39$$

The first member of the right side of eq. 39 becomes zero at infinite dilution; the activity coefficients approach unity as the concentrations approach zero. Accordingly if the left side of eq. 39 is plotted against the ionic strength, a very nearly straight line is obtained which can be extrapolated to zero ionic strength. The intercept at zero ionic strength is $-\frac{RT}{F} \ln K_a$ from which K_a can be computed.

Table 1 gives the ionization constants (thermodynamic constants) of some weak acids.

Temperature and Ionization. It would be expected that the influence of temperature on the ionization constants of acids could be described by the relation (see Chapter 2, Energetics, eq. 45)

$$\frac{d \ln K}{dT} = \frac{\Delta H^\circ}{RT^2} \qquad 40$$

However, the enthalpy (ΔH°) of the ionization of acids has a marked tendency to change with temperature and the ionization constants characteristically exhibit a maximum in relation to temperature. The relation between temperature and the ionization constant can be expressed empirically by

$$-RT \ln K = A - BT + DT^2 \qquad 41$$

and the maximum value of the ionization constant will occur at a temperature given by $(A/D)^{1/2}$ and at this temperature ΔH° is zero. For acids whose ΔH° is small at room temperature, the temperature at which a maximum occurs will be low whereas if ΔH° at room temperature is large, the temperature at which ΔH° becomes zero will be high.

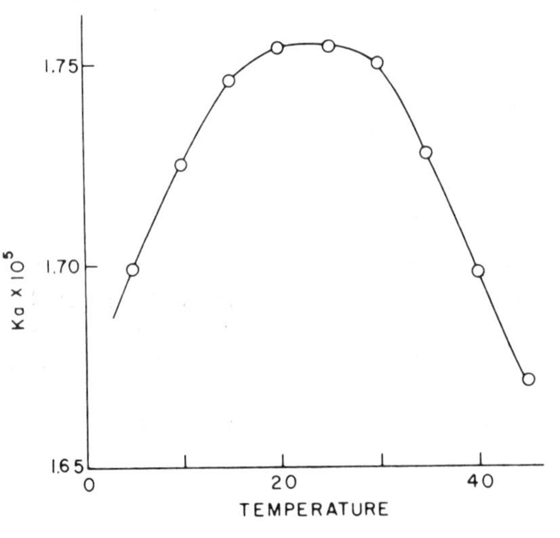

Fig. 4. Variation of K_a of acetic acid as a function of the temperature. Temperature expressed in degrees centigrade.

Shown in Figure 4 is a plot of the K value of acetic acid as a function of the temperature. The maximum in K_a (minimum in pK_a) for acetic acid occurs at 22.4°C. The heat of ionization is positive below this temperature and negative above.

This somewhat unexpected behavior of the ionization constants of weak acids in relation to temperature appears to arise primarily from two opposing effects. As the temperature is increased, due to kinetic energy, there is a natural and increasing tendency for the undissociated acid molecule to split into two parts, i.e., the proton and the anion. However, as the temperature increases, the dielectric constant decreases thus increasing the electrostatic force of attraction between the proton and the anion.

The carboxylic acids of the type R · COOH have low heats of ionization which approach zero at room temperature whereas acids of the type $R \cdot NH_3^+$ have high heats which are in the neighborhood of 10,000 calories per mole. This can be understood by contrasting the way in which these two ionize, for example

$$R{\cdot}COOH + H_2O \rightleftharpoons H_3^+O + R \cdot COO^- \qquad 42$$

but

$$R{\cdot}NH_3^+ + OH^- \rightleftharpoons R{\cdot}NH_2 + H_2O \qquad 43$$

Thus the ionization of R·COOH produces a hydronium ion. The heat of this process is low whereas the ionization of $R \cdot NH_3^+$ is coupled to the ionization of water; the heat of ionization of water is 13,519 calories per mole.

Buffer Capacity. It is obvious that the capacity of various aqueous solutions to resist pH changes with the addition of acids or of bases varies greatly. Van Slyke defines the buffer capacity as the reciprocal of the rate of change of pH of a solution relative to the addition of a strong base to the solution and

$$\text{Buffer capacity } (\beta) = \frac{dB}{dpH} \qquad 44$$

Solutions of strong acids and bases are excellent buffers if sufficiently concentrated. Consider, for example, a strong base which completely dissociates into metal and hydroxyl ions. Then the increment dB, in base, is equal to dOH^- and dB/dpH is equal to dOH^-/dpH. Since

$$pH = \log OH^- - \log K_w \qquad 45$$

dpH is equal to $d \log OH^-$ and

$$\frac{dB}{dpH} = \frac{dOH^-}{d \log OH^-} = 2.3\ OH^- \qquad 46$$

and similarly for a strong acid added to water

$$\frac{dB}{dpH} = \frac{-dH^+}{-d \log H^+} = 2.3H^+ \qquad 47$$

Adding the expressions for the buffer capacity of acids and bases

$$\frac{dB}{dpH} = 2.3(H^+ + OH^-) \qquad 48$$

In a solution of a weak acid in the presence of its salt of a strong base

$$K = \frac{H^+ \times B}{HA} \qquad 49$$

where B is the amount of base added and is equivalent to the salt as used previously in eq. 17. HA can be set equal to HA_0-B where HA_0 is the original concentration of acid before the addition of base. Substituting this equality in eq. 49 yields

$$K = \frac{H^+ \times B}{HA_0 - B} \tag{50}$$

and

$$B = \frac{K \times HA_0}{K + H^+} \tag{51}$$

Differentiating eq. 51 gives

$$\frac{dB}{dH^+} = -\frac{K \times HA_0}{(K + H^+)^2} \tag{52}$$

Multiplying both sides of eq. 52 by $-2.3H^+$ produces

$$\frac{dB}{dpH} = \frac{2.3K \times H^+ \times HA_0}{(K + H^+)^2} = \beta \tag{53}$$

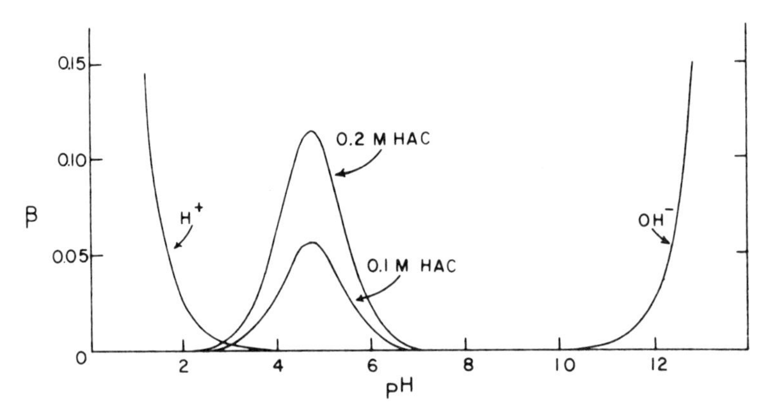

Fig. 5. Buffer capacity of strong acid (H^+); strong base (OH^-) and acetate buffers which are 0.1 M and 0.2 M respectively as functions of pH.

It is evident from eq. 53 that at constant H^+ concentration the buffer capacity is directly proportional to the buffer concentration. Figure 5 shows plots of the buffer capacities of a strong acid and a strong base as well as the capacity of an acetate buffer at two concentrations as functions of the pH. It is also evident from an inspection of Figures 2 and 5 that, for a weak acid in the presence of its salt of a strong base, the maximum buffer capacity occurs at a pH equal to the pK of the acid. This fact is also apparent from the following argument.

If eq. 53 is thrown into the log form and then differentiated in respect to β and H^+, the following is obtained

$$d \log \beta = d \log H^+ - 2d \log (K + H^+) \tag{54}$$

Remembering that dH^+ is equal to $-2.3H^+dpH$ and rearranging, eq. 54 gives

$$\frac{d\beta}{dpH} = \left(\frac{2H^+}{K + H^+} - 1\right) \beta \tag{55}$$

At the point of maximum buffer capacity $d\beta/dpH$ is zero, and from eq. 55, H^+ is equal to K. It is also evident from an examination of eq. 53 that when H^+ equals K (pH at which the buffer capacity is a maximum), β, the buffer capacity, is equal to $2.3\ HA_0/4$ or is $0.576\ HA_0$.

The purpose of a buffer is to resist changes in pH of a medium. The factors which will produce a change in the pH of a buffer are (1) addition of acids or bases, (2) dilution of the buffer, (3) increase in the concentration of neutral electrolyte and (4) temperature changes.

The buffer capacity which was discussed above is a measure of how successful the buffer will be in preventing excessive changes in pH due to the addition of acids and bases. The chemical mechanism of buffers composed of weak acids in the presence of their salts can be understood by considering the neutralization reactions. For example, suppose a strong acid such as HCl is added to an acetate buffer. The strong acid HCl has thus been replaced by the weak acetic acid and the addition of a strong base such as NaOH produces sodium acetate which is a weak base.

There are two effects produced upon the dilution of a buffer. The first is the effect of the water which itself is a weak acid and causes a shift towards pH 7 as dilution proceeds; it is obvious that if the buffer be diluted to such an extent that it no longer exists, the pH will have to equal that of pure water. The second is the effect of dilution on the activity coefficients. Dilution in the concentration ranges in which buffers are ordinarily used causes the activity coefficients to approach unity. Over the range of pH 4.5 to 9.5 in which neither the hydrogen ion concentration nor the hydroxyl ion concentration is comparable to the concentration of the buffer, the influence of dilution on the pH depends on the valence types and the buffer concentration.

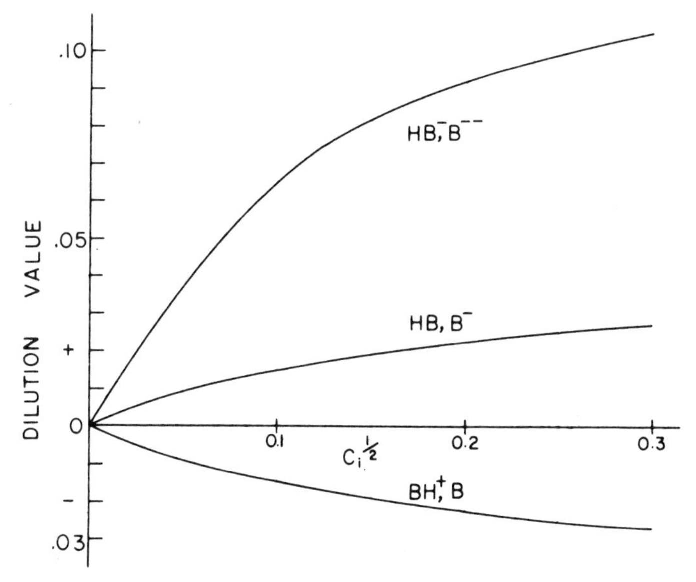

Fig. 6. Dilution values of buffers of the indicated ion types as functions of the square root of the initial molar buffer concentrations (C_i).

The dilution value of a buffer as defined by R. G. Bates is the increase of pH suffered by a solution of initial concentration C_i upon dilution with an equal volume of pure water to give a concentration $\frac{1}{2}C_i$. It is positive when the pH increases with dilution and negative when it decreases. Figure 6 shows plots of the dilution values of the indicated ion type buffers as functions of the square root of the initial buffer concentration (C_i). These values are valid in the pH range of 4.5 to 9.5.

The effect of adding neutral salts to a buffer system is in a sense the reverse of the dilution of a buffer and the pH changes are in an opposite sense to those experienced when the buffer is diluted. The effect here is essentially on the activity coefficient of the ions; the activity coefficient of the undissociated and uncharged acid molecules remains substantially unchanged.

Writing the expression for the ionization constant of a weak acid and neglecting the activity coefficient of the undissociated acid

$$K_a = \frac{H^+ \times A^-}{HA} \cdot \gamma_{H^+} \times \gamma_{A^-} \qquad 56$$

assuming that the concentration of the anions is equal to that of the salt

$$pH = pK_a + \log \frac{\text{salt}}{\text{acid}} + \log \gamma_{A^-} \qquad 57$$

Substituting from the Debye-Hückel theory the value of log γ_{A^-} (eq. 41, Chapter 3, Water and Electrolytes) at 25°

$$pH = pK_a + \log \frac{\text{salt}}{\text{acid}} - \frac{0.509 Z_1 Z_2 \Gamma/2^{1/2}}{1 + \Gamma/2^{1/2}} + 0.1\Gamma/2 \qquad 58$$

The influence of temperature on the pH of a buffer can be predicted from knowledge of the heat of ionization of the ionizing species; indeed the heat of ionization can be calculated from the change in pH of the buffer with temperature.

The Van't Hoff equation (Chapter 2, Energetics, eq. 45) is

$$\frac{d \log K}{dT} = \frac{\Delta H^\circ}{2.303 R T^2} \qquad 59$$

and substituting the value of the ionization constant of a weak acid

$$\frac{d \log \dfrac{H^+ \times A^-}{HA}}{dT} = \frac{\Delta H^\circ}{2.303\ R T^2} \qquad 60$$

and if the ratio of A^- to HA is fixed at unity

$$\frac{d \log H^+}{dT} = \frac{\Delta H^\circ}{2.303\ R T^2} \qquad 61$$

and

$$\frac{dpH}{dT} = -\frac{\Delta H^\circ}{2.303\ R T^2} \qquad 62$$

In using eq. 62, it must be kept in mind that the heats of ionization of acids change with temperature and, accordingly, a plot of pH against 1/T will not yield a straight line. The slope of the line at any point, however, will equal $\Delta H^\circ/4.574$.

Practical Buffers. The selection of the proper buffer system for a given experimental use is a common problem. In the extreme acid region, hydrochloric acid of the appropriate concentration is a suitable choice, the desired ionic strength being achieved by the addition of sodium or potassium chloride. Over the wide sweep of the range between pH 3 and 11 a buffer can be selected on the basis of the pK value of a weak acid (see Table 1); the pH of the maximum buffer capacity is equal to the pK. Usually, however, there are complicating factors to be considered such as solubilities, interaction of the buffer components with the experimental mixture, etc., so that the actual choice is limited. Shown in Table 2 are some suggested buffers with their effective ranges.

As noted in Table 2 the $KH_2PO_4 : Na_2HPO_4$ buffer covers the physiological pH region. Unfortunately, the phosphate system forms insoluble salts with calcium ions thus limiting the usefulness of this buffer. In addition, phosphate ions not infrequently interact with various enzyme systems and, lastly, due to the presence of a divalent anion ($HPO_4^=$), this buffer has large dilution and salt errors. There are few weak organic acids which have pK values in the range of that of the second hydrogen of phosphoric acid. The barbiturate buffer is a favorite for electrophoretic work, but this system has limited solubility. The organic amines have assumed considerable importance as buffers and of these tris-(hydroxymethyl) aminomethane, imidazole

TABLE 2. SUGGESTED BUFFERS WITH THEIR EFFECTIVE pH RANGES

System	pH *range*
HCl, NaCl	0– 2.5
Citric acid, Na citrate	2.2– 6.5
HCl, glycine	1.0– 3.7
Acetic, Na acetate	3.7– 5.6
KH phthalate, KNa phthalate	4.0– 6.2
KH_2PO_4, Na_2HPO_4	5.8– 8.0
Diethylbarbituric acid, Na salt	7.0– 9.0
HCl, Tris	7.2– 9.2
Boric acid, Na borate	8.0–10.0
Glycine, NaOH	8.2–10.1
$NaHCO_3$, Na_2CO_3	9.2–11.0
Na_2HPO_4, Na_3PO_4	11.0–12.0

and 4-amino-pyridine are solids at room temperature and can be crystallized and purified. Tris especially has come to the fore, the desired pH being achieved by the addition of HCl, and the ionic strength is equal to the sum of HCl and NaCl added. The heat of ionization of tris is large (see Table 1) and, accordingly, it has a high temperature coefficient, the pH falling with rising temperatures. The structural formula of tris is

```
              OH
              |
            H—C—H
     H        |        H
     |        |        |
 HO—C————————C————————C—OH          63
     |        |        |
     H       NH2       H
```

The Henderson-Hasselbalch equation is used to calculate the quantities of salt and weak acid to be used in the buffer of choice. In dealing with a monovalent anion the calculation of the ionic strength is simply equal to the anion concentration except in the more extreme acid or basic range where the ionic strength is equal to the sum of the anion concentration plus that of the hydrogen ion concentration. Divalent anions such as those of disodium phosphate and the citrates of higher charge present a minor complication which, however, is not difficult to resolve.

Several mixtures of buffers have been composed, the so-called universal buffers. Mixture of buffer substances will yield a constant buffering power over the working pH range if the consecutive pK values of the acids involved differ by not more than 1.2 pK-units.

Table 3 gives the National Bureau of Standards Standard Buffer Solutions.

TABLE 3. NBS STANDARD BUFFER SOLUTIONS (R. G. BATES) AT 25°C

Substance	pH
Potassium tetroxalate 0.05M	1.68
Potassium hydrogen-tartrate saturated at 25°	3.56
Potassium hydrogen-phthalate 0.05M	4.01
Potassium dihydrogen phosphate 0.025M Disodium hydrogen phosphate 0.025M	6.86
Borax 0.01M	9.18

Blood Buffers and Respiration. The important blood buffers are the hemoglobin and the bicarbonate-carbonic acid systems. The phosphate and the plasma proteins play relatively minor roles. The bicarbonate system acquires its importance from the rapidity and flexibility of action and from the fact that it is being constantly replenished due to the loss of CO_2 from the lungs.

The pH of the bicarbonate system in blood plasma at 38°C is given by the expression

$$pH = 6.1 + \log \frac{BHCO_3}{CO_2 + H_2CO_3} \qquad 64$$

where $BHCO_3$ represents the concentration of the bicarbonate and CO_2 plus H_2CO_3

represents the total dissolved carbon dioxide both hydrated and unhydrated. From Table 1 it is seen that pK_1 of carbonic acid is 3.88 and this pK_1 refers to the reaction

$$H_2CO_3 \rightleftharpoons H^+ + HCO_3^- \qquad 65$$

When CO_2 is dissolved in water only about 0.3 percent exists in the hydrated form (H_2CO_3) and the *apparent* K_1 of carbonic acid refers to the reaction

$$CO_2 + H_2O \rightleftharpoons H^+ + HCO_3^- \qquad 66$$

and the constant for this reaction at 25°C is 4.4×10^{-7}. The hydration of carbon dioxide is a comparatively slow reaction and in the body requires the presence of an enzyme, carbonic anhydrase, to increase the rate of reaction to make possible the rapid exchange of carbon dioxide necessary in the respiratory cycle.

The amount of gas dissolved in a liquid is proportional to the partial pressure of the gas. At 38°C, one ml. of blood plasma dissolves 0.51 ml. of CO_2 at 760 mm. of mercury pressure. The millimolar volume of carbon dioxide gas at 760 mm. Hg pressure and at 38°C is 22.26 ml. Accordingly, the concentration of dissolved carbon dioxide in plasma at 38°C is

$$CO_2(\text{mM/liter}) = \frac{0.51 \times 1{,}000}{760 \times 22.26} P_{CO_2} = 0.030\ P_{CO_2} \qquad 67$$

Equation 64 can then be written

$$pH = 6.1 + \log \frac{BHCO_3}{0.030\ P_{CO_2}} \qquad 68$$

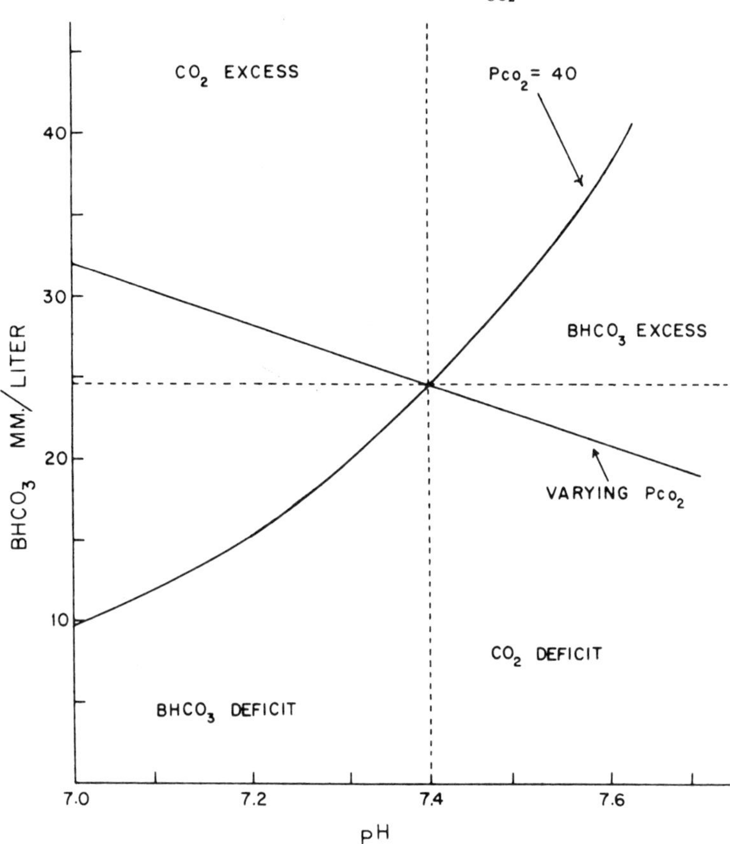

Fig. 7. $BHCO_3$ of human blood plasma as a function of pH. Normal values are pH 7.4 and a $BHCO_3$ concentration of 24mM./liter.

The pH of normal arterial blood is close to 7.40 and the partial pressure of CO_2 is about 40 mm. Hg. The concentration of dissolved CO_2 is then 1.20 mM. per liter, the $BHCO_3$ concentration is 24 mM. per liter and the ratio of the bicarbonate to the dissolved CO_2 is 20 to one. Obviously, if two of the three variables expressed in eq. 68 are known the third can be calculated. Shown in Figure 7 is a plot of the bicarbonate of plasma as a function of the pH of the plasma. Also shown is the bicarbonate concentration of normal plasma as a function of the pH at different P_{CO_2}; this amounts to a titration of normal plasma with CO_2.

An electrometric method has been devised for the determination of the P_{CO_2} of blood. The sample of blood is separated from a standard solution of potassium bicarbonate by a polyethylene or Teflon membrane; the membrane is easily permeable to CO_2 but to no other component in the system. The CO_2 of the blood comes to equilibrium with the potassium bicarbonate solution whose concentration is known. A pH determination on the bicarbonate solution, therefore, permits the calculation of the P_{CO_2}.

The control of body acidity in terms of the bicarbonate system lies in the exhalation of CO_2 from the lungs. The control of the so-called fixed acids lies ultimately in the operation of the kidneys.

Multiple Binding. An anion of a weak monobasic acid has one binding site for a proton, for example, an acetate ion. The binding reaction can be written

$$A^- + H^+ \rightleftharpoons HA \qquad 69$$

and the binding constant K is given by

$$K = \frac{HA}{A^- \times H^+} \qquad 70$$

The concentration of the anion A^- is evidently equal to $HA_0 - HA$ where HA_0 is the total amount of acid present, ionized as well as un-ionized. Substituting this information in eq. 70 and rearranging and solving for HA

$$HA = \frac{K \times H^+ \times HA_0}{1 + K \times H^+} \qquad 71$$

Taking the reciprocal of eq. 71 gives

$$\frac{1}{HA} = \frac{1}{K \times HA_0 \times H^+} + \frac{1}{HA_0} \qquad 72$$

If $1/HA$ is plotted against $1/H^+$ a linear relation will be found as shown in Figure 8.

To continue this argument, suppose the weak acid (say acetic) is attached to suspended inert particles. n_0 acetic acid molecules are adsorbed to each particle and

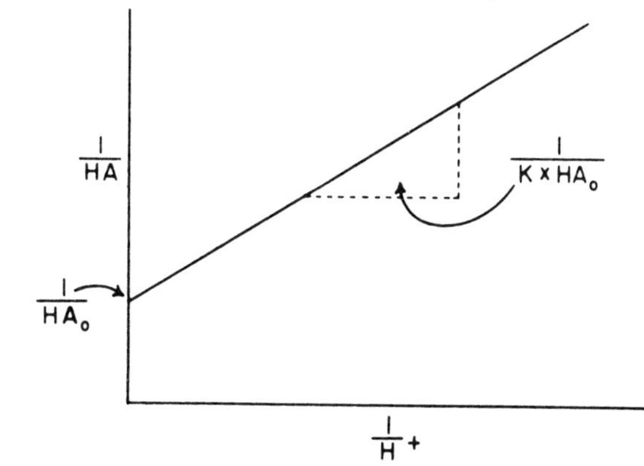

Fig. 8. Plot showing binding of a proton by a weak monobasic acid.

it is further assumed that the acetic acid molecules are sufficiently far apart on the surface of the particle so that they do not interfere with each other. Evidently, it will be impossible to distinguish between a titration curve of the acetic acid attached to the particles and that of a solution of acetic acid in the absence of the particles but containing the same number of acetic acid molecules.

The number of protons bound to each suspended particle having n_0 acetic acid molecules attached would be n. Clearly, in analogy to eq. 71

$$n = \frac{K \times n_0 \times H^+}{1 + K \times H^+} \tag{73}$$

and a plot of $1/n$ against $1/H^+$ would yield a straight line whose intercept on the y-axis is $1/n_0$ and whose slope is $1/K \times n_0$. Equation 73 is a proper expression for multiple binding of ions under the condition that the binding sites are far enough apart so that electrostatic fields do not interfere with each other. In fact, a linear plot of the kind indicated above is clear evidence that the binding sites are independent and further that there is only one intrinsic binding constant, i.e., all the sites are equivalent. A large variety of intrinsic binding constants, such, for example, as is encountered in the binding of water by a protein, introduces complications which can best be handled by other procedures (see Chapter 7, Osmotic Pressure and Related Topics). The equation for binding involving only two intrinsic constants is

$$n = \frac{n_0' \times K_i' \times C}{1 + K_i' \times C} + \frac{n_0'' \times K_i'' \times C}{1 + K_i'' \times C} \tag{74}$$

where n is the number of ions bound, n_0' and n_0'' are the total number of sites having intrinsic constants K_i' and K_i'' respectively. C is the concentration of the ion being bound. The plot of $1/n$ vs $1/C$ will not yield a straight line in this situation.

The question of multiple binding can be regarded in still another way. A dicarboxylic acid can exist as H_2A, HA^- and $A^=$ and the two binding constants (association constants) are

$$K_1 = \frac{H_2A}{H^+ \times HA^-} \quad \text{and} \quad K_2 = \frac{HA^-}{H^+ \times A^=} \tag{75}$$

Suppose there is no interference between the two binding sites. The ionic forms of the dicarboxylic acid can be diagramed as in Figure 9. In the ionized condition $A^=$, the chance of picking up a proton is twice as great as it is when one of the sites is occupied by a proton, HA^-. The chance of losing one proton when both sites are occupied by protons, H_2A, is twice as great as when only one site is occupied by protons, HA^-. It is, therefore, evident that

$$\frac{K_2}{K_1} = 4 \tag{76}$$

By the same line of reasoning, the relation between the association constants for a tricarboxylic acid (three independent binding sites per molecule) is K_1 equals 3 K_i, K_2 equals K_i and K_3 equals $K_i/3$ where K_i is the intrinsic association constant (reciprocal of the dissociation constant) of the acid as measured by the pH of the acid at half neutralization with a strong base. By an extension of the same line of reasoning as employed for two binding sites, it can be shown that the general relation between the intrinsic association constant, K_i, and any constant, K_n, for a polybasic acid containing n_o sites is

$$K_n = \frac{n_o - n + 1}{n} K_i \tag{77}$$

The above is a description of the so-called statistical binding. It is characteristic of statistical binding that the titration curve exhibits only one inflection corre-

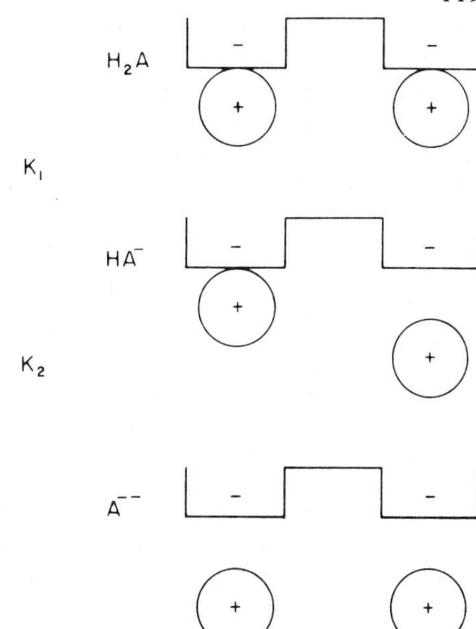

Fig. 9. Diagrammatic representation of the ionic forms of a dicarboxylic acid without interference of the binding sites for protons.

sponding to the intrinsic constant of the acid; there is complete overlap of the individual sites. In order to detect two acid groups in a molecule by titration, the ratio of their ionization constants must be considerably greater than 4 and the ratio must exceed about 60 before the titration curve will show a detectable inflection. Figure 10 shows a plot of log K_1/K_2 for a series of straight chain dicarboxylic acids against the number of CH_2-groups separating the carboxyl groups.

Evidently, the ratio of the first to the second dissociation constant appears to be approaching 4 as the distance between the two carboxyl groups is increased (see

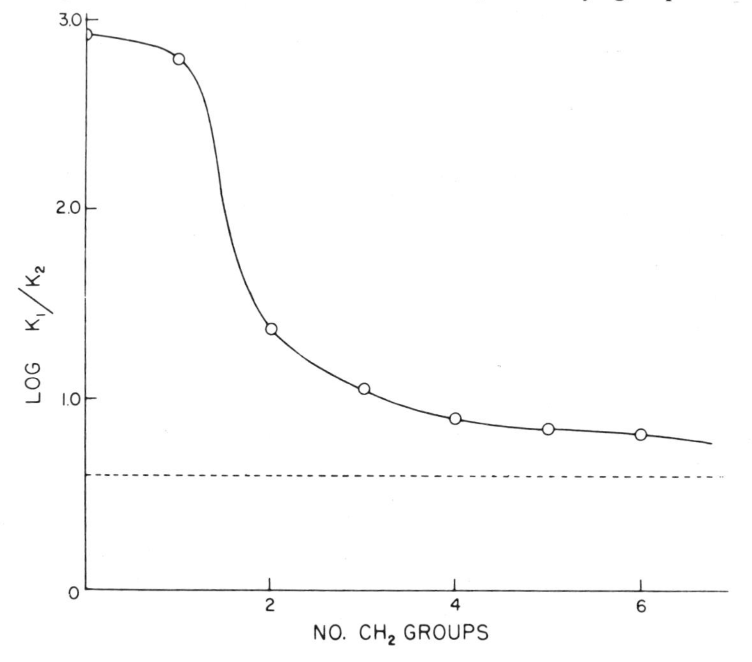

Fig. 10. Plot of log K_1/K_2 of a series of dicarboxylic acids against the number of intervening CH_2 groups. Horizontal dotted line corresponds to a ratio of K_1/K_2 of 4.

Fig. 10) but there is notable departure from this ratio and especially for the shorter chained acids.

The increase of K_1 relative to K_2 above the statistical ratio of 4 to 1 is attributed to electrostatic interaction. That is, the departure of a proton from a dicarboxylic acid leaves a net negative charge on the acid molecule and increases the affinity between the anion and remaining proton.

In addition to electrostatic effects, dicarboxylic acids can exhibit some curious anomalies in respect to dissociation. For example, the dissociation constants of fumaric acid in which the two carboxyls are trans to each other are 9.57×10^{-4} and 4.13×10^{-5} with a K_1/K_2 ratio of 23.2. Maleic acid, on the other hand, in which the two carboxyls are cis to each other has constants of 1.20×10^{-2} and 5.95×10^{-7} with a ratio of K_1/K_2 of 20,162. It seems likely that the monovalent anion of maleic acid is stabilized by internal hydrogen bonding and, accordingly, the second hydrogen is released with difficulty since it requires the rupture of a hydrogen bond to do so; no such internally hydrogen bonded structure is possible for the trans acid (fumaric). Internal hydrogen bonds of this type could occur in proteins and other biopolymers if the protogenic groups are properly spaced in respect to each other.

Electrostatic Effects. In the ionization of a proton from a large spherical central ion (for example, a positively charged protein ion) having n_0 available sites of which n sites are occupied by protons, the reaction is

$$P^{+n} \rightarrow P^{+(n-1)} + H^+ \qquad 78$$

According to eq. 73 the hydrogen ion concentration, H^+, is equal to $K_n/(n_0 - n)$ where K is the dissociation constant instead of the association constant as used in eq. 73. Taking logarithms and converting the hydrogen ion concentration to pH and K to pK and rearranging eq. 73 gives

$$\mathrm{pH} = \mathrm{pK} + \log \frac{n_0 - n}{n} \qquad 79$$

The free energy change for the ionization of the proton can be separated into the intrinsic part, ΔF_i, involving all the specific interacting factors and into an electrostatic part, ΔF_e, and

$$\Delta F = \Delta F_i + \Delta F_e \qquad 80$$

Attention is directed to the change of the ionization constant produced by the electrical field of the central ion. To calculate the change, it is necessary to estimate the work required to move a small ion from the solution to the central ion which is held at a constant potential. This work is simply $\epsilon\psi$ where ϵ is the electrostatic charge on the small ion and ψ is the potential of the central ion and its ion atmosphere. The free energy change for the process is

$$\Delta F = \Delta F_i + \epsilon\psi \qquad 81$$

Since the ΔF of ionization per molecule is equal to $2.3\ kTpK$ where k is the Boltzmann constant, eq. 81 becomes, after substitution and rearrangement,

$$\mathrm{pK} = \mathrm{pK}_i + \frac{\epsilon\psi}{2.3\ kT} \qquad 82$$

The potential of the large spherical central ion is given by the expression for two concentric spheres (Chapter 3, Water and Electrolytes) since the large ion is surrounded by an ion atmosphere of small ions and

$$\psi = \frac{Q(r_2 - r_1)}{Dr_1r_2} \qquad 83$$

where r_2 is the radial distance from the center of gravity of the ion atmosphere to the center of the large ion and r_1 is the radius of the central ion. D is the dielectric constant and Q is the charge on the large ion. From the discussion of the Debye-Hückel theory in Chapter 3, $1/\kappa$ is the distance from the center of gravity of the ion atmosphere to the surface of the large ion. Thus $1/\kappa$ is equal to $r_2 - r_1$. Substituting $1/\kappa$ for $r_2 - r_1$ in eq. 83 and rearranging there results

$$\psi = \frac{Q}{Dr_1\,(\kappa r_1 + 1)} \qquad 84$$

It is the practice to include in eq. 84 the radius of the small ions as being the closest approach of the ions to the large ion; the radius of the large ion is thereby increased by this distance. It appears that this correction in light of all the other approximations is something of an affectation and, accordingly, it is being omitted from the present discussion.

In eq. 84, Q, the charge on the large central ion, is equal to ϵZ where Z is the valence of the large ion. Substituting eq. 84 into eq. 82 along with the value of Q, there results per mole

$$\mathrm{pK} = \mathrm{pK_i} + \frac{Z\epsilon^2 N}{2.3\,RTD}\,\frac{1}{r_1(\kappa r_1 + 1)} \qquad 85$$

It is customary to substitute W in eq. 85 where W is defined as

$$W = \frac{\epsilon^2 N}{2DRT}\,\frac{1}{r_1(\kappa r_1 + 1)} \qquad 86$$

and eq. 85 is written

$$\mathrm{pK} = \mathrm{pK_i} + \frac{2\,WZ}{2.3} = \mathrm{pK_i} + 0.868\,WZ \qquad 87$$

Substituting eq. 79 into eq. 87 and rearranging gives

$$\mathrm{pH} - \log\frac{n_0 - n}{n} = \mathrm{pK_i} + 0.868\,WZ \qquad 88$$

The magnitude of W can be calculated by plotting the left side of eq. 88 against Z, the slope of the line being 0.868 W. The sign of WZ depends upon the ionization process being considered. If the small and large ions have the same sign of charge, WZ is negative whereas, if they bear opposite signs, WZ is positive. Evidently, the value of W will depend on the size and shape of the large central ion as well as on the ionic strength. Inspection of eq. 86 reveals that for a given valence, temperature and ionic strength, the larger the ion, the smaller is W. Since the effect of asymmetry would be to increase the surface of the particle and for a given charge the distances between the surface charge would increase, increasing the asymmetry of the large polyvalent ion also tends to decrease the value of W. Increasing the ionic strength also decreases W; the increase in electrolyte concentration provides greater screening for the central ion and decreases the electrostatic effects. Z represents the net charge on the central ion; if anions as well as protons are taken up by the central ion, Z is the difference between the bound protons and anions.

Shown in Figure 11 is a plot of the charge Z of a positively charged polyvalent ion (protein) resulting from the addition of a dilute solution of a strong acid to the protein solution as a function of the pH of the solution with and without electrostatic work at two different ionic strengths; evidently the electrostatic effect can be of considerable importance.

The discussion of the effect of the electrostatic fields on the proton uptake as presented is a very formal one and there are several complicating factors in any real situation. To begin with it has been necessary to assume regular geometrical shapes, i.e., spheres, rods and plane surfaces. If the actual shape departs from these, the

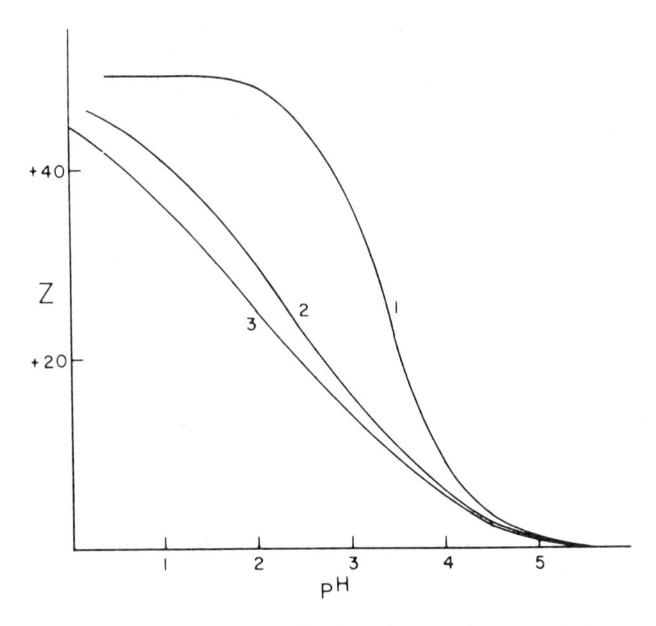

Fig. 11. Calculated proton binding of a spherical protein ion as a function of pH. Molecular weight 35,000; density 1.35 in water at 25°C; intrinsic pK_i is 3.40 and the maximum binding is 50. Curve 1, no electrostatic interaction. Curve 2, ionic strength 0.10, W is 0.05. Curve 3, ionic strength 0.05, W is 0.063.

argument fails in any rigorous sense. It is also necessary to assume a uniform distribution of charge on the large central ion and the influence of individual ion sites has been neglected. Tanford and Kirkwood have considered the situation in which the charges are not uniformly distributed over the surface of the central ion and conclude that the electrostatic work of charging is an important function of the configuration of the binding sites.

An important limitation to the electrostatic theory (so-called smeared site theory) outlined above is the necessity of assuming the existence of only one intrinsic ionization constant over the pH range covered by the plot of eq. 88 in the estimation of the electrostatic factor w. Two or more intrinsic constants spaced fairly close together can produce an increase in the calculated value of w and lead to significant error.

It has not been explicitly stated but the calculation of the electrostatic work is subject to the same limitations as are the Debye-Hückel expressions, i.e., the potential of the central ion must be small relative to kT and at room temperature should not exceed about 25 ordinary millivolts.

We will return to a discussion of the electrostatic factor in the section on protein titration.

Amino Acids. The amino acids, proteins, phospholipids and a few other substances form zwitter ions at the proper pH in which the molecule contains an equal quantity of negatively and positively charged groups. These charges have a discrete separation resulting in a high dipole moment of the molecule, although the dipoles in a protein tend to cancel each other resulting in a moderate overall dipole moment for these larger molecules. Consider the ionization of glycine as shown in Figure 12.

The ionization constants K_1 and K_2 can be written by inspection and

$$K_1 = \frac{H^+ \times {}^+A^-}{A^+} \text{ and } K_2 = \frac{H^+ \times A^-}{{}^+A^-} \qquad 89$$

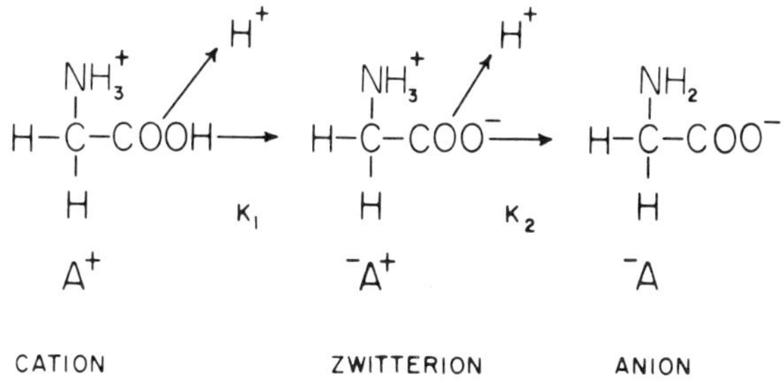

$$\underset{\text{CATION}}{\underset{A^{+}}{H-\overset{\overset{+}{NH_3}}{C}-COOH}} \xrightarrow[K_1]{H^+} \underset{\text{ZWITTERION}}{\underset{^{-}A^{+}}{H-\overset{\overset{+}{NH_3}}{C}-COO^-}} \xrightarrow[K_2]{H^+} \underset{\text{ANION}}{\underset{^{-}A}{H-\overset{NH_2}{C}-COO^-}}$$

Fig. 12. The sequence of ion forms of glycine starting from a solution of high acidity on the left and progressively adding alkali to give the ion forms shown to the right.

where A^+, $^-A^+$ and ^-A represent the concentrations of the various ion forms shown in Figure 12. The net charge on $^-A^+$ is zero and this is the zwitter ionic form; the pH at which this form has its maximum concentration is known as the isoelectric point. The pH of the isoelectric point (I.P.) can be calculated as follows: The product of the constants K_1 and K_2 is

$$K_1K_2 = \frac{(H^+)^2 \times A^-}{A^+} \tag{90}$$

At the isoelectric point A^- is equal to A^+ and rearranging eq. 90 gives

$$H^+ = (K_1 \times K_2)^{1/2} \tag{91}$$

Converting eq. 91 to pH and pK values yields

$$pH = I_P = \frac{pK_1 + pK_2}{2} \tag{92}$$

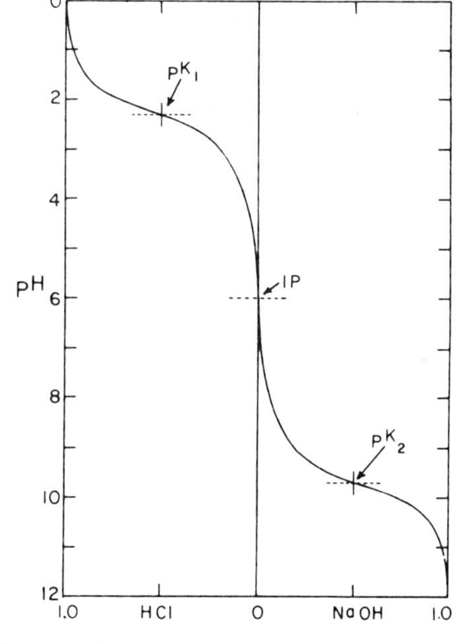

Fig. 13. Acid-base titration curve of glycine.

Shown in Figure 13 is a titration curve of the amino acid glycine. By setting the sum of the concentrations of all the ionic forms of the amino acid (A^+, $^-A^+$, ^-A) equal to unity and substituting the expressions for the ionization constants of the amino acids in eq. 89, the fraction of the amount of each ionic form of the amino acid can be calculated as a function of pH; the result of such a calculation is shown in Figure 14. It is to be noted from Figure 14 that the zwitter ionic form ($^-A^+$) predominates over a fairly wide range of pH; perhaps one should speak of an isoelectric zone instead of an isoelectric point.

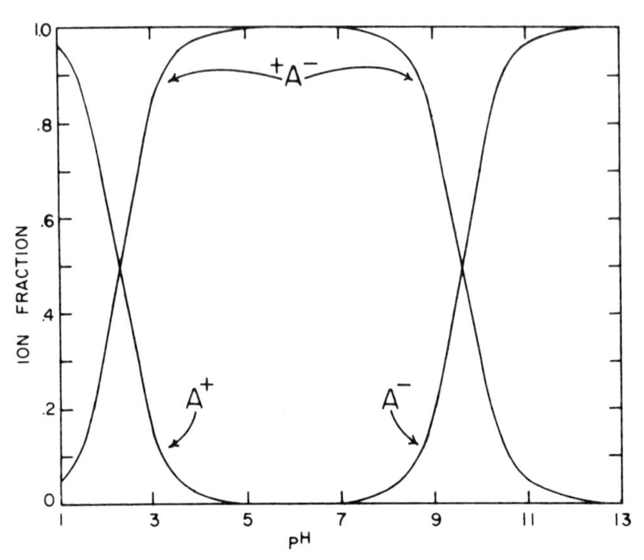

Fig. 14. Ionic forms of the amino acid glycine as a function of pH.

The amino acids lysine, arginine, histidine, glutamic acid, aspartic acid, tyrosine and cystine all have three groups capable of ionization and, accordingly, each has three pK values. Aspartic acid exhibits a titration curve shown in Figure 15. The ionic forms of aspartic acid are illustrated in Figure 16.

pK_1 and pK_2 of aspartic acid are fairly close together, being 1.88 and 3.65 respectively, and consequently there is an overlap between the release of these two protons. Inspection of Figure 15 does, however, reveal a distinct inflection between the two constants. The isolectric point of aspartic acid is given by $(pK_1 + pK_2)/2$.

The ionization of the sulfhydryl and phenolic groups represents a somewhat different ionization problem than do the amino groups. Both of these groups ionize in the alkaline region as does the amino group, but, unlike the ionization of the amino, their ionization produces a negative site on the amino acid.

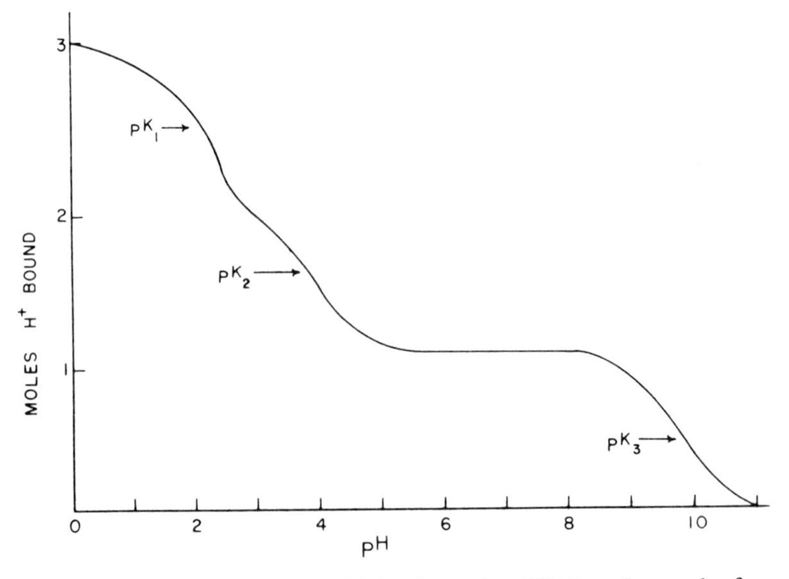

Fig. 15. Titration of aspartic acid showing moles of H^{+} bound per mole of amino acid as a function of pH.

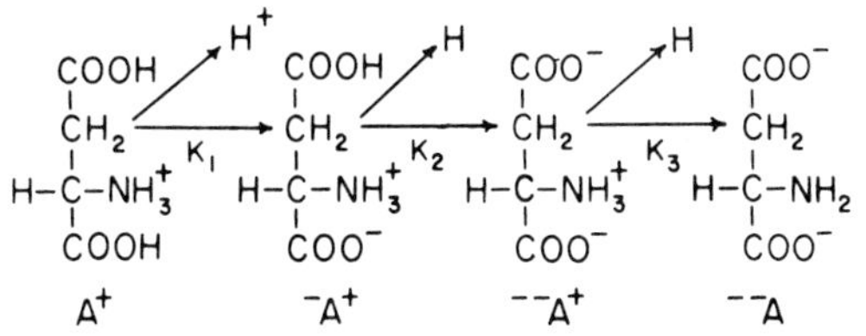

Fig. 16. Ionic forms of aspartic acid (compare with Fig. 12).

Benesch and Benesch studied the ionization of cysteine with the aid of ultraviolet absorption; the ionized sulfhydryl group shows strong absorption at a wavelength of about 240 mμ whereas the un-ionized group does not absorb in this region. As noted, cysteine has three dissociable protons and the various ionic forms of this amino acid bear a complex relationship to each other (see Fig. 17). Shown in Figure 18 are the ion fractions of the various forms of cysteine as calculated from the measured ionization constants. Analysis of the titration and spectrophotometric results yields the following provisional constants for the ionization of cysteine (see Fig. 17 for the identity of the constants): pK_1 1.8; pK_2 8.5; pK_3 8.9; pK_4 10.4; pK_5 10.0.

Table 4 shows the individual pK values for the amino acids.

TABLE 4. SUMMARY OF pK VALUES OF AMINO ACIDS

Amino Acid	pK_1	pK_2	pK_3
Aspartic	1.995	3.910	10.006
Glutamic	2.19	4.25	9.67
Histidine	1.82	6.00	9.17
Lysine	2.16	9.18	10.79
Arginine	1.81	9.01	12.5
Tyrosine	2.20	9.1(NH_2)	10.95(OH)
Cysteine	1.8	8.3	10.8

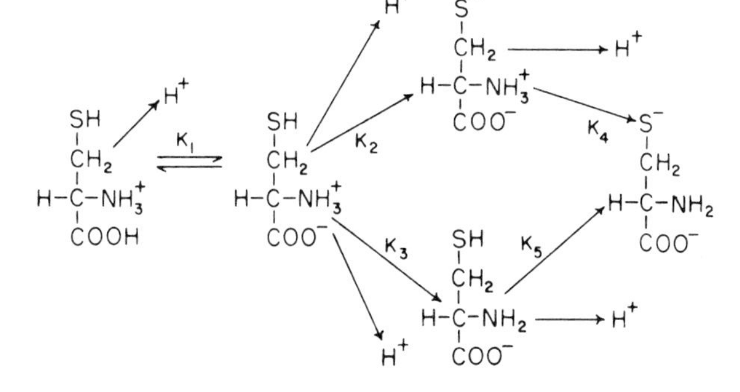

Fig. 17. Ionic forms of the amino acid cysteine.

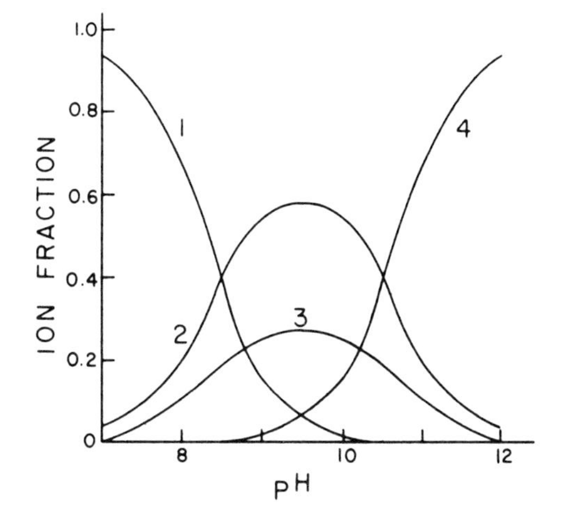

Fig. 18. Concentration of ionic forms of cysteine as a function of pH; calculated from ionization constants. Curve 1, HS-R-NH_3^+; Curve 2, $^-$S-RNH_3^+; Curve 3, HS-R-NH_2; Curve 4, $^-$S-R-NH_2. (R. E. Benesch and R. Benesch: J. Amer. Chem. Soc. *77:* 5877, 1955.

Protein Titration. Proteins consist of many different kinds of amino acid residues connected through peptide bonds to produce long polypeptide chains. Except for the ends of the chains the α amino and α carboxyl groups are involved in peptide bond formation and are unavailable for titration. Several of the amino acid residues, however, contain protogenic groups and these are the groups which participate in an acid-base titration of proteins. The residues responsible are listed in Table 5. In water soluble native protein molecules, the peptide chains have a highly folded specific arrangement leading usually to compact macromolecules with low degree of asymmetry.

It is usual to conduct a titration of a protein in the following manner: To a series of flasks is added a given volume of protein solutions of known concentration. The volume of each flask is then increased to a total and final volume by the addition of water, of the standard acid (usually HCl) or base (usually NaOH) plus the required amount of neutral salt such as NaCl to yield the desired ionic strength, the ionic strength being the same in all the flasks. The estimation of the ionic strength presents a degree of ambiguity, especially in the more acid or alkaline solutions, because a certain number of ions are bound by the protein. It is customary to compute the ionic strength from the sum of the concentrations of added electrolyte, the fact that some of the ions are bound being ignored. The protein solutions are brought to temperature and the pH measured with a glass electrode. A blank titration is then conducted, the protein being omitted from the solution (the ionic activities of the blank solutions can also be obtained from standard reference texts). Care must be exercised to exclude carbon dioxide in the alkaline regions from both the protein and blank solutions.

The number of equivalents of acid or base added are then plotted on a large scale graph against the pH of the solution. The differences in equivalents between the protein and the blank solutions at selected pH values are interpolated. These differences represent the protons bound by the protein as a function of pH; the activities of the small ions in solution at a given pH are assumed to be identical in the presence and absence of protein.

The titration curve of a protein reflects faithfully the ionogenic groups contributed by the amino acid residues; the titration curve of a protein can be predicted with fair accuracy from a knowledge of the chemical composition of a protein. Shown in Table 5 are the pK ranges and heats of ionization of the titratable groups in proteins.

The proton binding curves of many proteins have been measured; some with great care and attention to detail. Shown in Figure 19 is the titration curve of bovine serum albumin according to Tanford, Swanson and Shore. It is to be noted that increasing the ionic strength rotates the titration curve about the zero charge in a clockwise fashion; this is a reflection of the fact that the electrostatic work of charging the protein ion decreases as the ionic strength increases.

If all small ions are removed from a protein solution other than hydrogen or hydroxyl ions arising from water, the protein is said to be isoionic. The ions can be removed by dialysis or by an appropriate resin column. The isoelectric point is defined as the pH at which the net charge on the protein is zero. It is not meaningful to define the isoelectric point as the pH at which the electrophoretic mobility is zero because, as will be shown in Chapter 13, Electrophoresis and Electrokinetic Potentials, in general, zero electrophoretic mobility does not coincide with zero net charge. Perhaps one should speak of the isoelectrophoretic point as contrasted with the isoelectric point.

The addition of a neutral electrolyte such as NaCl will, in general, cause the pH of an isoionic protein solution to shift. If the sodium ions are bound, HCl is in effect released into solution and the pH drops, whereas if the chloride ions are bound the pH rises due to the NaOH produced. It is to be noted that the binding of either the cations or the anions changes the net charge on the protein and the isoelectric point

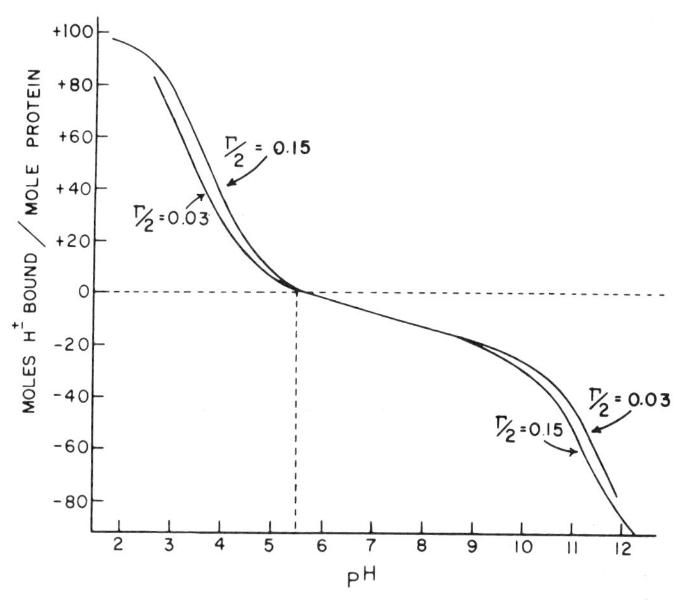

Fig. 19. Proton binding of bovine serum albumin at the indicated ionic strength and at 25°C. (C. Tanford, S. A. Swanson and W. S. Shore: J. Amer. Chem. Soc. *77*:6414, 1955.)

decreases with binding of anions and increases with the binding of cations.

TABLE 5. PK INTERVALS AND HEATS OF IONIZATION CONTRIBUTED BY AMINO ACID RESIDUE IN A PROTEIN

Titratable Groups	*Amino Acid Residues*	*pK Range*	*Heats of Ionization (cals/mole)*
α-Carboxyls	Any	3.0 to 3.6	−1,500 to 2,000
Distalcarboxyls	Aspartic and glutamic	3.0 to 4.7	−1,500 to 2,000
Imidazole	Histidine	5.6 to 7.0	7,000
α-Amino	Any	7.5 to 8.4	9,000 to 10,000
ε-Amino	Lysine	9.4 to 10.6	11,000 to 12,000
Guanidine	Arginine	12.0 to 13.0	12,000 to 13,000
Phenolic hydroxyl	Tyrosine	9.8 to 10.4	6,000
Sulfhydryl	Cysteine	9.1 to 10.8	6,500

A study of the shift of the pH of a protein solution with temperature permits the evaluation of the apparent heats of ionization of the protein as a function of pH as described previously (eq. 62). A comparison of the heats of ionization of the protein as determined by experiment along with a knowledge of the respective pK values of the available residues leaves little doubt as to the identity of the principal groups being titrated in a given pH range.

The net charge (Z) on the protein is obtained as the sum of all ions bound to the protein and

$$Z = H^+ + M^+ - A^- \qquad 93$$

where H^+, M^+, and A^- indicate the number of moles of protons, metal cations and anions respectively bound per mole of protein. To calculate the electrical work of charging the protein ion the net charge, Z, is to be used.

The net charge, Z, is the average charge on the protein. The protein ions, however, are in equilibrium with each other and at any given instant there are a variety of charged species present. Thus the square of the average value of Z differs from the square of Z for any given molecule and this difference is related to the slope of the titration curve by the expression

$$\frac{1}{2.3}\frac{dZ}{dpH} = Z_a^2 - Z_i^2 \qquad 94$$

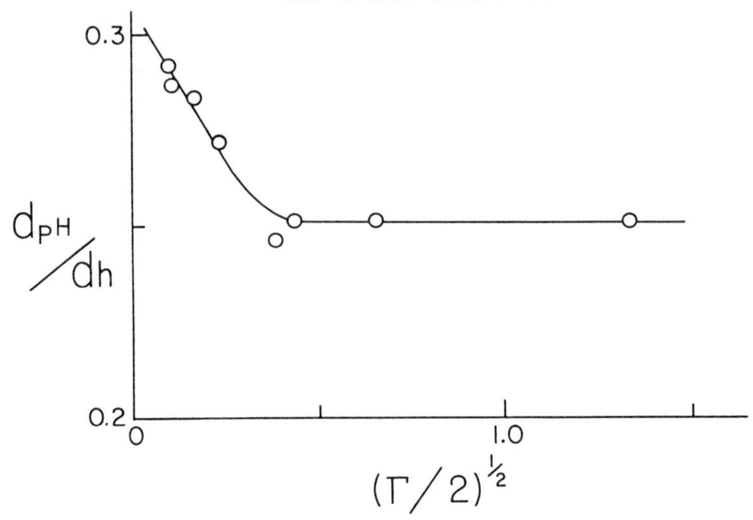

Fig. 20. Reciprocals of the slopes of the titration curves of ribonuclease at half neutralization of the carbonyl groups plotted against the square roots of the ionic strength. h is given in moles of hydrogen ions bound per mole of ribonuclease. (H. B. Bull and K. Breese: Arch. Biochem. Biophys. *110:* 331, 1965.)

where Z_a is the average net charge and Z_i is the net charge on any given molecule at a given time.

The effect of the electrical double layer (smeared site theory) around the central protein ion is to spread the titration curve of the protein and to decrease its slope. The comparison of the reciprocals of the slopes of the titration curves at a series of ionic strengths should, therefore, provide a measure of the importance of the electrostatic potential of the protein ion relative to the work of charging the central ion. Figure 20 shows such a plot for ribonuclease at half neutralization of the carbonyl groups.

Inspection of Figure 20 reveals that the slope of the titration curves becomes independent of the ionic strength at ionic strengths of 0.15 and above. It appears that, at this ionic strength and greater, the electrostatic potential of the central ion is no longer of importance as far as titrations are concerned and the smeared site theory is not applicable at higher ionic strengths; at and above 0.15, the approaching hydronium ion is concerned only with individual charged sites.

In the complete absence of electrostatic effects arising from the double layer of the protein ion and if there is only one intrinsic ionization constant, the buffer capacity of the protein should be given by (see eq. 53)

$$\beta_M = \frac{2.3\ K_i\ \mathrm{H}^+ n}{(K_i + H^+)^2} \qquad 95$$

where β_M is the molar buffer capacity, K_i is the intrinsic ionization constant and n is the number of groups being titrated per mole of protein with the intrinsic constant K_i. Figure 21 shows a plot of the molar buffer capacity of ribonuclease at an ionic strength of 1.80, at which the electrostatic effects for the central protein ion as a whole have surely been abolished.

The results shown in Figure 21 indicate a considerable dispersion of the apparent intrinsic carboxyl and, to a lesser extent, the imidazole constants of ribonuclease.

Provided the ionizing sites of a protein molecule are independent of each other, the dissociation curve of a protein should be given by the series

$$N = \sum_{n=1}^{n=n_o} \frac{K_n}{\mathrm{H}^+ + K_n} \qquad 96$$

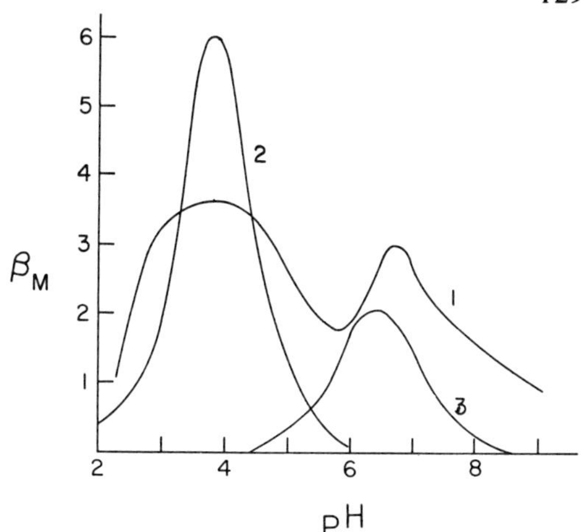

Fig. 21. Molar buffer capacity of ribonuclease as a function of pH. Ionic strength 1.80. Curve 1 measured from experimental titration curve; curve 2 calculated for 11 carbonyl groups $K_i = 3.80$; curve; 3 calculated for 4 imidaloze groups $K_i = 6.4$. (H. B. Bull and K. Breese: Arch. Biochem. Biophys. *110:* 331, 1965.)

where N is the number of sites which have released a proton, n_o is the total number of sites considered, K_n is the intrinsic ionization of the nth site. It is possible to obtain a numerical solution of the series given in eq. 96 by the use of a computer (this is an exercise in curve fitting). Shown in Figure 22 are the results of such an analysis of the titration curves of lysozyme, ribonuclease and of cytochrome C; evidently, there is a wide dispersion of the ionization constants of these proteins, wider than indicated in Table 5.

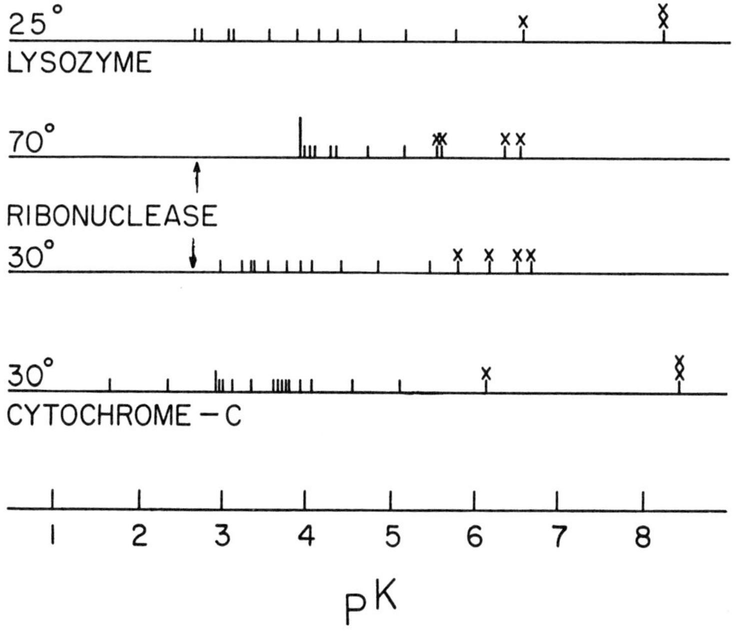

Fig. 22. pK values of lysozyme at an ionic strength of 0.15 and at 25°, of ribonuclease at an ionic strength of 0.20 at 30° and at 70°, and of cytochrome C at an ionic strength of 0.20 at 30°. Unmarked refer to carboxyl ionization. One asterisk, imidazole. Two asterisks, amino. Higher vertical lines indicate overlapping constants. (H. B. Bull and K. Breese: Arch. Biochem. Biophys. *117:* 106, 1966.)

The effect of the dispersion of intrinsic ionization constants is to cause an overestimation of W in eq. 88. In fact, a dispersion of the intrinsic ionization constants as shown in Figure 22 would give rise to an apparent W even in the complete absence of electrostatic effects. It is best to estimate W from eletrophoresis (see Chapter 13) rather than by a plot of eq. 88.

In many cases the titration curve of a protein is completely reversible over wide ranges of pH and furthermore the curve is independent of time. Hemoglobin, however, in the acid range yields an irreversible and time dependent course as shown by the work of Steinhardt and Zaiser (and illustrated in Fig. 23). The titration behavior of hemoglobin is probably related to the tendency of this protein to split into halves in an acid medium and to the dissociation of heme from the protein.

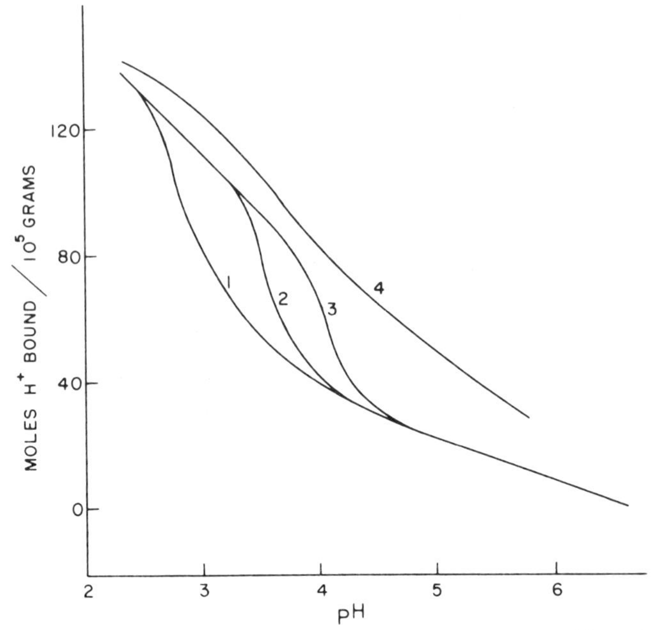

Fig. 23. Titration curve of carboxyhemoglobin. Total chloride 0.02 M at 25°C. Curve 1, 3 seconds; curve 2, 10 minutes; curve 3, 24 hours; curve 4, back titration. (J. Steinhardt and E. Zaiser: J. Biol. Chem. *190:* 197, 1951.)

Actually, pH dependent conformational changes in proteins frequently occur. Probably all proteins, if made sufficiently acidic, will undergo changes in structure. Conformational changes at more moderate acidification have been established for cytochrome C, bovine serum albumin and β-lactoglobulin. If the structural changes suffered by a protein are sufficiently drastic, the titration curve will be significantly altered. For example, the denaturation of a protein always leads to profound changes in the titration curve of the protein. The creation of a large net charge on the protein molecule leads to instability of the structure.

Titration of Nucleic Acids. The titration behavior of the nucleic acids has not been investigated in the depth that proteins have been studied in this respect. The titration of nucleic acids is usually accompanied by irreversible changes in the titration, thus strongly indicating profound structural changes in the molecules. The nucleic acids contain a smaller variety of titratable groups. DNA contains the purine and pyrimidine bases adenine, guanine, cytosine and thymine. These provide amino groups with pK near 4 and hydroxyl groups with pK near 11. The exposed phosphate group has a low pK and does not titrate. The charge on the nucleic acid in the neutral range is, therefore, negative and is due to the ionized phosphate. Table 6 gives the approximate pK values of bases (and phosphate) as they occur in nucleic acids. Figure 24 shows the forward and backward titration of DNA and RNA.

Ion Binding in General. Emphasis has been placed on proton binding by proteins. Both proteins and nucleic acids, however, bind many other ions both organic and inorganic and some of these interactions are of great interest and importance. In principle, the theory of proton binding applies equally well to ion binding in general; obviously, different experimental techniques must be employed.

TABLE 6. APPROXIMATE pK VALUES IN THE NUCLEIC ACIDS

Component	pK_1	pK_2
Phosphate	1.0	
Uracil	5.9 (OH)	9.4 (OH)
Thymine	4.0 (OH)	11.4 (OH)
Cytosine	4.2 (NH_2)	11 (OH)
Adenine	3.7 (NH_2)	
Guanine	2.4 (NH_2)	9.3 (OH)

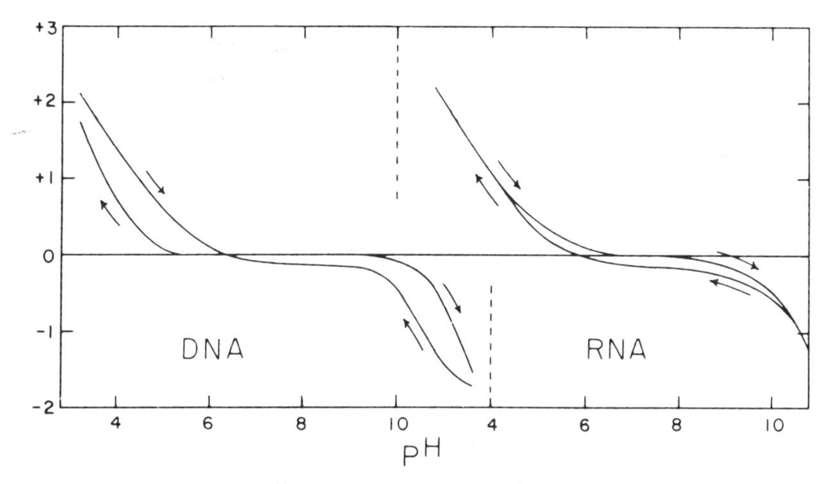

Fig. 24. Forward and back titration of herring sperm DNA ionic strength 0.05 and microsomal RNA in 0.1 M NaCl. Starting at or near pH 7.0. Ordinate indicates moles hydrogen ions bound for 4 gram-atoms of phosphorous. (DNA plot adapted from A. R. Peacocke and B. N. Preston: J. Chem. Soc. 2783, 1959; RNA plot adapted from U. Z. Littaver and H. Eisenberg: Biochim. et Biophys. Acta *32:* 320, 1959.)

If there is significant electrostatic interaction or if there is more than one intrinsic binding constant the plot of $1/n$ against $1/C$ will not be linear (see eq. 73). Scatchard has suggested a plot which reveals this departure from linearity more critically. Equation 73 is rearranged and C substituted for the concentration of hydrogen ions to give

$$\frac{n}{C} = K\,(n_0 - n) \qquad 97$$

The plot of n/C against n will give a straight line for statistical binding in the absence of interacting sites. This line will have a negative slope which is equal to $-K$ and the intercepts on the x and y axes will be Kn_0 and n_0 respectively. The merit to this plot is that the plot becomes more sensitive as the binding increases.

To measure ion binding, some method must be available to estimate the concentration of the ion in equilibrium with the protein or to measure directly the concentration of the protein-ion complex. There are a considerable number of methods available, some of which are more suitable than others for a given situation. The ones which are more apt to have wider utility are: (1) equilibrium dialysis, (2) ultrafiltration, and (3) change in light absorption spectrum.

In equilibrium dialysis the protein solution, enclosed in a sausage casing, is dialyzed against a solution of the ion under investigation. The protein molecules are too large to pass through the pores of the sausage casing membrane, whereas the smaller ions under investigation can and do pass. In due time the protein solution reaches equilibrium with the outside solution. With a knowledge of the total amount of protein and of small ion and, by analytical methods, the distribution of the small ion between the inside and outside of the sausage casing, the ion binding can be calculated. A correction has to be made for the Donnan equilibrium and for the amount of adsorption on the surface of the membrane.

In ultrafiltration, the solution in equilibrium with the dissolved protein is forced by hydrostatic pressure through a membrane whose pores are too small to permit the passage of the protein molecules. The ultrafiltered solution is analyzed and the ion binding calculated as is done in equilibrium dialysis. The method is subject to the same corrections as is the equilibrium dialysis method. A troublesome feature is the fact that the protein is concentrated relative to the small ion as filtration proceeds.

The absorption spectrum of a compound is apt to be shifted both in respect to intensity and wavelength when it is removed from solution and attached to a protein. This method is especially useful for compounds which absorb in the visible range (dyes). The method will receive further comment in Chapter 8, Solution Optics.

GENERAL REFERENCES

Bates, R. G.: Determination of pH. John Wiley and Sons, New York, 1964.

Steinhardt, J. and S. Beychok: Interaction of proteins with hydrogen ions and other small ions and molecules. *In* The Proteins, Academic Press, New York, 1964.

Steinhardt, J. and J. A. Reynolds: Multiple Equilibria in Proteins. Academic Press, New York, 1970.

Tanford, C.: Physical Chemistry of Macromolecules. John Wiley and Sons, New York, 1961.

Tanford, C.: The interpretation of hydrogen ion titration curves of proteins. Adv. Prot. Chem. *17*:69, 1962.

PROBLEMS

1. (a) Give in detail the steps involved in the calibration of the pH-scale.
   (b) What ambiguity or ambiguities are present in this calibration?

   Ans: See text.

2. (a) Derive the expression for the pH of an aqueous solution of a salt of a weak acid and a strong base in terms of the salt concentration.
   (b) Calculate the pH of 0.01 M sodium acetate.

   Ans: (a) See text; (b) 8.37.

3. (a) Calculate the grams of $NaH_2PO_4 \cdot H_2O$ and of $Na_2HPO_4 \cdot 12H_2O$ required to make one liter of buffer at pH 7.0 and with an ionic strength of 0.05.
   (b) Add 50 cc. of 0.05 M NaOH to the buffer in (a) and calculate the resulting pH. What would be the ionic strength of the new buffer?

   Ans: (a) Na_2HPO_4 4.06 grams, NaH_2PO_4 2.21 grams
   (b) pH 7.16, ionic strength 0.0524.

4. The amount of phosphate, expressed as HPO_4, in a urine sample of a patient was found to be 450 milligrams percent. The volume of the 24 hour sample was 1500 cc. and the pH of this sample was 6.21. Calculate the base conserved by the body expressed in ml. of normal alkali.

   Ans: 37.8 ml. alkali.

5. The pH of a given sample of blood plasma after exposure to a certain partial pressure of carbon dioxide was found to be 7.60. The total CO_2 expressed in volume percent was 87. Calculate the pressure of CO_2 used. What are the circumstances which could have produced this condition in the patient?

   Ans: (a) $P_{CO_2} = 39.9$ mm. Hg; (b) ingestion excess $NaHCO_3$.

6. (a) The pK_a values for glutamic acid are 2.19, 4.25 and 9.67. Assume a 0.1 M solution of glutamic acid and draw an approximate titration curve of this acid with 0.1 M NaOH.
   (b) Identify the pK_a values in terms of the groups responsible for proton release.

   Ans: See text.

7. 50 cc. of 0.1 N HCl were added to 100 cc. of 0.1 N tris at 25°C and the pH was found to be 8.38. The heat of ionization of tris is known to be 11,700 calories. What is the pH of this solution at 37°C.

   Ans: 8.05.

8. The pK_a values for arginine are 2.17, 9.04, and 12.48. Calculate the relative amounts of the ionic forms of this amino acid at pH 9.04.

   Ans: 0.5 each of $^{-}A^{+}$ and $^{-}A^{2+}$.

9. The pK values for the dissociation constants of citric acid are 3.13, 4.76, 6.40 respectively. Calculate and plot the distribution of the ionic forms of citric acid as a function of pH.

   Ans: Too extensive to give.

10. Derive expressions for the ionization of acid groups on a plane surface and for two concentric cylinders* taking into consideration the electrostatic charge.

    Ans: See text. *See T. L. Hill: Arch. Biochem. Biophys. *57:* 229, 1955.

11. Assume that a spherical protein molecule whose molecular weight is 35,000 and whose density is 1.35 is able to bind 50 protons when fully titrated and the intrinsic pK value of the groups being titrated is 3.40. The titration is carried out at an ionic strength of 0.05. Draw the titration curve with and without electrostatic interaction at 25°.

    Ans: See Figure 11.

12. The following is the partial (but sufficient for your purposes) amino acid analysis of β-lactoglobulin whose molecular weight is 35,000 and whose isoionic point is at 5.20.

Amino Acid	*Percent*	*Molecular Weight*
Leucine	15.6	131
Glycine	1.4	75
Arginine	2.88	174
Histidine	1.58	155
Lysine	11.4	146
Aspartic	11.4	133
Glutamic	19.5	147
Amide	1.31	17
Serine	5.00	105

Calculate the number of residues of each amino acid per mole of protein and construct a titration curve for this protein. This protein contains two peptide chains.

Ans: Too extensive to give. However, see experimental curve in R. K. Cannan, A. H. Palmer, and A. C. Kibrick: J. Biol. Chem. *142:*803, 1942. Adjust molecular weight of protein.

CHAPTER 6

Biopolymers

There occurs in living tissue a vast array of high molecular weight compounds which are appropriately called biopolymers. The biopolymers perform numerous physiological functions such as: structure of cells and tissues, catalysis (enzymes), food source, osmotic gradients, locomotion (muscles), hormonal control, immune reactions (γ-globulin), storage of genetic information (nucleic acids) and probably others not yet identified.

The structures of nucleic acids and of proteins are directly controlled genetically and are highly specific. By contrast all other biopolymers are synthesized and degraded in tissue through the action of enzymes leading to simpler structures with a wide molecular weight dispersion.

A polymer is a multiple of a monomer and, if the multiplication of the monomer is sufficiently large, it is a high polymer indicating a considerable molecular weight and a high degree of polymerization; thus rubber is a high polymer of isoprene and starch is a high polymer of glucose. A copolymer is a high polymer consisting of two or more chemically different monomers; a protein is a copolymer of amino acids.

Polymers can be linear with the monomeric units arranged in chains as in cellulose or, by cross linking, the chains can form network polymers as are formed by the keratin chains in fingernails. The polymers may have many branched chains, for example, glycogen and amylopectin; these are known as branched polymers. Obviously, the form which the polymer assumes (linear, network or branched) has a profound effect upon the properties of the polymer; amylose, which is the linear starch polymer, forms stiff gels, whereas amylopectin, the branched polymer, forms only viscous solutions.

A linear polymer of high molecular weight and with complete freedom of rotation about valence bonds will assume the shape of a random coil. The random coil is the most probable shape of the molecule and is analogous to the shape described by the random flight of an insect; it is also the shape of the volume element defined by the random translational motion of a molecule as the result of Brownian motion (heat motion). The only restriction here is that whereas an insect can occupy a position which he has previously occupied, obviously no two or more parts of a flexible molecular chain can be in the same place simultaneously. Shown in Figure 1 is a two-dimensional projection of a random coil.

The configuration of a random coil is a matter of probability and there are many structures simultaneously present in a solution. The average distance r between the

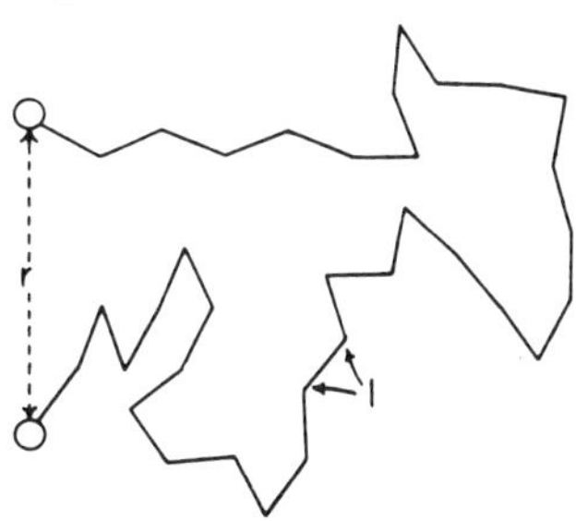

Fig. 1. Two dimensional projection of a random coil. l is the length of each segment and r is the distance between the two ends of the flexible molecule.

ends of such a polymer chain is given by

$$(\bar{r^2})^{1/2} = ln^{1/2} \quad 1$$

or

$$\bar{r^2} = nl^2 \quad 2$$

where $(\bar{r^2})^{1/2}$ is the root mean square of the distance between the ends of the polymer chain. l is the length of each link (segment) in the polymer and n is the number of bonds per molecule. The relationships shown in eqs. 1 and 2 yield the probable distance between any two segments in a freely jointed chain if the number of chain segments between them is sufficiently large.

Another important measure of the effective size of a randomly coiled molecule is the root-mean-square distance of the elements of the chain from its center of gravity. This quantity is designated by $(\bar{S^2})^{1/2}$ and is referred to as the radius of gyration of the molecule and for freely jointed chains

$$(\bar{S^2})^{1/2} = \left(\frac{\bar{r^2}}{6}\right)^{1/2} \quad 3$$

or

$$\bar{S^2} = \frac{nl^2}{6} \quad 4$$

The random coil represents the most probable state of a molecule and, accordingly, in this condition the entropy of the molecule is at a maximum. To stretch the molecule requires that entropy of the polymer be decreased and energy must be expended to do this.

No real valence bonds show complete freedom of rotation, thus even polymethylene exhibits favored positions of the carbon atoms relative to each other. The potential energy passes through 3 equal maxima and 3 equal minima for each complete rotation of the valence bond. For ethane the energy humps are about 3,300 calories each. Figure 2 is a diagrammatic representation of the energy changes accompanying rotation about a carbon-carbon bond.

If there are double bonds in the polymer chain or if massive groups are attached along the chain, the rotations about the valence bonds are severely restricted. This kind of restriction amounts to decreasing n, the number of links and of increasing l, the length of each link. As can be seen in eq. 1, the effect of hindered rotation, whatever may cause it, is to increase the value of r for a given molecular weight of the polymer; the random coil opens up; if it opens up completely it becomes a rod.

A random coil is thus subject to a number of modifying factors. A good solvent, i.e., a solvent which tends to attach itself to the polymer, tends to swell the coil making $\overline{r^2} > nl^2$, whereas a poor solvent tends to shrink the coil; an "ideal" solvent leaves the coil unchanged.

Rubber comes closest of any of the biopolymers of fulfilling the conditions for a random coil. Rubber consists of long chains of 500 to 5,000 isoprene units polymerized in the cis configuration about the repeating double bonds. It contains no polar groups or massive side groups to hinder rotation about single valence bonds,

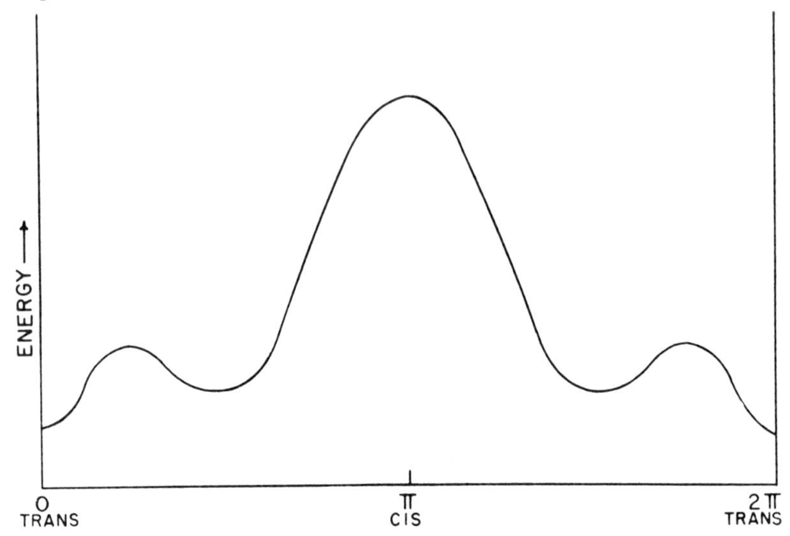

Fig. 2. Diagrammatic representation of the energy change for the rotation around a carbon-carbon bond between two adjacent segments. The trans configuration is more stable than the cis.

and such chains possess great flexibility. When the rubber chains are cross-linked at wide intervals by disulfide bonds (vulcanization) to prevent slippage of the isoprene chains relative to each other, the rubber is capable of long range and reversible extension under load. Upon stretching, the entropy of the rubber is decreased and, accordingly, its contraction is a spontaneous process. The study of rubber-like elasticity has greatly increased the understanding of the molecular basis of the elasticity of living tissue.

The stiffness of the linear polymer chain leads to departure from a random coil as noted above. There are other and more important causes for departure from a random coil. Intramolecular interactions such as those which arise from electrostatic forces and hydrogen bonding can and do produce tremendously specific structures in which randomness of configuration is absent such, for example, as occurs in native proteins and nucleic acids. There can be, however, significant interaction between parts of a flexible chain without the formation of a specific structure. Such a structure results when proteins and nucleic acids are denatured; there is a considerable measure of randomness in the folding of the polymer chains. They do not, however, form random coils; the entropy of configuration is not maximal. It is best to refer to them as disordered coils as distinguished from random coils.

Mention has been made of the forces causing departure from randomness and it is perhaps worthwhile to review briefly the nature of these molecular forces.

Hydrogen Bonds. One of the more interesting of chemical bonds as far as the biochemist is concerned is the hydrogen bond. The significance of this bond arises not from its strength but from its weakness. The small bond energy along with the low activation energy for its formation and rupture makes the hydrogen bond of particular interest in reactions and molecular structures at biological temperatures.

Hydrogen bonds result from the attraction of two electronegative atoms for a proton. Accordingly, only the most electronegative atoms are capable of forming hydrogen bonds and, furthermore, the strength of the bond increases with the electronegativity of the two bonded atoms. The important hydrogen-bond-forming atoms are fluorine, oxygen and nitrogen, although chlorine, bromine and sulfur may be able to form weak hydrogen bonds. The hydrogen bond is essentially electrostatic in character, but, unlike electrostatic interactions between ions, the bond has direction. The bond is frequently nearly linear but deviations of up to about 30° from linearity have been observed without a significant decrease in strength.

There are numerous examples of hydrogen bonds. Fatty acid crystals are strongly stabilized by hydrogen bonds between two carboxyl groups. It is significant, however, that acetic acid when dissolved in water exhibits practically no dimer formation; the hydrogen bonds have been dissolved by water. Hydrogen bonds determine the magnitude and nature of the mutual interaction between water molecules and are consequently responsible for the striking physical properties of this uniquely important substance. The structure of water is discussed in more detail in Chapter 3.

Intramolecular hydrogen bonds are of frequent occurrence as, for example, in simpler molecules such as nitrophenols, ethyl acetoacetate and maleic acid. Intermolecular hydrogen bonding is sensitive to dilution whereas intramolecular hydrogen bonding is not. Intramolecular hydrogen bonding is of great importance in stabilizing the fine structure of proteins and of nucleic acids, and these topics are to be considered later.

The strength usually assigned to hydrogen bonds is in the order of 3 to 5 kilocalories per mole. The energy of the hydrogen bond in ice is estimated to be 4 kilocalories per bond. The length of hydrogen bonds lies between 2.5 and 3.0A; that in water is 2.76A and the NH··O bridge in peptides is estimated to be 2.75 to 2.8A; hydrogen bond lengths are considerably greater than those of covalent bonds; there is an inverse relation between bond strength and bond length. Table 1 gives the approximate bond strengths and lengths of some hydrogen bonds.

TABLE 1. APPROXIMATE BOND STRENGTHS IN KILOCALORIES PER MOLE AND BOND LENGTHS IN ANGSTROM UNITS FOR SOME HYDROGEN BONDS

Bond	*Energy*	*Length*
F-H-F	6.7	2.26
N-H-F	5	2.63
N-H-O	5	2.78
O-H-O	4.5	2.76
N-H-N	1.3	3.38
O-H-Cl	3.9	

The detection and estimation of the extent of hydrogen bonding are frequently difficult and depend more often than not on inference and on indirect methods. For example, if the distance R(H···B) is less than $[r_H - r_B - 0.2A]$ where r_H and r_B are the Van der Waal radii for atoms H and B, the existence of a hydrogen bond is indicated. Of the various methods for the detection and estimation of hydrogen bonds, infrared absorption is the most definitive. The stretching of the hydrogen valence bond incidental to the formation of a hydrogen bond results in a decrease in the normal frequency vibration with a resulting shift in the absorption band; however, the bending mode of the bond is shifted to frequencies above that in the absence of H-bonding. Not only can the extent of hydrogen bonding be estimated from infrared studies but the direction of the hydrogen bond can be determined with the use of polarized infrared; this technique has been especially useful in the study of the configuration of synthetic peptides.

Nuclear magnetic resonance and inelastic neutron scattering have been employed for the study of hydrogen bonding.

Hydrophobic Interactions. Hydrocarbons are characteristically insoluble in water and water does not wet an oily surface. These substances are said to be hydrophobic or water hating. If such groups as $-COOH$, $-OH$ and $-SO_3H$ be substituted into a hydrocarbon, they tend to make the hydrocarbon water soluble and the hydrocarbon now becomes hydrophilic or water loving. The above groups all have the ability to form hydrogen bonds with water. The presence of both hydrophobic and of hydrophilic groups in the same molecule produces a polar molecule in the Langmuir sense but not necessarily in the sense of separation of electrostatic charge (dipole moment); the dipole moments of the fatty acids are very nearly independent of the length of the hydrocarbon chain.

Whereas the word hydrophobic has been used, it must not be imagined that there is an actual repulsion between water and a hydrocarbon chain. Indeed there exist fairly large attractive forces between water and hydrocarbons, but it so happens that these forces are smaller than those existing between the water molecules themselves. The paraffin molecules are unable to break the water-water bond and the hydrocarbons are insoluble in water. It is possible to calculate from the studies of W. D. Harkins the total energy in ergs required to separate an interface one square centimeter in area formed between paraffin and paraffin, paraffin and water and water and water—the work of cohesion and of adhesion. The results of such calculations are shown in Table 2.

Table 2. COHESION AND ADHESION ENERGIES IN ERGS PER SQ. CM. AT 20°

Combinations	*Energies in ergs/cm²*
n-butane:n-butane	92
n-octane:n-octane	100
n-octane:water	107
water:water	233

It is to be noted from Table 2 that the attraction of water for n-octane is greater than is that of octane for octane. The energy of the "bonds" between n-octane molecules can be approximately estimated from the results given in Table 2 and turn out to be about 6 kilocalories per mole—a not inconsiderable energy term. It is to be noted from Table 2 that if an octane-water surface is replaced by an octane-octane and by a water-water surface there is a gain of 226 ergs per square centimeter.

Popular model systems for the discussion of hydrophobic interactions (some prefer to designate this interaction as hydrophobic bonding) are solutions of hydrocarbons in water. The reaction considered is

$$\text{Hydrocarbon (in non-polar solvent)} \rightarrow \text{Hydrocarbon (in } H_2O) \qquad 5$$

The entropy change at 25° for the transfer of a mole of aliphatic hydrocarbon from a hydrophobic environment to the water phase is negative to the extent of about 20 entropy units and the enthalpy change is negative to about 2,000 calories per mole. Accordingly, the free energy change for the transfer is about 3,000 calories positive. Thus, whereas the enthalpy change favors the solubility of hydrocarbon in water, the decrease in entropy is so large as to reduce the solubility of the hydrocarbon in water to a low level. It is postulated that the negative entropy change arises from the ordering of the water in the immediate vicinity of the hydrocarbon. As a consequence of the negative entropy change, the free energy change becomes more strongly positive with increasing temperature; the solubility of the hydrocarbon in water decreases with increasing temperature.

If a flexible long chain molecule containing both hydrophilic and hydrophobic centers is dissolved in water, the molecules as a result of hydrophobic interactions should arrange themselves in such a fashion that the hydrophobic portions of the molecules are sequestered in the interior of a collapsed coil with the hydrophilic groups on the outside of the coil and directed towards the water phase.

Hydrophobic interactions alone, however, cannot provide for specificity in the folding of the flexible molecule. Since the energy of hydrophobic interaction increases with increasing temperature, it can be concluded that if hydrophobic interaction is significantly more important than other types of interaction, the collapsed structure should become more stable with increasing temperature.

Van der Waal Forces. Gases tend to depart from the ideal gas laws; this departure is due to the volume occupied by the gas molecules as well as to attractive forces between the gaseous molecules. Such forces are known as Van der Waal or London or dispersive forces. It was noted above that the energy of cohesion between hydrocarbon molecules (n-octane) can be considerable. Pure normal long chain hydrocarbons form hard stable crystals with high melting points indicating the existence of fairly strong attractive forces. These forces arise from induced dipoles. Aliphatic hydrocarbons have no permanent dipole moment; nevertheless, on a time average the electrons in atoms will suffer a temporary displacement, giving rise to a momentary dipole. Such a transitory dipole is capable of inducing a dipole in a neighboring atom and gives rise to an attractive force. The tendency of an atom to form temporary dipoles is related to the polarizability of the atom which in turn is related to the index of refraction. The theoretical treatment of Van der Waal forces is fairly complex and their magnitudes cannot, in general, be calculated exactly. They vary inversely as about the sixth power of the distance between the atoms. The high exponential dependence on distance insures as close a fit between molecules as possible. Thus Van der Waal forces, unlike hydrophobic interaction, can give rise to a high degree of specificity.

Electrical Effects. It has been noted in Chapters 3 and 5 that the electrical work of charging a spherical ion large or small is

$$W = \frac{Z^2\epsilon^2}{2D}\left(\frac{\kappa}{1+\kappa a}\right) \qquad 6$$

and it is evident that the free energy of charging, when the charge is progressively increased from zero to some finite quantity given by $Z\epsilon$, must always be positive. Thus the charge will always lead to instability of the large ion and will tend to make it swell and depart from its specific structure.

A protein molecule tends to become unstable if the pH is shifted a sufficient distance from the isoelectric point; the net electrostatic charge on the molecule has been increased. At the isoelectric point of the protein there are an equal number of negative and positive sites on the surface of the molecule; the net charge is zero and the effect noted above does not exist. However, any considerable departure of the negative and positive sites from a uniform distribution would lead to local instability which would require adjustment in the molecular structure with a restoration of the uniform distribution of charged sites.

Evidently, if the polyelectrolyte is flexible with charges of like sign distributed along its length and, further, the coil can be penetrated by water, the coil will swell as a result of charging the polymer. This is very much the situation experienced in the swelling of a gelatin gel as the pH is shifted away from its isoelectic point. Such swelling can be understood in terms of a Donnan equilibrium with an increase of osmotic pressure inside the coiled molecule.

The important electrical effects on the structure of biopolymers are probably a good deal more subtle than indicated above. The various covalent bonds found in proteins and nucleic acids have permanent dipole moments, for example, that for the peptide bond is 3.7 Debye units. There is a complex electrostatic interaction of the dipoles. The forces of interaction depend on the magnitude of the dipole moments, on the distance between the dipoles, on the relative orientation of the dipoles and on the dielectric constant of the medium. The dielectric constant inside a protein molecule is assumed to have a value in the range of 3 to 5.

Peptide Structure. Polypeptides can be formed by the condensation of amino acids usually employing some variation of the original Leuch's synthesis which uses the N-carboxyanhydrides of amino acids. Synthetic polypeptides have been of great interest and their structures in the solid state as well as in solution have been the subject of much investigation.

Stretched peptide chains were shown by X-ray diffraction methods to exist in silk fibroin and later were identified in many of the synthetic polypeptide preparations. Such a chain with its approximate dimensions is diagramed in Figure 3.

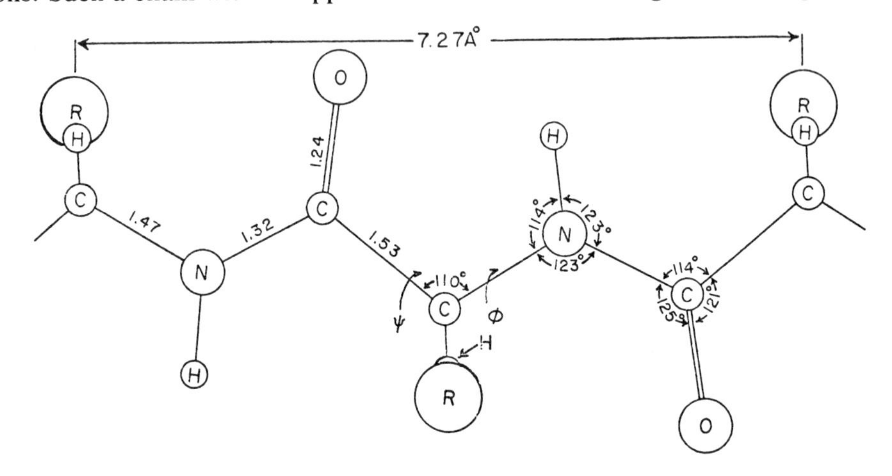

Fig. 3. A polypeptide chain in the stretched β-keratin configuration.

It was evident from the early work of Astbury and Street on wool keratin that peptide chains can form folded structures when conditions are suitable. In fact, hair and wool both exist in nature as folded structures. Many suggestions were made concerning the configurations of folded peptide chains but it remained for Pauling and Corey to delineate the problem. In a series of important papers these authors explored various polypeptide structures. They made use of X-ray diffraction studies on simpler compounds containing the peptide bond followed by the construction of accurate atomic models, and proposed definite possible structures of the peptides. The most interesting of these structures was the so-called α-helix. This structure contains 3.6 amino acid residues per turn of the helix and is stabilized by the formation of hydrogen bonds between neighboring carboxyl and imino groups adjacent to the peptide bonds as the chain coils back on itself. Central to their ideas was the realization that, due to resonance, the peptide bond has substantial double bond character thus preventing rotation about this bond; this limited the number of possible configurations available to the peptide chain. By means of X-ray diffraction, the α-helix has been shown to occur in synthetic polypeptides as well as in several proteins. It should be possible to have either a right or left handed helix in the same sense that one can have a right or left handed screw. It has been found, however, that the α-helix invariably occurs in right handed form, both in polypeptides and in proteins.

Figure 4 is a diagrammatic sketch of the α-helix.

Whereas the structure of polypeptides as determined by X-ray diffraction constitutes far more impressive evidence for the existence of the α-helix in the solid state, there is good presumptive evidence (mostly from optical rotatory dispersion and circular dichroism studies on peptides and proteins) of the existence of α-helixes in solution. Those solvents which tend to break hydrogen bonds, such as formamide, also disrupt the α-helix, and other solvents, such as carbon tetrachloride, chloroform and 2-chloroethanol, tend to stabilize the helical form; the ability to form an

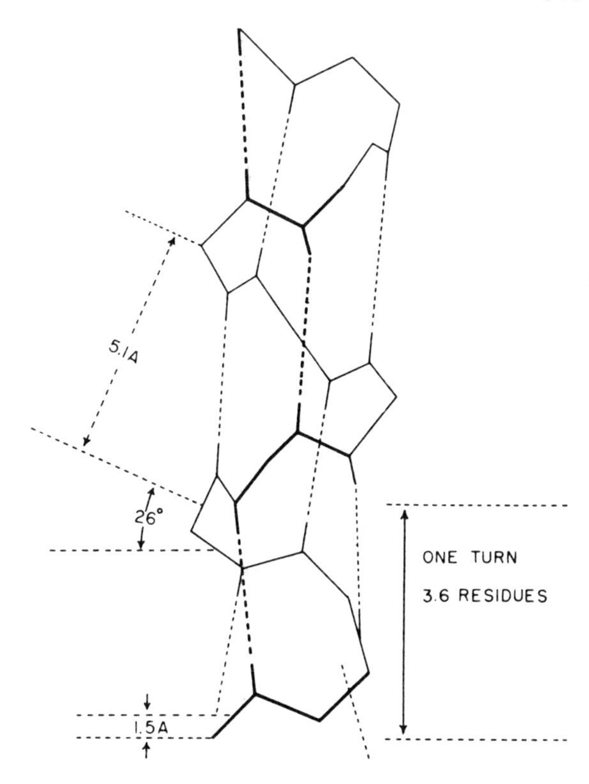

Fig. 4. Skeleton of α-helix. Dotted lines in skeleton indicate location of H-bonds which serve to stabilize the helix. The α-helix is approximately a cylindrical rod about 15.3A units in diameter.

α-helix depends on the amino acids which have been polymerized.

Amino acids can be grouped into two categories: those amino acids whose homopolymers can form helixes in solution and those whose homopolymers cannot. The helix-forming acids are: alanine, arginine, aspartic acid, cysteine, glutamic acid, leucine and lysine. The non-helix formers are: glycine, phenylalanine, proline, serine, threonine, tryptophane and tyrosine. Although poly-L-proline will form a helix, it is not an α-helix. Ionization of the amino acid residues would be expected to disrupt the helix. Shown in Figure 5 is an approximate plot of the percent α-helix as determined from optical rotatory studies of poly-L-glutamic acid as well as of poly-L-lysine as a function of pH.

There appears to be considerable tendency for a linear polymer with regularly spaced interacting sites to form molecular helixes. Thus, both nucleic acids and proteins contain helixes. A crystal form of amylose starch (linear variety) obtained by precipitating starch from a paste with alcohol gives good X-ray diffraction patterns and these can be interpreted as indicating a helical structure with 6 glucose residues per turn. A notable property of amylose starch but not of amylopectin (the branched variety) is to yield an intense blue color with iodine. The suggestion has been made that the iodine atoms occur in the core of an amylose helix and that the amylose helix is stabilized in aqueous solution by the presence of iodine.

Molecular Model Building. Much qualitative and sometimes quantitative information can be had by building molecular structures using space filling atomic models. It was partially such construction that led Pauling and Corey to their α-helix configuration for the peptide chain. If one builds a molecular model of a peptide chain with complete freedom of rotation about all valence bonds, one quickly becomes discouraged with the enormous number of allowable configurations. There are, however, some restrictions which can be applied. For example, due to resonance

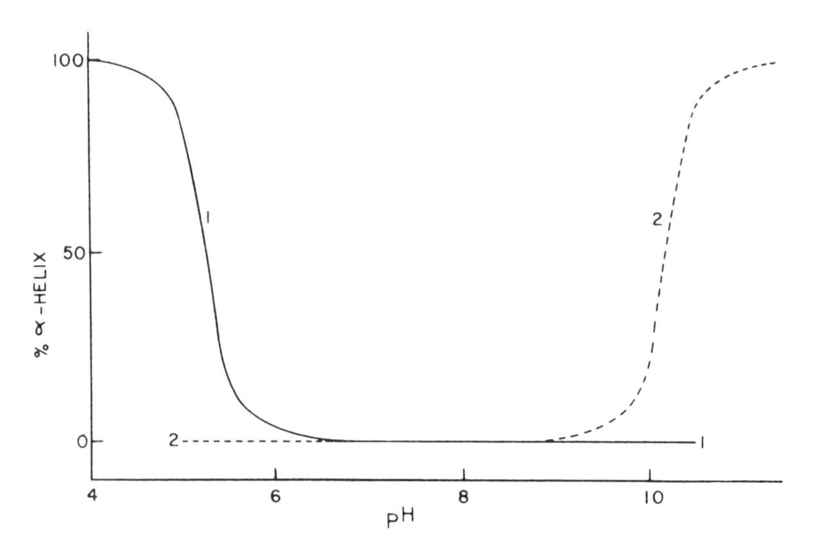

Fig. 5. Approximate percent of α-helix as function of pH. Curve 1, poly-L-glutamic acid. Curve 2, poly-L-lysine. (Calculated from data of P. Doty, K. Imahori and E. Klemperer: Proc. Nat. Acad. Sci., U.S. *44:* 424, 1958.)

there is no rotation of the peptide bond and, furthermore, the groups about the peptide bond are planar. As a result of the double bond character of the peptide bond, cis and trans isomers are possible. It has been found that the cis configuration does not occur in open peptide chains; the trans form is energetically favored and the barrier to the interconversion of the cis and trans conformers exceeds 20 Kcals per mole. Given these restrictions, it is evident that in a peptide chain only rotation about the $C_\alpha - C'$ and $C_\alpha - N$ bonds can occur (see Fig. 3) and, further, the configuration of the peptide chain is specified by these angles of rotation. The angle of rotation about the $C_\alpha - C'$ bond is known as ψ and that about the $C_\alpha - N$ bond as ϕ. The rotations are performed in a clockwise sense; both ψ and ϕ are zero in the standard configuration as shown in Figure 3. As the valence bonds $C_\alpha - C'$ and $C_\alpha - N$ are rotated relative to each other, it will be discovered that certain combinations of angles are incompatible due to interference between the atoms attached to the peptide chain. Contact distances for various atomic combinations occurring in a peptide are shown in Table 3.

TABLE 3. CONTACT DISTANCES IN A-UNITS FOR VARIOUS ATOMS

Contact type	*Distance*	*Contact type*	*Distance*
H----H	2.0	O----C	2.8
H----N	2.4	O----N	2.7
H----C	2.4	N----C	2.9
H----O	2.4	N----N	2.7
O----O	2.7	C----C	3.0

Contact distances are not exactly defined and those in Table 3 can be slightly less than those given. It is possible to construct a graph of the angle ψ against ϕ both extending from 0° to 360°. Certain areas of this map will include allowable configurations and other areas will be forbidden. The angle $C' - C_\alpha - N$ which is given as 110° in Figure 3 is capable of small variations; the allowable structures as functions of ψ and ϕ are critically dependent on the $C' - C_\alpha - N$ angle.

The relation between the angles ψ and ϕ can be programmed on a computer without the use of space-filling atomic models; this is the modern way of building molecular models.

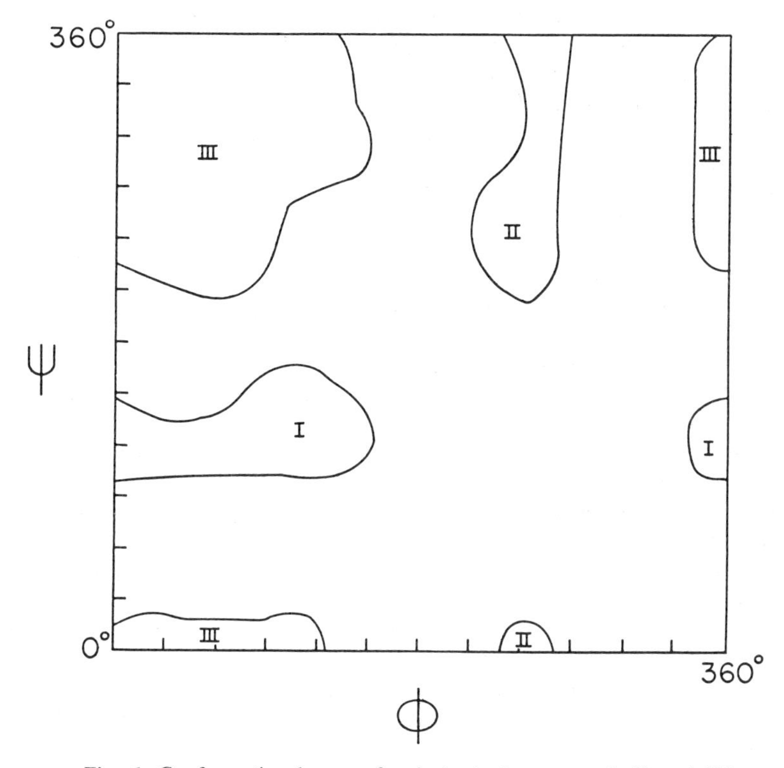

Fig. 6. Conformational map of poly-L-alanine; areas I, II and III are allowed configurations. (Adapted from D. A. Brant, W. G. Miller and P. J. Flory: J. Mol. Biol. *23:* 47, 1967.)

Whereas models based on contact distances between atoms can yield information about allowable configuration, the inclusion of forces of interaction between the atoms as well as the contact distances produces more realistic models and still further restricts the probable configurations of the peptide chains. These forces are of the same kind as discussed earlier in this chapter and the conformational potential energy of an infinite polypeptide chain can be expressed as

$$E = E_v + E_t + E_e + E_h \qquad 7$$

where E_v is the Van der Waal energy of interaction, E_t is the energy associated with rotation about the $C_\alpha - C'$ and $N - C_\alpha$ bonds and exists because of torsional barriers to rotation; there is not complete freedom of rotation about these valence bonds. E_e expresses the electrostatic interaction energy between the dipoles. E_h represents the contribution arising from hydrogen bonding. Figure 6 shows a conformational map for poly-L-alanine; the areas represented by I, II, III are the allowed configurations. The original figure contained contour lines of the conformational energies; these have been omitted for the sake of clarity.

Results of two-angle-rotation investigations on peptides can be summarized in general terms as follows: rotations are nearly independent of rotations of neighboring residues, but are greatly influenced by the nature and size of amino acid chain residues. Any restrictions on the rotation of the valence bonds tend to expand the coiled peptide chains.

It is believed that the native conformation of a protein under conditions approaching physiological conditions is the stable configuration whose potential energy is minimal and, further, this conformation is completely determined by the sequence of the amino acid residues in the peptide chain. It is hoped by some that the structure of the native molecule can be assigned through computer programs which will reduce the configurational energy of a protein to a minimum. This is a formidable task and it is far from certain that this endeavor will be completely successful. It is, however, not necessary that such a program meet with complete success to yield much of interest.

Native Protein Structure. It is convenient to consider protein structure on four levels. The primary structure refers to the sequence of the amino acid residues in the peptide chain. Peptide chains are linear without branching and the sequence of the residues is genetically controlled through nucleic acids. The lengths of the peptide chains vary considerably depending on the individuality of the protein; thus the A-chain of insulin contains 21 residues whereas a protein such as bovine serum albumin has over 500 residues in a single peptide chain. The secondary structure involves the folding of the linear chains into helixes of regular or irregular character or into twisted and bent linear segments.

Three arrangements of peptide chains have been observed to occur in native proteins: (1) stretched parallel chains in the β-form stabilized by interchain hydrogen bonds; (2) the α-helix or some closely related helical form; and (3) a specific order in which the peptide chains are neither stretched nor in helical form. It would be expected that this last arrangement would give rise to the specific properties which the protein might exhibit.

Some proteins (myoglobin, hemoglobin) exhibit a high percentage of α-helix (about 80 percent) whereas others (lysozyme) have a moderate amount and still others have practically none (ribonuclease, chymotrypsinogen). The tertiary structure pertains to the three-dimensional folding of the folds of the peptide chains frequently giving rise to a water-soluble globular protein. Disulfide formation between half cystine residues can have an important effect on the secondary as well as on the tertiary structure. Disulfide bonds (a kind of vulcanization) can occur between folds of a single peptide chain or between two peptide chains and restrict the conformations which the peptide chain or chains can assume. There is a strong tendency for the hydrophobic amino acid residues to be in the interior of the molecule and for the hydrophilic ones to be on the surface of the globular protein molecule. This separation is, however, not complete. A large fraction of the surface of a typically globular protein in solution is hydrophobic in the sense that the surface is occupied by atoms which have no ability to form hydrogen bonds with water. Assuming, for example, that the horse heart cytochrome C molecule is spherical and allowing 25 sq. A-units per hydrophilic group, and further assuming that all of the hydrophilic residues emerge on the surface, it turns out that only about half of the surface is occupied by hydrophilic groups. The shapes of globular proteins are irregular with many surface hills and valleys. Proteins with compact and specific structures have an anhydrous interior with the exception of the hemoglobin molecule which apparently has a small pool of water inside.

Lastly, two or more of the globular units can associate without covalent bonding, as is true of hemoglobin, to give rise to a specific structure; this is the quaternary level. It is difficult to overemphasize the importance of water on the secondary, tertiary and quaternary protein structure; water, although outside of the protein molecule, is just as much part of the fine structure of proteins as are the amino acid residues.

The structures of native proteins are vastly more complex than those of the synthetic polypeptides. The synthetic polypeptide is usually a polymer of one or perhaps two amino acids whereas the proteins contain 15 to 20 different amino acids with residue groups differing greatly in size and in functionality. Thus, results and ideas derived from polypeptide study, while valuable, need careful scrutiny before application to structural problems encountered in proteins.

The tertiary structures of a number of proteins (whale myoglobin, hemoglobin, lysozyme, ribonuclease, carboxypeptidase, papain and chymotripsin) have been successfully studied by means of X-ray diffraction technique. This success depended upon the use of the method of isomorphous replacement with heavy atoms to determine phase angles. For example, two of the four cysteines of horse hemoglobin combine with paramercuribenzoate to form six different isomorphous atom compounds.

The Fourier synthesis showed that hemoglobin consists of 4 subunits in tetrahedral array and each subunit closely resembles the sperm whale myoglobin. It is significant that the peptide chains of both proteins showed strong evidence for extensive folding in the α-helix configuration (about 80 percent).

Examination of the tertiary structures of those proteins whose structures have been determined reveal no regulating structural principles and it would have been totally impossible to have predicted these structures either on theoretical grounds or from a knowledge of protein chemistry, although, as noted earlier, currently there are efforts to arrange computer programs to minimize the potential energy of protein molecules and to arrive at the configuration of the native structure. In the known protein structures the peptide chains appear extensively convoluted and display no discernible order, although it is clear each molecule of a given protein must be folded in exactly the same specific manner as the other molecules. Otherwise, the detailed X-ray diffraction pictures could not have been obtained. A large number of different proteins have been investigated by a variety of techniques and a wealth of information is available. In order to confine this discussion, hemoglobin has been selected for more extensive comment.

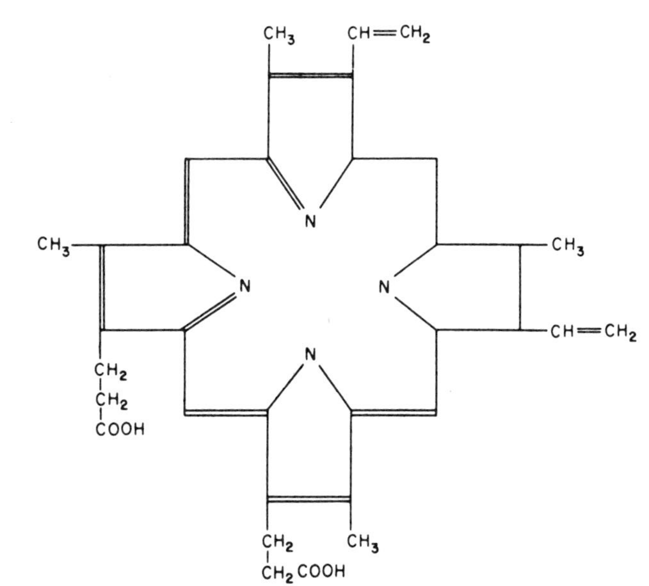

Fig. 7. Protoporphyrin attached to hemoglobin.

Hemoglobin. Due to its physiological importance and ready availability, hemoglobin is probably the most investigated of all proteins; it is a very complex molecule. The molecular weight of human hemoglobin is 64,450 and all mammalian hemoglobins resemble each other closely. As noted above, its tertiary structure is known down to a 2.3A unit level as a result of the impressive X-ray diffraction studies of the Perutz-Kendrew group. The molecule is compact, somewhat irregular in shape, with a low order of asymmetry. There appear to be many hills and valleys on its surface. The amino acid sequence of human hemoglobin has also been determined. The molecule is composed of four hemes attached to the protein part (globin). Attention will first be directed to the hemes and oxygen uptake followed by a brief comment on globin.

The porphyrin of hemoglobin, protoporphyrin, the structural formula of which is shown in Figure 7, is a plate-like structure and like other porphyrins forms metal complexes not only with iron but also with copper, magnesium, manganese and zinc ions. The complex formed between protoporphyrin and ferrous iron is called heme.

Heme has a magnetic moment (paramagnetic) corresponding to four unpaired electrons characteristic of electrons involving ferrous iron. In heme, therefore, the ferrous iron must be attached to the pyrrole nitrogen, the ionic bonds forming, with the replacement of two protons, with porphyrin. The two electrons remaining on the porphyrin may be considered to be equally distributed by resonance among the nitrogen atoms of the four pyrrole rings.

When heme becomes oxygenated, it becomes diamagnetic indicating loss of ionic bonds; however, the iron is not oxidized to the ferric state. When hemoglobin binds oxygen, the binding is covalent. Each heme binds one molecule of oxygen at saturation so that a molecule of hemoglobin having four porphyrin rings binds four molecules of oxygen. The structures of the binding sites of oxygen on hemoglobin are diagramed in Figure 8; included also is the configuration of the methemoglobin complex.

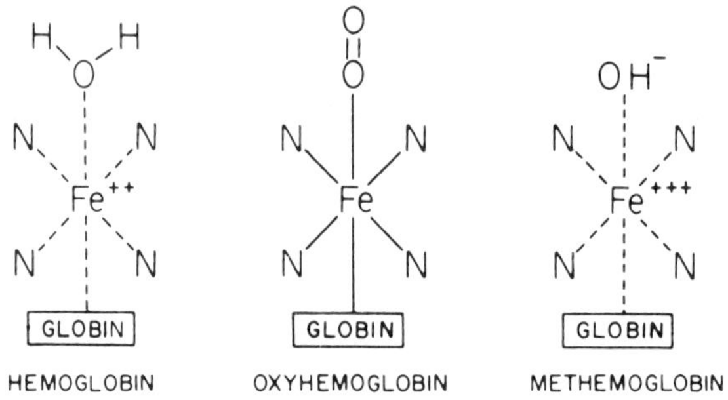

Fig. 8. Porphyrin-iron complexes associated with globin; predominantly ionic bonds are indicated by dotted lines; predominately covalent bonds by solid lines.

The oxidation of the ferrous iron of hemoglobin to the ferric state produces methemoglobin in which the binding of iron is again essentially ionic (Fig. 8) and which forms salts such as chloride and acetate. When heme is split from hemoglobin it quickly oxidizes to hemin (hematin) in which the iron is ferric. Heme is dissociated from hemoglobin by the acidification of a hemoglobin solution. It is not known how the hemoglobin manages to keep the heme in the reduced state but it is speculated that the heme is partially buried in a pocket formed by the hydrophobic residues of the globin leading to an environment of low dielectric constant, and this retards the irreversible oxidation of the ferrous to the ferric state.

The combination of hemoglobin with oxygen follows an S-shaped curve when the moles of oxygen bound are plotted against the partial pressure of oxygen (see Fig. 9). The reaction for the combination of oxygen with hemoglobin can be written

$$Hb + 4O_2 \rightleftharpoons Hb(O_2)_4 \qquad 8$$

and if the four sites (iron atoms) in the hemoglobin molecule functioned independently, the oxygen uptake could be expressed by the use of the mass action law with one intrinsic binding constant. Figure 9 shows such a plot (dotted curve) with the intrinsic constant set equal to that observed at half saturation of the hemoglobin. Evidently, there is strong interaction between the binding sites in the hemoglobin molecule; after the first oxygen is bound it becomes easier to bind the other oxygen molecules. The nature of this interaction is still unclear and is particularly difficult to understand in light of the fact that the heme groups in hemoglobin are as shown by X-ray diffraction about 25 A-units apart. Nuclear magnetic resonance studies in-

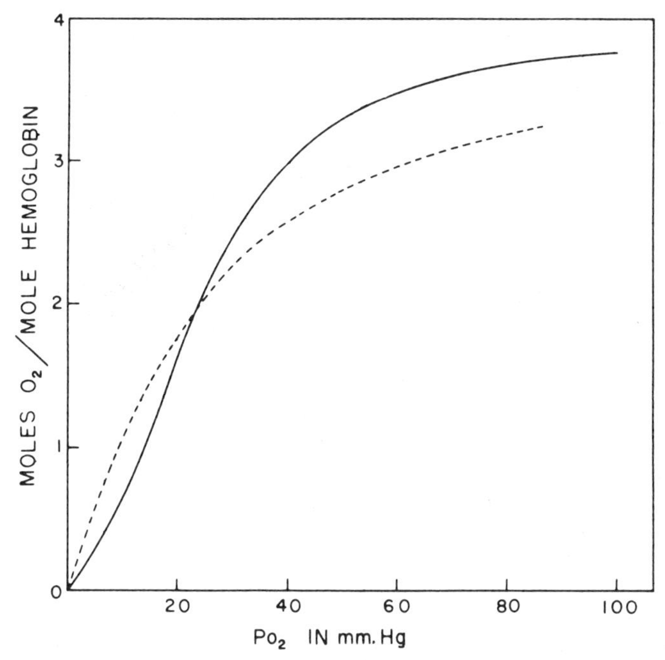

Fig. 9. O_2 uptake curve for hemoglobin, pH 7.4, 38°C. Dotted curve, calculated O_2 uptake with non-interacting sites. P_{CO_2} about 40 mm. Hg.

dicate that there is no heme-heme interaction and that the increase in oxygen affinity with increased oxygen uptake resides in structural changes in the protein part of the molecule. Indeed X-ray diffraction studies show that the subunits roll upon each other upon the addition of oxygen and a different quaternary structural arrangement results.

A. V. Hill proposed the following equation to describe the interaction between hemoglobin and oxygen

$$\log Y = n \log X + K \qquad 9$$

where Y is the percentage saturation of hemoglobin with oxygen, X is the partial pressure of oxygen, n is the slope of the plot of log Y against log X and reflects the number and extent of interaction between reactive sites. Mammalian hemoglobins have four interacting sites, but n is found experimentally to be about three, indicating less than total interaction. Sigmoidal dependence of activity on substrate concentration was originally thought to be unique to hemoglobin, but such behavior has since been reported for a number of enzymes. Such effects have become known as allosteric.

The advantages of such heme-linked groups in the respiratory cycle is of special importance because it permits near saturation of the hemoglobin in the lungs where the P_{O_2} is high (about 90) and the release of a large fraction of the oxygen in the capillary beds in which the P_{O_2} is much lower (about 20).

Hemoglobin becomes a stronger acid upon being oxygenated and the pK of one group, probably a histidine residue in the hemoglobin molecule, is shifted from 7.64 to 6.70 as a result of oxygenation. This shift is accompanied by a release at pH 7.4 of 0.7 mole of hydrogen ions. There is a reciprocal relationship between the oxygen uptake of hemoglobin and the acidity and this relationship is known as the Bohr effect. In capillary beds the pH of the blood is slightly lower due to the release of carbon dioxide into the blood. This shift towards the acid region along with the increase in carbon dioxide concentration decreases the affinity of the hemoglobin for oxygen and, accordingly, increases the extent of oxygen release where it is needed.

Globin, the protein part of hemoglobin, can be prepared by the careful acidification of a hemoglobin solution, whereupon the heme dissociates from the globin and can be extracted with an organic solvent. The native globin can be combined with oxidized heme (hemin), the hemin reduced back to heme and the native hemoglobin regenerated. Globin is much more unstable than is hemoglobin; heme lends stability to the molecule.

Globin consists of two pairs of identical peptide chains of about equal size. The peptide chains of normal adult human hemoglobin are known as α and β chains respectively. There are many different globins, the differences between the globins usually consisting of the replacement of one or a few amino acid residues in the peptide chains at a given sequence in the chain by other residues. To a cursory examination, the alteration in the peptide chains appears trivial but nevertheless it can have a profound effect on the properties of the hemoglobin.

Hemoglobin acidified to about pH 6 splits symmetrically into half molecules, each half being represented by $\alpha\beta$. There is a further split into monomers. Upon neutralization, the fragments recombine and indeed it is possible to dissociate mammalian hemoglobins from two different species in the presence of each other and upon recombination, after neutralization, hybrid hemoglobin molecules are formed.

Examination of the amino acid content of hemoglobin reveals an unusually high content of histidine, giving hemoglobin considerable buffer capacity in the physiological pH range. Hemoglobin also contains more than ordinary amounts of amino acids with hydrophobic residues (residues incapable of hydrogen bond formation).

Protein Denaturation. One of the outstanding properties of proteins is that of undergoing denaturation when subject to relatively mild treatment. During denaturation, many or all of the specific biological and some of the chemical properties of proteins are greatly altered. Agents which bring about protein denaturation are heat, short wavelength radiation, surface forces, strong acids and bases, alcohols, urea, guanidine · HCl, long chain alkyl sulfates, and others not mentioned. The typical changes suffered by a globular protein upon denaturation are: viscosity increase of the solution, decrease of diffusion rate, increase in levo optical rotation, decrease in solubility, loss of ability to crystallize, exposure of groups undetectable or partially undetectable in the native configuration such as tyrosyl hydroxyls, sulfhydryl and amino groups, increased digestability by proteolytic enzymes, modification of immunological properties, and loss of enzymatic properties if the protein is an enzyme.

Heat denaturation is no doubt the oldest known reaction of proteins; people have been cooking proteins for a long time. It was the first protein reaction to be subject to careful physical measurements (H. Chick and C. J. Martin, 1910). Typically, a protein heated in solution undergoes profound conformational changes usually between 60° and 70°, although the temperature of denaturation varies with the physical conditions.

Some proteins exhibit reversibility in respect to heat denaturation, notably pepsinogen, chymotrypsinogen, trypsin, and it is, accordingly, possible to calculate the thermodynamic parameters of the reaction. Table 4 shows the ΔH, ΔS and ΔF for the heat denaturation of these proteins. Heat denaturation is characterized by a large positive entropy and enthalpy changes; the driving force for the reaction is primarily entropic.

TABLE 4. ΔH AND ΔF IN KCALS PER MOLE AND ΔS IN ENTROPY UNITS PER MOLE FOR THE HEAT DENATURATION OF PROTEINS

Protein	*Conditions*	ΔF	ΔH	ΔS
Pepsinogen	pH 7, 45°	1.3	31	93
*Chymotrypsinogen	pH 2, 43.6°	0.3	99.6	316
Trypsin	.01 NHCl, 46°	0	67.6	213

*See Figure 12, Chapter 2.

The kinetics of heat denaturation have been extensively investigated. The reaction is first order in respect to protein concentration and is characterized by very large heats of activation (the rate increases drastically within a few degrees rise in temperature). The heat of activation frequently exceeds 100 Kcals per mole, indicating that a large number of chemical bonds need to be activated simultaneously to achieve denaturation.

The energies of activation depend greatly on the environment and it is noticed that the maximum energy of activation occurs at or near the isoelectric point of the protein, falling off as the solution is made either acidic or basic in reference to the isoelectric point. It is likely that a large fraction of the observed heat of activation at the isoelectric point arises from the heat of ionization of a number of ionogenic groups of the protein, and it is only after the protein has a substantial net electrostatic charge on it that thermal denaturation occurs. It appears that coulombic repulsion combined with increased kinetic energy resulting from the increase in temperature brings about denaturation.

The considerable role of water in heat denaturation should be noted. For example, egg albumin samples containing varying quantities of water were heated for 10 minutes at 76.5°, and the amount of denatured protein as judged by solubility in quarter saturated Na_2SO_4 determined. These results are shown in Figure 10.

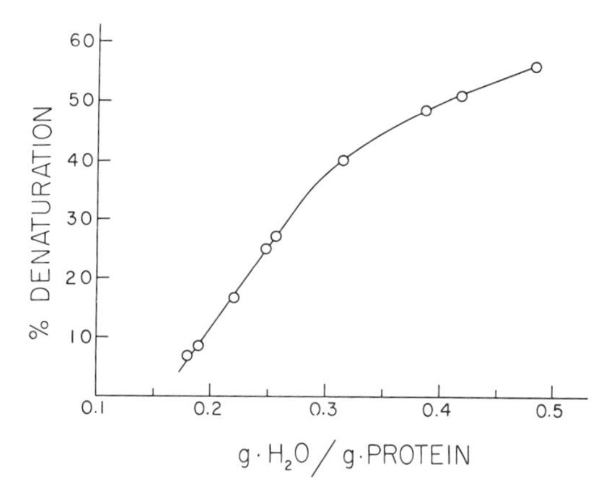

Fig. 10. Percent denaturation of egg albumin as a function of water content. 10 minutes at 76.5°. (Adapted from H. B. Bull and K. Breese; Arch. Biochem. Biophys. *128:* 488, 1968.)

After about half of the water of hydration has been removed from egg albumin, the protein becomes very stable towards heat; at elevated temperatures, water is a powerful protein denaturant.

Shifting the pH of protein solutions away from the isoelectric point of the protein decreases the stability of the protein; acid-base denaturation of proteins has the characteristics of heat denaturation at room temperature.

Shown in Figure 11 is the percent denaturation of ribonuclease as judged by optical changes at pH 6.83, 3.16 and 0.89 as a function of temperature; these changes are reversible.

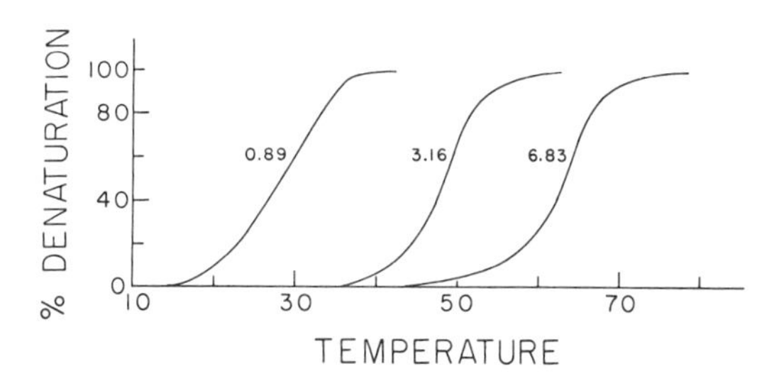

Fig. 11. Percent denaturation of ribonuclease as a function of the temperature at pH 0.89. 3.16 and 6.83. (Adapted from J. Hermans and H. A. Scheraga: J. Amer. Chem. Soc. *83:* 3283, 1961.)

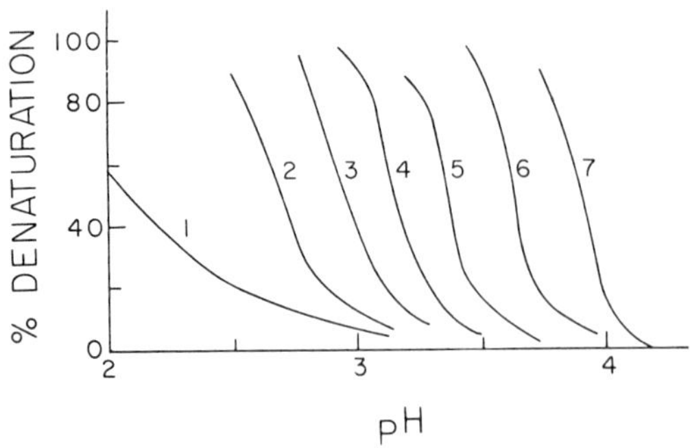

Fig. 12. Percentage of egg albumin denatured as a function of pH. 1, HCl; 2, acetic; 3, propionic; 4, *n*-butyric; 5, *n*-valeric; 6, *n*-caproic; 7, *n*-heptanoic. 0.05 MKCl, 20 hours 30°. (Adapted from H. B. Bull and K. Breese: Arch. Biochem. Biophys. *120:* 309, 1967.)

The short chain fatty acids are strong protein denaturants and the longer the aliphatic chain the more potent they are. Not only is a simple pH effect present but, more importantly, the hydrocarbon portion of the acids are interacting with the hydrocarbon parts of the amino acid residues. Figure 12 shows the denaturation of egg albumin (as judged by solubility) by fatty acids as a function of pH.

Many years ago H. Wu proposed that protein denaturation consisted essentially in the transformation of the highly specific structure of the native protein to the disordered denatured molecular structure without rupture of covalent bonds. As a general description of the process, Wu's ideas are still valid. The heat denaturation of protein bears a strong resemblance to a melting of the native structure. Urea denaturation can be looked upon as a partial dissolution of the peptide chains of the native structure. During the process of denaturation, chemical groups which were buried in the interior of the protein molecule become exposed to the solution and the biological specificity, which depends on the specific arrangement of the peptide chains, is lost.

It appears that the native structure is always the stable structure at or close to the conditions of the native environment. As a result of a change in the solvent, temperature, etc., there is a corresponding change in the structure of the native molecule; the molecule becomes denatured if the environmental changes are sufficiently drastic. If the original conditions of the solvent are restored, there is a strong tendency for the protein to return to its native structure; denaturation is reversed. Mention has been made of the ability of denatured protein to polymerize, and if such polymerization occurs it is impossible for the individual protein molecules to return to their native configuration; protein denaturation cannot be reversed if protein association occurs during denaturation.

The term cooperative structure is frequently used in reference to proteins. The essential meaning of this term is that the structure assumed by a protein molecule depends upon all of its parts and the disruption of a portion of the structure leads to the disruption of the entire structure.

Solubility of Proteins. Many factors influence the solubility of biopolymers; among these factors may be mentioned: temperature, electrolytes, nature of the polymer, and nature of the solvent.

The process of solution can be regarded in a sense as melting of the solute, and the variation of the solubility with temperature should be closely related to the heat of solution. If heat is evolved on solution, then the solubility will decrease as the temperature is increased. The heat of solution is not always constant with the change in temperature and can indeed change sign over a sufficient temperature change. For example, Sörensen in his classic work on the solubility of egg albumin in the presence of 2.14 M ammonium sulfate found the following solubilities for this protein in grams per liter at the indicated temperatures: 0°, 3.18 grams; 12°, 2.09 grams; 20°, 1.81 grams; 29°, 2.24 grams.

Comment was made in Chapter 3, Electrolytes and Water, under a discussion of the Hofmeister series, on the effect of salts on the solubility of proteins. The influence of salts on the solubility of proteins is complex. In general, dilute salt solutions increase the solubility of proteins. Indeed, euglobulins are insoluble in pure water but become soluble in dilute-salt solutions. β-lactoglobulin from cow's milk is also insoluble in pure water but shows complete miscibility in a dilute solution of a neutral salt. As the salt concentration is increased beyond a certain point, however, protein solubility decreases. The classical methods of protein isolation and purification depend upon differential solubilities in salt solutions of different concentrations.

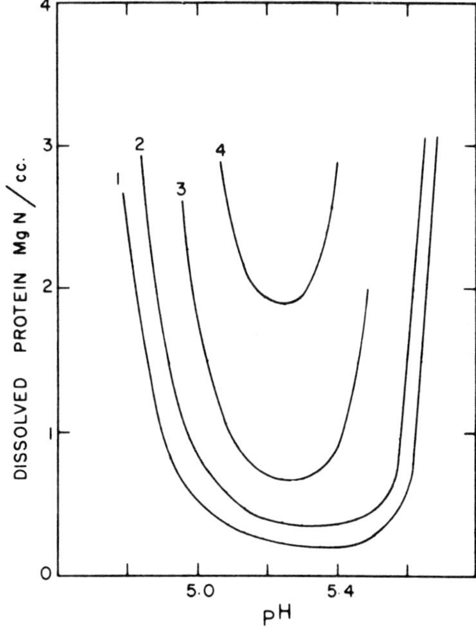

Fig. 13. Solubility of β-lactoglobulin as a function of pH. Curve 1, 0.001 ionic strength; Curve 2, 0.005 ionic strength; Curve 3, 0.01 ionic strength; Curve 4, 0.02 ionic strength. (A. Grönwall: Comp. rend. Trav. Lab. Carlsberg *24:* 185, 1942.)

The solubility of proteins is a marked function of the pH; the solubility being least in the isoelectric zone and increasing on both the acid and the basic side. Figure 13 shows the variation of the solubility of β-lactoglobulin as a function of pH.

It is easy to offer a qualitative explanation of the influence of pH on the solubility of a protein. The solubility is the resultant of the balance of the attraction of the solute molecules for each other, which tends to prevent solution, and the attraction of the solvent molecules for the solute, which tends to promote solution. At the isoelectric point the attraction of the protein molecules for each other is maximal. If the pH is shifted away from the isoelectric point, the protein acquires a net charge. This decreases the attraction of the protein molecules for one another, and consequently the solubility of the protein tends to increase.

Protein solubility at constant pH, temperature, pressure and salt concentration is an important but seldom used method for the detection of heterogeneity of a protein sample. There are several tests for the purity of a protein sample such as electrophoresis, chromatography, sedimentation, and solubility. Crystallinity of a protein is a deceptive measure of purity; a protein crystal can be grossly impure in respect to a single protein component.

According to Gibbs' phase rule, at constant temperature and pressure the solubility of a pure chemical compound is independent of the amount of solid phase present. Accordingly, if the amount of protein in a salt solution is progressively increased, the protein added all dissolves so that the amount in solution is equal to the amount added and a straight line plot with a slope of unity is obtained. Finally, as more protein is added, the solution becomes saturated with protein and, if the protein is pure, the amount of protein in solution is independent of the quantity of protein added. The possible situations to be encountered are shown in Figure 14.

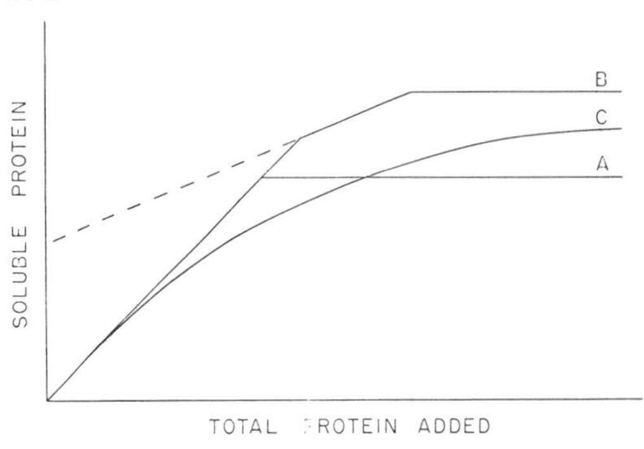

Fig. 14. Diagrammatic solubility curves. *A*, Solubility of a single pure protein. *B*, Solubility curve of a mixture of two proteins. *C*, Solubility of a preparation containing several protein components.

The success of using the solubility of a protein to characterize the protein depends considerably on the technique employed. In the first place, an amorphous protein preparation is, in general, significantly more soluble than is a crystalline preparation of the same protein. Further, the solubility curve must be reversible in respect to solution and precipitation and, accordingly, ample time for equilibration must be allowed.

A variation of the solubility method outlined above is to study the solubility of a protein as a function of the salt concentration at constant temperature and pH. Wong and Foster have used this method to detect microheterogeneity in bovine serum albumin. The method they employed was to dissolve the protein in 25 weight percent ammonium sulfate and to dialyze samples of this solution against a series of solutions of ammonium sulfate of progressively increasing salt concentrations extending from 26 to 32 percent. After a 20- to 24-hour equilibration, the protein in solution was measured and expressed as the fraction of the total protein present; this value expressed as f_s is shown plotted against the weight percent ammonium sulfate in Figure 15. It appears that the fraction of protein in solution depends only on the salt concentration and not at all on the quantity of protein present. This result clearly demonstrates that the samples of protein are not homogeneous and, further, the samples employed differed significantly in respect to their heterogeneity.

Nucleic Acids. One of the most fascinating developments of modern times has been the elucidation of the structures of the nucleic acids and the relation of these structures to protein synthesis and to genetics.

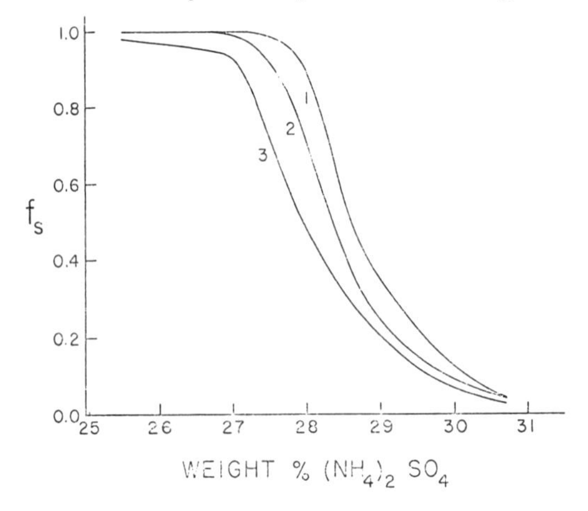

Fig. 15. Soluble fraction, f_s, of bovine serum albumin as a function of the weight percent concentration of $(NH_4)_2SO_4$. Curve 1, most soluble subfraction; curve 2, unfractionated protein; curve 3, least soluble subfraction. Each subfraction constituted approximately one-third of the protein. 25°. (Adapted from K. Wong and J. F. Foster: Biochem. *8:* 4096, 1969.)

There are two large classes of nucleic acids referred to respectively as deoxyribonucleic acid (DNA) and ribonucleic acid (RNA). These are high molecular weight unbranched linear polymers. The chains can be written in their simplest

form as

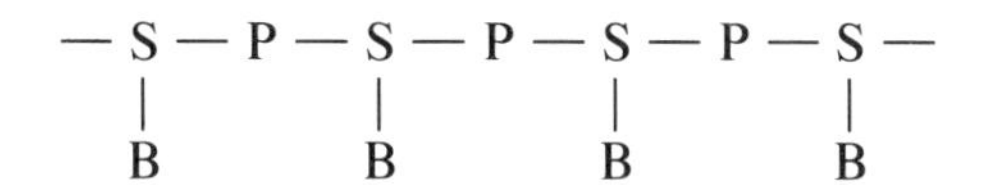

where S represents the sugar, P the phosphoric acid residue and B the various nitrogen bases (purines and pyrimidines). The nucleic acid chains are somewhat analogous to peptide chains; the sugar and phosphoric acid constitute the backbone of the chain and the nitrogen bases correspond to the amino acid side chains; their sequence in the nucleic acid chains gives individuality to the chains. The combination of phosphoric acid, sugar and nitrogen base is known as a nucleotide, which is the monomeric unit. The monomers are polymerized through phosphate ester bonds with the sugar.

DNA and RNA resemble each other closely differing only in the pentose sugar (deoxyribose contains one less hydroxyl group than does ribose) and by the replacement of the pyrimidine thymine in DNA by the closely related pyrimidine, uracil, in RNA. DNA codes itself as well as RNA. However, in coding for RNA, only one strand of the two available strands of DNA is used. The other strand along with the information it contains is apparently thrown away. RNA codes for protein structure and determines the sequence of the amino acid residues in the peptide chain. DNA, unlike RNA, lends itself to X-ray diffraction studies in the solid state, a fact which enabled Watson and Crick to propose their well-known two stranded helical structure for this molecule shown diagrammatically in Figure 16.

The diameter of the DNA helix is about 20A and each strand makes a complete turn every 34 A. Since there are 10 nucleotides on each chain per turn of the helix the distance per nucleotide is, accordingly, 3.4 A. The core of the helix consists of the flat purine and pyrimidine bases stacked on top of each other, and the helix is stabilized by hydrogen bonds between the nitrogen bases located on the two strands. There is a specificity in respect to which bases can pair; thus adenine can hydrogen bond with thymine and guanine with cytosine. This arrangement requires the two chains to be complementary and to run in opposite directions. The interaction between the nitrogen bases is shown in Figure 17.

It will be noticed from Figure 17 that the combination of thymine with adenine involves the formation of two hydrogen bonds whereas the cytosine-guanine interaction gives rise to three such bonds. If DNA is heated in solution at about 90°, the DNA denatures; it "melts," the absorbance at 260 $m\mu$ greatly increases and the two strands of the helix separate. It is found that DNA containing a higher ratio of cytosine plus guanine to thymine plus adenine exhibits a higher melting temperature than does that with a lower ratio; the extra hydrogen bond tends to increase the stability of the helix. In addition to hydrogen bonding as indicated above, there is also considerable Van der Waal and dipole interaction between the stacked bases. The dipole interaction of the stacked bases leads to hypochromism, that is, the absorbance at 260 $m\mu$ of the intact DNA helix is considerably less than that expected from the sum of the individual nitrogen bases. Hypochromism is further discussed in Chapter 8. Upon denaturation and disruption of the helix, hypochromism disappears. Whereas DNA usually exists as double stranded helix, it can exist and operate as single strands or even as circular strands.

As noted above, the sequence of the bases in the DNA determined the sequence of the bases in new DNA molecules as well as in the RNA chains. This is accomplished through base pairing. However, for this synthesis an energy source is needed because the free energy for the hydrolysis of the phosphate ester bonds is negative to the extent of about 6 Kcals per mole. It appears that ATP transfers its phosphate groups to the nucleosides of the other bases (a nucleoside is the combination of the sugar with a base, deoxyribose for DNA and ribose for RNA) thus converting the

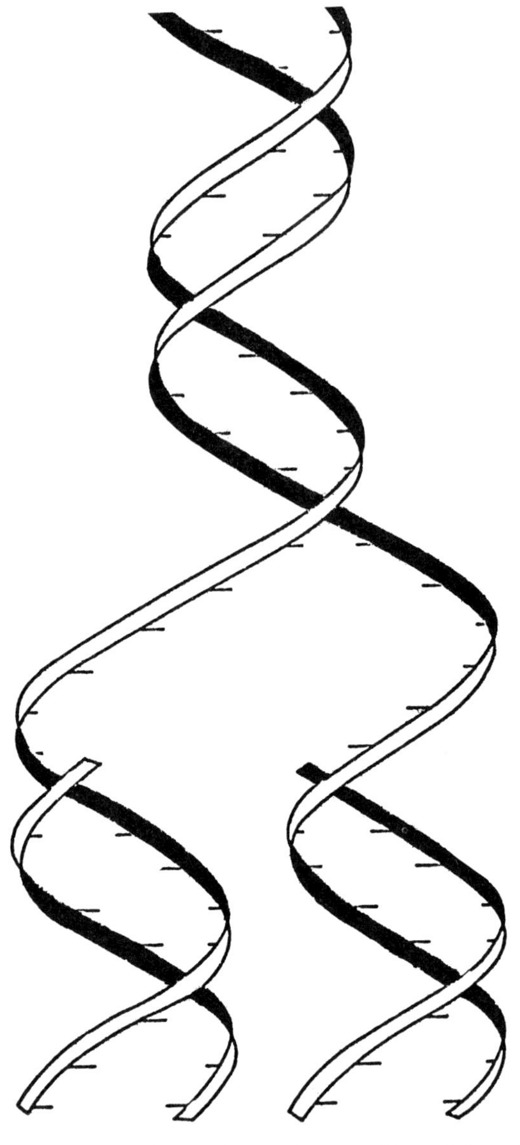

Fig. 16. Watson-Crick double stranded helical structure of DNA. Also represented is the unwinding of the helix with a start on the replication of two new strands of DNA.

nucleosides into triphosphates. The triphosphate then attaches itself to the growing nucleic acid chain at the correct point and simultaneously two phosphate groups are released. The actual formation of the required ester bond is mediated through the action of an enzyme, DNA polymerase.

The functions and structures of RNA are more complex than are those of DNA. To start with there are at least three types of RNA: transfer RNA (tRNA sometimes called soluble RNA and written SRNA), messenger RNA (mRNA), and ribosomal RNA (rRNA).

As noted above, RNA codes for the sequence of the amino acids in proteins but the amino acids themselves have no specific affinity for RNA. The coding is accomplished through a series of fairly complex reactions. ATP combines with the carboxyl group of the amino acids thus activating them; two phosphate groups are simultaneously released. An enzyme is required to catalyze this reaction and the amino acid adenosine monophosphate remains bound to the enzyme. The enzyme then transfers the amino acid to the terminal adenylic acid of the specific tRNA

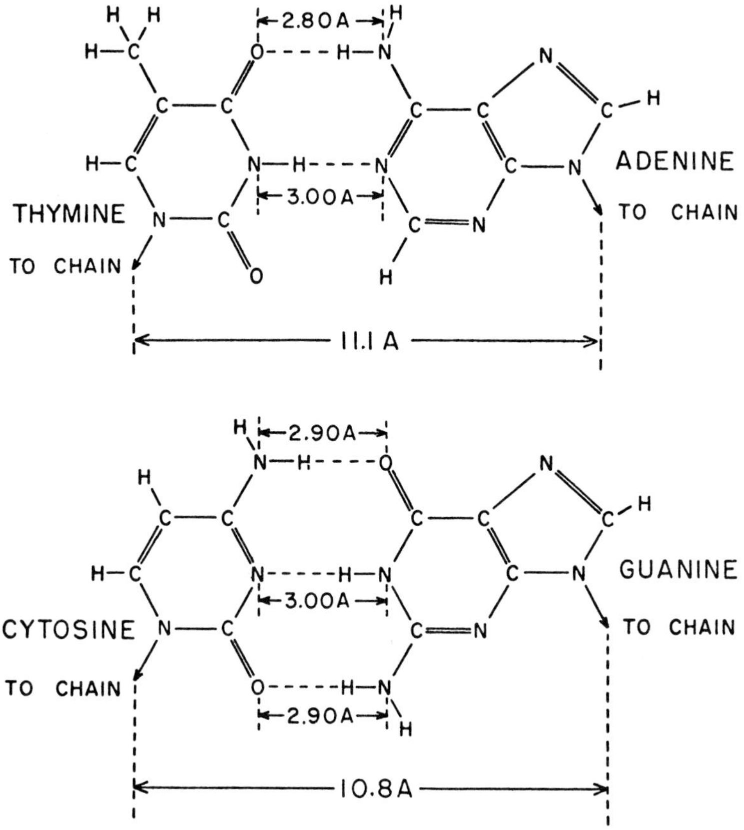

Fig. 17. The hydrogen-bonded base pairs in DNA.

releasing adenosine monophosphate and giving a covalent bond between the carboxyl group of the amino acid and the ribose component of tRNA. The activating enzyme is thus able to recognize both the side chain of the amino acid as well as the tRNA specific for a given amino acid. tRNA has a molecular weight of about 25,000 containing about 80 nucleotides in a single chain terminating at one end by the sequence cytidylic acid, cytidylic acid, adenylic acid with the opposite end terminating in guanylic acid. Each amino acid has a specific tRNA and, accordingly, there are at least 20 different tRNAs.

Protein synthesis itself occurs on the ribosomes which are spherical particles with molecular weights of about three million and are partly protein and partly rRNA. rRNA comes in two classes in regard to size. One of these has a molecular weight of about one million and the other has about half this weight. Apparently, however, rRNA plays no direct role in protein synthesis, the active template being instead mRNA. mRNA binds to the surface of the ribosome and selects the specific tRNA with its amino acid attached and orients them in the proper sequence corresponding to the sequence of its own bases. Stepwise peptide chain growth begins with the amino terminal end and successive peptide bonds are formed by the release of tRNA; enzymes and an energy source are required. The energy source is not adenosine triphosphate in this instance but is instead guanosine triphosphate. About ten seconds are required to synthetize the average size protein molecule and once the peptide chain is formed it spontaneously folds into the tertiary structure specific for the amino acid sequence in the peptide chain. The size of the mRNA molecule depends on the number of amino acid residues in the peptide chain being manufactured; the number of nucleotides is equal to three times the number of amino acid residues.

The genetic information is carried in the DNA molecule in the sequence of the four bases adenine, cytosine, guanine and thymine. This information is delivered to RNA and is coded in terms of the four bases adenine, cytosine, guanine and uracil. Evidently, the number of sequence permutations is 4^n where n is the number of nucleotides in a given nucleic acid molecule; the amount of information which can be conveyed is enormous – enough and more to literally describe an elephant.

A sequence of three bases is required for each amino acid and with four bases there are 64 possible three-letter codons as they are called. This number is more than enough to take care of the twenty amino acids. In fact, a single amino acid may have more than one codon. By a series of ingenious investigations, the genetic code has been broken and the proper codons assigned to the correct amino acid.

It is clear that the secondary structure of the nucleic acids depends primarily on hydrogen bond formation and the so-called hydrophobic bonds play minor or no roles at all; nucleic acids appear to be highly hydrated. It is possible that the water molecules themselves lend stability.

Both DNA and RNA have a strong tendency to complex with proteins, especially if the protein is very basic, for example, histones and protamines. Such complexes are usually dissociated at higher pH values as well as in relatively concentrated solutions of neutral salt; this behavior indicates the importance of electrostatic interaction.

A number of plant viruses have been crystallized which are complexes of RNA and a specific protein. The nucleic acid in the nucleoprotein is surrounded by a protective coating of protein and the complex is arranged in a definite pattern as revealed by X-ray diffraction studies. The RNA in nucleoproteins typically has a very high molecular weight of the order of a few million, whereas the soluble RNA fraction unassociated with protein has a more moderate molecular weight of the order of tens of thousands. Reversible dissociation of the nucleoprotein into subunits readily occurs.

The plant virus about which more is known than perhaps any other is tobacco mosaic virus (TMV). The native TMV molecules are rods about 3,000A long and about 180A in diameter and with a molecular weight of about 40 million; the molecules are easily visible in the electron microscope. TMV can be dissociated into RNA and free protein. The protein can be further dissociated into units whose molecular weight is about 17,000. The native nucleoprotein is in the form of a giant helix with about 2,000 protein subunits associated with a single RNA chain which is also in the form of a helix. The protein is on the outer periphery with the RNA chain towards the center of the rod. The structure resembles in a crude way a corncob with the corn kernels in place; the corn kernels represent the protein subunits.

GENERAL REFERENCES

Dayhoff, M. O.: Atlas of Protein Sequence and Structure. Nat. Biomed. Res. Found., Silver Springs, 1969.

Dickerson, R. E. and I. Geis: The Structure and Action of Proteins. Harper and Row, New York, 1969.

Joly, M.: A Physico-Chemical Approach to the Denaturation of Proteins. Academic Press, New York, 1965.

Ramachanchan, G. N. and V. Sasisekharan. Conformation of polypeptides and proteins. Adv. Prot. Chem. *23*:283, 1968.

Rich, A. and N. Davidson: Structural Chemistry and Molecular Biology. W. H. Freeman, San Francisco, 1968.

Tanford, C.: Protein denaturation. Adv. Prot. Chem. *23:* 121, 1968; *24:* 1, 1970.

Watson, J. D.: Molecular Biology of the Gene. Benjamin, New York, 1965.

PROBLEM

1. Distinguish between a random coil, a disordered coil and a cooperative structure.

   Ans: See text.

CHAPTER 7

Osmotic Pressure and Related Topics

The properties of solutions which depend primarily on the number of solute molecules present rather than on the kinds of particles are called colligative. These properties embrace vapor pressure lowering, boiling point elevation, freezing point depression, osmotic pressure and, at very low concentrations, surface tension lowering.

Vapor Pressure. A solution containing two kinds of spherical particles of identical size in equilibrium with the vapor phase is diagrammed in Figure 1. If both of the molecular species (Fig. 1) are volatile and the solution is ideal, the relative concentration of the two kinds of molecules in the vapor must equal that in the liquid phase and N_1-vapor is equal to N_1-liquid where N_1 is the mole fraction of component 1. The partial pressure P_1 of component 1 in the gas phase is

$$P_1 = P_0 N_1 \qquad 1$$

where P_0 is the total gas pressure exerted by components 1 and 2. Evidently, P_0 is also the vapor pressure of the pure liquid component 1. Equation 1 can also be expressed as (N_1 plus N_2 is unity)

$$\frac{P_1}{P_0} = 1 - N_2 \qquad 2$$

or

$$\frac{P_0 - P_1}{P_0} = N_2 \qquad 3$$

Equations 1, 2 and 3 are expressions of Raoult's law.

Whereas the escaping tendency of a real gas dissolved in a real liquid is proportional to its mole fraction, the proportionality constant is not in general P_0 but is Henry's constant K and

$$P_2 = KN_2 \qquad 4$$

where P_2 is the partial pressure of dissolved gas usually expressed in millimeters of

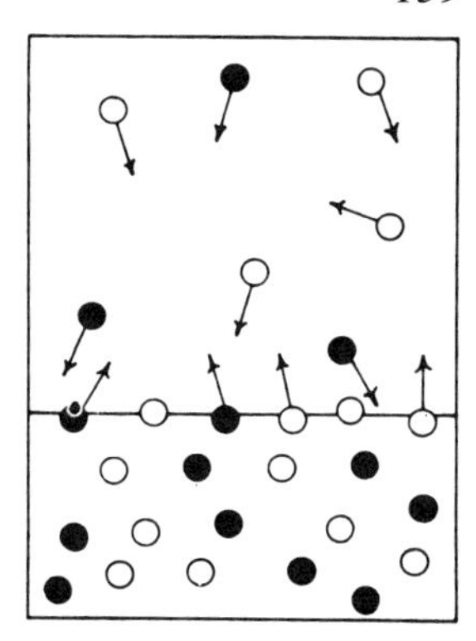

Fig. 1. Diagrammatic representation of a solution composed of two kinds of particles (open and closed circles) in equilibrium with the vapor mixture.

mercury and N_2 is the mole fraction of the dissolved gas. For example, 0.168×10^7, 7.51×10^7 and 4.04×10^7 are the Henry constants respectively for carbon dioxide, for nitrogen and for oxygen dissolved in water at 38°C. Therefore, per 1,000 grams of water at 38° and at a pressure of 760 mm. mercury, 0.0251 moles of carbon dioxide, 0.0005 moles of nitrogen and 0.00104 moles of oxygen would be dissolved. One thousand grams of physiological saline will dissolve 0.02417 moles of carbon dioxide under the same physical conditions.

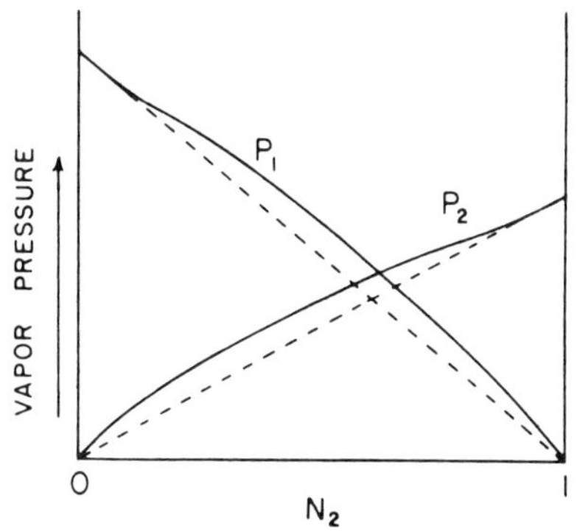

Fig. 2. Hypothetical vapor pressure relations of a solution of two volatile components. Dotted lines represent Raoult's law.

Figure 2 diagrams the vapor pressure relations of a solution of two volatile components with positive deviations from Raoult's law. It is to be noted that the "solvent" tends to obey Raoult's law as the concentration of the solute becomes very small, whereas the solute obeys Henry's law.

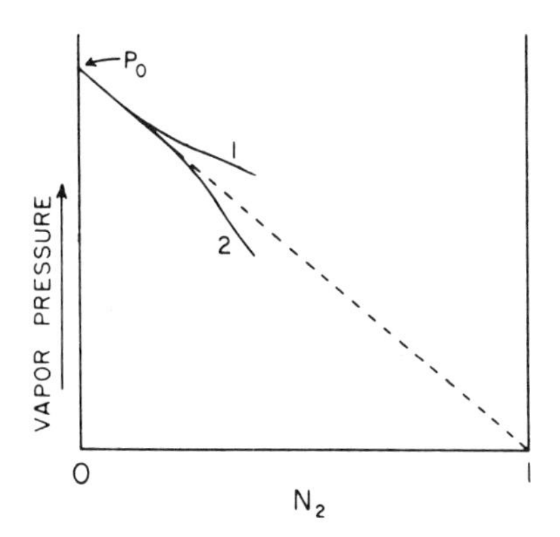

Fig. 3. Diagrammatic representation of vapor pressure of a volatile solvent containing a non-volatile solute. Curve 1 represents a positive deviation from Raoult's law; Curve 2, a negative deviation.

Figure 3 shows the vapor pressure of a volatile solvent containing a non-volatile solute. The solution approaches ideal behavior as the concentration of the solute is decreased.

Equation 3 can be written for dilute solution with $n_1 >> n_2$, as

$$\frac{P_0 - P_1}{P_0} = \frac{n_2}{n_1} = \frac{w_2}{M_2} \times \frac{M_1}{w_1} \qquad 5$$

where w_1 and w_2 are the respective weights of components 1 and 2 and M_1 and M_2 are their molecular weights respectively. The vapor pressure lowering of water produced by dissolving 10 grams of a protein whose molecular weight is 10,000 in 1,000 grams of water at 25°C would be from eq. 5

$$P_0 - P_1 = \Delta P = \frac{23.756 \times 18 \times 10}{1{,}000 \times 10{,}000} = 0.000427 \text{ mm Hg} \qquad 6$$

where 23.756 is the vapor pressure, P_0, of pure water at 25°C. Evidently, vapor pressure lowering would be an inappropriate method for the evaluation of the molecular weight of proteins and other high molecular weight substances; the mere trace of a low molecular weight substance such as glucose would lead to error.

The activity of a volatile solvent such as that of water is simply

$$\text{Activity} = \frac{P_1}{P_0} \qquad 7$$

where P_1 is the partial pressure of the solvent vapor in equilibrium with the solution and P_0 is the vapor pressure of the pure solvent at the same temperature.

There is an explicit relation between the activities of all the components of a system; it is impossible to change the activity of any one component without altering the activities of all other components. In Chapter 2 it was shown that for a two-component system at equilibrium

$$\mu_1 \, dn_1 + \mu_2 \, dn_2 = 0 \qquad 8$$

where μ_1 and μ_2 are the chemical potentials of the solvent and solute respectively, and n_1 and n_2 are the number of moles of each. For a solution at constant temperature and pressure

$$n_1\mu_1 + n_2\mu_2 = \Delta F \qquad 9$$

which upon differentiation gives

$$n_1 d\mu_1 + \mu_1 dn_1 + n_2 d\mu_2 + \mu_2 dn_2 = \Delta F \qquad 10$$

and at equilibrium eq. 10 becomes

$$n_1 d\mu_1 + \mu_1 dn_1 + n_2 d\mu_2 + \mu_2 dn_2 = 0 \qquad 11$$

Comparison of eqs. 11 and 8 gives

$$n_1 d\mu_1 + n_2 d\mu_2 = 0 \qquad 12$$

Since the activity, a, is related to the chemical potential by

$$\mu = \mu_0 + RT \ln a \qquad 13$$

$$n_1 d \ln a_1 + n_2 d \ln a_2 = 0 \qquad 14$$

Replacing n_1 and n_2 by mole fractions and remembering that the sum of the mole fractions must be unity, eq. 14 reduces to

$$N_1 d \ln a_1 + (1 - N_1) d \ln a_2 = 0 \qquad 15$$

which shows that in a two-component system the change of the activity of one component in respect to composition completely determines how the activity of the other component will change with composition.

Measurement of Vapor Pressure. The vapor pressure of a liquid or of a solution can be measured quite simply by determining the height of a column of mercury or of Cenco Hyvac pump oil (density of 0.895 at 25°C) which the vapor will support as diagramed in Figure 4.

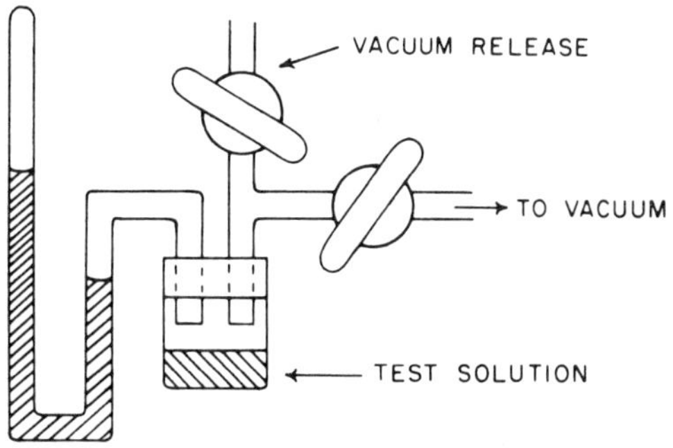

Fig. 4. Simple manometric method for vapor pressure.

There are many variations of the manometric method for vapor pressure. The vapor pressure of volatile liquids or of gas adsorbed on solids can be determined by this technique. For example, the McBain balance makes use of a delicate quartz spiral spring mounted inside a vertical arm of the vacuum line. Attached to the end of the spring is a platinum bucket containing the solid. The amount of adsorption on the solid as a function of the vapor pressure of the volatile compounds can thus be measured.

Isopiestic Method. Vapor will distill from a solution of higher vapor pressure and condense into the one having the lower pressure if given the opportunity to do so. At equilibrium, both solutions exert the same vapor pressure and, if the two solutions contain the same volatile solvent but different non-volatile solutes are analyzed, the vapor pressure of the solvent as a function of the concentration of the solutes is determined. It is further necessary that the vapor pressure of one of the solutions be known and serve as a standard.

Figure 5 shows the relative water vapor pressure at 25° of solutions of sorbitol, urea and several electrolytes as a function of the molal concentration of the solute.

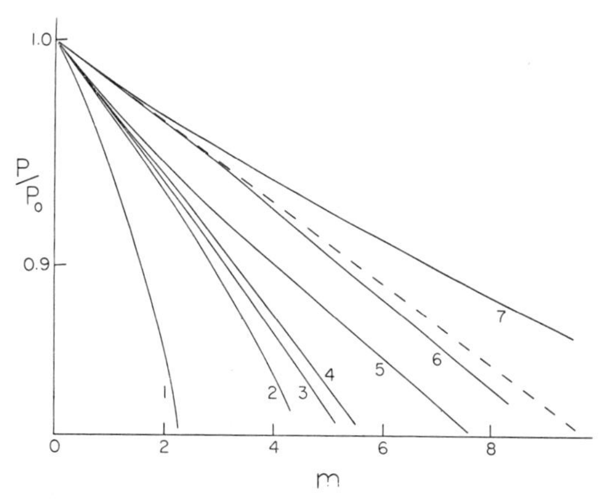

Fig. 5. The relative water vapor pressure P/P_0, plotted against the molal concentration of solutes. 1, $CaCl_2$; 2, NaCl; 3, KCl; 4, CsCl; 5, guanidine · HCl; 6, sorbitol; 7, urea. Dashed line is ideal curve for undissociated solute, 25°.

The isopiestic method is especially appropriate for the study of the hydration of biopolymers. The reference solutions of known water vapor pressure are placed in the bottoms of small vacuum dessicators along with magnetic stirring bars. The protein or other biopolymer, either in the solid state or in concentrated aqueous solution, is placed in low form weighing bottles and the weighing bottles floated in the reference solutions in the jars. The dessicators are evacuated and placed in a constant temperature bath and the reference solutions stirred with magnetic stirrers. After the samples have reached equilibrium, they are weighed and reweighed after drying to constant weight. The loss in weight represents the water uptake by the samples. Figure 6 shows the grams of water adsorbed per 100 grams of dry protein as a function of the relative vapor pressure of water at 25°C. Figure 7 shows similar relations for water uptake by calf thymus DNA, histone and one-to-one mixtures of DNA and histone.

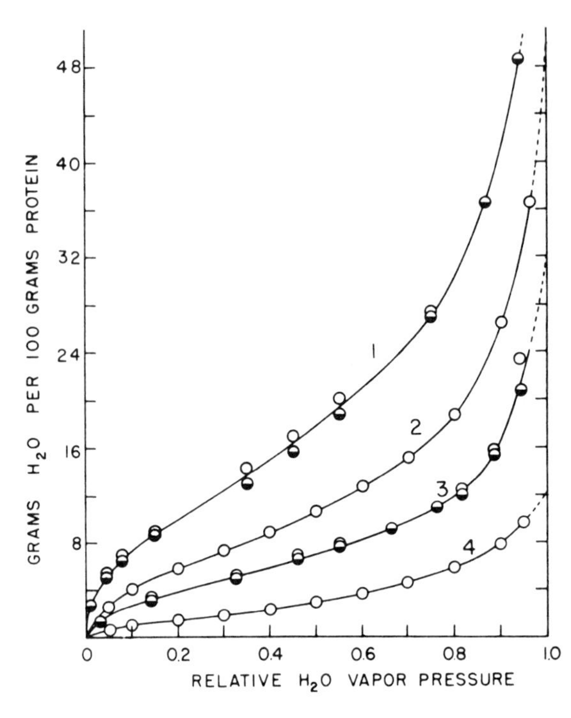

Fig. 6. Water adsorption curves for unstretched nylon (4), for wet (open circles) and dry (half circles) silk (3), for unlyophilized egg albumin (2), and for wet (open circles) and dry (half circles) collagen (1) at 25°C. (H. B. Bull: J. Amer. Chem. Soc. *66*:1499, 1944.)

The free energy of hydration can be calculated in the following manner. It was noted in Chapter 2, Energetics, that the free energy change for the transfer of a mole of a substance from activity a_1 to activity a_2 at constant temperature is

$$\Delta F = RT \ln \frac{a_2}{a_1} \qquad 16$$

Correspondingly, to transfer a mole of water from pure water (unit activity) to the protein where the water activity is P/P_0, P being the vapor pressure of water in equilibrium with adsorbed water and P_0 the vapor pressure of pure water, is

$$\Delta F = RT \ln \frac{P}{P_0} \qquad 17$$

and to transfer dn moles of water

$$dF = RT \ln \frac{P}{P_0}\, dn \qquad 18$$

Equation 18 can be rewritten as

$$dF = RTd\left(n \ln \frac{P}{P_0}\right) - RTn\, d \ln \frac{P}{P_0} \qquad 19$$

which on integration gives

$$\Delta F = nRT \ln \frac{P}{P_0} - RT \int_0^P \frac{n\, d\, P}{P} \qquad 20$$

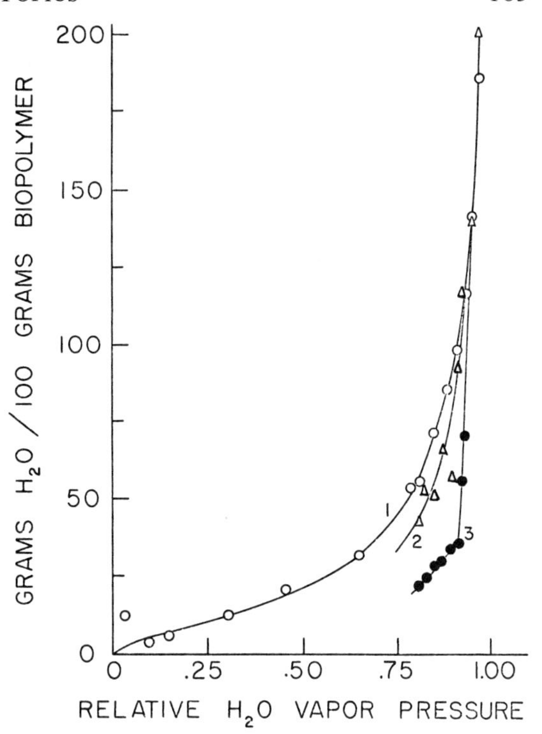

Fig. 7. Water adsorption curves for: 1, calf thymus DNA (2.86×10^{-4} mole NaCl per g. DNA); 2, 1:1 histone–DNA mixture (2.21×10^{-4} mole NaCl per g. mixture); 3, calf thymus histone (2.05×10^{-4} mole NaCl per g. histone). 25°. (Private communication from D. K. Chattoraj.)

The first term on the right side of eq. 20 gives the free energy change for the expansion of water vapor from P_0 to P and the second term gives the free energy of expansion from P, the vapor pressure in equilibrium with the adsorbed water to zero vapor pressure. Thus the process of hydration consists of two steps: (1) the vapor pressure of pure water is expanded to a pressure P and (2) the water vapor at pressure P is transferred to the dry material in infinitesimal increments until n moles of water have been adsorbed and P is the vapor pressure in equilibrium with the adsorbed water, pure water and completely dry protein being the standard reference state. The first term on the right side of eq. 20 can be evaluated directly by substituting the appropriate values and the second term is estimated by plotting n/P against P and determining the area under the curve. It will be noticed from eq. 20 that the energy of hydration of protein in equilibrium with pure water is

$$\Delta F = -RT \int_0^{P_0} \frac{n\, dP}{P} \qquad 21$$

Table 1 gives the free energies of hydration of 100 grams of dry protein at 25°C by water vapor.

TABLE 1. FREE ENERGIES (ΔF) AND HEATS (ΔH) OF HYDRATION OF 100 GRAMS OF DRY PROTEIN AT 25°C.

Protein	ΔF_{25}	ΔH
Nylon, unstretched	− 254	− 510
Nylon, stretched	− 244	− 540
Silk	− 656	−1270
Wool	− 975	−1950
Salmin	−1450	−1800
Elastin	−1000	−2250
Collagen	−1620	−3640
Egg albumin, native	− 956	−1350
Egg albumin, heat coagulated	− 814	−1400
β-Lactoglobulin, lyophilized	− 917	−1610
β-Lactoglobulin, crystals	−1020	− 940
Serum albumin, horse	−1034	−1450

With some modifications and within certain limits, the isopiestic method provides a means by which the binding of water and of added small molecular weight solute molecules by proteins or other biopolymers can be measured simultaneously.

There is one dependent variable, i.e., the amount of water in the sample at equilibrium. There are four independent variables: (1) temperature, (2) the amount of protein, (3) the amount of solute added to the sample and (4) the water vapor pressure of the reference solution.

The temperature can be held constant and all results can be referred to a constant amount of protein (one gram), and, accordingly, we are left with two independent variables, the concentration of the reference solution and the amount of solute added to the protein. These two can be reduced to one independent variable by holding one constant while varying the other.

A simple theory of the addition of small solute molecules to protein can be formulated along the following lines. Dry protein is added to an aqueous solution containing n_1 moles of water and n_2 moles of low molecular weight solute, for example, urea. The protein binds Δn_1 moles of water and Δn_2 moles of urea. The protein solution is now allowed to come to equilibrium with an isopiestic reference solution containing n_1' moles of water and n_2' moles of urea. If the protein is sufficiently dilute so that it makes no contribution to the lowering of the water vapor pressure, the following statement is correct

$$\frac{n_2 - \Delta n_2}{n_1 - \Delta n_1} = \frac{n_2'}{n_1'} \qquad 22$$

Equation 22 can be rearranged in a variety of ways; a useful form is

$$n_1 m_2' - 55.506\ n_2 = m_2' \Delta n_1 - 55.506\ \Delta n_2 \qquad 23$$

n_1, Δn_1, n_2 and Δn_2 are expressed in moles, n_1' is equal to the moles of water in 1,000 grams and, accordingly, n_2' becomes the molality of the reference solution expressed as m_2'. If Δn_1 and Δn_2 are indeed constant, a plot of $[n_1 m_2' - 55.506\ n_2]$ against m_2' should be linear, and from such a plot Δn_1 and Δn_2 can be obtained out of hand. It is to be expected that, over at least part of the range of the concentration of the solution, Δn_1 and Δn_2 will not be constant and no linear relation will be obtained by the above plot; however, eq. 23 will still be valid.

Even if Δn_1 and Δn_2 are dependent on m_2', it is possible to draw some useful qualitative conclusions from a plot of eq. 23. For example, the surface mole fraction of the solute bound to the protein must be less than, equal to, or greater than the solute mole fraction in the reference solution if the function $[n_1 m_2' - 55.506\ n_2]$ is positive, zero, or negative respectively. It is also clear that the slope of the plot can never be negative if Δn_1 and Δn_2 are constant and independent of m_2'. Any portion

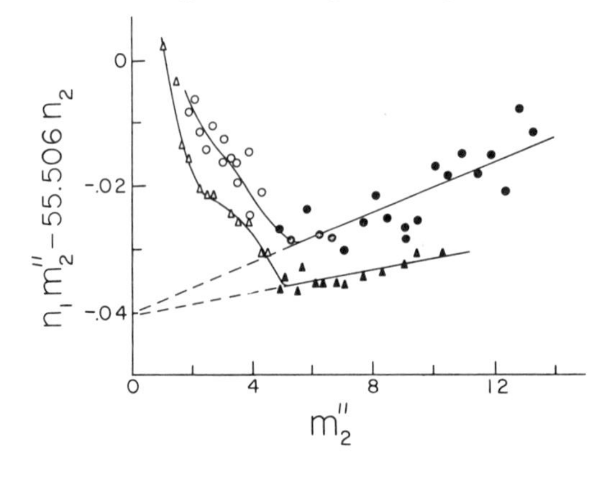

Fig. 8. The function $[n_1 m_2' - 55.506 n_2]$ plotted against the molal concentrations of the reference solutions. Circles, urea; triangles, guanidine · HCl. Solid circles and solid triangles have been used to calculate least square lines as shown. 25°. (H. B. Bull and K. Breese: unpublished results.)

of the plot suspected of linearity cannot extrapolate to a positive value at zero concentration of the reference solution. In keeping with the above theory, bound solute and solvent molecules are defined as those which cannot participate in the colligative properties of the protein solution. Adsorption from solution can be positive, zero, or negative whereas binding can be zero or positive, but never negative. Figure 8 shows plots of the function $[n_1 m_2' - 55.506\ n_2]$ against the molal concentration of the reference solutions of urea and of guanidine · HCl in the presence of egg albumin.

The values of Δn_1 and Δn_2, estimated from the linear portions of the plots shown in Figure 8, are 89 moles of water and 33 moles of urea bound per mole of egg albumin. 41 moles of water and 33 moles of guanidine · HCl are bound per mole of egg albumin. The molecular weight of egg albumin is 45,000. It is to be recalled that both urea and guanidine · HCl are potent protein denaturants.

Heat of Vaporization. The heat of vaporization of a liquid may be calculated quite simply by measuring the vapor pressure of the liquid at equilibrium with the liquid at two different temperatures and then substituting the results so obtained in the Clausius-Clapeyron equation. The process is to be visualized as follows: The conversion of a liquid to a vapor may be represented by a reaction

$$\text{Liquid} \rightleftharpoons \text{Vapor} \qquad 24$$

The equilibrium constant can be represented by P_1 the vapor pressure of the liquid. According to the Van't Hoff equation which was discussed in Chapter 2, the equilibrium constant is related to the heat of the reaction by

$$\frac{d \ln K}{dT} = \frac{\Delta H}{RT^2} \qquad 25$$

Then substituting the value of P for the equilibrium constant, there results

$$\frac{d \ln P}{dT} = \frac{\Delta H}{RT^2} \qquad 26$$

which is an expression of the Clausius-Clapeyron equation where ΔH is the heat of vaporization of the liquid.

The integration of eq. 26 leads immediately to

$$\ln \frac{P_2}{P_1} = -\frac{\Delta H}{R}\left(\frac{1}{T_2} - \frac{1}{T_1}\right) \qquad 27$$

which permits the calculation of the molar heat of vaporization from the vapor pressure P_2 at temperature T_2 and the vapor pressure P_1 at temperature T_1.

The heats of hydration of a protein or other biopolymer can also be calculated by the use of the Clausius-Clapeyron equation, although it is best to use graphical integration for this purpose. The partial molal heat of hydration can be written

$$-q = \frac{d\Delta H}{dn} = \frac{RT_1T_2}{T_2 - T_1} \ln \frac{x_1}{x_2} \qquad 28$$

where x_1 and x_2 are the relative vapor pressures at temperatures T_1 and T_2 respectively which produce the same extent of hydration of the protein. If q is plotted against n and the area under the curve measured, the integral heat of hydration is found (see Table 1).

The Hill method for the determination of the vapor pressure of biological fluids has several advantages. The apparatus consists of a thermopile of many thermocouples wound on a thin flat insulator with the junctions of one kind on one side and the junctions of the other kind on the reverse side of the flat insulator. On one side is placed a piece of filter paper saturated with a standard solution and on the other a piece of filter paper saturated with the test solution (biological fluid). The rate of evaporation of the water is proportional to its vapor pressure; accordingly, the solution with the higher vapor pressure is cooler and this is recorded as a difference in potential between the two sides of the thermopile.

The Mechrolab Inc. has produced a commercial instrument utilizing the Hill principle under the deceptive name of Osmometer which is very effective for relatively small vapor pressure depressions. Use is made of thermistors instead of thermocouples.

Freezing Point Depression. It is observed that, if a solute is added to a solvent, the freezing point of the solution is less than that of the pure solvent. The depression of the freezing point by the solute may be regarded in the following way: The conversion of the solid, frozen solvent into a liquid (melting) can be expressed as a reversible reaction

$$\text{Solid} \rightleftharpoons \text{Liquid} \tag{29}$$

the equilibrium constant of which is

$$K = \frac{N_1}{N_s} \tag{30}$$

where N_1 is the mole fraction of the liquid solvent in solution and N_s is the mole fraction of the frozen solvent which is unity. Then K is equal to N_1 and the integrated Van't Hoff equation can be written

$$\ln N_1 = -\frac{\Delta H}{R}\left(\frac{T_0 - T}{TT_0}\right) \tag{31}$$

where T_0 is the melting point of the pure solvent and T is the melting point of the solution. ΔH is the heat of fusion of the solvent and T and T_0 are very nearly equal. N_1 can be replaced by $(1 - N_2)$; then eq. 31 becomes

$$\ln (1 - N_2) = -\frac{\Delta H \Delta T}{RT_0^2} \tag{32}$$

Expanding the log term and rejecting all terms except the first and realizing that $n_1 >> n_2$, eq. 32 can be written

$$\Delta T = \frac{w_2 M_1 R T_0^2}{M_2 w_1 \Delta H} \tag{33}$$

where w_1 is the weight of the solvent, M_1 is its molecular weight and w_2 is the weight of the solute and M_2 is its molecular weight. The freezing point depression of 1,000 grams of water produced by the addition of one mole of the solute is

$$\Delta T = \frac{1.987 \times 18.016 \times (273.15)^2}{1436.13 \times 1{,}000} = 1.860 \tag{34}$$

where 1436.13 is the molar heat of fusion of water.

If 10 grams of protein whose molecular weight is 10,000 is dissolved in 1,000 grams of water, the freezing point depression produced by the protein would be 0.00186°C which is so small that it could hardly be detected much less measured with accuracy.

Provided a sensitive thermometer is available, it is possible to construct a simple and inexpensive freezing point depression apparatus. The arrangement for such equipment is diagramed in Figure 9.

Freezing point depression studies have long been of interest and of importance in biochemical and physiological investigations. Not only is the method appropriate for the determination of the molecular weights of small molecules but the method

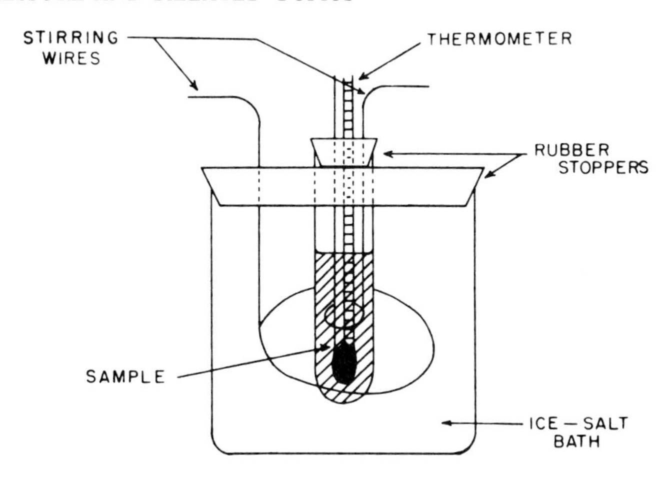

Fig. 9. Freezing point depression apparatus.

has also led to much that is known about the electrolyte and water balance of the body.

Early studies on the freezing point depression of tissue seemed to show that the freezing point of tissue was considerably below that of blood plasma. Such large differences in freezing point would indicate tissue to be very hypertonic relative to its extracellular fluid. There are good reasons for believing that the lower freezing point of tissue is due to autolysis of the tissue. Maffly and Leaf have approached this problem in a different way; they have studied the melting point of water in tissue and in blood serum. Tissues were removed from anesthetized animals and dropped immediately into liquid nitrogen. The frozen material was warmed at a constant rate and the increase in the temperature of the suspension was measured with a thermistor. There was no significant difference between the melting curves of rat liver and of rat blood serum. Similar results were obtained with skeletal muscle, heart and brain. Thus the water activity of mammalian cells is the same or very nearly the same as that of the extracellular fluids.

Osmotic Pressure. Consider the distillation of a solvent from pure solvent into a solution as diagramed in Figure 10. A cellophane membrane supports the solution

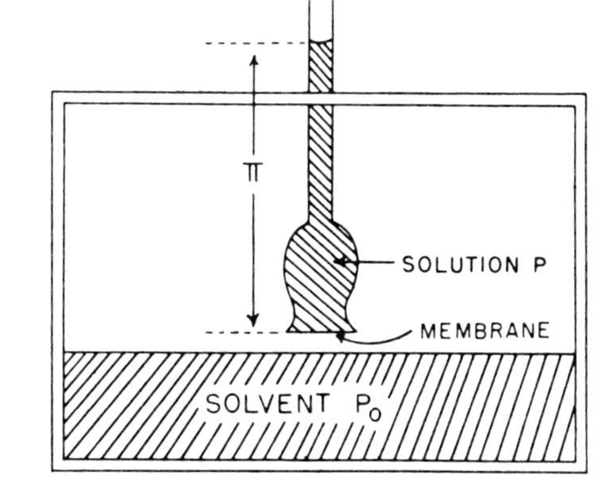

Fig. 10. Distillation of solvent into the solution with the production of an osmotic pressure, II.

and water molecules can diffuse through this membrane; the membrane is attached to an inverted tube ending in a capillary. The solvent and solution are enclosed in an air-tight box filled with an inert gas at atmospheric pressure.

Since the vapor pressure of the pure solvent is greater than that of the solution, solvent will distill into the solution until the hydrostatic pressure built up inside the tube containing the solution prevents the further distillation; the system is now at equilibrium and the hydrostatic pressure is known as the osmotic pressure (Π).

The vapor pressure of the pure solvent is P_0 and that of the solution is P. If one mole of the solvent is transferred from the pure solvent to the solution, the free energy gained is

$$\Delta F = RT \ln \frac{P}{P_0} \qquad 35$$

and the work required to bring about this transfer is $\Pi\overline{V}_1$ where $\overline{V}_1$ is the partial molar volume of the solvent. At equilibrium, the total free energy change is zero and

$$\Pi\overline{V}_1 + RT \ln \frac{P}{P_0} = 0 \qquad 36$$

or

$$\Pi = -\frac{RT}{\overline{V}_1} \ln \frac{P}{P_0} = -\frac{RT}{\overline{V}_1} \ln a_1 \qquad 37$$

Since P/P_0 is equal to the activity of the solvent (a_1), it is to be noted from eq. 37 that the osmotic pressure is closely related to the activity of the solvent. From Raoult's law P/P_0 is equal to $\left(1 - \frac{n_2}{n_1 + n_2}\right)$ or

$$\Pi = -\frac{RT}{\overline{V}_1} \ln \left(1 - \frac{n_2}{n_1 + n_2}\right) \qquad 38$$

Expanding the logarithm and rejecting higher terms

$$\Pi = \frac{RT}{\overline{V}_1} \frac{n_2}{n_1 + n_2} \qquad 39$$

Since, in dilute solutions, $n_1 >> n_2$ and $\overline{V}_1$ is very nearly equal to M_1/ρ_1, where M_1 and ρ_1 are the molecular weight and the density respectively of the solvent

$$\Pi = \frac{\rho_1 RT n_2}{M_1 n_1} \qquad 40$$

Since $M_1 n_1$ equals the total weight (w_1) of the solvent in solution,

$$\Pi = \frac{\rho_1 RT n_2}{w_1} \qquad 41$$

Since n_2 is the number of moles of the solute dissolved in weight w_1 of the solvent and if w_1 is set equal to 1,000, eq. 41 becomes

$$\Pi = \frac{\rho_1 RT w_2}{1{,}000\ M_2} \qquad 42$$

where w_2 is the grams of solute dissolved in 1,000 grams of solvent and n_2 is equal to w_2/M_2.

If the osmotic pressure is expressed in centimeters of water, R has the value

$$R = \frac{22{,}414 \times 76.0 \times 13.597}{273.15} = 8.48 \times 10^4 \text{ cc. cm. } H_2O \text{ per mole per degree} \qquad 43$$

Substituting the values for the constants at 25°C into eq. 42 there results for aqueous solutions

$$\Pi = \frac{2.528 \times 10^4 \rho_1 w_2}{M_2} \qquad 44$$

If 10 grams of protein whose molecular weight is 10,000 are dissolved in 1,000 grams of water at 25°C the resulting osmotic pressure should be 25.3 cm. of water pressure. Such a pressure can be accurately measured with an appropriate manometer and, accordingly, osmotic pressure, unlike the vapor pressure lowering and the freezing point depression, can be used to measure the molecular weight of proteins and other biopolymers. Indeed, in careful and skilled hands, osmotic pressure measurement in the range of molecular weights of 15,000 to 100,000 is the most unambiguous method available for molecular weight determination.

An osmotic pressure measurement, in effect, counts the number of particles which cannot pass through a membrane; a particle (molecule or ion) which can pass through the membrane registers no osmotic pressure unless its passage is constrained in some way (Donnan equilibrium). As long as the ideal solution laws are obeyed or very nearly obeyed, osmotic pressure can reveal no specific property of the molecule other than its molecular weight. Departure from ideal behavior does, however, depend, albeit in a complex manner, upon the specific nature of the solute and solvent molecules.

Osmotic pressure measurements on a polydispersed solution of macromolecules give a number average moleculear weight defined by the relation

$$\frac{N_1M_1 + N_2M_2 + N_3M_3}{N_1 + N_2 + N_3} \text{ etc} = \frac{\Sigma N_iM_i}{\Sigma N_i} = M_n \qquad 45$$

where N_i is the number of moles of the macromolecule whose molecular weight is M_i. Obviously, if there is only one species of macromolecule present, then M_n is equal to M_i.

The relation between the vapor pressure lowering, the freezing point depression and the osmotic pressure for ideal solutions is summarized in Table 2.

TABLE 2. RELATION BETWEEN VAPOR PRESSURE LOWERING (ΔP), FREEZING POINT DEPRESSION (ΔT) AND OSMOTIC PRESSURE Π FOR IDEAL SOLUTION

Vapor Pressure		*Freezing Point*		*Osmotic Pressure*
$\Delta P = \frac{n_2}{n_1} P_0$		$\Delta T = \frac{n_2}{n_1} \frac{RT_0^2}{\Delta H_f}$		$\Pi = \frac{n_2}{n_1} \frac{RT}{\overline{V}_1}$
$\frac{\Delta P}{P_0}$	=	$\frac{\Delta T \Delta H_f}{RT_0^2}$	=	$\frac{\Pi \overline{V}_1}{RT}$

Osmotic Pressure Measurements. Over the years there has been considerable experimental work on osmotic pressure. Industrial laboratories have used the method to characterize polymers of industrial interest. Biochemists have confined most of their measurements to protein solutions and among these studies should be mentioned the early and now classical work of Sörensen on egg albumin, of Adair on hemoglobin and on other proteins. The papers of Scatchard can be read with profit.

Figure 11 illustrates a simple osmometer which has been widely used and modified. A capillary correction has to be applied to the indicating column of n-decane. The technique of osmotic pressure measurements has been greatly advanced by the development of high speed osmometers which are commercially available. These instruments employ a feedback circuit which imposes a counter pressure to maintain the level of the meniscus of the protein solution constant. With proper preparation of the apparatus and the membrane, measurements can be completed in a few minutes.

Equation 44 is valid only for ideal solutions, but all solutions behave in an ideal manner at infinite dilution. Accordingly, if the ratio π/w_2 be extrapolated to zero concentration (w_2 equals zero), the molecular weight can be calculated unambiguously from the intercept on the y-axis. Figure 12 shows such extrapolations for a series of native proteins in 0.1M buffers at or near their respective isoelectric points. The results leave little doubt as to the correct molecular weight. Figure 13 exhibits

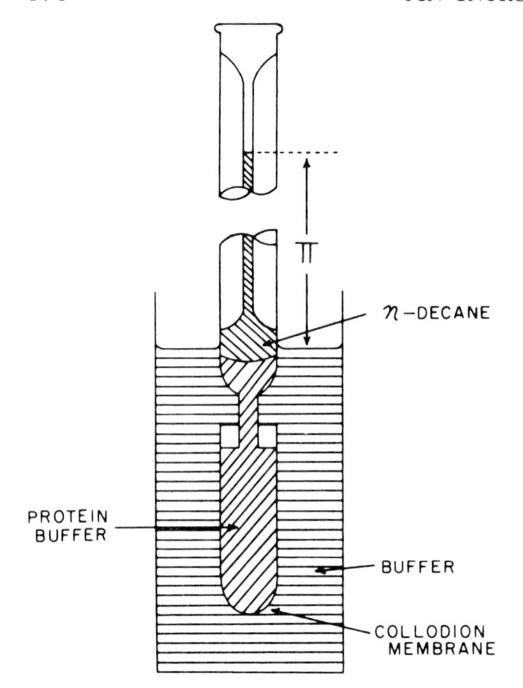

Fig. 11. Static osmometer.

similar plots for the same series of proteins dissolved in 6M guanidine hydrochloride; 6M guanidine hydrochloride is a potent protein denaturant. Evidently, denatured proteins show much greater departure from ideality with increasing protein concentration than do the native proteins.

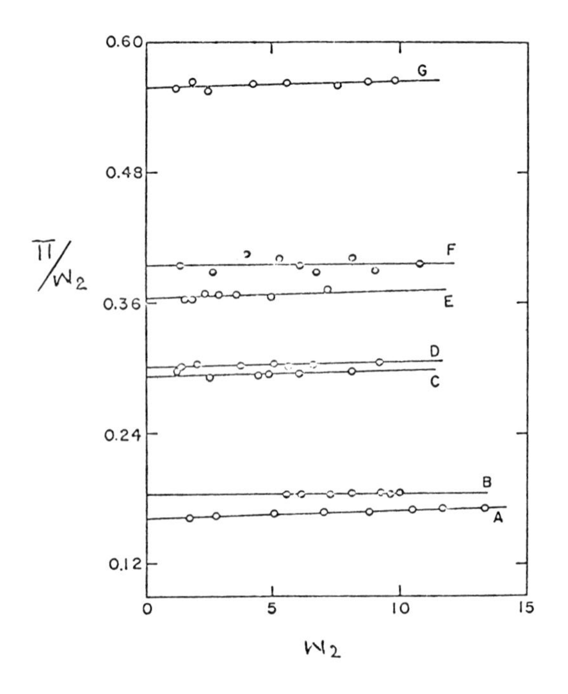

Fig. 12. Plots of π/w_2 vs w_2 for native proteins. π is expressed in centimeters of water, w_2 in grams per liter. (A) Aldolase, (B) lactate dehydrogenase, (C) enolase, (D) alcohol dehydrogenase, (E) serum albumin, (F) methemoglobin, and (G) egg albumin. (F. J. Castellino and R. Barker: Biochem. *7:* 2207, 1968.)

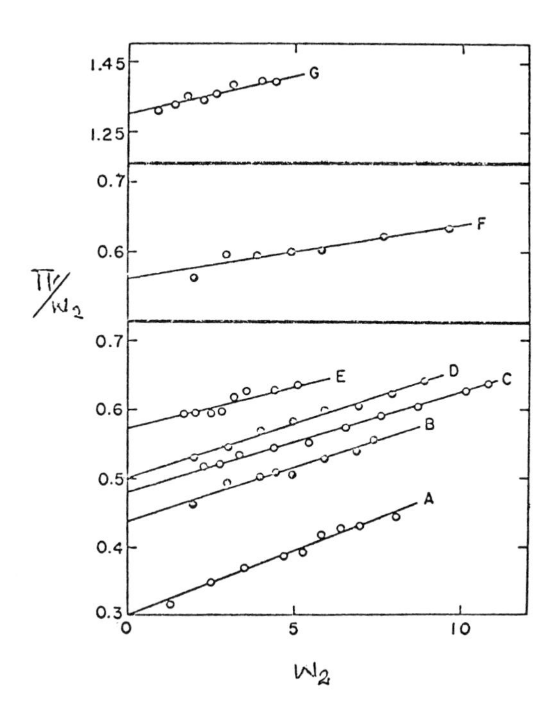

Fig. 13. Plots of π/w_2 vs w_2 for proteins dissolved in 6M guanidine hydrochloride. π is expressed in centimeters of water, w_2 in grams per liter. (A) Serum albumin, (B) egg albumin, (C) aldolase, (D) alcohol dehydrogenase, (E) lactate dehydrogenase, (F) enolase, and (G) methemoglobin. (F. J. Castellino and R. Barker: Biochem. *7:* 2207, 1968.)

Departure from Ideality. As the concentration of the solute (protein) increases, progressive departures of the measured osmotic pressure from ideality are to be expected. Such departure is shown in Figure 13. A clearer statement of this situation is shown diagrammatically in Figure 14 in which Π/w_2 is plotted against w_2 where w_2 is the weight concentration of the biopolymer. The results shown in Figure 14 can be expressed by the equation

$$\frac{\Pi}{w_2} = \frac{\rho_1 RT}{w_1 M_2} + RTBw_2 \tag{46}$$

where B is the so-called interaction constant which expresses the deviation from ideality and into whose nature we will now inquire.

As noted in Chapter 2, for any isothermal process at constant pressure

$$\Delta F = \Delta H - T\Delta S \tag{47}$$

and the partial molar quantities of the solvent can be written

$$\Delta\overline{F}_1 = \Delta\overline{H}_1 - T\Delta\overline{S}_1 \tag{48}$$

The free energy change $\Delta\overline{F}_1$ for the transfer of one mole of the solvent in a reversible process from the solution to the solvent is $-\Pi\overline{V}_1$ and substituting this information into eq. 48, there results

$$\Pi = \frac{T\Delta\overline{S}_1}{\overline{V}_1} - \frac{\Delta\overline{H}_1}{\overline{V}_1} \tag{49}$$

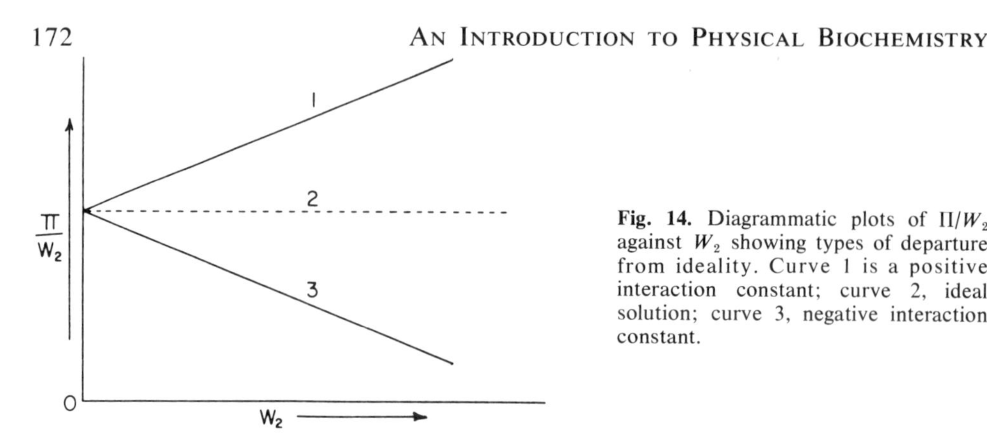

Fig. 14. Diagrammatic plots of Π/W_2 against W_2 showing types of departure from ideality. Curve 1 is a positive interaction constant; curve 2, ideal solution; curve 3, negative interaction constant.

The deviation of the osmotic pressure or, for that matter, the freezing point depression or the vapor pressure lowering could, therefore, be due to the entropy term or to the heat term or to both; these terms correspond to the entropy and heats of mixing of the solute and the solvent. If the heat term is small and can be neglected, eq. 49 reduces to

$$\Pi = \frac{T\Delta\overline{S}_1}{\overline{V}_1} \quad 50$$

The entropy of mixing can be calculated from the number of ways the solute and solvent molecules can be arranged in solution, i.e., the thermodynamic probability of the possible configurations in solution. The total number of available configurations will depend on the number, size and shape of the solvent and solute molecules. The configurational entropy of mixing for spherical solute and solvent molecules is

$$\Delta S = -R\,[n_1 \ln V_1 + n_2 \ln V_2] \quad 51$$

where n_1 and n_2 are the number of moles of the solvent and solute respectively and V_1 and V_2 are the corresponding volume fractions of the solvent and solute. Since the volume fractions are always less than unity, the entropy of mixing is positive and the process is spontaneous; the entropy of mixing favors solution stability.

If the solute and solvent molecules are spheres of equal size, eq. 51 reduces to

$$\Delta S = -R\,[n_1 \ln N_1 + n_2 \ln N_2] \quad 52$$

For $n_1 >> n_2$ and per mole of solvent, eq. 52 becomes

$$\Delta S = -R \ln N_1 \quad 53$$

From eqs. 50 and 53

$$\Pi = -\frac{RT}{\overline{V}_1} \ln N_1 \quad 54$$

which is identical with eq. 38 and can be looked upon as an expression of Raoult's law.

The nature of the interaction constant is complex and depends not alone on the configurational entropy as indicated above but also upon the heat of mixing as well as upon the entropy terms arising from the interaction between solute molecules. A study of the variation of the osmotic pressure with temperature could provide a basis for deciding which part of the interaction constant arises from the entropy and from the heat of mixing. B is the effective molecular volume of the solute. The volume is related to the co-volume or excluded volume and under most circumstances is a complicated function of the actual volume. In the case of hard, non-interacting spheres, the excluded volume is four times the molecular volume. Efforts have been made to interpret the co-volumes of proteins in aqueous solutions in terms of protein hydration but there appears to be no way in which the interaction constant B can be meaningfully related to hydration although hydration is undoubtedly involved.

As noted in Figure 14, B may be positive or negative; Table 3 summarizes the conditions which determine the sign of B.

TABLE 3. CONDITIONS INFLUENCING THE SIGN OF THE INTERACTION CONSTANT B

Positive Sign	*Negative Sign*
1. Solute molecules large relative to solvent molecules	1. Poor solvent
2. Elongated solute molecules (rigid rods or flexible chains)	2. Interaction between solute molecules
3. Good solvent	3. Positive heat of mixing
4. Negative heat of mixing	
5. Net electrostatic charge on solute molecules (Donnan equilibrium)	

Associating Systems. The reversible association of biopolymers is an important problem and frequently enters into and confuses measurements made on such systems by osmotic pressure, by sedimentation, by light scatter and by electrophoresis. This is not a simple problem. If the various non-diffusable components behaved in an ideal manner, the equilibrium constant for the association could be calculated from osmotic pressure by solving the appropriate simultaneous equations. Such calculations are, however, surprisingly awkward. The reversible reaction is

$$nB \rightleftharpoons B_n \qquad 55$$

where B is the monomer which polymerizes to form B_n. The equilibrium constant for this reaction is per unit volume

$$K = \frac{N_2}{(N_1)^n} \qquad 56$$

where N_2 is the number of moles of polymer and N_1 refers to the monomer. The moles of monomer plus polymer (found from osmotic pressure assuming ideal behavior) is

$$N = N_1 + N_2 \qquad 57$$

The following relation is also true

$$N_o = N_1 + n\, N_2 \qquad 58$$

where N_o is equal to w/M_1, w is the total weight of the biopolymer, and M_1 is the molecular weight of the monomer. Combining eqs. 56, 57 and 58 gives

$$K = \frac{(n-1)^{n-1}\,(N_o - N)}{(nN - N_o)^n} \qquad 59$$

If n is 2, the plot of $(N_o - N)$ vs $(2N - N_o)^2$ should be linear with a slope of K. Correspondingly, if n is 3, the plot of $(N_o - N)$ vs $(3N - N_o)^3$ should give a straight line with a slope of $K/4$.

R. F. Steiner has suggested for an ideal system involving the following reversible associations

$$B_1 + B_1 \overset{K_1}{\rightleftharpoons} B_2 \qquad 60$$

$$B_2 + B_1 \overset{K_2}{\rightleftharpoons} B_3 \qquad 61$$

The association constants K_1 and K_2 can be estimated as follows: The molar concentration of the monomer C_1 is obtained from the ratio C_1/C where C is the total molar concentration of the biopolymer. C is equal to w/M_n, w being the total weight of the biopolymer and M_n the number average molecular weight calculated from the osmotic pressure measurements assuming ideal behavior. The ratio C_1/C is found by graphical integration using the following relation

$$\ln C_1/C = \int_o^c \left(\frac{M_1}{M_n} - 1\right) d \ln C \qquad 62$$

where M_1 is the molecular weight of the monomer.

The association constants are given by

$$\frac{d}{dc_1}\left(\frac{C - C_1}{C_1}\right)_{C_1 \to 0} = K_1 \qquad 63$$

and by

$$\frac{d}{dC_1}\left(\frac{C - C_1 - K_1C_1^2}{C_1^2}\right)_{C_1 \to 0} = K_1K_2 \qquad 64$$

Thus the slope of a plot of $(C - C_1)/C_1$ against C_1 is equal to K_1 as C_1 approaches zero and correspondingly a plot of $\frac{(C - C_1 - K_1C_1)^2}{C_1^2}$ against C_1 as C_1 approaches zero gives K_1K_2 as the slope. Whereas association or dissociation can no doubt be detected by means of osmotic pressure, the actual numerical values of the constants are to be regarded with reservation. The only proper way to measure the molecular weight unambiguously from osmotic pressure measurements is to extrapolate a plot of Π/w_2 to zero value of w_2.

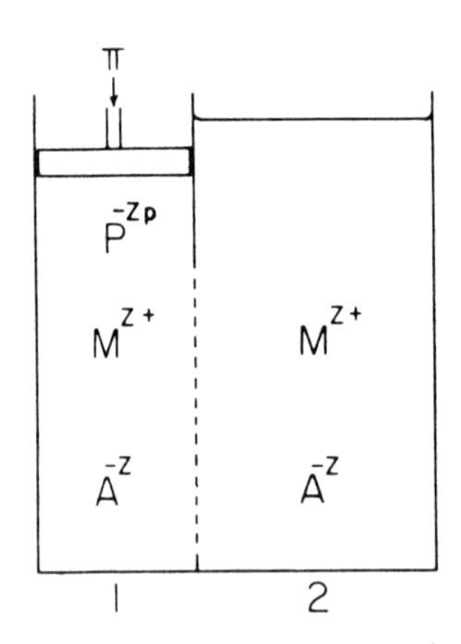

Fig. 15. Two compartments (1 and 2) separated by a semipermeable membrane (dotted line) permitting passage of all molecules and ions except the protein anion (P^{-z_p}). M^{z+} represents the diffusible cation of valence $z+$ and A^{z-}, the diffusible anion of valence $z-$. Π is the osmotic pressure.

Donnan Equilibrium. The net electrostatic charge on a biopolymer ion has a profound effect upon its behavior as was first realized by Donnan. When a charged particle is restrained a Donnan equilibrium results. Figure 15 shows a semipermeable membrane through which all constituents except the charged biopolymer (protein) ion can pass. When equilibrium has been attained and the pressure exerted on compartment 1 is equal to the osmotic pressure, the free energy of transfer of one mole of neutral salt from compartment 2 to compartment 1 at constant temperature and pressure is

$$\Delta F = n_M RT \ln \frac{M_1}{M_2} + n_A RT \ln \frac{A_1}{A_2} \qquad 65$$

where n_M and n_A are the number of moles of cation and anion respectively. Since, for a neutral salt, n_M is equal to z^- and n_A is equal to z^+, where z^- and z^+ are the valences of the anion and cation respectively, and, further, since at equilibrium ΔF is zero, eq. 65 can be rearranged to give

$$z^- \ln \frac{M_1}{M_2} = z^+ \ln \frac{A_2}{A_1} \qquad 66$$

from which

$$\left(\frac{M_1}{M_2}\right)^{\frac{1}{z^+}} = \left(\frac{A_2}{A_1}\right)^{\frac{1}{z^-}} \qquad 67$$

Providing z^+ and z^- are both unity, eq. 67 becomes

$$\frac{C_{M_1}}{C_{M_2}} = \frac{C_{A_2}}{C_{A_1}} = r \qquad 68$$

where r is the ion ratio characterizing the distribution of diffusible ions across the membrane. Since electrical neutrality must be maintained in both compartments,

$$C_{M_2} = C_{A_2} \qquad 69$$

and

$$C_{M_1} = C_{A_1} + Z_p C_p \qquad 70$$

where C_P is the molal concentration of the protein or of other ions to which the membrane is impermeable and Z_P is the valence of the large ion. At equilibrium, all the ion concentrations are interrelated and r may be expressed by any two of the concentrations. Thus by the proper substitutions and rearrangements of eqs. 68, 69, 70 (M and A are both univalent)

$$r = \left(\frac{C_{M_1}}{C_{M_1} + Z_P C_P}\right)^{1/2} \qquad 71$$

$$r = \left(1 + \frac{Z_P C_P}{C_{A_1}}\right)^{1/2} \qquad 72$$

$$r = \frac{Z_P C_P}{2\, C_{M_2}} + \left(1 + \frac{Z_P^2 C_P^2}{4\, C_{M_2}^2}\right)^{1/2} \qquad 73$$

To obtain an idea of the magnitude of r, suppose compartment 1 of Figure 15 contains 10 grams of protein per 1,000 grams of solvent. The molecular weight of the protein is 40,000. Further, suppose the protein has a negative valence of 10 in one series of experiments and a negative 40 in another series. Shown plotted in Figure 16 are the values of r corresponding to the equilibrium concentrations of M_2 for the two assumed valences of protein.

The osmotic pressure due to the difference in concentrations of the diffusible ions on the two sides of the membrane is

$$\Pi_i = RT\,(C_{M_1} + C_{A_1} - C_{M_2} - C_{A_2}) \qquad 74$$

Again making the restriction that M and A are univalent, the value of Π can be expressed in terms of $Z_p C_p$ and the concentration of any one of the diffusible ions by the use of eqs. 68, 69, and 70 to give

$$\Pi_i = 2\, RTC_{M_1}\left[1 - \left(1 - \frac{Z_p C_p}{C_{M_1}}\right)^{1/2} - \frac{Z_p C_p}{2\, C_{M_1}}\right]_{C_p \to 0} = \frac{RTZ_p^2 C_p^2}{4\, C_{M_1}} \qquad 75$$

$$\Pi_i = RTC_{A_1}\left[1 - \left(1 + \frac{Z_p C_p}{C_{A_1}}\right)^{1/2} + \frac{Z_p C_p}{2\, C_{A_1}}\right]_{C_p \to 0} = \frac{RTZ_p^2 C_p^2}{4\, C_{A_1}} \qquad 76$$

$$\Pi_i = 2\,RTC_{M_2}\left[\left(1 + \frac{Z_p^2 C_p^2}{4\,C_{M_2}^2}\right)^{1/2} - 1\right]_{C_p \to 0} = \frac{RTZ_p^2 C_p^2}{4\,C_{M_2}} \qquad 77$$

The expansion of the square root terms in eqs. 75, 76, and 77 and the retention of only the first term of the series in each case gives the expression on the extreme right. If the equivalent concentrations of the protein ion become comparable to that of any of the diffusible ions, the expansion series do not converge rapidly enough to be of help in numerical calculations and, in general, it is best to use the unexpanded equations. It is evident from eqs. 75, 76, and 77 that as C_p approaches zero Π_i/C approaches zero and the protein solution approaches ideal behavior in spite of the presence of a net charge. It can also be noted that at sufficiently high concentrations of diffusible ions Π approaches zero irrespective of the equivalent concentration of protein.

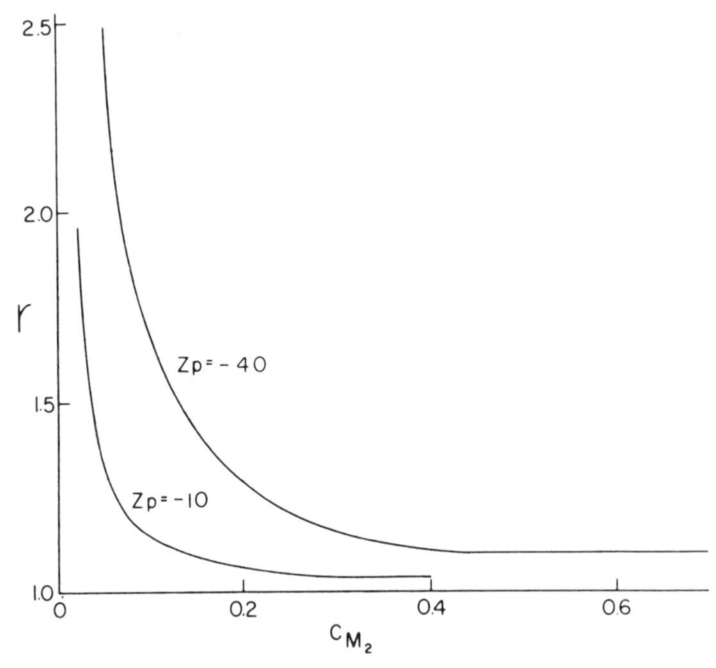

Fig. 16. Calculated values (see eq. 73) of r as functions of the salt concentration in compartment 2 (see Fig. 15). Ten grams of protein of molecular weight 40,000 dissolved in 1,000 grams of solvent. Valence (Z_p) of protein is -10 or -40 as indicated.

In the equation

$$\Pi = \frac{\rho_1 RTw_2}{M_2 w_1} + RTBw_2^2 \qquad 78$$

Π_i can be identified with RTB provided causes other than the Donnan equilibrium for departure from ideality can be neglected. Under these circumstances, RTB is Π_i/w_2^2 where w_2 as used previously (see eq. 42) is the grams of protein per 1,000 grams of water.

Numerical calculations are of aid in obtaining a notion of the importance of the Donnan equilibrium. Let it be assumed that the molecular weight of the protein is 40,000 and its valence is -10 (see Fig. 15). This imaginary osmotic pressure experiment is conducted in such a fashion that the salt concentration in compartment 2 is maintained constant first with C_{M2} equal to 0.01M and then for another series of

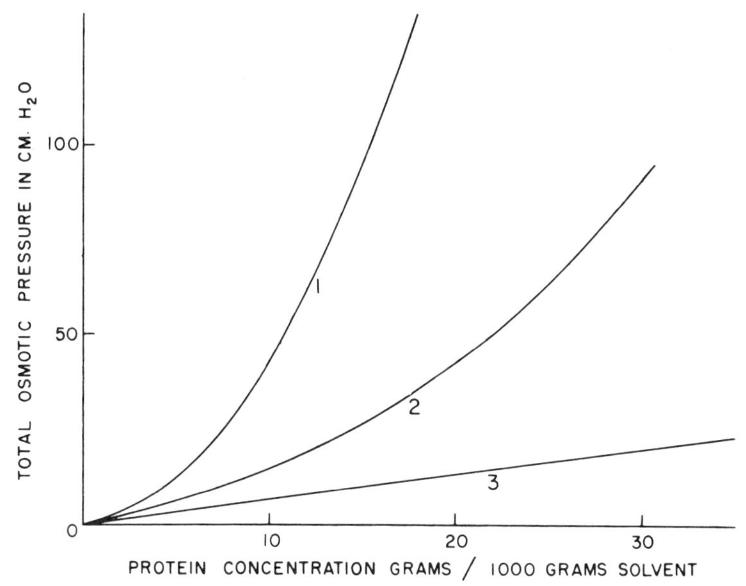

Fig. 17. Calculated osmotic pressures. Curve 1, C_{M2} equals 0.01M; curve 2, C_{M2} equals 0.05M; curve 3, ideal solution without Donnan effect. Z is 10 and molecular weight of protein is 40,000, 25°C.

experiments at 0.05 M, varying the protein concentration in both series. The temperature is to be at 25°C. Experimentally, this procedure would be awkward; it is much more easily conducted on paper. Shown plotted in Figure 17 are the total osmotic pressure values (the ideal pressure plus Π_i) calculated as a function of the protein concentration.

From an inspection of Figure 17 it is evident that the Donnan equilibrium can increase the osmotic pressure considerably above the ideal pressure; the Donnan equilibrium can lead to substantial departure from ideality.

Within the framework of their derivation, the Donnan equilibrium equations derived above are exact, but several factors have been neglected and these must now be considered. The departure of the osmotic pressure of a high molecular weight substance from ideality is due to several causes and these have been mentioned already. It is to be remembered that an osmotic pressure measures directly the logarithm of the activity of the solvent. The activities of all the components of a system are related and, accordingly, the activities of the diffusible ions as well as the activity of the biopolymer influence the activity of the solvent. The activities of the diffusible ions and that of the biopolymer are also mutually dependent. If the interaction between a diffusible ion and the biopolymer is sufficiently strong, the ion is considered to be bound to the biopolymer; the extent of ion binding can be estimated by means of the Donnan equilibrium. For example, it is possible to measure the pH of the inside and the outside compartments and thus obtain values of r. The concentration of the anion (say chloride) is measured in compartment 2 by accepted analytical methods and the concentration of chloride in compartment 1 is calculated from the known and corresponding values of r. The difference between the total chloride in compartment 1 and that in solution (as obtained from r) is the chloride ion bound by the protein.

Using radioactive Na^{24} as a tracer, Saifer and Steigman report satisfactory agreement between the r values calculated from pH measurements and the sodium ion distribution. Bovine serum albumin was the protein employed. Figure 18 shows a plot of some of their results.

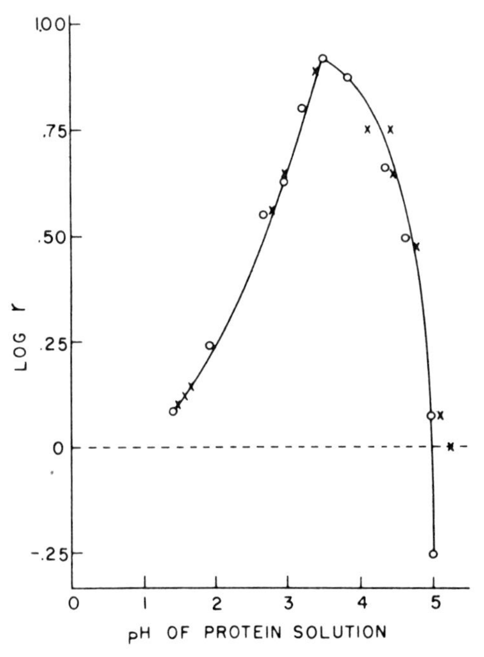

Fig. 18. Comparison of r-values from pH measurements and from sodium ion distributions. Five per cent bovine serum albumin. x – log Na_1/Na_2, o – pH_2 – pH_1. (A. Saifer and J. Steigman: J. Phys. Chem. *65*:141, 1961.)

Incidentally, when a solution of a biopolymer is concentrated by whatever means, the molecules of the biopolymer are brought closer together and the osmotic pressure is the pressure which must be exerted to concentrate such a solution. It has been useful to view an osmotic pressure as arising from a distillation of the solvent from one compartment to another, but there is some merit in regarding the osmotic pressure as arising from forces of repulsion between the biopolymer molecules. Osmotic pressure measurements, therefore, can provide a method by which, in principle, it should be possible to calculate the forces of repulsion between the biopolymer molecules.

The Donnan Potential. On the face of it, there should exist a potential difference between the two sides of the membrane in a Donnan equilibrium and this potential difference should be given by (see Fig. 15 and assuming the diffusible ions to be univalent)

$$E = \frac{RT}{F} \ln \frac{M_1}{M_2} = \frac{RT}{F} \ln \frac{A_2}{A_1} \qquad 79$$

But the system is in equilibrium and in this condition no work can be done by the system and, accordingly, if reversible electrodes be inserted into the two compartments, the potential observed must be zero. With another experimental arrangement, however, a potential difference can be measured; the reversible electrodes can be connected to the two compartments by salt bridges and a potential difference observed.

The distribution of diffusible ions and the osmotic pressure resulting can be treated by the methods of thermodynamics, but the potential difference is extrathermodynamic in origin and depends primarily on the difference in mobilities of the cations and anions in the salt bridges leading to the compartments. Even if the mobilities of the diffusable ions are identical in a pure solution of the salt, a potential difference will still be observed. For example, in compartment 1 the large protein ion has limited mobility and there is a deficiency of diffusible anions which is matched by the equivalent charge on the protein; there is in effect a liquid junction potential at the tip of the salt bridge which may be of considerable magnitude depending on the condition of the Donnan equilibrium. The flow of current in a Donnan equilib-

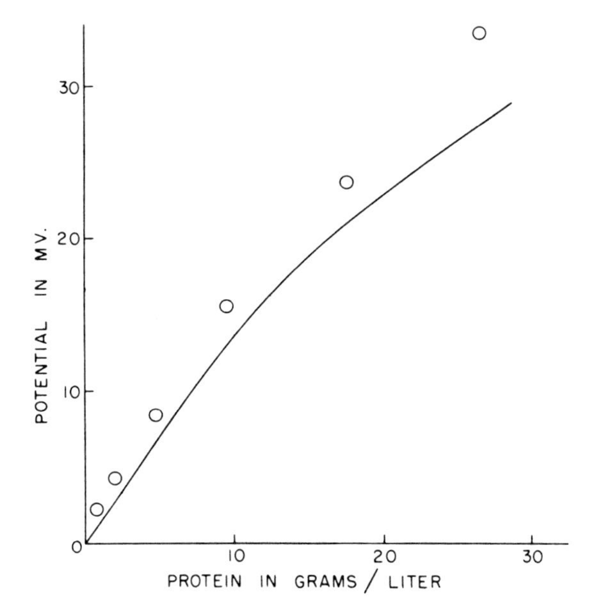

Fig. 19. Comparison between experimental (circles) and computed (line) Donnan potentials. Bovine serum albumin with Z equal to -19.4. Activity of KCl in compartment 2 was 1.919 x 10^{-3}. (W. J. H. M. Möller: Electrical Transport Properties of Alkali Albuminates. Thesis, University of Utrecht, 1959.)

rium disturbs the equilibrium conditions which cannot be exactly restored by reversing the current. Liquid junction potentials will be considered in Chapter 12, Conductance and Ion Transport.

Shown in Figure 19 is the plot of the potential difference between the two compartments at equilibrium as determined experimentally with salt bridges and as calculated by eq. 79.

Osmosis in Biology. About two thirds of the osmotic pressure of blood serum under physiological conditions can be predicted from the known molecular weights of the serum fractions using the expression for ideal solution behavior; the departure from ideality is due principally to the Donnan effect.

The blood plasma proteins are of great importance in maintaining the water balance between the blood and interstitial fluid and, in spite of the many complications which have since been recognized, the original conception of Starling has considerable merit. This idea is diagramed in Figure 20. The hydrostatic pressure resulting from the pumping of the heart on the arterial side of the capillary bed is greater than the difference between the osmotic pressure due to the plasma and that of the interstitial proteins and, accordingly, water and small solute molecules are forced out of the capillaries and into the tissue spaces. On the venous side of the capillary bed, the osmotic pressure difference exceeds the hydrostatic pressure and, accordingly, water diffuses from the tissue spaces and into the capillaries thus establishing water and solute exchange between the circulatory system and the tissue spaces. If the protein plasma level is depleted, water is not withdrawn from the tissue spaces and interstitial water accumulates and edema results.

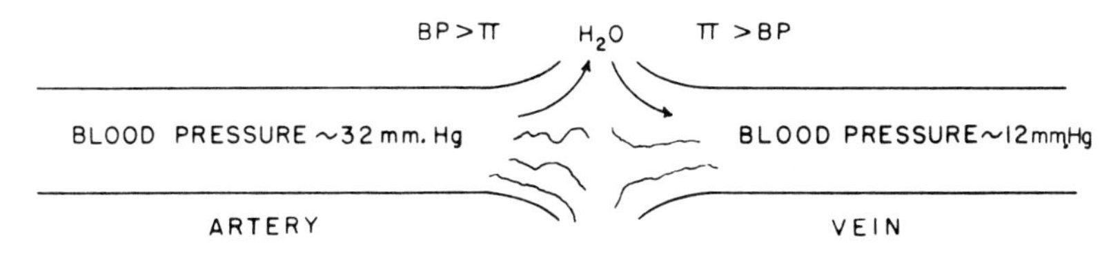

Fig. 20. Relation between blood pressure and osmotic pressure and water balance between blood plasma and the interstitial fluid (Starling).

Osmotic Behavior of Small Ions. It has long been known that red blood cells neither swell nor shrink when placed in a solution containing about 0.90 percent sodium chloride. Such a solution is said to be isotonic with red cells and is 0.154 molar in respect to sodium chloride. In a sufficiently dilute salt solution, the cells swell and hemolize. If the sodium chloride concentration is made significantly larger than 0.154 molar, the cells lose water and become creanated. Since sodium chloride in solution ionizes to yield one ion each of sodium and of chloride, the total ion concentration is 0.308 molar. Thus a 0.308 molar solution of glucose is also isotonic with blood plasma.

Evidence indicates that the passage of water across cell membranes involves, in general, no active mechanism and water transfer is osmotic in character. There appears to be a quantitative relation between the osmotic pressure of the solution bathing living cells and the uptake or loss of water by the cells. This relation can be expressed by the equation

$$\Pi_1 (V_1 - b) = \Pi_0 (V_0 - b) \qquad 80$$

where Π_1 is the known osmotic pressure of the external solution, V_1 is the expected equilibrium volume of the cell, Π_0 is the original osmotic pressure of the external medium, V_0 is the original volume of the cell. b is a constant which is in the nature of a correction for the volume occupied by the osmotically inactive material in the cell. For sea urchin cells, b is about 12 percent of the initial volume of the cell and for mammalian red blood cells it is about 45 percent. The rate of water uptake by cells in response to osmotic changes is rapid and complete adjustment usually requires only a few minutes. The osmotic pressure of the external solution can be controlled by electrolytes and small organic molecules such as glucose and eq. 80 is obeyed. This indicates that the rate of water penetration through the cell membranes greatly exceeds that of the electrolytes and small organic molecules.

In spite of what has been written above, several examples can be cited of water transfer in biological material "uphill" against osmotic gradients. The secretions of salivary glands, the passage of water through frog skin, the flow of water in trees and the excretion of dilute urine in the kidneys all involve the transfer of water from a concentrated to a more dilute solution. In the case of the kidneys, the urine is diluted by reabsorption of many of the dissolved molecules by active mechanisms in the kidney tubules. In principle, this may very well be the mechanism by which water is secreted against osmotic gradients in other tissues. Water can and does move from concentrated to dilute solution through inert membranes as a result of the differential rate of penetration of inorganic ions through the membrane; this is known as negative anomalous osmosis and is now fairly well understood and will be discussed at a later point.

Of considerable physiological significance is the swelling of mitochondria. Active swelling of mitochondria is induced by calcium ions, phosphate, thyroxine or reduced glutathione. Swelling of mitochondria loosens the coupling between phosphorylation and respiration and ATP is no longer produced from ADP and inorganic phosphate. ATP itself causes the swollen mitochondria to contract. Such problems as these, while osmotic in character, cannot be expressed by a Donnan equilibrium.

Osmotic Coefficients. Note has been made of isotonic solutions in respect to swelling of cells and tissues. A more exact way of dealing with such phenomena is through a consideration of osmotic coefficients.

It is possible to express deviations of the osmotic pressure of a solution from ideality in several ways. For example, instead of introducing the interaction constant B, eq. 46 could be expressed as

$$\Pi = \frac{\rho_1 RT w_2 \Phi}{1{,}000 M_2} \qquad 81$$

where Φ is the osmotic coefficient and takes into account the departure from ideality and would be a function of the concentrations of various constituents of the osmotic system. Electrolytes dissociate into ions and each ion produced has an osmotic effect and to cover the more general case eq. 81 is changed to

$$\Pi = \frac{vRTM_1 m\Phi}{1{,}000\ \bar{V}_1} \tag{82}$$

where v is the number of particles (ions) produced by the dissociation of the solute and m is its molality. The molal osmotic coefficient is expressed exactly as

$$\Phi = -\frac{1{,}000\ \ln a_1}{vm\mathrm{M}_1} \tag{83}$$

Shown in Figure 21 are the osmotic coefficients of several solutes dissolved in water at 25° as functions of the molality of the solute. The results shown in Figure 21 are to be interpreted in terms of eqs. 82 and 83.

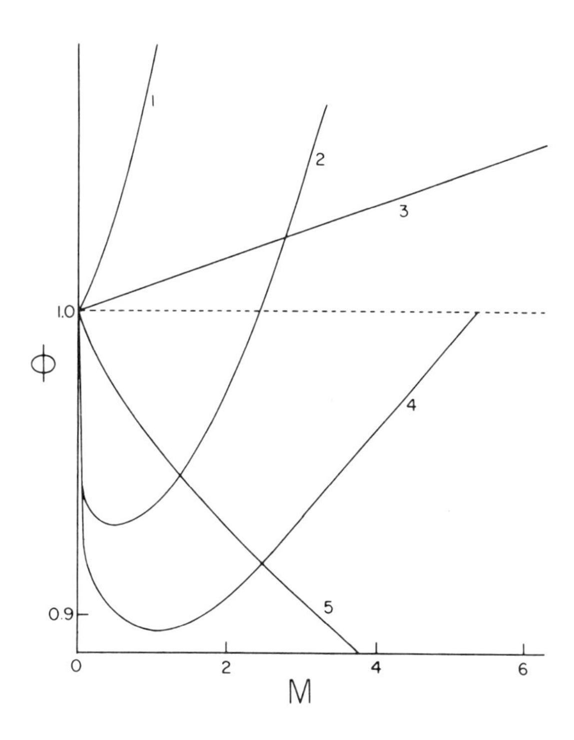

Fig. 21. Osmotic coefficients (ϕ) as functions of the molal concentrations in water at 25°C. Curve 1, sucrose; curve 2, NaCl; curve 3, glycerol; curve 4, KCl; curve 5, urea. (G. Scatchard, W. J. Hamer and S. E. Wood: J. Amer. Chem. Soc. *60:* 3061, 1938.)

GENERAL REFERENCES

Kupke, D. W.: Osmotic Pressure. *In* Advances in Protein Chemistry, Vol. 15. Academic Press, New York, 1960.

Overbeek, J. Th. G.: The Donnan Equilibrium. *In* Progress in Biophysics, Vol. 6. Pergamon Press, New York, 1956.

Steiner, R. F.: Reversible association processes of globular proteins. Arch. Biochem. Biophys. *49:*400, 1954.

PROBLEMS

1. The following table shows the grams of water bound per 100 grams of dry egg albumin as a function of the relative water vapor pressure both at 25° and at 40°C.

Relative Vapor Pressure	*Grams of* H_2O *Adsorbed per 100 Grams Dry Protein*	
	25°C	*40°C*
0.05	2.65	2.53
0.10	3.91	3.63
0.20	5.86	5.50
0.30	7.56	7.14
0.40	9.28	8.84
0.50	10.95	10.40
0.60	12.72	12.12
0.70	15.26	14.35
0.80	18.97	18.00
0.90	27.00	24.90
0.95	36.20	32.00

Calculate the free energy change and the heat change at 25°C for the hydration of 100 grams of dry protein.

Ans: ΔF, -956; ΔH, -1350.

2. Nitrogen mustards have been found to yield some relief in the treatment of Hodgkin's disease. Nitrogen mustard is not well tolerated, but its worst effects can be avoided by intravenous injection of a solution of the sulfur containing amino acid cysteine. Problem: You are given pure crystalline cysteine hydrochloride and you are asked to make up one liter of solution containing the maximum quantity of cysteine at pH 7.4 and which is isotonic with blood and suitable for immediate intravenous injection into a human being. Give calculations and describe procedure for the preparation of the solution. Note: Cysteine is exceedingly unstable in weak alkali and you cannot autoclave this solution without decomposition.

Ans: 16.187 grams cysteine HCl, 8.629 grams $NaHCO_3$.

3. In respect to osmotic pressure, what is the meaning and the nature of the interaction constant B? What factors tend to make this constant positive and what factors tend to make it negative? Suggest reasons why the plot of the osmotic pressure of egg albumin against concentration yields very nearly a straight line, whereas a similar plot for serum globulin has a pronounced curvature.

Ans: See text.

4. Explain the occasion for a Donnan equilibrium, and derive the expression for the ratio of the diffusible ions across a semipermeable membrane in terms of the original concentrations of the ions and of the equivalent concentration of protein. Let the volumes of the two compartments be equal to each other.

Ans: See text.

5. The osmotic pressure of human hemoglobin in 0.2 M phosphate has been measured at 3.0°C by H. Gutfreund with the following results:

Conc. Grams per 100 ml.	*Osmotic Pressure cm.* H_2O	*Conc. Grams per 100 ml.*	*Osmotic Pressure cm.* H_2O
0.47	1.6	3.25	12.8
0.56	2.0	3.47	13.2
0.60	2.2	3.75	14.9
1.29	4.7	4.34	17.6
1.66	6.5	4.83	19.9
2.39	9.5	5.40	22.6
3.09	11.3	7.01	30.6

Calculate the molecular weight of human hemoglobin and the B-constant.

Ans: Mol. wt. 67,300; B 0.53×10^{-4} ml/g.

6. 1.427 grams of β-lactoglobulin in 100 grams of 0.5 M NaCl (density of 1.0206) at the isoelectric point of this protein yielded an osmotic pressure of 10.95 cm. of water at 25°C against 0.5 M NaCl. The same weight of this protein was adjusted to pH 6.5 by the addition of NaOH. NaCl and water were added to give 100 grams of solvent which was 0.01 M in respect to NaCl. The osmotic pressure of this solution against an equal volume of 0.01 M NaCl solution was 14.65 cm. of water at 25°C.
   (a) Calculate the mol. wt. of β-lactoglobulin (β-lactoglobulin can be assumed to behave ideally at its isoelectric point).
   (b) Calculate the approximate charge on the protein at pH 6.5.
   (c) What would be the freezing point depression of the inside and outside solutions?

   Ans: (a) 33,600 (b) 5.90 (c) outside 0.0394°: inside 0.0405°

7. The osmotic pressure of solutions of bovine serum albumin was measured. The pH of the solutions was adjusted by the addition of dilute HCl or of NaOH. NaCl was added to give the desired ionic strength. The following results were obtained

W_2	pH	z	Cl_1^-	Cl_2^-	Π
60.31	4.22	22.9	.1587	.1530	17.55
58.85	4.36	21.5	.1566	.1517	17.09
59.03	4.52	16.1	.1549	.1507	17.46
57.23	4.79	8.9	.1524	.1494	17.82
57.42	4.98	5.3	.1523	.1488	18.93
27.28	5.28	1.5	.1506	.1489	8.35
17.69	5.42	− 1.1	.1498	.1488	5.07
57.71	6.19	− 6.5	.1512	.1482	21.48
56.17	6.64	− 9.2	.1518	.1491	21.40
50.15	7.06	− 12.2	.1537	.1501	19.27
66.88	7.28	− 13.9	.1543	.1501	29.45
59.76	7.54	− 16.1	.1554	.1514	25.24
57.33	7.97	− 20.5	.1541	.1522	24.34
57.11	8.15	− 22.9	.1542	.1511	24.46

W_2 = grams isoelectric protein per kilogram of water
pH = of protein solution
z = valence of the protein as determined by acid-base titration whose molecular weight is 69,000
Cl_1^- = the total molal chloride concentration of protein solution both in bound and unbound form
Cl_2^- = molal chloride concentration of outside solution
Π = is given in mm. Hg at 0°C

(a) Calculate the B-constant from the charge on the protein, its molecular weight and the salt concentration.
(b) Calculate the B-constant using the observed osmotic pressure and the molecular weight of the protein.
(c) Compare the B-values obtained by these two methods by plotting against Z. Explain why the two methods of calculation do not agree.

The answers are too extensive to give; however, see Figure 14, Chapter 8, Solution Optics, for a plot of the B-Values.

CHAPTER 8

Solution Optics

Colorimetric analytical methods because of their rapidity, ease of application, and sensitivity have been long a source of comfort to the biochemist. Optical rotatory studies have also played important and historical roles in the development of biochemistry. As the years have passed, solution optics have become more and more sophisticated both in theory and in application. To deal with the total subject in anything like a comprehensive manner would be an heroic undertaking and would require more knowledge and space than the author has at his disposal. The aim of this chapter is set at a modest level: the general and elementary theory is to be reviewed along with a discussion of the refraction and polarization of light; light scatter will be summarized; the absorption of electromagnetic radiation will be outlined and a short section on photochemical reactions is included.

Elementary Theory. In certain respects an alternating electric current passing through a conductor bears a strong resemblance to the passage of light through space; both are accompanied by electrical and magnetic fields of force and both arise from the acceleration of charged particles. Figure 1 shows the electric and magnetic fields associated with plane or linearly polarized electromagnetic radiation. Ordinary light vibrates in all planes and only if special restraints are imposed does light become polarized; the necessary conditions will be discussed presently. In ordinary spectroscopy the electric field is of primary interest rather than the magnetic field, although in the radio and microwave regions attention is centered on the magnetic part of the radiation.

The equation for the wave shown in Figure 1 is that of undamped harmonic motion and is

$$Y = Y_0 \text{ Cos } 2\pi v\left(t - \frac{x}{V}\right) \qquad 1$$

where Y is the amplitude of the wave at any time, t, v is the frequency in oscillations per second and V is the velocity in centimeters per second. Equation 1 can be expressed also in the form (*see* Chapter 1, Mathematical Review)

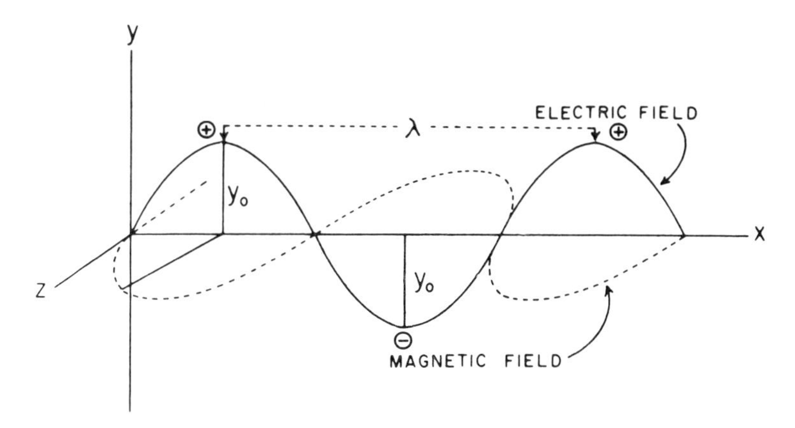

Fig. 1. Diagrammatic representation of a plane polarized light wave with its electric vector in the plane of the paper. Note alternations in the sign of the electric vector. Y is the amplitude of the electric field, Z is the amplitude of the magnetic vector, and λ is the wavelength.

$$Y = \frac{Y_0}{2}\left(e^{i\theta} + e^{-i\theta}\right) \qquad 2$$

where i is the square root of minus one and θ is equal to $2\pi v\left(t - \frac{x}{V}\right)$

Evidently, the velocity of the propagation of the wave is equal to the frequency multiplied by the wavelength and the energy of the wave at a given frequency is proportional to the square of the amplitude. The wavelengths of radiation are expressed in units such as in Angstrom units (10^{-8} cm.), in mμ. (10^{-7} cm.), in nanometers (nm, 10^{-7} cm.), in μ (10^{-4} cm.), in cm., in meters (10^{2} cm.), etc., depending on their magnitudes. Wavelengths can also be expressed in wave numbers which refer to the number of waves per centimeter. Thus a wavelength of 10μ is equivalent to 1,000 cm.$^{-1}$; the wave number is denoted by $\bar{v}$.

Electromagnetic radiation covers a vast spectrum of wavelengths extending from very short wavelengths of cosmic rays to long radio waves. The visible spectrum occupies only a small segment of the total spectrum. Figure 2 exhibits the spectrum in logarithmic scales of wavelengths, of frequency and wave numbers, the units being centimeters and seconds. Also shown is the energy in kilocalories per mole.

As noted above, visible light spans only a small part of the electromagnetic spectrum. Shown in Table 1 are the wavelengths in mμ corresponding to the various colors.

TABLE 1. WAVELENGTHS IN mμ OF THE COLORS

Color	Wavelength
Ultraviolet	< 400
Violet	400-450
Blue	450-500
Green	500-570
Yellow	570-590
Orange	590-620
Red	620-750
Infrared	> 750

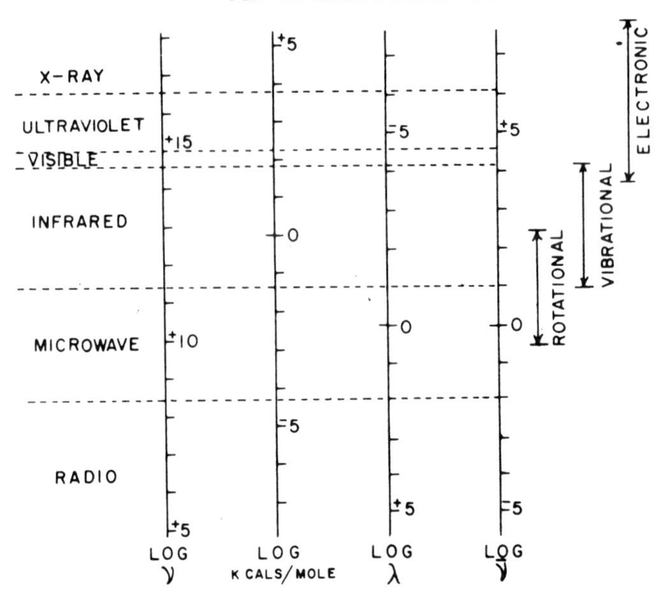

Fig. 2. Logarithmic scales of electromagnetic radiation. Comparison of scales in a vacuum. Designation into regions should overlap somewhat. ν frequency in cycles per second, λ in centimeters and $\bar{\nu}$ in cm^{-1}. The origin of the spectra is shown on the right.

Refraction of Light. An electromagnetic wave passing through a medium interacts with the electrons in the atoms of the medium as the electrical field oscillates from positive to negative. The electrons absorb energy from the light because they are displaced and release the energy back to the light beams as the phase changes. The net effect is to decrease the wavelength of the light and to slow the velocity of the beam without change in frequency. The ratio of the velocity of light in a vacuum to that in the medium is the index of refraction and

$$n = \frac{C}{V} \qquad 3$$

where C is the velocity of light in a vacuum (2.998×10^{10} cm. per second) and V is the velocity in the medium.

The slowing of the light when passing into a medium of higher refractive index is, as noted above, due to polarization of the electrons in the material. There should, accordingly, be a relation between the dielectric constant at high frequencies (frequencies of visible light) and the index of refraction of the medium since the dielectric constant also depends on the degree of polarization. The dielectric constant at frequencies of visible light is equal to the square of the index of refraction of the medium.

It is evident that if the wave front of incident light enters a medium of higher index of refraction at an angle other than normal to the surface, the first part of the wave which enters the medium will be slowed relative to the rest of the wave front. The light beam will be bent (refracted) through an angle which is closer to the normal to the surface. Light entering a medium of smaller index of refraction will be refracted in the opposite sense. The relation between the sines of the angles and the index of refraction is given by

$$\frac{\sin i}{\sin r} = \frac{n_2}{n_1} \qquad 4$$

where i is the angle of incidence and r is the angle of refraction. n_1 and n_2 are the indices of refraction of the incident medium and the refracting medium, respectively. If the incident medium is a vacuum, n_1 is unity (the index of refraction of air is very nearly unity). The index of refraction of a given substance decreases with increasing wavelength of the light (blue light is refracted more than is red, except in the region of an absorption band); it also decreases with increasing temperature.

The index of refraction of a liquid can be measured with an Abbe refractometer or with a dipping refractometer. Both instruments employ prisms and the angles of refraction are determined directly. The interferometer provides two paths for a split beam of monochromatic light. One of these beams passes through a liquid (water) and the other through the solution. The split beams are then recombined and the degree of interference measured. One of the split beams has been slowed relative to the other and, accordingly, the two beams are out of phase and destructive interference of the light results.

The index of refraction of a solution is directly proportional to the concentration of the solute over a wide range of concentration. The specific refractive increment of a solution is given by the relation

$$\delta = (n_{\text{soln}} - n_{\text{solvent}}) / C \qquad 5$$

where C is expressed in grams per 100 mls. of solution. Index of refraction has long been used to determine the concentration of solutions of various materials. The refractive increments of all the proteins lie fairly close together although there is sufficient individuality so that a common value is not available and each protein has to be treated separately.

Since the index of refraction is proportional to the concentration, index of refraction gradients also describe the concentration gradients which exist between a solution and a solvent. Optical methods are available for the plotting of the gradients in refraction as functions of distance through the solution-solvent boundary. These methods are used extensively in electrophoresis, diffusion, chromatographic analysis and ultracentrifugation, and will receive additional discussion in Chapter 11, Diffusion.

Double Refraction. Suppose an electron could oscillate in one plane more easily than it could in a second plane perpendicular to the first. Evidently the substance would have two indices of refraction. The situation is crudely diagramed in Figure 3. If light is passed through a calcite ($CaCO_3$) crystal in the proper direction, the beam of light is broken into two beams. This can be demonstrated by viewing an object through a rhomb of calcite; a double vision is observed to appear. On

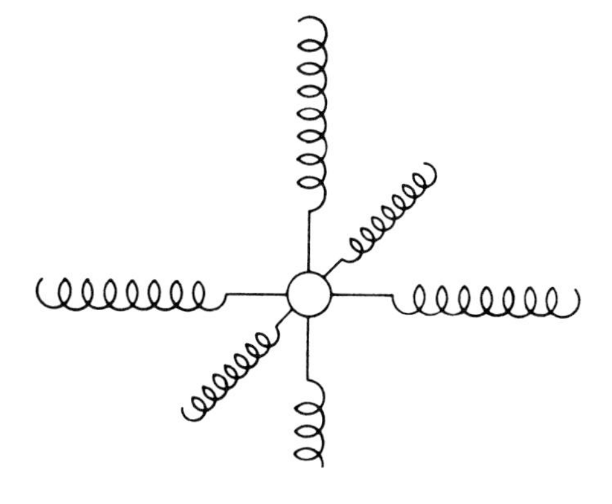

Fig. 3. An asymmetrically bound electron as in an optically anisotropic crystal.

rotation of the rhomb one of the images behaves in an extraordinary manner; it rotates around the other image. This is called the extraordinary ray. The other ray is the ordinary ray. It will be discovered that there is one direction in which light can be passed through a calcite crystal without being broken into two beams. This direction corresponds to the optic axis of the crystal. Calcite, like a number of other crystals, is anisotropic, the index of refraction being different in two directions through the crystal. Both beams of light emerging from such a crystal are plane polarized, and the planes of vibration of the two beams are at right angles to each other. The magnitude of the double refraction is measured by the difference between the index of refraction of the ordinary ray (n_0) and that for the extraordinary ray (n_e). Light whose plane of vibration is parallel to the optic axis is called the extraordinary ray; and light whose plane of vibration is perpendicular to it is known as the ordinary ray. If $n_e > n_0$, the double refraction is positive; if $n_e < n_0$, it is negative.

If a calcite rhomb is cut diagonally at the proper angle, the cut surfaces polished and reunited and cemented with balsam cement, the ordinary ray strikes the balsam at an angle greater than the critical angle. This ray is then totally reflected out of the optical path. The extraordinary ray, however, which is substantially undeviated, is transmitted through the balsam layer with little loss of intensity and so emerges through the exit face of the crystal. A calcite crystal prepared in this manner is known as a nicol prism and is a common means of producing plane polarized light.

Double refraction of an object can be detected by placing it between two crossed nicol prisms. Light passing through the first nicol (the polarizer) is plane polarized. When it enters the anisotropic object the plane polarized beam is broken into two rays, one vibrating parallel to the optic axis of the object and the other at right angles to it. If white light has been used, the field, as viewed through the second nicol (the analyzer), is seen to be beautifully colored and there will be no position of the second nicol at which all the light can be extinguished. Each wavelength of light has its own index of refraction through a medium, and the velocities of the various wavelengths in the two beams of light through the double refracting object are all different. On emergence of the light, some of the wavelengths will differ in phase and their intensity will be diminished, whereas others will be in phase and will be reinforced. The net effect is that the object appears highly colored and the color varies as the analyzer is rotated.

A particle subject to a single simple harmonic motion vibrates in one plane; this is the situation with plane polarized light. The projection of the motion of the particle in a plane perpendicular to the direction of propagation will describe a straight line. A second harmonic motion, perpendicular to the plane of the first, introduces complications in the motion of the particle. If the two harmonic motions are in phase, the projection of the motion of the particle will still be that of a straight line but the direction of the line will be inclined at an angle between the two planes of harmonic motion, the angle depending on the amplitude of the two waves. If the two harmonic motions are out of phase by 90° and the amplitudes are unequal, the particle will describe an ellipse and the light is said to be elliptically polarized. If the two waves (still 90° out of phase) have the same amplitude, the path of the particle will be circular. The rotation of a centrifuge can be resolved into two vibrations of equal amplitude at right angles to each other. This is the situation with circularly polarized light.

Mica is a doubly refracting mineral and sheets of mica can be prepared in almost any desired thickness by careful cleavage of the sheets. Light passed through the mica sheets will emerge broken into two rays polarized at right angles to each other. Due to the fact that the two indices of refraction differ, the two rays have a phase difference depending on the thickness of the mica. By a suitable choice of thickness, it is possible to slow one ray relative to the other by a quarter wave or by a half wave. These thicknesses of mica are spoken of as quarter wave plate or half wave plate, respectively.

Plane polarized monochromatic light passed through a half wave plate will be completely out of phase in the two perpendicular planes. Destructive interference would result if the two harmonic motions were in the same plane. Two rays plane polarized at right angles produce plane polarized light rotated through 45°. Plane polarized monochromatic light passed through a quarter wave plate has maximum amplitude in one wave while that of the other is zero; the result is circularly polarized light. The insertion of another quarter wave plate in the optical path converts the circularly polarized light back into plane polarized light.

There are other ways of producing plane polarized light besides that of double refraction. For example, light scattered from a solution (Rayleigh scatter) will be partially polarized and this topic will be discussed presently. Reflected light will also be polarized, the plane of polarization being parallel to the reflecting surface; the extent of polarization is a function of the angle of reflection. Some crystals have the property of absorbing those vibrations which are perpendicular to the crystal axis and of transmitting the vibrations which are parallel to this axis; this behavior is known as dichroism. Crystals of methemoglobin are strongly birefringent with a high dichroic ratio. It is, therefore, concluded that the hemin groups which are responsible for the absorption are parallel to each other in the hemoglobin molecule. Sheets of polyvinyl alcohol complexed with iodine and subjected to high stress orient the crystals of iodine and produce polarized light, the unwanted rays being absorbed by the iodine. These are known as polaroids and have proved extremely useful.

Birefringence (double refraction) of a substance can be due to intrinsic birefringence or to form birefringence. Polarizability of chemical bonds is pronouncedly directional and, if the material is arranged in a crystal such that the chemical bonds show a predominate orientation, the substance will show intrinsic birefringence. Form birefringence results when isotropic rods or discs are arranged parallel to each other; obviously, a substance may exhibit intrinsic and form birefringence simultaneously. The two can be distinguished by varying the index of refraction of the suspending medium. If the index of refraction of the medium is the same as that of the material, the form birefringence vanishes. It is impossible to abolish the intrinsic birefringence by varying the index of refraction of the medium.

Many biological structures such as nerves and muscles exhibit birefringence and a study of the sign and magnitude of the birefringence of these objects can give valuable information concerning their structure.

It is also possible to produce birefringence in a solution or suspension of elongated molecules or particles by any technique which will bring about orientation of the particles or molecules. If a suspension of long asymmetric particles, such as those of vanadium pentoxide, is placed between crossed nicols, the field will be completely dark. If, however, streaming motion is produced in the suspension, the suspension becomes double refracting as is evidenced by a light field which persists as long as the suspension is in motion. A strong static electric field also induces double refraction not only in a suspension of particles but also in a liquid such as nitrobenzene. The production of double refraction by a static field is known as the Kerr effect. Both stream birefringence and the Kerr effect can be used to determine the rotary diffusion coefficient of the particles or molecules and can thus yield information regarding the size and shape of the particle. The rotary diffusion coefficient will be discussed under the general topic of diffusion.

Optical Rotation. A beam of plane polarized light can be resolved into a right and a left circularly polarized component of equal amplitude. It has been noted previously that if one of these components is slowed relative to the other, a phase difference results. Upon recombination of the two components after passage through the medium it will be observed that the plane of the emergent light will have been rotated through an angle, α'. A diagrammatic representation of the combination of the right and left vectors of intensity is shown in Figure 4.

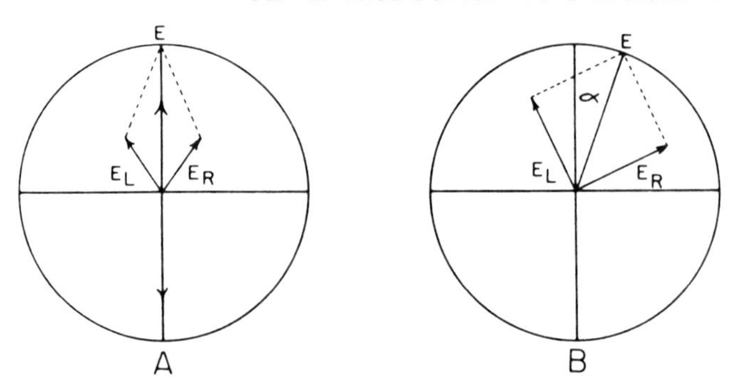

Fig. 4. Resolution of plane polarized light into two circular components E_L and E_R. *A*, Before passage through an optically active medium. *B*, After passage through an optically active medium with a phase difference between the two components.

The angle α^1, through which rotation has taken place, is given by the relation

$$\alpha' = \frac{\pi}{\lambda} (n_L - n_R) \qquad 6$$

where λ, the wavelength of the light, is that in a vacuum. n_L and n_R are the two indices of refraction of the birefringent medium. α' has the dimensions of radians per centimeter. Conventionally, rotatory power is expressed in terms of specific rotation $[\alpha]$ which is defined as

$$[\alpha] = \frac{\alpha}{L \times C} \qquad 7$$

where α is the observed rotation in degrees, L is the length of the optical path through the solution in decimeters, and C is the concentration in grams of optically active material per cubic centimeter. Since the optical activity is a function both of the wavelength of light and of the temperature, both of these factors have to be designated. It is the custom to employ the D-line of sodium. Since α' is expressed in radians per centimeter, the relation between α' and $[\alpha]$ is given by

$$[\alpha] = \frac{1{,}800\alpha'}{\pi C} \qquad 8$$

where C again is expressed in grams per cubic centimeter. Suppose an observed rotation was 10° or 0.1745 radian. The D-line of sodium has been used (5.892×10^{-4} cm.), then from eq. 6, $n_L - n_R$ is 3.27×10^{-6}. By convention, if the analyzer must be turned clockwise to obtain transmission of the plane polarized light, the substance is said to be dextrorotatory or +; if the analyzer is turned counterclockwise the substance is levorotatory or −. From eq. 6 the sign of the rotation depends upon the relative magnitudes of n_L and n_R.

The molar rotation is defined by the relation

$$[M] = \frac{[\alpha]\, M}{100} \qquad 9$$

where M is the molecular weight of the optically active material.

It has long been known that optical activity is related to molecular asymmetry. Molecules which do not have a plane or center of symmetry exhibit optical activity. Such molecules can exist in two stereochemical forms which are mirror images of each other. It can be said that such molecules are birefringent and in one stereochemical configuration n_L is larger than n_R (dextrorotatory) and in the other n_R is larger than n_L (levorotatory), but the absolute difference between n_R and n_L is the same in both cases. On first thought it might appear that the extent of rotation would depend upon the orientation of the molecule; a dextrorotatory molecule would become levorotatory if it got turned around to expose the opposite face. This, however, is not true; a right-hand screw is right-handed no matter how you look at it.

Whereas it is possible for compounds other than those containing carbon to show optical activity (quartz, for example), interest has centered mostly on carbon compounds. Any carbon compound in which the four valences of carbon are bonded to four different groups will be optically active. Consider, for example, the amino acid alanine which is optically active. If the formula of alanine is written in the plane of the paper as follows

$$\begin{array}{c} CH_3 \\ | \\ H—C—NH_2 \\ | \\ COOH \end{array}$$

it would appear that the groups around the carbon could be arranged in six different ways. However, if a three dimensional structure is inspected, it will be discovered that there are in fact only two different arrangements which are possible. (examine Fig. 5.)

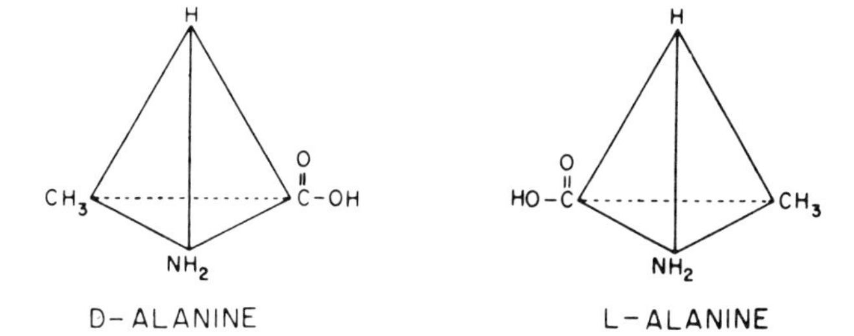

Fig. 5. Stereochemical isomers of alanine.

The tetrahedral representation of groups around the asymmetric carbon follows the convention introduced by Emil Fischer who evidently had two possible choices and by chance selected the one which subsequently has been demonstrated to be correct by X-ray diffraction studies. There exists, however, no simple relationship between the stereochemical configuration and the direction or magnitude of the optical rotation. All amino acids occurring in proteins are of the L-series structurally speaking, but their optical activity can be either positive or negative and, further, the specific rotation varies with pH and temperature. For example, L-alanine in water has a specific rotation of + 2.4 degrees and in 6 N HCl this becomes + 14.5 degrees. To say that the relation between chemical structure and optical activity is complex would be a gross understatement. Not alone is the structure of the molecule involved, but the interaction with the solvent can have profound effects.

Optical rotatory power being dependent upon the refractive index of the solvent, it is necessary to correct for the influence of the refractive index and to reduce the rotation to unit refractive index. Such a correction leads to the reduced rotation defined by

$$[\acute{\alpha}]\lambda = \frac{3}{n^2 + 2}[\alpha]\lambda \qquad 10$$

where n is the refractive index of the solvent at wavelength λ.

The optical rotatory power of a polymeric substance is essentially a function of the individual residues rather than of the total molecule and, accordingly, the reduced mean residue rotation is given by

$$[m']_\lambda = \frac{M\ Rw}{100} [\alpha']_\lambda \qquad 11$$

where $M\ Rw$ is the mean residue weight of the momeric substance and for most proteins (amino acid residues) is about 115.

Rotatory Dispersion. It has already been noted that optical rotation is dependent on the wavelength of the light. This dependence over a considerable range of wavelengths can be expressed by a single term Drude equation of the form

$$[m']_\lambda = \frac{A_c \lambda_c^2}{\lambda^2 - \lambda_c^2} \qquad 12$$

where A_c and λ_c are empirical constants; λ is the wavelength at which the measurement has been made. A_c and λ_c can be obtained from the experimental data by use of an appropriate plot. For example, $[m']\lambda^2$ can be plotted against $[m']_\lambda$ and λ_c obtained from the slope and A_c from the intercept. Systems which obey the single term Drude equation are said to give plain dispersion curves. However, some systems do not give a linear relation when plots of the type described above are constructed. These are complex dispersion curves and a two-term equation in the form

$$[m']_\lambda = \frac{a_o \lambda_o^2}{\lambda^2 - \lambda_o^2} + \frac{b_o \lambda_o^4}{(\lambda^2 - \lambda_o^2)^2} \qquad 13$$

has to be employed to represent the data. a_o, b_o and λ_o are empirical constants which, however, can be related to molecular structure. Equation 13 is known as the Moffitt equation and originally was derived on theoretical grounds but is now regarded more as a convenient empirical equation. A plot of $[m']_\lambda\ (\lambda^2 - \lambda_o^2)$ vs $1/(\lambda^2 - \lambda_o^2)$ should yield a linear relation from which the constant terms can be estimated. The value of λ_o can be adjusted until a linear relation is observed. The plot of $[m']_\lambda$ against wavelength sometimes gives maxima and minima. Such anomalous behavior is known as a Cotton effect. If $[\alpha]$ becomes more negative with decreasing wavelength and then passes through a minimum, the behavior is known as a negative Cotton effect.

Some compounds do not show the Cotton effect; $[\alpha]$ simply increases positively or negatively with decreasing wavelength; it is likely that the measurements have not been extended to sufficiently short wavelength to enter the region of absorption.

In a Cotton effect the two circular components of the plane polarized light not only travel at unequal velocities through the optically active medium, they are also absorbed to different extents. Under these circumstances, the two components are not only out of phase but their amplitudes differ and the light becomes elliptically polarized. The differences in absorption of the two components are usually not large and the resulting ellipse is extremely elongated; the major axis of the ellipse rather than the plane of polarization has been rotated.

To obtain a Cotton effect, it is necessary that the displacement of the electron have not only a linear component but likewise a circular one in the direction of the applied field. Such a displacement produces a magnetic moment in the direction of the electric field of the light wave as well as an electric moment perpendicular to that direction. The intensity of absorption depends only on the electric dipole transition moment, but the intensity of circular dichroism depends on both the electric and magnetic dipole transition moments. It is thus possible to have an optically active as well as an optically inactive absorption band.

The difference in optical density for the two circular polarized components is measured directly and expressed in terms of extinction coefficients. The molecular ellipticity is defined as the angle whose tangent is the ratio of the minor to the major axis of the elliptical polarized emergent light which was originally plane polarized before passing through unit length of the substance at unit concentration. The molecular ellipticity is given by

$$[\theta]_\lambda = 2{,}303 \frac{(4{,}500)}{\pi} (\epsilon_L - \epsilon_R) \qquad 14$$

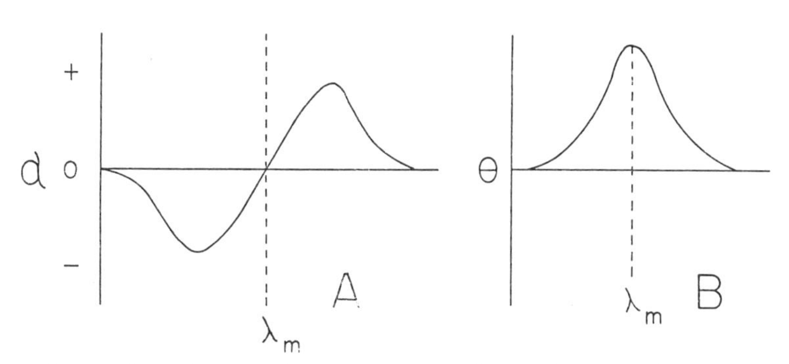

Fig. 6. *A*, Optical rotation in the neighborhood of an optically active absorption band. *B*, Ellipticity as a function of the wavelength. λ_m is the wavelength of maximum absorption (idealized plot).

where ϵ_L and ϵ_R are the molecular extinction coefficients for left and right circularly polarized light. Figure 6 shows an idealized representation of the relation between rotation (positive Cotton effect) and ellipiticity involving a single optically active absorption band with maximum absorption at a wavelength λ_m.

The ellipticity of a peptide reflects quite directly the conformation of the peptide. For example, Figure 7 shows a plot of the ellipticity of poly-L-lysine in an α-helical, a β-conformation and in a random conformation as a function of the wavelength.

As noted above, ϵ_L and ϵ_R do not differ greatly and it is necessary to measure optical density differences of the order of 1×10^{-4} or less. It is only recently that instruments have become available which are capable of accurate work at wavelengths down to about 190 mμ.

The relation between circular dichroism and optical rotatory dispersion is an intimate one. Optical activity at any wavelength arises because of optically active absorption bands, but these bands are frequently far removed from the wavelength at which a measurement is made; the region of measurement may be completely transparent. In principle, the circular dichroic behavior can be calculated from optical rotatory dispersion and conversely the circular dichroic curve can be obtained from optical rotatory dispersion. However, optical rotatory dispersion values can arise in a complex manner from a number of widely separated optically active absorption bands whereas circular dichroic measurements are confined to one band or closely associated bands at a time. It is thus apparent that circular dichroism is a sharper tool than is optical rotatory dispersion. Since circular dichroic equipment of good quality is now available, it is likely that it will displace optical rotatory dispersion as the method of choice in the study of molecular structure.

The above discussion of rotatory dispersion and of the Cotton effect has been simplified as much as possible; the actual situation is frequently complex and these very complexities have proved of great value in investigations on, for example, the structure of steroid molecules.

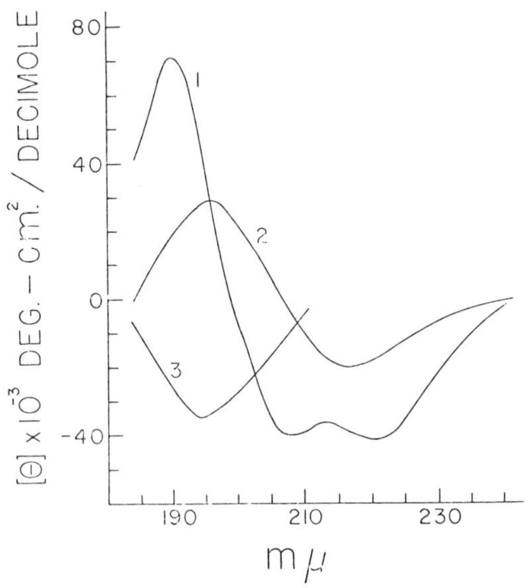

Fig. 7. Ellipticity plotted against the wavelength for poly-L-lysine. Curve 1, α-helical; curve 2, β-conformation; curve 3, random conformation. (R. Townend, T. F. Kumosinski, S. N. Timasheff, G. D. Fasman and B. Davidson: Biochem. Biophys. Res. Com. *23:* 163, 1966.)

In earlier years, optical rotation played a significant role in the investigation of the structure of sugars and mention has just been made of the use of rotatory dispersion in the study of complex steroid molecules. Another field of molecular structure which has been illuminated by the use of polarized light is the structure of polypeptides and of proteins.

The specific rotation $[\alpha]_D$ of most native proteins is between $-30°$ and $-60°$ and, on denaturation, the specific rotation increases and lies between $-80°$ and $-120°$. The sum of the rotation of the L-amino acids in a protein is about $-100°$ and, accordingly, it is only when the protein is denatured that its optical rotary power approaches that to be expected from constituent amino acids. The conclusion is drawn that the structure of the native protein molecule makes a contribution to optical rotation.

The relation between rotatory dispersion and the conformation of peptide chains both in synthetic polypeptides and more specifically in proteins is far from simple and is presently the subject of intensive investigation. Whereas theory has played some part in these investigations, its role has been secondary and primarily the problem has been approached empirically through studies on synthetic polypeptides whose conformation is either known or presumed to be known.

The Moffitt equation (eq. 13) has been extensively employed in the interpretation of optical rotatory dispersion curves of polypeptides and proteins. a_0 is interpreted as a complex constant representing both the intrinsic residue rotation as well as interactions within the helix and would be expected to vary with the environment. λ_0 is the wavelength at which the chromophore group absorbs and was estimated by Moffitt to be at 200 mμ for a helically arranged peptide chain with two absorption bands. For a right handed α-helix, Moffitt estimated b_0 should be -580; a left handed helix would give b_0 of the same magnitude but of opposite sign. The λ_0 determined experimentally for synthetic polypeptides believed to be in the form of α-helixes was 212 and the b_0 values were all close to -630. Thus, the theory was apparently confirmed. In principle, when the α-helix is converted into a random coil, b_0 should be zero since the b_0 term arises from interaction within the helix. Actually, b_0 is not zero for non-helical peptides but its value is much less than the 630 observed for peptides in helical form. Large deviations of λ_0 from 212 have been observed for randomly coiled peptides.

A helical synthetic peptide does not obey a single one term Drude equation and

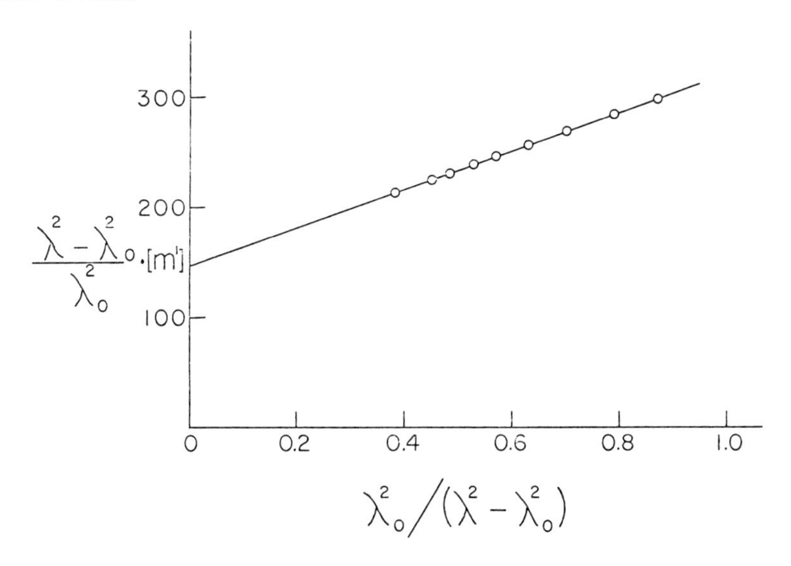

Fig. 8. Linearized Moffitt plot for egg albumin.

a two term equation must be used. Random coiled peptides as well as globular proteins, however, follow a single term Drude equation but the Moffitt constants can still be evaluated for random coiled peptides and for proteins both native and denatured. Since the constants a_0 and b_0 are believed to reflect the conformation of the peptide chains in proteins, it is possible to estimate from the measured constants for a given protein the helical content of the protein.

Figure 8 shows a Moffitt plot for egg albumin in which a_0 is 150 and b_0 is -159. The amount of α-helix present was estimated from the empirical relation

$$\% \ \alpha\text{-helix} = \frac{-(b_o - 100)}{8} \qquad 15$$

to be 32 percent; λo was assumed to be 212 mμ. Cotton effects have now been observed for a number of proteins in the wavelength range of 180 to 240 mμ. For example, shown in Figure 9 is the Cotton effect exhibited by phosphoglucomutase. There are three optically active absorption bands in this region associated with peptide bonds arranged in right-handed helical configuration. The longest wavelength band occurring at about 222 mμ is the weakest one of the three and is due to the promotion of an electron from an oxygen non-bonding atomic orbital to an antibonding molecular orbital involving oxygen, carbon and nitrogen atoms. The splitting of a single absorption band gives rise to the other two optically active absorption bands occurring at about 190 and at 205 mμ respectively. The 205 mμ band is polarized perpendicular to the helix axis. It is possible through a study of the Cotton effects shown by a protein to estimate the helical content of the protein.

In addition to the peptide bond, there are several chromophores which absorb below 230 mμ, and also at longer wavelengths. Optical activity in the range of wavelengths from 240 to 310 mμ for proteins without prosthetic groups must be due to histidine, to cystine or to the aromatic residues.

Light Scatter. It is a common observation that a beam of light becomes visible when passed through a smoky atmosphere; the light is scattered from the smoke particles and this scattering is known as a Tyndall cone.

The problem of light scatter by particles whose radii are small compared with the wavelength of light was first investigated by Lord Rayleigh, who derived an

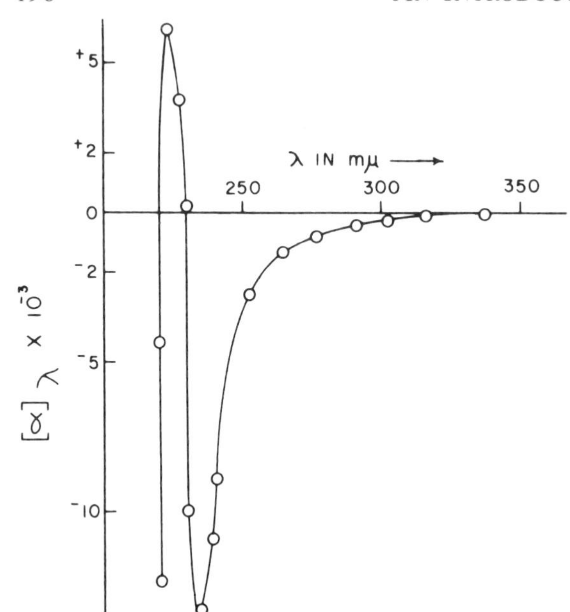

Fig. 9. The Cotton effect of phosphoglucomutase. The specific rotation has a minimum of −14,700° at 232 mμ. (B. Jirgensons: Biochem. *1*:917, 1962.)

equation expressing the turbidity of an ideal gas of low density. In effect, Lord Rayleigh considered the particles to act as linear electric oscillators and that the oscillating electric field of the incident light induces an oscillating electric moment in the particles; the particle becomes a broadcasting station and reradiates the energy received without change of frequency. This is an example of forced harmonic motion; the forcing frequency is not equal to the resonance frequency of the electrons. Shown in Figure 10 is a diagrammatic representation of a light wave forcing an oscillation on an electron with scatter in the Z-direction; note the attenuated amplitude of the scattered radiation. The oscillating electron cannot broadcast in the Y-direction because it is vibrating in the $\pm Y$-directions. It does, however, broadcast in the X- and Z-directions; this is the basis for the polarization of scattered light.

It was noted earlier that the intensity of light is proportional to the square of its amplitude. That this is true can be seen from the following argument: the velocity of the oscillatory motion at constant frequency and wavelength must increase as the amplitude increases; the particle has further to go in a given time interval. Since the kinetic energy of a particle is proportional to the square of its velocity, the intensity of the scattered light is proportional to the square of the amplitude of the light.

The polarizability of a medium is related to the ease of displacement of the electrons in a medium and, accordingly, the amplitude of the scattered light is proportional to the polarizability. In turn, the intensity of the scattered light is proportional to the square of the polarizability. The polarizability is related to the optical dielectric constant (the dielectric constant at the frequency of a light wave). It will be recalled that the optical dielectric constant is equal to the square of the refractive index of a medium.

These relations are expressed by the equation

$$\alpha = \frac{D - D_0}{4\pi N D_0} = \frac{n^2 - n_0^2}{4\pi N n_0^2} \qquad 16$$

where α is the polarizability, D_0 is the dielectric constant of the medium and D is that of the particle. N is the number of particles per unit volume. n_0 is the refractive

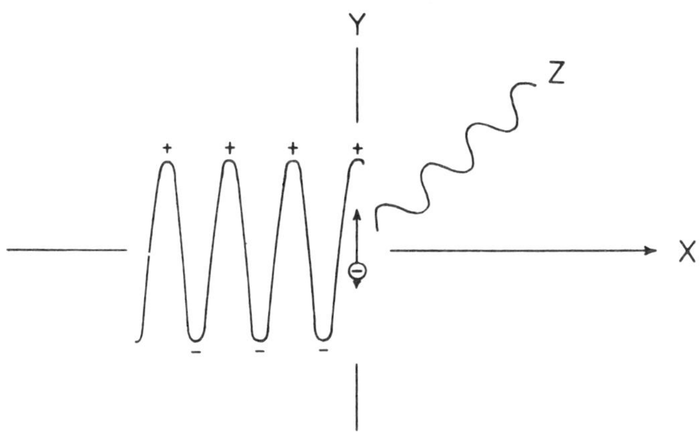

Fig. 10. The forced oscillation of an electron by a light wave and reradiation of the light in Z-direction.

index of the medium and n that of the particles. As noted above, the intensity of the scattered light is proportional to the square of the polarizability and, accordingly, from eq. 16 there results

$$\frac{I_s}{I_0} = k\left(\frac{n^2 - n_0^2}{4\pi N n_0^2}\right)^2 = k'\left(\frac{n - n_0}{N}\right)^2\left(\frac{n + n_0}{N}\right)^2 \cong k''\left(\frac{n - n_0}{N}\right)^2 \qquad 17$$

where I_s is the intensity of the scattered light and I_0 is the intensity of the incident light. k and k' are proportionality constants. The term $(n - n_0)/N$ is clearly proportional to the specific refractive increment and, accordingly.

$$\frac{I_s}{I_0} = k''\left(\frac{dn}{dc}\right)^2 \qquad 18$$

It is well known that the intensity of the scattered light is dependent on the wavelength of the light, blue light being scattered more than red light. The dependence of the scatter or wavelength can be determined from the following dimensional analysis. For a small particle, the amplitude of scattered light is equal to the sum of the amplitudes of all the electric oscillators in the particle and the number of such oscillators in the particle is proportional to the mass of the particle and to the volume of the particle. From the above discussion, the intensity of the scattered light is proportional to the square of the mass (volume) of the particle or to L^6. The intensity of the scattered light is also proportional to the reciprocal of the square of the distance of the observer from the source of the light or to L^{-2}. The ratio of the scattered light to the incident light is a dimensionless numeric and, accordingly, in the proportionality between scatter intensity and the dimensions of length, there must exist a length raised to the −4 power. This length is the wavelength of light.

The turbidity is equivalent to the extinction coefficient of an absorbing system and

$$I = I_0 e^{-\tau L} \qquad 19$$

or

$$\tau = \frac{1}{L}\ln\frac{I_0}{I} \qquad 20$$

where L is the length of the light path through the scattering medium and I_0 and I are the intensities of the incident and transmitted light respectively. The turbidity of a suspension is evidently proportional to the number of scattering particles, N, per

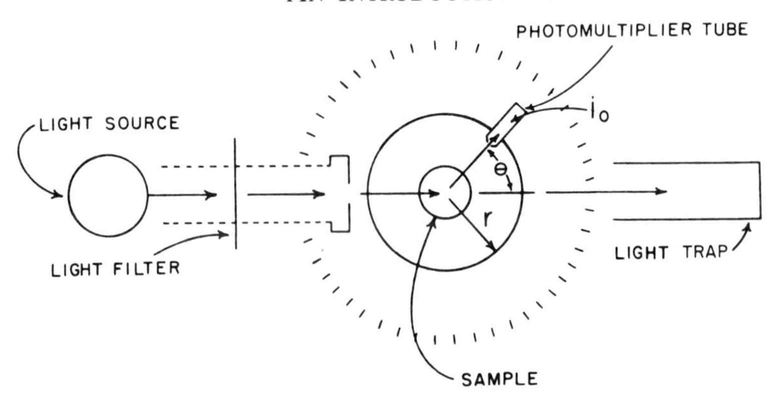

Fig. 11. Diagrammatic representation of apparatus for the measurement of light scatter from a solution. Photomultiplier can be rotated around sample container and the intensity of scatter scanned as a function of the angle.

unit volume. The turbidity is also proportional to the mass of the particles squared. These proportionalities can be expressed as

$$\tau = H'Nm^2 \qquad 21$$

where H' is a proportionality constant and m is the mass of the individual particles. Nm is the weight concentration of the particles and m is proportional to the molecular weight of the particles. Using this information eq. 21 becomes

$$\tau = HCM \qquad 22$$

where C is the weight concentration and M is the molecular weight. H is a new proportionality constant. Equation 22 can be rearranged to give

$$\frac{HC}{\tau} = \frac{1}{M} \qquad 23$$

It is clear from eq. 23 that the larger the molecular weight of the particles, the greater is the light scatter for a given weight concentration.

Except for suspensions of larger particles, the turbidity to be measured is usually so small that the problem can better be approached by determining the intensity of the scattered light; the turbidity can then be calculated from the intensity of the scattered light. An apparatus suitable for the measurement of the intensity of scattered light is diagramed in Figure 11.

The relation between τ and i_0 is obtained by an integration of the observed scattered intensity over a sphere of radius r and

$$\tau = \frac{8\pi}{3} i_\theta \frac{r^2}{I_0} = \frac{16\pi}{3} \frac{i_{90°} r^2}{I_0} \qquad 24$$

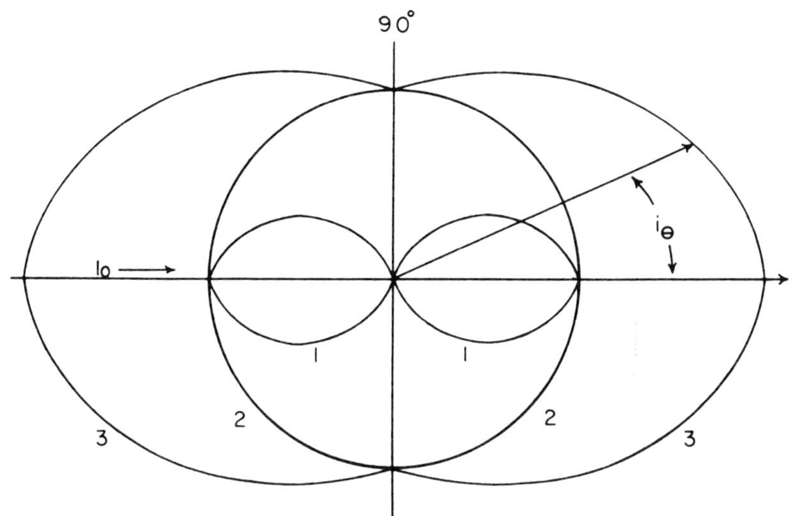

Fig. 12. Intensity envelope of scattered light. Curve 1, horizontally polarized component. Curve 2, vertically polarized component. Curve 3, total scatter intensity.

The factor $i_\theta r^2/I_0$ in eq. 24 is known as Rayleigh's ratio and is denoted by R_θ.

If the incident light is unpolarized and the scattering particles are isotropic and small compared with the wavelength of the light, the scattered light at 90° to the direction of the incident beam is completely plane polarized. The plane of polarization is vertical to the plane formed by the direction of the incident beam and the direction of the observation. At all other angles, the total intensity of scattered light is made up of two incoherent plane polarized components, a vertical one which is spherically symmetrical and a horizontal component which varies with the angle of observation and is proportional to $\cos^2 \theta$, the angular dependence of R_θ being given by the relation

$$R_\theta = R_{90} (1 + \cos^2 \theta) \qquad 25$$

A plot of eq. 25 will yield the intensity of the total scatter as a function of the angle of scatter. Such an intensity envelope is shown in Figure 12.

Concentration Fluctuations. In addition to the treatment of scatter as being the sum of that from the individual particles as discussed above, there is another approach which is based on the microscopic concentration fluctuations of a solute in a solution.

Light passed through an optically clear liquid will be scattered to a small extent due to fluctuations in the density of small volume elements (thermal fluctuations). Additional scatter from solutions arises because of fluctuations in concentration of small volume elements and if the concentration of the solute is significant and the molecules are large compared with those of the solvent, the extent of scatter will greatly exceed that of the pure solvent. Figure 13 shows a diagrammatic sketch of light scatter due to concentration fluctuations. In a solution at equilibrium, the change in the mean value of the concentration in a small volume element involves a change in free energy and the probability of such a change occurring is given by Boltzmann's distribution equation. The magnitude of the scattering effect is related to the index of refraction of the medium and to the index of refraction increment in respect to the solute concentration. The product of the magnitude of the scatter and the probability of the fluctuations in concentration of the solute gives an expression for the intensity of light scatter for particles less than 1/20th of the wavelength of the

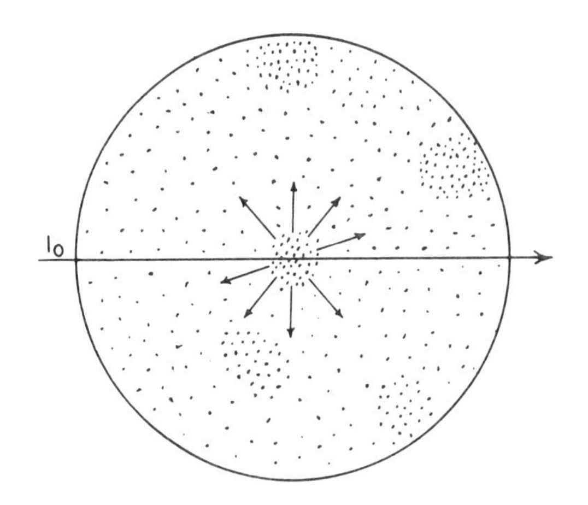

Fig. 13. Diagrammatic sketch of light scatter from a solution due to concentraction fluctuations.

light. Whereas the derivation of this equation requires a number of steps, in principle, the derivation is a simple one and leads to

$$R_{90} = \frac{2\pi^2 RTC\overline{V}_1}{\lambda^4 N \left(\dfrac{dF}{dC}\right)} n^2 \left(\frac{dn}{dC}\right)^2 \qquad 26$$

where n is the index of refraction of the solution and n_0 is that of the solvent; N is Avogadro's number. F is the free energy and since ΔF is equal to $\Pi_0 \overline{V}_1$, where Π_0 is the osmotic pressure and $\overline{V}_1$ is the partial specific volume of the solvent, eq. 26 can be written

$$R_{90} = \frac{2\pi^2 RTC}{\lambda^4 N \dfrac{d\Pi_0}{dC}} n_0^2 \left(\frac{dn}{dC}\right)^2 \qquad 27$$

The osmotic pressure of a solution can be represented by (see Chapter 7)

$$\Pi_0 = \frac{RTC}{M} + RTBC^2 \qquad 28$$

and

$$\frac{d\Pi_0}{RTdC} = \frac{1}{M} + 2BC \qquad 29$$

Combining eqs. 27 and 29 and replacing Rayleigh's ratio R_{90} by $3\tau/16\pi$, yields

$$\frac{\Pi C}{\tau} \cdot \frac{32\pi^2 n_0^2}{3\lambda^4 N} \left(\frac{dn}{dC}\right)^2 = \frac{1}{M} + 2BC \qquad 30$$

The constant terms can be collected and combined into one constant H and the appearance of eq. 30 is simplified to give

$$\frac{HC}{\tau} = \frac{1}{M} + 2BC \qquad 31$$

A plot of HC/τ against C should yield a straight line; the intercept on the y-axis is $1/M$ and the slope of the line is $2B$. It should be noted that since light scatter is proportional not only to the number of particles but likewise to the size of the particles, a weight average molecular weight is obtained. It will be recalled that osmotic pressure at low concentrations is proportional only to the number of particles and hence gives a number average molecular weight defined as

$$M_n = \frac{\Sigma M_i N_i}{\Sigma N_i} \qquad 32$$

where N_i is the number of molecules with a molecular weight of M_i. The weight average molecular weight is defined as

$$M_w = \frac{\Sigma N_i M_i^2}{Z N_i M_i} = \frac{\Sigma C_i M_i}{\Sigma C_i} \qquad 33$$

Only if the solute is monodispersed does M_n equal M_w. The molecular weights of a number of purified proteins have been determined and found to be in satisfactory agreement with those measured by other methods provided careful experimental procedure is followed. The presence of extraneous dust and debris in the experimental solution requires special attention.

Edsall et al. have demonstrated a close correspondence between the B-values calculated from the osmotic pressure measurements of Scatchard et al. on solutions of bovine serum albumin and B-values obtained from light scatter measurements under the same conditions. Shown in Figure 14 is a plot of these B-values as functions of the charge on the bovine serum albumin molecule obtained from the titration curve of this protein.

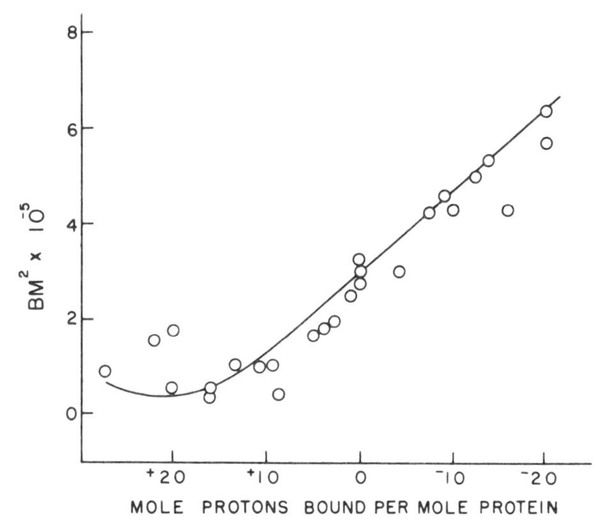

Fig. 14. B-values as a function of the number of protons bound by a mole of bovine serum albumin. Solid line from osmotic pressure. Circles from light scatter. (J. T. Edsall, H. Edelhoch, R. Lontie and P. R. Morrison: J. Amer. Chem. Soc. *72:* 4641, 1950.)

Whereas for simple well-behaved systems the plot of eq. 31 yields satisfactory straight lines as demanded by theory, there are situations which present complications. As pointed out by Timasheff and Kronman three cases can be described in which the plot of HC/τ against concentrations does not yield straight lines. These are (1) associating protein, (2) isoionic protein in pure water, and (3) protein at high charge in ion free solution.

For example, in Figure 15 is shown a plot of HC/τ against the concentration of equimolecular mixtures of egg albumin and pepsin. Provided the excess departure from ideality with increasing protein concentration can be attributed to complex formation, the equilibrium constant for the formation of the complex between egg albumin and pepsin can be calculated. In many ways, light scatter should offer a means of study for interacting protein systems of all kinds, but it should be noted that there are many complicating factors.

Larger Particles. If the solute molecules have a dimension which is greater than about 1/20th of the wavelength of light, the light scattered from different parts of the molecule suffers destructive interference. The envelope of the scattered inten-

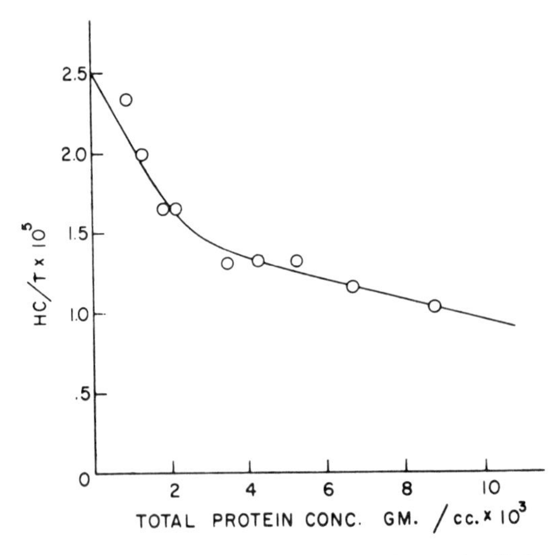

Fig. 15. HC/τ plot against concentration of equimolecular mixtures of egg albumin and pepsin; ionic strength 0.15. (D. S. Yasnoff and H. B. Bull: J. Biol. Chem. *200:* 619, 1953.)

sity is no longer symmetrical since the light scattered in the forward direction is not subject to interference but the difference in phase increases with increasing angle, θ, leading to greater destructive interference in the backward direction. The intensity of the 90° scattering is also less than that for the corresponding particle of same molecular weight but of smaller size. The difference between the backward and forward scatter for larger particles can be understood by an examination of Figure 16 which shows two scatter centers on the same particle separated by a distance comparable to the wavelength of the light. The light waves of the forward scatter will evidently be in phase and no destructive interference results. The right angle scattered light will be partly out of phase due to the difference of optical distance between the two scatter centers and the backward scatter has still greater opportunity for destructive interference. Evidently, the exact state of affairs will depend on the ratio of the distance separating the scatter centers and the wavelength of the light.

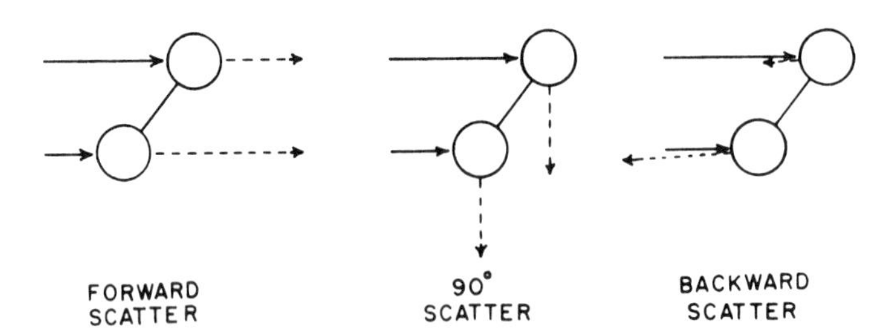

Fig. 16. Illustrating the difference in optical path of forward, right angle and backward scatter from two scattering centers on the same particle separated by a distance comparable to the wavelength of the light.

A correction for this loss in intensity due to interference must therefore be applied before the molecular weight can be calculated by the use of eq. 31. The wavelength of light to be considered is that in the medium. For example, the wavelength of the mercury green line in air is 546.1 mμ and in water it is about 546.1/1.33 or is 409.0 mμ. One twentieth of this value is about 200 A and represents the largest dimension which will yield a symmetrical light scatter intensity envelope. Figure 17 is a diagrammatic representation of the light scatter intensity envelope given by particles which are about the same size as the wavelength of the light.

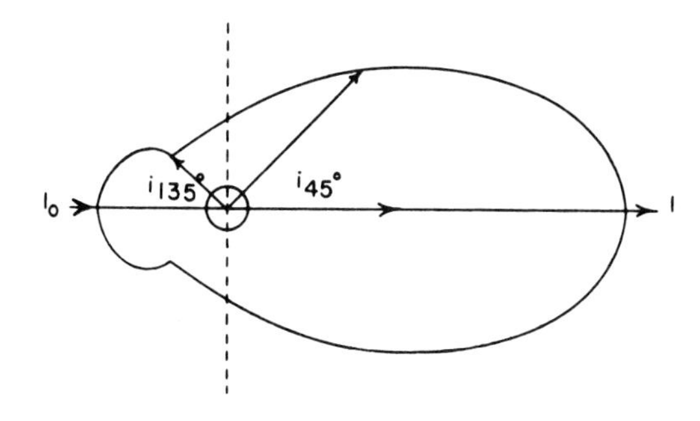

Fig. 17. Intensity scatter envelope of light whose wavelength is about the same as the diameter of suspended spheres.

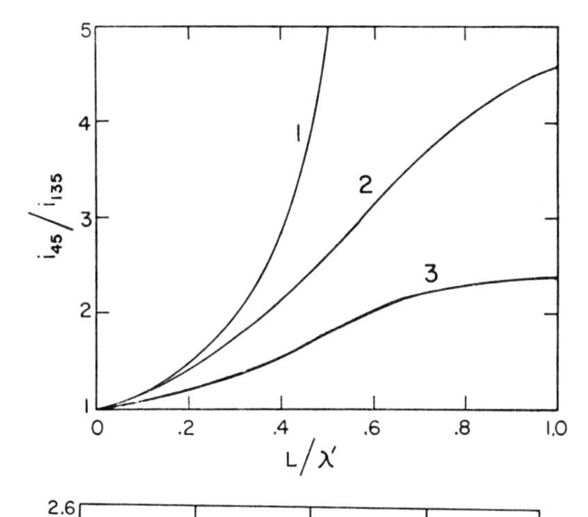

Fig. 18. Ratio of intensity of scatter at 45° to that at 135° as function of L/λ'. Low relative refractive index. Curve 1, spheres; curve 2, coils; curve 3, rods.

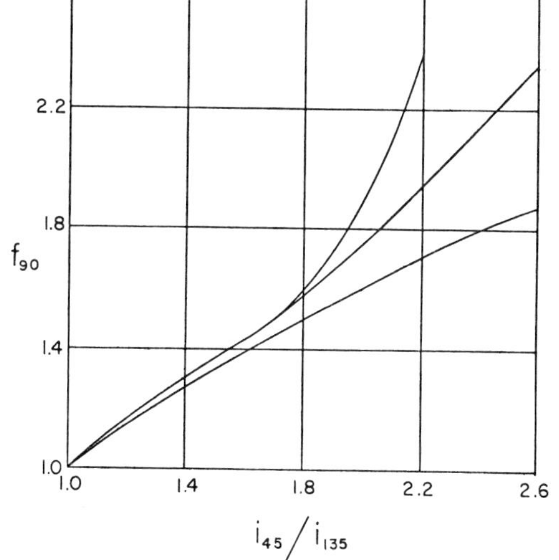

Fig. 19. Correction factors (f_{90}) for scatter at 90° plotted against the dissymmetry ratio; (1) rods, (2) coils, and (3) spheres.

The dissymmetry of the intensity envelope will not only depend upon the size of the particle but likewise on its shape. The dissymmetry is usually characterized by the ratio of the intensity of the scattered light at 45° to that at 135°. Shown in Figure 18 are the long dimensions of spheres, random coils and rods divided by the wavelength and plotted against the dissymmetry ratios.

From the dissymmetry ratio as defined by i_{45}/i_{135} and a knowledge of the shape of the particle, it is possible to obtain the correction factor by which to multiply the scatter at 90° to obtain the true τ and to calculate the molecular weight by means of

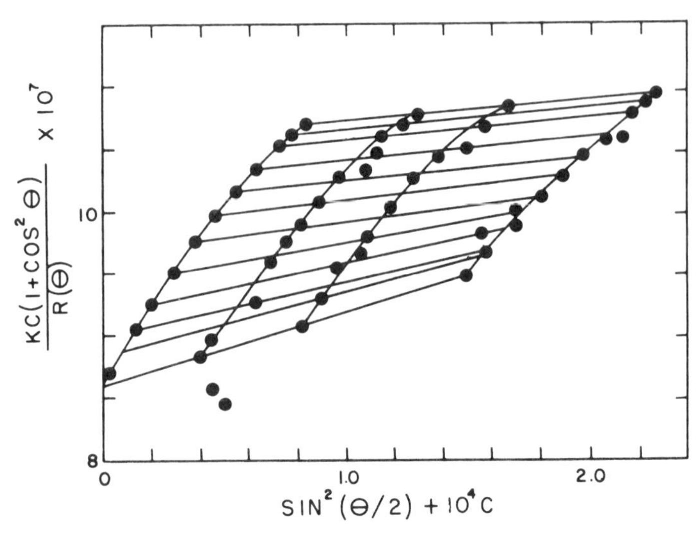

Fig. 20. Zimm plot for RNA from bacteriophage MS2 in 0.05 M NaCl + 0.03 M Tris pH 7.0 at 6°C. Some experimental points omitted to give clearer figure. The extrapolation on y-axis corresponds to a molecular weight of 1.15 x 10^6. (J. H. Strauss, Jr., and R. L. Sinsheimer: J. Mol. Biol. *7:* 43, 1963.)

eq. 31. Figure 19 shows the correction factors plotted against the dissymmetry ratios.

Since there is no destructive interference of the scattered light in the forward direction, and the total scatter at i_0 is exactly twice what it is at i_{90} for a symmetrical scatter envelope, a determination of i_0 should permit the calculation of the molecular weight of the particles. Such an extrapolation is best performed by means of a Zimm plot in which HC/τ_θ is plotted against $\sin^2(\theta/2) + KC$ where K is an arbitrary constant to spread the X-scale. Smoothed lines are drawn through the experimental points at constant angle and at constant concentration, thereby forming a series of grid lines. Each of the lines is then extrapolated to zero angle or zero concentration. Lines are then drawn through these extrapolated points; the two lines should meet at the same intercept of HC/τ_0 at zero concentration. The reciprocal of the intercept is equal to the molecular weight if the incident light is vertically polarized or is twice the molecular weight if unpolarized light has been used. Such a Zimm plot for RNA is shown in Figure 20.

Still Larger Particles. It has been noted that as the particles in suspension become larger than about 1/20th the wavelength of light, the backward scatter decreases progressively with increasing particle size whereas the intensity of the forward scatter continues to increase with the square of the particle size. At a particle radius which is one-fourth of the wavelength of the light, there is complete destructive interference of the backward scatter and its intensity becomes zero.

As the particles become still larger, extinction occurs at intermediate angles to the direction of the beam and reinforcement begins to appear in the backward direction. This leads to maxima and minima in the intensity of scattered light as the suspension is scanned at various angles to the incident light. The position of these maxima and minima is related to the particle size, the wavelength of the light and the difference of refractive index between the particle and the medium. This phenomenon has been made use of to measure the size of living cells in suspension and to study their osmotic behavior. An ordinary light scatter apparatus can be modified to accommodate these measurements and the position of the first minimum as the scanning phototube moved from zero angle noted. For spherical particles whose diameters extend from about 0.2 to 4 microns, the particle diameter is given by

$$d = \frac{Z\lambda_m}{\sin \frac{\theta}{2}} \tag{34}$$

where λ_m is the wavelength of the light in the medium, θ is the angle at which the first minimum occurs, and Z is a function of the index of refraction of the particle and of the medium. Between n_1/n_2 equal 1 and 1.55 where n_1 is the index of refraction of the particle and n_2 that of the medium, Z is a linear function of n_1/n_2 and

$$Z = 1.062 - 0.347 \frac{n_1}{n_2} \tag{35}$$

Dilute suspensions must be used and the calculated particle diameter should be independent of the wavelength of the light; dependence of d on wavelength indicates polydispersity.

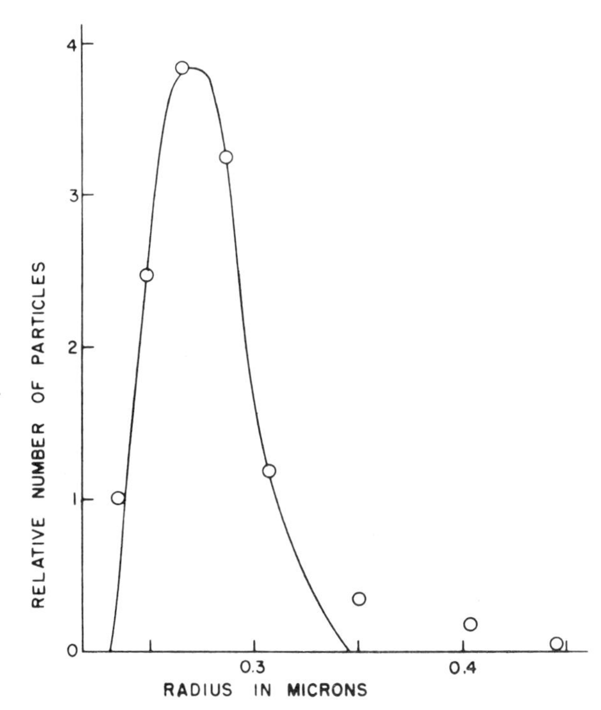

Fig. 21. Particle size distribution of n-octadecane emulsion after centrifugation for 5 minutes at 2,000 RPM in water at 30°. Experimental points from Coulter counter. Solid line from light scatter at wavelength of 409.3 $m\mu$. (S. Bandyopadhyay: Private communication.)

If the particles in suspension are sufficiently heterogeneous in respect to particle size, no maxima or minima will be observed as the angle of scanning is varied; the dispersion in particle size causes the maxima and minima to average out. Serious effort has been made to use light scatter to obtain the particle size distribution of larger spherical particles. The situation is somewhat unsatisfactory, however, because there is no explicit expression available to describe such a distribution in terms of light scatter. There exist several procedures by which the problem can be approached. The essential information required is the variation of the turbidity of the suspension as a function of concentration and of the wavelength of the light. The index of refraction of the particles and of the medium are also required. It is also necessary to make an assumption regarding the nature of the particle size distribution. Figure 21 compares the particle size distribution as obtained by light scatter and by the Coulter counter of a *n*-octadecane emulsion in water prepared by sonification.

Even for large particles such as encountered in emulsions and living cells, it is possible to extrapolate the scatter intensity to zero angle by means of a Zimm plot and to eliminate the effects of destructive interference of the light and to calculate the average molecular weight for the suspension with use of eq. 31.

Frequent use is made of turbidity measurements to estimate the concentration of bacterial suspensions, emulsions, etc. If the particle sizes do not change from one situation to another this is an acceptable procedure, but it is evident that light scatter is not only proportional to concentration; it is also a function of particle size.

For more concentrated solutions and suspensions, difficulty is encountered with multiple scatter. Light scattered from one particle is scattered by another particle, etc. This phenomenon becomes of importance when the turbidity exceeds about 0.2 and will confuse any light scatter treatment; it is necessary to work with more dilute suspensions.

Spectroscopy. The reflection, refraction, diffraction, and polarization of light can be described and understood in terms of the wave theory of light, i.e., that electromagnetic radiation consists of transverse waves as described at the beginning of the chapter. However, at the turn of the century, Planck concluded from his studies on the radiant energy of black bodies at high temperatures that energy was radiated in discrete packages. Many investigations have confirmed the dual nature of electromagnetic radiation as particles and as waves. It is concluded that radiation can be emitted only in multiples of $h\nu$ where h is Planck's constant whose dimensions are action and whose numerical value is 6.625×10^{-27} erg sec. and ν is the frequency in cycles per second. $h\nu$, therefore, has the dimensions of energy. The quantity of energy $h\nu$ is a quantum of radiation; it is also spoken of as a photon. In a sense, the treatment of light as a wave is a mathematical fiction; the detection and measurement of light intensity always requires light to behave as particles.

Not only do electromagnetic radiations have the character of particles, but particles in motion can be diffracted as though they have a wavelength λ given by the expression

$$\lambda = \frac{h}{mV} \qquad 36$$

where m is the mass of the particle and V is its velocity (mV is the momentum of the particle). This statement has been amply confirmed in the study of electron diffraction.

For an adequate presentation and understanding of emission and absorption spectroscopy one runs head-on into quantum theory and wave mechanics; without a knowledge of these tools, one is severely handicapped in relating spectroscopy to molecular structure.

When a solution is exposed to electromagnetic radiation a number of things can happen separately and together: (1) the radiation can be reflected, (2) it can be refracted without absorption, (3) it can be scattered (Rayleigh scatter) and the frequency of the radiation is unchanged, (4) it can give rise to a Raman spectrum in which the light is scattered at longer and shorter wavelengths than the incident light, (5) the radiation can be absorbed and the energy directly released as heat (the more common form of absorption), (6) the radiation is absorbed and then released after a brief interval at somewhat longer wavelengths; this behavior includes both fluorescence and phosphorescence (it is difficult to observe phosphorescence except in the solid state), (7) the energy can be absorbed and gives rise to a photochemical reaction. The ultimate fate of all radiation is conversion into heat. The kinetic energy of a free molecule can vary in a continuous manner but the energy associated with periodic motion of the molecule as a whole and with the constituent atoms relative to one another as well as that of electronic transitions can assume only discrete levels, the so-called quantitized energy levels which characterize stationary states.

The energy change of an electron, atom, or molecule is related to the frequency of the absorbed electromagnetic radiation by

$$\Delta E = E - E_0 = h\nu \quad 37$$

where E is the energy of the higher energy state and E_0 is that of the lower and ν is the frequency of the absorbed radiation. Only certain energy levels of an atom or molecule are possible and, accordingly, the frequency of the absorbed radiation occurs at the corresponding energy levels and gives rise to the characteristic absorption spectra.

In describing spectroscopy, the choice has been made to present in sequence radiofrequency, infrared, visible and ultraviolet spectroscopy. The first involves the interaction of magnetic fields with electrons and nuclei, the second the rotation and twisting of parts of molecules, and visible and ultraviolet spectra require electronic transition. The procession is from low to higher energy transitions.

Radiofrequency Spectroscopy. A group of methods which embraces nuclear magnetic resonance, electron spin resonance, and microwave spectroscopy is included under the term, radiofrequency spectroscopy. In these methods, the frequencies compared to visible light are very low, and the energy transitions are small.

Electrons and nuclei of atoms are charged bodies with angular spin and, accordingly, they produce magnetic fields of their own; their magnetic moments are inversely proportional to the mass of the particle. Since the mass of the proton is 1837 times that of the electron, the magnetic moments that are associated with electron spin are much larger than those of atomic nuclei. Both electron spin resonance as well as nuclear magnetic resonance depend for their success on the interaction of an external magnetic field with the magnetic moments of the electrons and the nuclei respectively. Both methods have great potential interest for biochemistry and, accordingly, brief summaries of the methods will be presented.

Electron Spin Resonance. This method is also known as paramagnetic resonance. Electrons have a spin and in an ordinary covalent bond the two electrons spin in an opposite sense so that they give rise to no net magnetic moment. In fact, when such substances with paired electrons are placed in a magnetic field a magnetic moment is induced which opposes the applied field and they tend to move out of the field. This is the general state of matter and these substances are said to be diamagnetic. On the other hand, a substance with unpaired electrons has a permanent magnetic moment and tends to become orientated in line with the applied magnetic field. These substances are paramagnetic. It is evident that magnetic studies on free radicals are valuable in characterizing oxidation-reduction reactions. (See Chapter 4, Oxidation-Reduction Potentials.)

Magnetic susceptibilities were formerly studied with some form of the Gouy balance (the Gouy balance is still used for specific purposes) in which the substance in question was suspended between the poles of an electromagnet from a sensitive analytical balance and the magnetic force acting on the substance was measured. Unfortunately this method is not capable of detecting small quantities of a free radical, and very concentrated solutions have to be used. More recently, the method of electron spin resonance has been employed with great success.

When a free radical is exposed to a magnetic field there exists an energy difference in respect to the orientation of the electrons, i.e., whether or not the magnetic field of the odd electron spin is parallel or antiparallel to the direction of the applied field. When the energy of the applied magnetic field is equal to the energy of the transition of the electron from antiparallel to parallel orientation such a transition occurs and electromagnetic radiation is absorbed whose energy per quanta corresponds to the energy of transition (see Fig. 22).

The energy difference ΔE of the transition of the electron spin is equal to $h\nu$ where h is Planck's constant and ν is the frequency of the absorbed radiation. The microwave energy supplied to the system is usually set at about 3 cm. wavelength;

such a wavelength corresponds to a low frequency and to a small energy per quanta. The energy difference, ΔE, is also equal to BgH where B is the magnetic moment of the electron spin, g is the gyromagnetic ratio which is the ratio of the magnetic moment to the spin angular momentum of the electron, and reflects the interactions of the electron's magnetic moment with the electron's environment. For organic radicals, g varies about the free electron value of 2.0023, in a narrow range of about 2.0010 to 2.0070. A given free radical exhibits characteristic g-values within this range. H is the external field expressed in Gauss. Experimentally, the magnetic field is increased, the frequency of the microwave remaining constant until resonance

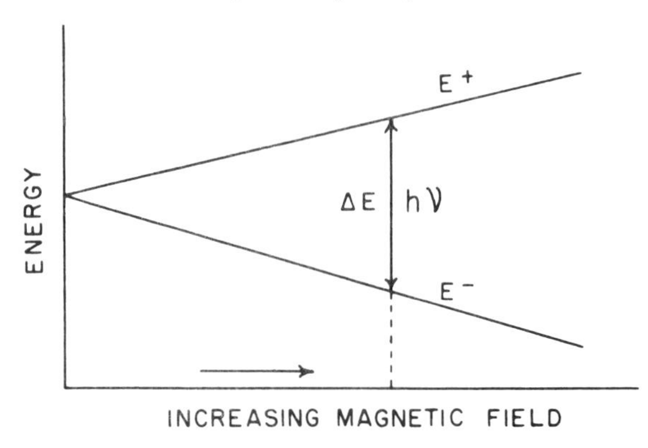

Fig. 22. Plot of the two permitted energy levels (E_+ and E_-) of an electron spin as a function of the applied magnetic field. Transition occurs at a critical magnetic field strength.

with the radiation is reached at which point absorption of the radiation is recorded by the detecting system. A diagrammatic representation of the experimental arrangement is shown in Figure 23. The electron spin resonance spectrometer signal is in the form of a differential absorption curve. The differential curve is integrated and the area under the integrated curve is proportional to the concentration of the free radical. The detectable concentration of free radical with presently available equipment is about $5 \times 10^{-7}M$.

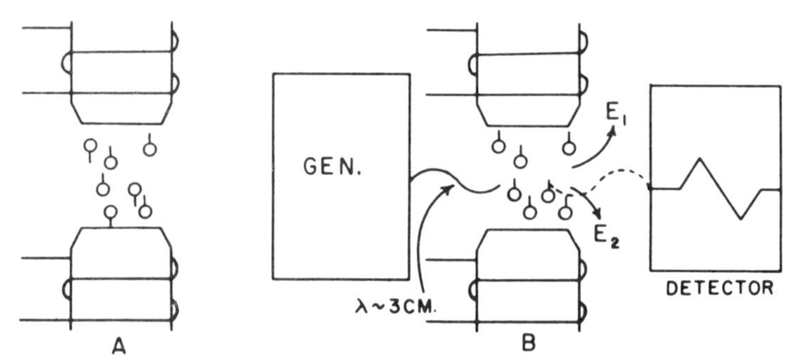

Fig. 23. Diagrammatic experimental arrangement for electron spin resonance. *A*, Magnetic field not in resonance with electron spin. *B*, Magnetic field in resonance with electron spin and with energy absorption which is recorded as a differential spectrum.

Owing to the interaction of the magnetic field produced by the electron spin of the unpaired electrons with the magnetic moment of certain atomic nuclei, the absorption spectrum reflects the fine structure of the whole molecule. Electron spin resonance has been used primarily in the study of free radicals and in biological oxidation; free radicals have been detected in living tissue.

Nuclear Magnetic Resonance. Electron spin resonance and nuclear magnetic resonance resemble each other fairly closely; nuclear magnetic resonance is, however, somewhat more complex. Atomic nuclei which have odd mass numbers or odd numbers of protons in the nucleus have magnetic moments. The nuclei which show magnetic moments in decreasing order of usefulness are *H, F, P, N,* O^{17}, and C^{13}.

When the nuclear magnets are placed in a magnetic field, they precess about the direction of the magnetic field in the same sense as a wobbly spinning top precesses in the gravitational field of the earth. The frequency of precession, ν, of the wobble is proportional to the strength of the magnetic field, H, and

$$\nu = \frac{\gamma}{2\pi} H \qquad 38$$

where γ, the gyromagnetic coefficient, is the ratio between the magnetic moment and the angular momentum spin of the nucleus.

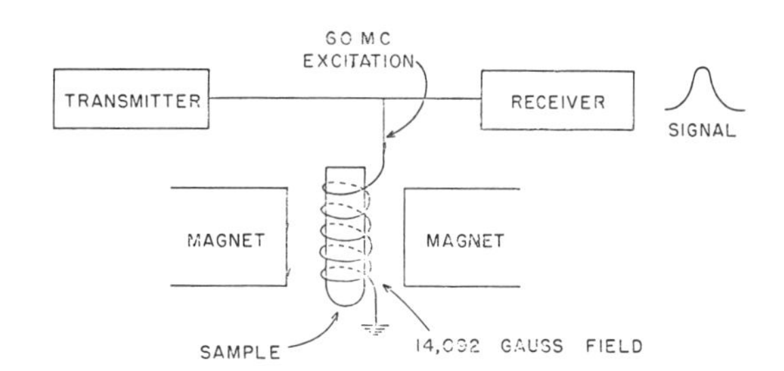

Fig. 24. Diagrammatic sketch of nuclear magnetic resonance apparatus.

A proton placed in a magnetic field has two possible orientations relative to a static external magnetic field. The low energy orientation has the nuclear magnetic moment aligned parallel and the high energy orientation antiparallel (opposed) to the external magnetic field. There is only a small energy difference between these two orientations and, accordingly, the number of nuclei in the low energy state exceeds those in the higher energy state by only a small margin. The nuclear transition corresponds to the promotion of nuclei from the low to the higher energy state. This transition occurs when a magnetic field with its vector rotating in a plane perpendicular to the main magnetic field has the same frequency as the frequency of the processing nuclei; the two are in resonance and absorption of energy by the nucleus occurs. For a proton in a magnetic field of about 14,000 Gauss, the resonance frequency is about 60 megacycles per second (60 MC). The nuclei in the upper spin state return to the lower spin state by relaxation. The shorter the relaxation times, the broader are the spectral peaks. Broadening of the spectral peaks is so pronounced in solids and in viscous liquids that NMR is substantially useless for their study. Broadening also occurs if the external magnetic field is not uniform. The experimental arrangement for NMR measurements is diagrammatically sketched in Figure 24.

Nuclear magnetic resonance is a powerful tool for the investigation of the structure and the identification of organic molecules; the NMR spectra of a large number of compounds have been catalogued. In an aqueous solution, the water protons contribute so enormously to the spectrum as to obscure the details of that arising from the solute. It is customary to use deuterium oxide as a solvent rather than water. Whereas deuterium nuclei do have spin and an associated magnetic moment, their absorption is far removed from that of the protons and, accordingly, is not seen. When exposed to the same effective magnetic field, all protons process at the same frequency. However, protons in an organic molecule are shielded from the applied magnetic field to different extents and, accordingly, each proton which is differently shielded will yield a separate and distinct signal. The shielding arises from an induced circulation of the electrons about the nucleus. The electrons produce a magnetic field which opposes that of the applied field. The magnitude of the effective field is

$$H_{eff} = H_o (1 - \sigma) \qquad 39$$

where H_o is the strength of the applied magnetic field, and σ is the shielding factor. The separation of resonance frequencies due to different shielding is known as the chemical shift

The chemical shift parameter, δ, is defined as

$$\delta = \frac{H_r - H_s}{H_r} \qquad 40$$

where H_s and H_r are the field strengths producing resonance for a particular nucleus in the sample (H_s) compared with the reference (H_r). δ is dimensionless and is expressed in parts per million (ppm). Tetramethylsilane (TMS) is the reference usually selected because it has a single sharp peak which occurs at a higher frequency than that of protons usually found in other compounds. The frequency scale is expressed relative to that of TMS by the symbol τ where τ is equal to $10 - \delta$; the larger τ, the greater is the magnetic shielding. The area under a given absorption peak is proportional to the number of protons having the same environment in the molecule.

At low resolution (low external magnetic field) single peaks will be observed for each set of protons in a molecule. For example, ethanol at low resolution gives three peaks corresponding to protons associated with CH_3, CH_2 and OH and further the areas under these peaks are in the ratio 3: 2: 1, respectively. With increasing magnetic field it is seen that the individual peaks begin to split due to spin-spin interactions, that is, the behavior of an individual proton is influenced by other protons which it "sees." The spin coupling permits the identification of near neighbor protons in a molecule.

Effort has been made to examine the molecular structure of polypeptides and proteins using NMR. There are, however, difficulties associated with such studies and these center around the following: (1) Assignment of the spectral peaks to the chemical groups responsible may be ambiguous. (2) Peak overlap may occur because in a large molecule there will be numerous protons with nearly the same environment. (3) In a rigid macromolecule such as a native protein, the slower translational and rotational motions give rise to peak broadening just as extreme peak broading is observed in viscous and solid samples. There are two approaches to the above problems. One is to reduce the number of protons by exchange with deuterium, thus decreasing the number of protons which have to be identified. The other expedient is to increase the flux of the magnetic field. Increasing the intensity of the magnetic field does not result in peak broading, and the height of the peak is proportional to the intensity of the magnetic field, thus leading to sharper spectra and better resolution.

Infrared. Molecules absorb in the infrared region because of changes in the vibrational frequencies of atoms and groups of atoms in a molecule. In contrast to the spectra in the visible and ultraviolet, the electronic configuration is not altered; instead the dipole moment of the molecule shifts due to the movement of the nuclei of the bonded atoms. Spectra of interest in the infrared arises from the movement along the valence bonds or normal to them and these are bond stretching and bond deformation respectively. Bond deformation or bending vibrations can assume a variety of forms. For example, in a saturated hydrocarbon the deformation vibrations of C-H bonds in methylene groups fall into four classes with the somewhat descriptive names of rocking, wagging, twisting, and scissoring.

There are two factors that cause a vibrational frequency to be high: a small mass of the participating atoms and a high value of the restoring force. The frequencies due to stretching are higher than those due to deformation. In Chapter 1, Mathematical Review, it was shown that the natural frequency of a harmonic oscillator is given by $\frac{1}{2\pi}(k/m)^{1/2}$ where k is the force constant (elastic coefficient) and m is the mass of the particle. In dealing with two bonded atoms, the reduced mass of the two atoms is used. The reduced mass is equal to $m_1m_2/(m_1m_2)$ where m_1 and m_2 are the masses of the two bonded atoms in grams. The approximate value of the stretching frequency in reciprocal centimeters is given by

$$\nu = \frac{1}{2\pi}\left[\frac{R\ (m_1 + m_2)}{m_1m_2}\right]^{1/2} \qquad 41$$

where R, the force constant, is approximately 5, 10, and 15 $\times$ 10^5 dynes/cm for single, double, and triple bonds respectively.

The infrared spectrum extends from the end of the visible to the beginning of the microwave region corresponding to wavelengths from about 0.75 μ to about 1,000 μ. Only a comparatively short span of the spectrum is of immediate interest to the biologist since the vast majority of the spectra which are of significance to molecular structures are found between the wavelength limits of 2.5 to 15 μ.

The measurement of infrared absorption requires special apparatus. Glass cannot be used for lenses or for prisms since glass is opaque to infrared. Concave mirrors are employed in place of lenses and prisms are made of salts such as NaCl; the prisms can be replaced by diffraction gratings. The choice of the solvent also presents difficulties since water has a rich and intense infrared absorption spectrum of its own over much of the working range.

A non-linear molecule containing n atoms has $3n - 6$ possible fundamental vibrational modes that can produce absorption of infrared light. Methane is indeed a simple molecule, but in principle it can have nine fundamental absorption bands. Additional non-fundamental absorption bands of greatly reduced intensity may also occur because of the presence of harmonics. Thus it is seen that infrared spectra can be very complex.

In a large molecule such as a sterol the infrared absorption spectrum is highly elaborate with considerable intricate detail. Infrared spectroscopy provides a means of identification that is free of ambiguity. A great many sterol-like compounds have been "finger-printed" and the infrared spectroscopy has been of great help in the investigation of sex hormones and hormones of the adrenal cortex.

The various valence bonds have rather characteristic frequencies of vibration and accordingly, under favorable conditions, it is possible to obtain information concerning the types of bonds in a molecule from its infrared spectrum. It should, however, be noted that coupled frequencies between the various valence bonds can occur and with resulting shifts of the expected frequencies. It is true, however, that the frequency of vibration of bonds connecting hydrogen atoms with heavier atoms is not greatly influenced by changes made elsewhere in the molecule; most of the motion occurs in the light hydrogen atoms. For example, molecules containing aliphatic CH_2-groups show absorption bands at about 2860 cm.$^{-1}$ and at 2930 cm.$^{-1}$ as a result of bond deformation and of bond stretching, respectively.

It is possible to detect and to estimate the extent of hydrogen bond formation from infrared spectra. In the formation of the bond

$$\text{A—H --- A}'$$

the force constant of the A—H bond is diminished by the hydrogen bond. The modes of vibration of this system consist of (1) the motion of the hydrogen atom in which the A—H bond is stretched; (2) the hydrogen atom is moved perpendicularly to the A—H bond (deformation), and (3) the A—H moves relative to A′. The first mode because of the weakening of the A—H bond incidental to the formation of the hydrogen bond will be at a lower frequency (the unbonded A—H will yield a frequency corresponding to about 3,000 cm.$^{-1}$). The second mode (deformation) has a higher frequency because the resistance to deformation has been increased as a result of hydrogen bond formation. Because of the weakness of the hydrogen bond itself and the large masses of A—H and of A′, the third mode of vibration is very low and not often seen. Not only is there a decrease in the stretching frequency of the A—H bond upon hydrogen bond formation, but the integrated absorption is greatly increased.

Infrared studies have been most useful in the investigation of polypeptides and the employment of polarized infrared has been especially rewarding. It has become possible to synthesize long chain polypeptides of relatively high molecular weight such as poly-L-alanine, poly-L-glutamic acid and a variety of other peptides. The regions of the infrared spectrum of interest for polypeptide study is in the vicinity of 1600 cm.$^{-1}$ and in the range of 3,000-3,500 cm.$^{-1}$. There is a frequency at about 1550 due to N—H deformation, one at 1640 due to C=O stretching. The 3,000-3,500 cm.$^{-1}$ range embraces stretching frequencies for the C—H, N—H, and O—H bonds as well as activity from the CH_2 groups. The formation of hydrogen bonds of the type N—H--O has been found to shift the stretching frequency from 3450-3570 to 3050-3330. Since the CH stretching frequency falls in the range of interest for hydrogen bond formation, confusion of identity can arise. This confusion can be relieved by exchanging the H-atoms with D-atoms; such an exchange will not occur with a C—H bond but will with a N—H bond. The ratio of the A—H stretching frequency to the A—D stretching is about 1.36.

Atoms whose mode of vibration is parallel to the vector of polarized infrared will absorb more intensely than those which are perpendicular. This provides a means by which the direction of the various bonds in a specimen can be determined. The directional effect in the absorption of light is known as dichroism and the ratio of the optical density measurement first with the plane of vibration of the light parallel to the fiber axis then perpendicular to it is known as the dichroic ratio. The orientation of the hydrogen bonds in samples of synthetic polypeptides has been studied with polarized infrared. It has been found that, in stretched peptides (β-keratin form), the hydrogen bonds are directed mostly perpendicularly to the fiber axis; these are interpeptide chain hydrogen bonds. Polypeptides folded in the α-helical form have their hydrogen bonds directed mostly parallel to the fiber axis; these are intrapeptide chain bonds and serve to stabilize the α-helix.

Transitions in peptides can be studied with infrared. Figure 25 shows the infrared spectra of poly-L-proline at 31° and at 65°. Notable, at the lower temperature, are the absorption band at 1624 cm^{-1} which is predominantly a carbonyl-stretching motion and the band at 1456 cm^{-1} which is predominantly a C-H bending motion of the pyrrolidine ring. At the higher temperature, the poly-L-proline precipitates and the infrared spectra change. As can be seen from Figure 25 the 1624 cm^{-1} band is shifted to higher and the 1456 cm^{-1} band to lower frequencies as the temperature is raised. The shift of the 1624 cm^{-1} band can be interpreted as resulting from the rupture of deuterium bonds between D_2O and the carbonyl with a consequent dehydration. The 1456 cm^{-1} shift is probably directly related to helical transitions occurring at higher temperatures.

Electronic Configuration. As has been noted, the absorption of radiation in the visible and the ultraviolet region is due to electronic transition. In general, the situation is inherently more complex than absorption in the infrared.

The formation of a co-valent chemical bond is accompanied by the sharing of the valence electrons of the two atoms; the shared electrons occupy molecular orbitals in pairs and with opposite spin. For example, in a saturated hydrocarbon, there are two electrons shared between each of two carbon atoms with the electron density concentrated between the two carbons; the electrons occupy σ-orbitals. It is difficult to promote such electrons to higher energy levels and saturated hydro-

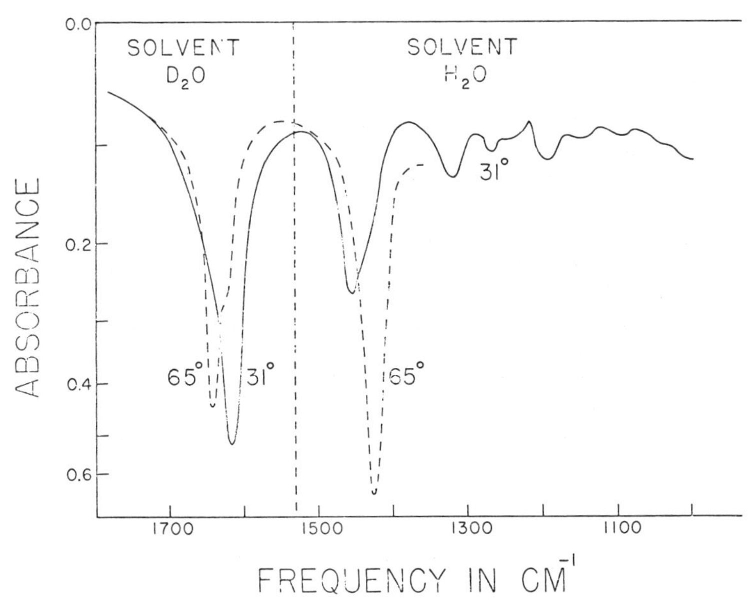

Fig. 25. Infrared spectrum of a one percent aqueous solution of poly-L-proline at 31° and at 65°. Solvents: D_2O 1800–1530 cm^{-1}; H_2O, 1550–1000 cm^{-1}. (Adapted from C. A. Swenson and R. Formanek: J. Phys. Chem. *71:* 4073, 1967.)

carbons only absorb in the far ultraviolet, at wavelengths less than about 180 mμ.

The introduction of a double bond between two carbon atoms such as in ethylene requires the sharing of four electrons between the two carbon atoms. Two of these electrons occupy the σ molecular orbital as described above for the saturated hydrocarbons, but the other two electrons occupy the π orbital; the electron spins of the two electrons in the π orbital are also opposite. Actually, there are three σ orbitals associated with each carbon atom. Two of these orbitals contain the electrons bonding the two hydrogen atoms and the third forms the carbon-carbon linkage. The three σ orbitals are co-planar. The maximum charge density of the π orbitals occurs in two regions, one above and one below the plane containing the carbon and hydrogen nuclei; the molecular plane itself is a region of zero charge density for the π orbital. The binding of the electrons to the nuclei in the π bond is weaker than in the σ bonds and the π electrons are excited by radiation of a lower frequency than are the σ electrons. In a molecule containing conjugated double bonds, the individual π orbitals overlap to form a set of π orbitals and these orbitals extend over the entire conjugated region of the molecule; thereby the wavelength of absorption is increased. If the conjugation is sufficiently extensive, absorption may occur in the visible.

Upon excitation, the position of the molecular orbital electrons is shifted relative to the nuclei in such a manner that the electron density between the atoms decreases to a level which is less than that for two non-bonded atoms the same distance apart. The excited orbital is then said to be anti-bonding and is usually denoted by an asterisk such as σ^* or π^* depending upon the orbital involved. The presence of anti-bonding electrons can lead to dissociation of the molecule and a photochemical reaction.

Also important in molecular spectroscopy in the ultraviolet and visible range are the non-bonding valence electrons, the so-called lone pair electrons. These electrons are mostly localized in specific atoms such as oxygen, nitrogen, halogen atoms, or sulfur. Some of these lone pair electrons can be excited by relatively low energy corresponding to wavelengths above 200 mμ.

As noted, electrons in occupied orbitals can be excited to unoccupied ones by absorbing light. If, during this excitation process, the electron's spin does not change, there will be no net angular momentum, and the ground and excited states are called singlets. It is possible, however, that the spin of the promoted electron or the one remaining in the ground state is reversed. The molecule will then have an electronic magnetic moment and the energy level will be split into three levels in a magnetic field; this is a triplet state.

Raman Spectra. Light can be scattered without change in frequency. This scatter is known as Rayleigh scatter and has been discussed. There is also the Raman scatter in which the light causes forced oscillations in the molecule, but the disturbance is insufficient to produce an electronic transition. There is, however, a shift in the vibrational level of the molecule. When the molecule returns to its ground state, radiation is emitted the wavelength of which is different from that of the original light. The difference in energy between the vibrational levels of the molecule and the difference in frequency of the incident and scattered light is related by

$$\Delta E = h(\nu_i - \nu_r) \qquad 42$$

where ΔE is the energy difference between the vibrational levels, ν_i is the frequency of the incident light and ν_r is that of the scattered light. The energies associated with vibrational levels are not large and correspond closely with the energies of infrared absorption. Thus if ultraviolet or visible light is used as a source, the scattered light is still in the ultraviolet or visible, and accordingly the problem of light absorption by water, which is one of the technical difficulties met with in infrared, is completely avoided. The Raman spectra usually bear a strong resemblance to infrared spectra of a molecule, but there is no necessity for them to be identical.

Fluorescence. There are two other ways by which radiant energy can be scattered and these are known as phosphorescence and fluorescence. The mechanism of scatter is, however, very different than it is for the Rayleigh and Raman scatter. In both phosphorescence and fluorescence, an electron is promoted to a higher energy level by absorbed radiation and on returning to the ground state emits radiation. Both the Rayleigh and Raman scatter will accommodate incident light of any frequency, whereas phosphorescence and fluorescence exhibit characteristic spectra both for absorption and for radiation.

Whereas a direct transition of the excited electron to the ground state is possible, and in which case the emitted radiation has the same frequency as that absorbed, more usually, some energy is lost by molecular collisions and the electron reduced to another permitted level but with less energy. The return of the electron to the ground state is then accompanied by the emission of light of a longer wavelength than that absorbed. In phosphorescence, the initial loss of energy by the electron places the electron on a level from which a transition to the ground state is forbidden. It can, however, slowly return to the ground state with the release of radiation. There is thus characteristically a difference in the time constants for fluorescence and for phosphorescence, fluorescence being the faster by several orders of magnitude (the half life of the activated molecule usually being in the order of $10^{-9} - 10^{-7}$ seconds). Phosphorescence is more often observed in a highly viscous or solid medium.

In terms of the electronic transitions, fluorescence results from the decay of an excited singlet state, whereas if the excited singlet state undergoes a transition to the related triplet state and then returns to the ground state the result is phosphorescence (see Fig. 26).

All molecules which absorb light energy should fluoresce but not all do. This means that the fluorescence efficiency of most absorbing molecules is very low.

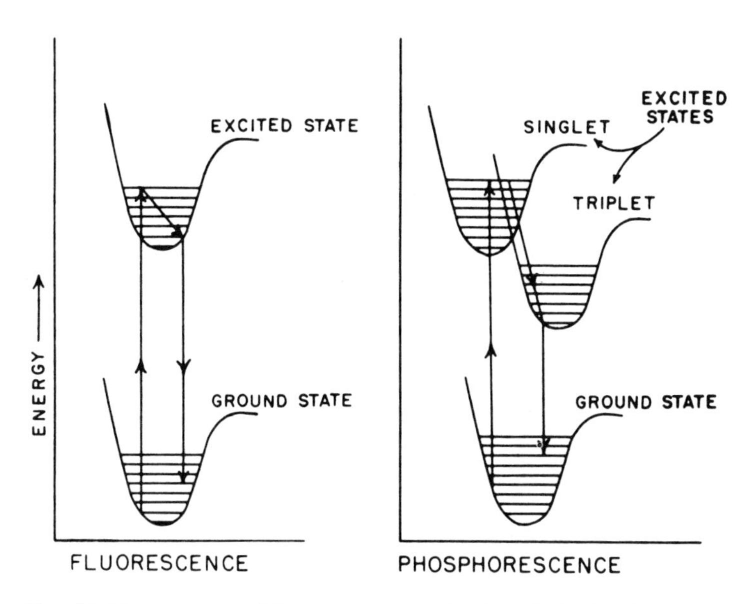

Fig. 26. Electron potential-energy curves showing distinction between fluorescence and phosphorescence.

Fluorescence can be quenched either through interaction with the solvent or with another solute. In principle, the quantitative measurement of fluorescence is simple. A strong beam of monochromatic light of the desired exciting wavelength is allowed to impinge on the sample and the sample is scanned at right angle to the incident beam. Fluorescent techniques are widely used in analytical biochemistry; also, fluorescent antibody tracers are used to locate the corresponding antigens in tissues or cells.

Fluorescent probes attached to specific sites on protein molecules can give information about the surface characteristics of the protein. Whereas phenylalanine, tyrosine and tryptophane residues which occur in proteins exhibit weak fluorescence, it is the usual practice to insert unnatural, strongly fluorescent chromophore probes on the protein either by co-valent bonding or by specific absorption.

A fluorescent group having a higher dipole moment in the excited state than in the ground state tends to orient water molecules about itself to a greater extent in the excited state. Some of the hydration energy is lost upon return to the ground state and, accordingly, the emitted light suffers a red shift relative to a non-polar solvent. It is also observed that the quantum yield is higher (the intensity of the emitted light is greater) in a non-polar solvent. A fluorescent probe inserted on a binding site of a protein can thus yield information concerning the polarity of the environment of the binding site from the wavelength maximum of the emitted light.

It is possible for a chromophore to absorb energy and to release the energy to another chromophore with singlet-singlet transfer through dipole interaction across distances up to about 70 A. The efficiency of transfer is dependent on the orientation of the two chromophores relative to each other and varies as the reciprocal of the sixth power of the distance separating the chromophores.

Finally, the rotary diffusion coefficient of a protein as well as the flexibility of the site of attachment can be determined through a study of the depolarization of fluorescent light (see Chapter 11).

Absorption of Light. An excited molecule can lose its excess energy by emitting radiation, by undergoing a chemical reaction, or by transferring the energy to another molecule as a result of a molecular collision. In this last process the kinetic

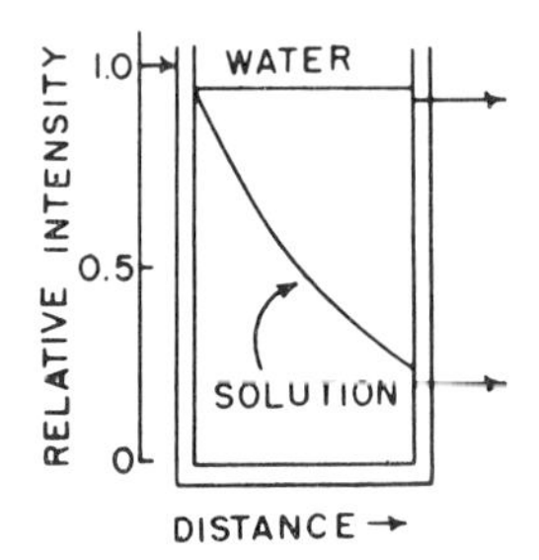

Fig. 27. Decrease of light intensity with distance through a solution. Note the drop in intensity due to reflection from the walls of the container.

translational energy of the system is increased and there is an increase in the temperature. This is the more usual, immediate, fate of absorbed radiation and is the ultimate one for all radiation and the one which is now to be considered.

A close relation exists between the extent of absorption and the concentration of the absorbing substance. The change of the intensity of light with distance through the solution can be written

$$\frac{dI}{dx} = -KCI \tag{43}$$

where C is the concentration of the absorbing substance in solution, I is the intensity of the light at the distance x and K is a constant characteristic of the wavelength of the light and of the substance. The negative sign indicates that the intensity of the light decreases with increasing distance. The decrease of light intensity with distance as it impinges on a transparent cell containing a solution of absorbing material is shown in Figure 27.

Rearranging eq. 43 and integrating between the limits of x equals zero where I is the intensity of the incident light and denoted by I_0 and x equals the distance, d, gives

$$\ln\frac{I}{I_0} = -KCd \tag{44}$$

Equation 44 can be expressed in logarithms to the base 10 and

$$\log\frac{I_0}{I} = \epsilon Cd = \text{optical density} \tag{45}$$

The quantity $\log\frac{I_0}{I}$ is known as the optical density or the absorbance. ϵ is the molar extinction coefficient with the units liter mole^{-1} cm^{-1}. On the other hand, if the concentration is expressed in grams per cubic centimeter, ϵ becomes the specific extinction coefficient. Equation 45 is an expression of Lambert-Beer's law. The percent transmission is defined as $100I/I_0$ and the relation between the percent transmission and the optical density is evidently

$$OD = \log\frac{100}{\%_t} = 2 - \log \%_t \tag{46}$$

where OD is the optical density and $\%_t$ is the percent transmission. The two coefficients which are sufficient to describe the interaction of radiation with a medium are the refractive index and the absorption coefficient.

According to eq. 45, the plot of the optical density against the concentration should yield a straight line; it frequently does, but not always. If the light filter or monochromator of the spectrophotometer is not sufficiently discriminating so that the light used covers a wide band of wavelengths, deviations from Beer's law can occur. At low concentrations the decrease in the light transmitted is a function primarily of

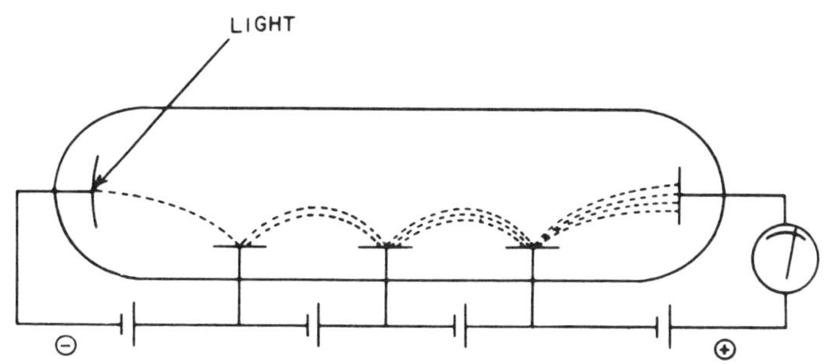

Fig. 28. Photomultiplier tube (diagrammatic).

the peak absorption with a characteristic extinction coefficient. As the concentration is increased, the light at the peak absorption has been so attenuated that the instrument response is shifted to those wavelengths which have been less absorbed and whose extinction coefficient is less. This is accompanied by a decreased slope of the plot of the optical density vs the concentration as the concentration is increased.

Other causes for deviations from Beer's law can be grouped under the general heading of chemical effects, i.e., the chemical character of the solute changes with increasing concentration and the degree of association of the solute in solution can undergo changes with solute concentration. These factors can have very considerable effects upon the extinction coefficient of the solute.

Measurement of Absorption. There are many "black boxes" on the market which are used to measure the optical densities of solutions; some of these are very elegant instruments. It is usually necessary to measure only the intensity of the transmitted light relative to that of the incident light and this is much easier than the measurement of the absolute intensity.

Whereas the human eye is one of the most sensitive instruments for the detection of radiation in the visible, it is not very well suited for quantitative measurements. In the earlier years of biochemistry, colorimeters in which visual comparisons were made between a standard and an unknown were used extensively; colorimeters have become museum pieces.

Relative light intensities are ordinarily measured by means of photoelectric colorimeters and spectrophotometers. Photoelectric colorimeters are of several designs and makes, but they all have in common the use of color filters to select the desired wavelengths in the visible. Ordinarily, they employ photovoltaic cells consisting of a transparent layer of metal covering a layer of a semi-conductor (crystalline selenium is most often the choice) on a steel backing. The absorption of an incident photon leads to the formation of an electron and a hole pair in the semi-conductor. This in turn leads to a charge difference between the semi-conductor and the transparent metal layer which induces a current flow in the external circuit. Photovoltaic cells undergo fatigue on exposure to light and the stability of such cells is greatly improved by employing matched cells. One of the cells is the balancing cell which receives light directly from the tungsten lamp. The other is the measuring cell and is exposed to the light which has passed through the colored solution. The two cells are connected to a potentiometer circuit and a galvanometer is used as a null indicator. The spectral response of the selenium cell extends from about 250 to 750 mμ with a maximum at about 570 mμ; it is suited primarily for the visible region. Photoelectric colorimeters are employed exclusively in analytical procedures.

Spectrophotometers are much more versatile than are colorimeters. Either prisms or diffraction gratings serve to isolate the wavelengths desired and phototubes typically are the receptors. The phototube has a sensitized cathode surface which emits electrons when light strikes its surface (photoelectric effect). The emitted electrons are attracted to a positive anode and produce a current flow in an ex-

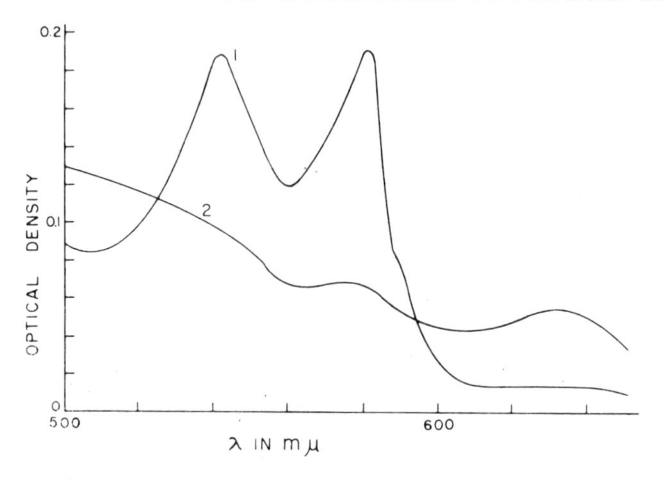

Fig. 29. Curve 1, Oxyhemoglobin. Curve 2, Methemoglobin. Solutions are 0.025 percent, pH 6.32; ionic strength 0.05 (bovine).

ternal circuit. The tube is evacuated. The current flow which is small lends itself to amplification because of the high internal resistance of the tube. There is no fatigue response in the phototube.

The photomultiplier tube has built-in amplification. Electrons released due to the light beam in turn cause the emission of additional electrons from the anode. The electrons are directed to another anode and this process is continued through a number of steps, the final emission being multiplied many fold over the original, but is still proportional to the intensity of the light beam. The photomultiplier tube is diagramed in Figure 28.

Electronic Absorption Spectra. Many molecules of biological interest absorb in the visible and ultraviolet. A very partial list of such substances would include the pyridine nucleotides, the cytochromes, myoglobin, hemoglobin, purines, pyrimidines, nucleic acids, and aromatic amino acids and proteins containing residues of these acids, sterols and carotenoids.

Of great aid in the study of hemoglobin are the distinctive spectra in the visible of the various forms of this molecule—differences which reside in the porphyrin nucleus. A comparison of the spectrum of oxyhemoglobin and of methemoglobin is shown in Figure 29. Oxyhemoglobin has the iron in the ferrous state whereas in the methemoglobin the iron has been oxidized to ferric iron. This change gradually and spontaneously occurs in a stored solution of purified oxyhemoglobin. Frequently, it is of interest to determine the extent to which a preparation of oxyhemoglobin has deteriorated to the methemoglobin. The specific extinction coefficients of oxyhemoglobin at a wavelength of 500 mμ is about 340 and at 578 mμ about 803. The corresponding values for the methemoglobin are 532 and 213 respectively. Then the optical densities of a solution containing a mixture of oxyhemoglobin and methemoglobin are

$$OD_{500} = 340\ C_{HbO_2} + 532\ C_{\text{met}\ Hb} \qquad 47$$

$$OD_{578} = 803\ C_{HbO_2} + 213\ C_{\text{met}\ Hb} \qquad 48$$

where the concentrations C_{HbO_2} and $C_{\text{met}\ Hb}$ are given in grams per cubic centimeter of solution. It is easy to solve the above simultaneous equations for C_{HbO_2} and $C_{\text{met}\ Hb}$ providing the optical densities have been determined at wavelengths of 500 mμ and 578 mμ.

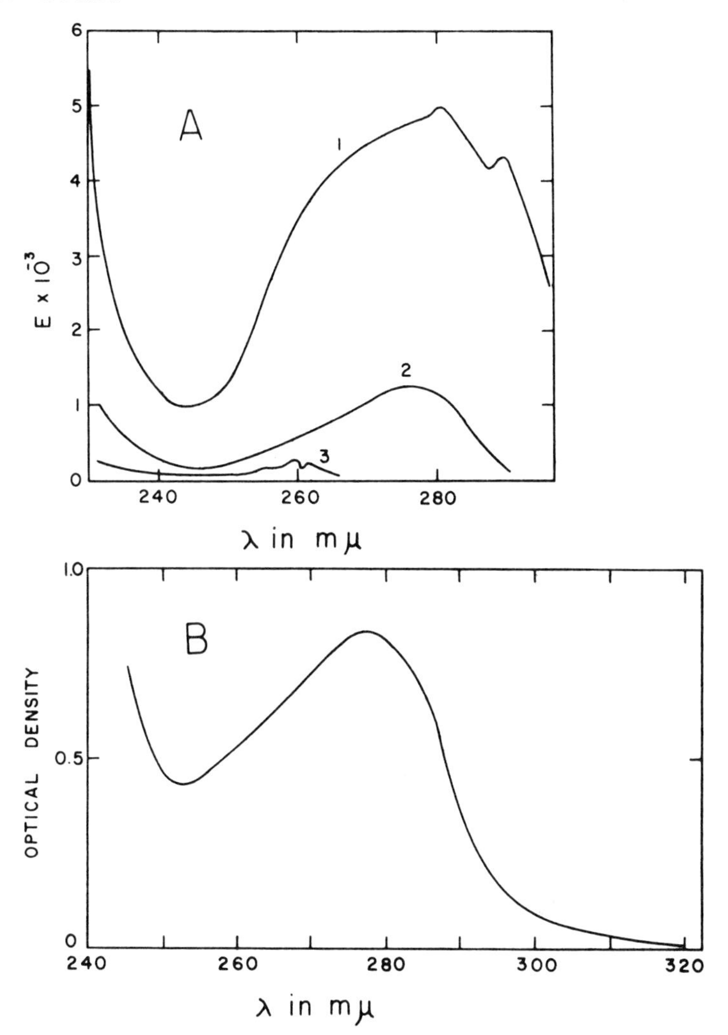

Fig. 30. *A*, Curve 1, tryptophane; curve 2, tyrosine; curve 3, phenylalanine. *B*, Bovine serum albumin, 0.125 percent, in neutral pH range.

The peptide bond of a peptide (or protein) absorbs in the neighborhood of 220 mμ. A protein usually shows strong absorption at about 280 mμ in a rather broad peak. Absorption in this region occurs because residues of phenylalanine, tyrosine and tryptophane are present in most proteins; the groups responsible for the absorption by these amino acids are phenyl, phenol, and indole respectively. Figure 30*A* shows the molar extinction coefficients of phenylalanine, tyrosine, and of tryptophane plotted against the wavelength. Figure 30*B* is the optical density of a 0.125 percent bovine serum albumin solution as a function of wavelength. Whereas the intensity of absorption by tryptophane is much greater than the other two residues, owing to its greater abundance in protein, tyrosine is usually responsible for most of the absorption in the 280 mμ region. Once the specific extinction coefficient of a given protein has been determined at or about a wavelength of 280 mμ, the concentration of the protein in solution can be rapidly and accurately measured. Under the appropriate conditions this is the method of choice for the assay of a protein.

It should be noted immediately that the absorption spectra are, in general, sensitive to the environment provided by the solution in respect to the polarity of the solvent, temperature and, more importantly, the degree of ionization of the absorbing

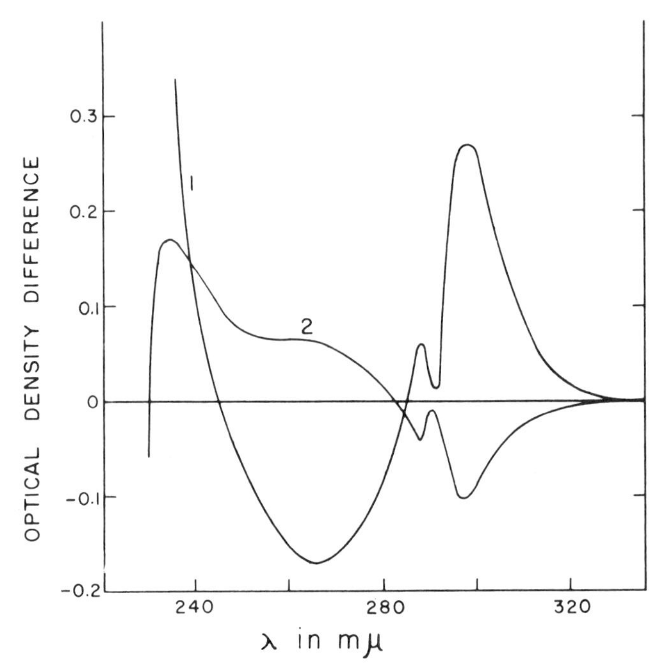

Fig. 31. Difference spectra of 3.05×10^{-4} M tryptophane; curve 1, pH 1 vs pH 6 change due to repression ionization COOH; curve 2, pH 13 vs pH 6 change due to loss of proton by amino group. (J. Hermans, J. W. Donovan and H. A. Scheraga: J. Biol. Chem. *235:* 91, 1960.)

molecule. A polar solvent tends to broaden the spectrum and to abolish the fine structure. Lowering the temperature increases the discreteness of the absorption bands. Ionization of a group in the neighborhood of the chromophore usually produces a marked shift in the absorption maximum. For example, the ionization of the phenolic hydroxyl of tyrosine produces a red shift with a maximum at about 290 mμ instead of 275 mμ of the un-ionized group (see Fig. 30). The change in color of an acid-base indicator upon ionization is a familiar example of the influence of ionization on the absorption spectrum in the visible region.

To bring out the finer details of the change of an absorption spectrum produced by a change in environment, it is helpful to obtain a difference spectra. That is, the solution in some selected standard state and concentration is used as a blank in the spectrophotometer. Another solution of the same concentration of absorbing material but with shifted pH or with some other change of environment is then compared with the blank over a range of wavelengths. The customary technique used to obtain difference spectra is to employ two optical cells containing two different solutions in tandem as the blank. The two solutions are then mixed and the same length of the mixed solution compared with the total optical length of the two unmixed solutions.

Figure 31 shows the difference spectra of tryptophane at pH 1 and at pH 13 compared with that at pH 6. This figure serves not only to illustrate difference spectra, but likewise the influence of pH on the spectrum of tryptophane; both the carboxyl and the amino groups are separated from the chromophore (indole) by two carbon atoms; the indole group itself does not ionize. A study over the complete pH range would permit an accurate determination of the ionization constant and indeed the authors have carried out such experiments and calculations.

There have been a number of studies of ion binding of proteins based upon spectral methods. Frequently when an organic ion with a characteristic absorption

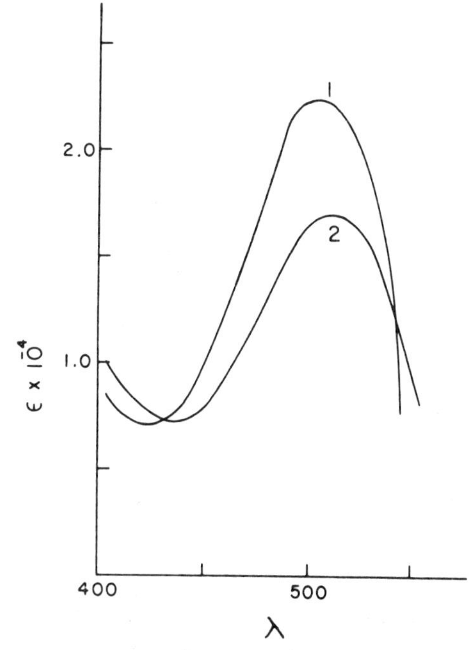

Fig. 32. Absorption spectrum of azosulfathiazole at pH 6.8; curve 1, in buffer only; curve 2, in 0.155 percent bovine serum albumin. The concentration of the dye in both curves 1 and 2 was 1.546 x 10^{-5} M. (I. M. Klotz, H. Triwash and F. M. Walter: J. Amer. Chem. Soc. *70:* 2935, 1948.)

spectrum is bound to a protein there is a shift both in the wavelength of maximum absorption as well as of the extinction coefficient. Provided the extinction coefficient at a given wavelength for both the free dye as well as that of the complexed dye is known, it is a simple matter to calculate the dye binding from a single measurement of the optical density of a solution of the dye and the protein; the total concentration of each must, of course, also be known. Figure 32 shows the molar extinction coefficient of azosulfathiazole at pH 6.8 both in the presence and absence of bovine serum albumin.

Optical methods have been extensively employed in the study of nucleic acids. The purine and pyrimidine bases, because of the conjugated double bond system they contain, show intense and characteristic absorption spectra in the region extending from about 240 mμ to about 290 mμ with maxima around 260 mμ. Naturally the nucleic acids also exhibit strong absorption in this range of the ultraviolet spectrum; their specific extinction coefficients are some fifty times larger than those of the tissue proteins.

The absorption spectra of both the bases as well as of the nucleic acids are marked functions of the pH of the solution and this variation of the spectra with pH is related to the ionization of the bases and provides a good method for the measurement of their ionization constants.

The ultraviolet light absorption of an intact polynucleotide is much less than the sum of absorption of the constituent nucleotides. This effect is known as hypochromicity. Any agent which disturbs the structure of the nucleic acid molecule such as extreme pH, heat, hydrolysis, etc., causes an increase in the molar absorption coefficient. Tinoco ascribes hypochromism in polynucleotides solely to coulombic interaction between electrons in the bases. The dipoles induced by the absorption of light interact with each other and their interaction depends upon their orientation relative to each other. If the groups are randomly oriented, there is no net effect upon the spectrum. End to end orientation leads to enhanced absorption, whereas parallel stacking of the moments causes a decrease in absorption, i.e., hypochromism; examine Figure 33.

Oriented nucleic acid preparations exhibit dichroism in the ultraviolet, the dichroism of DNA being more pronounced than that of RNA. To produce dichro-

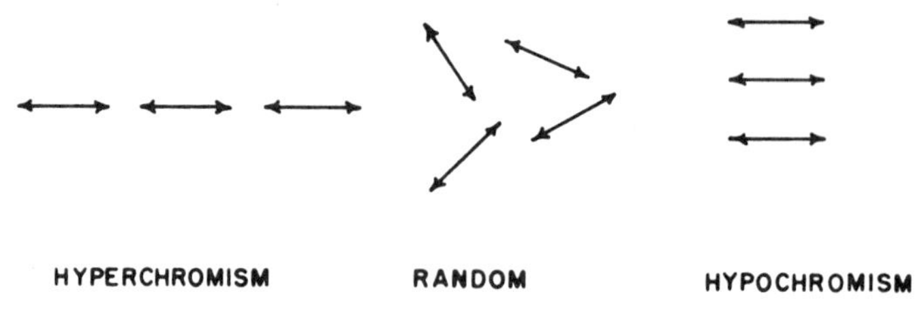

Fig. 33. Orientation of induced dipoles leading to hyper- and to hypochromism. (I. Tinoco: J. Amer. Chem. Soc. *82:* 4785, 1960.)

ism, a viscous gel of the nucleic acid is sheared and the extent of dichroism which can be obtained depends critically on the water content; the dichroism increases with increasing humidity up to 90 percent water above which the specimen becomes unstable and changes to an isotropic form. The electric vector of the principal absorption is normal to the direction of shear of the gel. This indicates that the ring planes of the bases are oriented perpendicular to the long axis of the nucleic acid; this is the situation which would be expected to lead to hypochromism as noted above.

Absorption spectroscopy has been a powerful adjunct in the study of cellular metabolism. For example, reduced nicotinamide adenine nucleotide shows a strong absorption maximum at about 340 mμ and absorption at this wavelength is completely abolished when the nucleotide is oxidized causing the nucleotide to lose its aromatic character. Even more useful have been the characteristic differences in the absorption spectra of oxidized and reduced cytochromes which have led to rapid measurement of metabolic events in living cells.

Solvent Perturbation Spectra. A change in the index of refraction of the environment of a chromophore alters the interaction between dipoles of the medium and those of the ground and excited state of the chromophore, which in turn leads to an alteration in the absorption spectrum of the chromophore. Not alone is there a shift in the wavelength of maximum absorption but also the extinction coefficient can be enhanced or depressed. As noted above, the absorption of most proteins in the region of 280 mμ is largely due to tyrosine residues. It is found that if a protein is dissolved in 20 percent ethylene glycol the absorption spectrum is shifted from that of the protein in water solution; the new spectrum is known as the solvent perturbation spectrum. The perturbation spectrum corresponds to a shift of the absorption of a certain number of the tyrosyl (or tryptophanyl) residues of the total present in the protein. It is thus possible from such measurements to calculate the number of tyrosyl residues exposed on the surface of the protein molecule as well

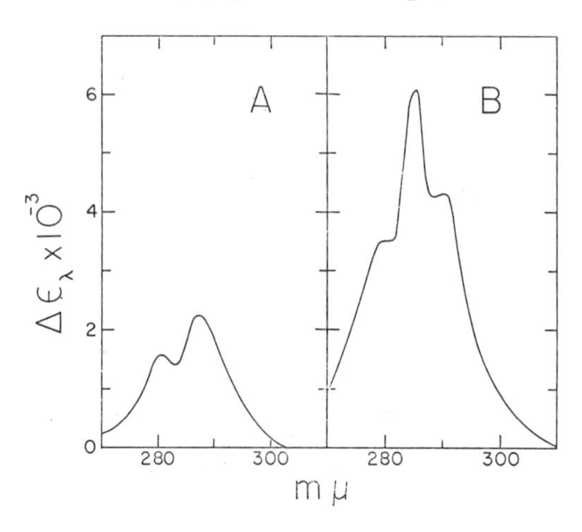

Fig. 34. Solvent perturbation difference spectra of horse heart ferricytochrome C normalized against that of the heme moiety. *A,* In 20 percent ethylene glycol. *B,* In 8 M urea. (Adapted from E. Stellwagen and S. van Rooyan: J. Biol. Chem. *242:* 4801, 1967.)

as the number buried inside the protein molecule which, accordingly, do not enter the ethylene glycol environment.

Figure 34*A* shows the solvent perturbation difference spectrum of horse heart ferricytochrome C in 20 percent ethylene glycol as contrasted with the spectrum in the absence of ethylene glycol. The observed difference spectrum corresponds to two tyrosyl and no tryptophanyl residues on the surface of the native protein molecule. Figure 34*B* shows the solvent perturbation difference spectrum of cytochrome C in 8 M urea (urea at this concentration is a powerful protein denaturant). The difference spectrum obtained indicates the complete exposure of four tyrosyl and one tryptophanyl residue; urea exposes all of the tyrosyl and the one tryptophanyl residues in horse heart cytochrome C.

Photochemical Changes. The absorption of radiant energy may or may not lead to a chemical reaction. In case the radiation does produce a chemical reaction, the energy per mole which is called an Einstein is

$$\mathrm{E} = Nh\,\nu = \frac{\mathrm{NhC}}{\lambda} \tag{49}$$

where N is Avogadro's number and C is the velocity of light in a vacuum. Inserting the numerical values of the constants in eq. 49, there results per mole

$$E = \frac{2.854 \times 10^4}{\lambda} \tag{50}$$

where λ is expressed in mμ and E in kilocalories. One mole of absorbing substance should decompose for every $2.854 \times 10^4/\lambda$ kilocalories of radiation absorbed and the photochemical equivalent of one kilocalorie is $\lambda/2.854 \times 10^4$ moles. The results of photochemical experiments are expressed in terms of the quantum yield which is defined as the number of moles decomposed by each quantum of radiation absorbed. It is evident from an inspection of eq. 49 that the energy per quantum is directly proportional to the frequency and, in a given medium, inversely proportional to the wavelength (see Fig. 2). The energy per mole in the near infrared ($\lambda \sim 15\mu$) is about 2 kilocalories and in the ultraviolet the range is from 70 to 200 kilocalories. The energy yield in the infrared is, in general, too small to support a chemical reaction. The shorter wavelengths associated with ultraviolet and especially with X-rays mean that the energy per quanta is so large that its absorption is apt to lead to destructive reactions such as the denaturation of proteins, destruction of tissue and the mutation of genes. Only in the narrow range of the visible and the near ultraviolet are the energy yields per quanta commensurate with moderate chemical activity.

Exposure of proteins to ultraviolet light leads to profound changes in the properties of the proteins; the protein becomes denatured and, if the protein is an enzyme, the enzymatic activity is lost. McLaren and Luse conclude that the destruction of the cystyl and tryptophanyl residues is primarily responsible for the loss of activity of chymotrypsin, of lysozyme, of ribonuclease and of trypsin when exposed to ultraviolet light whose wavelength was 253.7 mμ. From the quantum yields of the free amino acids they were able to predict the quantum yield of the proteins. Neither peptide bond rupture nor the destruction of the tyrosine residues is involved.

Only light which is absorbed can produce a chemical change; the chemical change produced can, however, be indirect. For example, Weil and Seibles found that the irradiation of a solution of ribonuclease and of methylene blue with light whose wavelength was about 670 mμ destroyed the histidine residues in ribonuclease; the maximum absorption of methylene blue is at 670 mμ. What happens is that the light-excited dye acts as a hydrogen acceptor removing a hydrogen from the histidine residue. The methylene blue is reduced to the colorless leuco form and the histidine residue is oxidized. The methylene blue is reoxidized by oxygen and thus the photochemical reaction continues.

The release of heat by chemical reactions is familiar. In the appropriate systems the energy derived from chemical reactions can be released as light and give rise to luminescence. There are several examples of bioluminescence including that of the North American firefly. In this reaction, luciferin, which is a fairly complicated substance supplied by the firefly, combines with ATP in the presence of the enzyme luciferase. This enzyme-substrate complex upon oxidation by molecular oxygen radiates light with a maximum peak at 562 mμ. In this reaction pyrophosphate is released.

There are two photochemical reactions in biology of overriding importance, namely photosynthesis and the perception of light by the eye leading to vision.

Photosynthesis. Photosynthesis activity is associated with the chloroplasts of the leaves of green plants. The chloroplasts which are about 5μ in diameter have a complex structure and contain among other things the grana. The grana have a fine and well-organized layer structure in which are found the chlorophyll pigments. The absorption spectrum and the chemical structure of chlorophyll are shown in Figure 35. Phytol attached to the porphyrin is a long chain optically active aliphatic alcohol. Note the similarity of chlorophyll to heme in hemoglobin. Carotenoids and

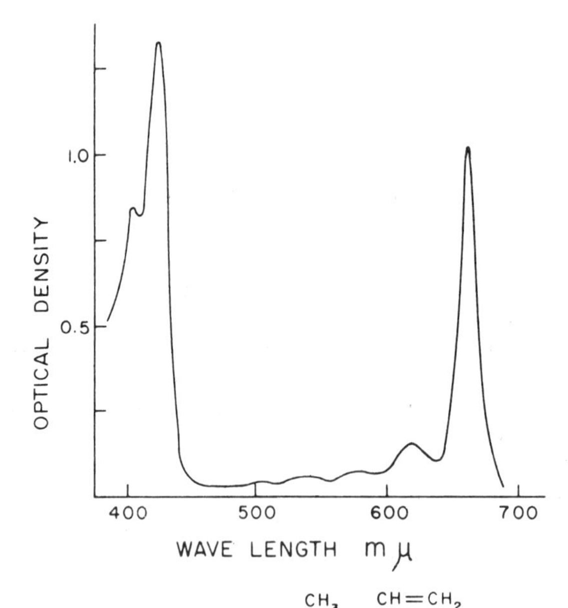

Fig. 35. Absorption spectrum and structural formula of chlorophyll *a*.

plycobilins accompany the chlorophyll pigments and probably play significant supporting roles in photosynthesis. Chlorophyll b differs from chlorophyll a in having a methyl group replaced by a formyl group. Whereas chlorophyll a and b absorb strongly in the violet (420 mμ), photosynthetic activity occurs as a result of absorption of red light in the neighborhood of 670 mμ which corresponds to an energy of about 40 kilocalories per mole.

There has been intensive investigation of photosynthesis and much valuable information has accumulated, but the essential nature of the initial photochemical reaction is elusive. Early work concerned itself primarily with the reduction of carbon dioxide with the formation of formaldehyde and various carbohydrates. It has long been known that the fixation of carbon dioxide by the plant can occur in the absence of light (dark reaction) but its reduction requires the absorption of light; during the photosynthetic reaction, gaseous oxygen is released. More recently, attention has been directed to phosphorylation with the production of ATP from ADP. and the reduction of carbon dioxide and the synthesis of carbohydrates are

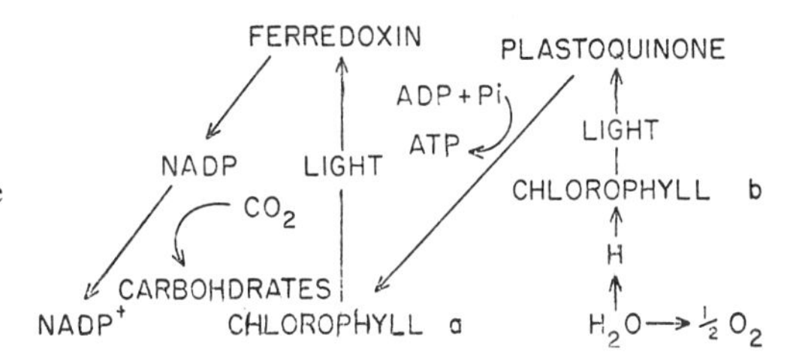

Fig. 36. Schematic representation of the reactions occurring in photosynthesis.

regarded as removed from the photochemical step.

Apparently, the conversion of radiant energy into chemical energy involves two photoreactions. The first is the transfer of electrons and protons from water to plastoquinone through the action of light on chlorophyll b. The passage of electrons through the plastoquinone and electron carriers results in the formation of ATP from ADP and inorganic phosphate. The electrons are then transported to chlorophyll a. Another light reaction occurs resulting in the passage of electrons to ferredoxin. The electrons then reduce the oxidized nicotine adenine dinucleotide phosphate ($NADP^+$) leading to the reduced form (NADP). The reduced NADP is responsible for the reactions which reduce carbon dioxide leading to the production of carbohydrates. Figure 36 gives a schematic representation of these reactions.

Ferredoxin is a water soluble protein found in the chloroplast of green leaves. It is an iron containing oxidation-reduction enzyme with a strongly negative (reducing) oxidation-reduction potential (about −0.43 volts at pH 7).

Vision. The human eye is indeed an important and sensitive detector of electromagnetic radiation having a suitable wavelength. In the back of the eye is the retina, the thickness of which is between 0.2 and 0.3 mm. The retina is a complex structure containing the light-sensitive receptors along with the connecting nerve fibers and the necessary veins and arteries. The light-sensitive receptors are of two kinds, rods and cones. The rods are by far the more sensitive and participate in night vision and cannot distinguish color; all cats look gray in the dark. The cones provide color vision. The rods and cones also differ in their distribution in the retina—the cones are concentrated around the center of the retina with the rods being more numerous towards the peripheral areas of the retina; it is easier to observe an object in the dark by not looking directly at it.

Rod vision is maximally sensitive at about 510 mμ (the cones are maximally

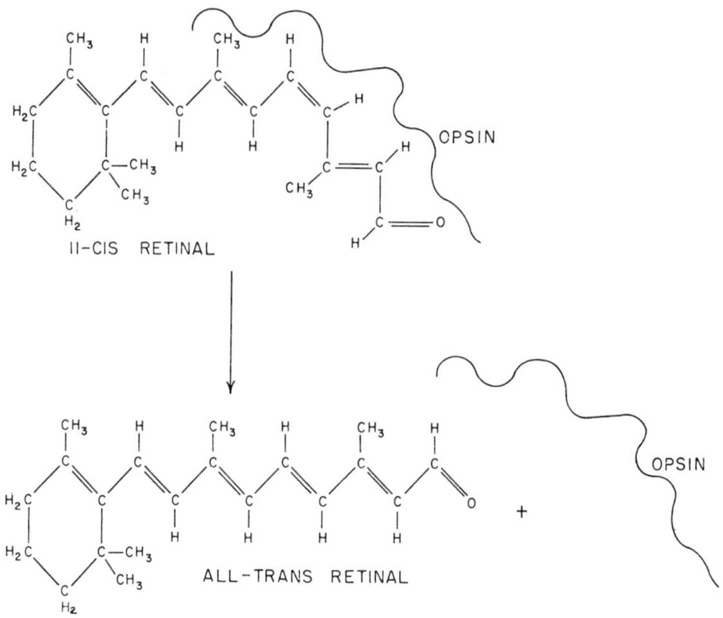

Fig. 37. Conversion of retinal 1 to all-trans retinal 1 by light and the dissociation of the all-trans retinal 1 from the protein opsin.

sensitive at 540 mμ) and at this wavelength will respond to about 10 quanta corresponding to about 4×10^{-11} erg. At the dark adapted threshold for rod vision, a single rod absorbs a quantum about once on the average in 38 minutes. It appears that a single quantum absorbed by a single molecule of the visual pigment is sufficient for visual excitation.

There is a time constant involved in vision and the eye's ability to integrate a total signal does not extend beyond about a tenth of a second. The time constant of the eye is related to the critical flicker frequency—the frequency of the light flicker below which the eye is able to perceive flicker. The flicker frequency increases with increasing illumination of the retina and the cones have a higher critical flicker frequency than do the rods.

The aldehydes derived from the oxidation of vitamin A_1 and A_2 by alcohol dehydrogenase are known as retinals (formerly called retinenes). Retinals 1 and 2 have identical structures except that retinal 2 contains one more double bond in the ring than does retinal 1. The retinals when combined with the specific proteins called opsins produce the visual pigments rhodopsin in the rods and iodopsin in the cones. The essential photochemical reaction is the conversion of the 11-cis retinal to the all-trans retinal. The 11-cis retinal is bound to the opsin but the all-trans is not. Figure 37 shows the relation between the 11-cis retinal and the all-trans retinal.

The initial photochemical reaction is followed by a series of complex reactions in which the dissociated all-trans retinal is converted back into vitamin A and then into 11-cis retinal followed by binding to the opsin. Under physiological conditions the equilibrium between vitamin A and retinal is far in the direction of the vitamin A, so far that oxidation to retinal occurs only to the extent that the retinal is trapped by combination with opsin. The combination of the retinal with opsin seems to depend on the combination of the aldehyde group of retinal with an amino group of the opsin. After the retinal is bound by the opsin, the resulting pigment is again ready to receive a light signal. How the photochemical reaction produces a nerve response is unknown.

Normal human vision requires four different opsins: one in the rods to produce rhodopsin and three in the cones to produce the necessary red, green and blue pigments. The retinals are common to the rods and cones.

GENERAL REFERENCES

Barrow, G. M.: Introduction to Molecular Spectroscopy. McGraw-Hill, New York, 1962.
Beychok, S.: Circular dichroism of biological macromolecules. Science *154:* 1288, 1966.
Dyer, J. R.: Applications of Absorption Spectroscopy of Organic Compounds. Prentice-Hall, Englewood Cliffs, 1965.
McLaren, A. D. and D. Shugar: Photochemistry of Proteins and Nucleic Acids. Pergamon Press, Oxford, 1963.
Schellman, J. A. and C. Schellman: The Conformation of Polypeptide Chains in Proteins. *In:* The Proteins, Vol. 2. Academic Press, New York, 1964.
Strong, J.: Concepts of Classical Optics. W. H. Freeman, San Francisco, 1958.
Udenfriend, S.: Fluorescence Assay in Biology and Medicine. Academic Press, New York, 1962.
Van de Hulst, H. C.: Light Scattering by Small Particles. John Wiley and Sons, New York, 1957.

PROBLEMS

1. 10 ml. of thermally insulated benzene are exposed to a beam of ultraviolet light with a wavelength of 280 mμ for 10 minutes and it is found that the temperature of the benzene has increased from 25.00°C. to 25.10°C. How many quanta of light have been absorbed by the benzene?

   Ans: 2.0×10^{18}

2. Three grams of powdered pyrex glass were added to 10 ml. of a bovine serum albumin solution at pH 5.05 in 0.05 ionic strength acetate buffer. The initial concentration of the protein was 0.185 grams per 100 ml. of solution. After adsorption had occurred, the suspension of glass was centrifuged and to a volume of the supernatant solution was added an equal volume of the acetate buffer. The optical density of this solution was 0.559 at a wavelength of 278 mμ and 0.062 at 320 mμ. The optical densities of the original solution before adsorption but after the same dilution were 0.618 at 278 mμ and 0.001 at 320 mμ. The ratio of the optical densities of a dilute suspension of pyrex glass at 278 mμ to that at 320 mμ is 2.10. Calculate the milligrams of protein adsorbed per gram of glass. Note: It is impossible to remove the suspended glass completely from the protein solution, hence the correction for the optical density of the glass.

   Ans: 1.87 mg. protein.

3. The following results were obtained in the study of light scattering by a protein solution (the scatter due to the solvent has been subtracted).

Protein conc. g/cc. $\times 10^3$	*Turbidity in* 1/cm. $\times 10^5$
3.48	160.7
4.87	211.9
6.95	308.8
8.11	385.3
11.59	557.2

   H has the value 0.95×10^{-5} sq. cm. per gram squared. Calculate the molecular weight of this protein and the value of the interaction constant B. What would be the osmotic pressure of a one percent solution of this protein?

   Ans: Mol. wt. 43,500; interaction constant -1.37×10^{-4}; osmotic pressure, 5.46 cm. H_2O.

4. Suppose you wished to determine the shape and the molecular weight of a sample of nucleic acid by means of light scatter. What kind of measurements and calculations would you need?

   Ans: See text.

5. A 2 percent solution of serum albumin (mol. wt. 69,000) is mixed with an equal volume of a 1 percent solution of egg albumin (mol. wt. 45,000). Assume the interaction constant to be zero and calculate the average molecular weight of the mixture by (a) osmotic pressure and (b) by light scatter.

   Ans: (a) 58,585; (b) 61,000.

6. The absorption of radiant energy by a molecule leads to an excited molecular configuration. In due time, all of the molecules will have been excited and absorption should stop. Explain why, in most absorptions, there is no change of absorption with time.

   Ans: See text.

7. Bromoeosine solutions are yellowish-red to transmitted light with a greenish fluorescence. Describe the fate of all visible radiation falling upon a solution of this substance.

   Ans: See text.

8. A spectrophotometer operates with a ± 5 percent uncertainty in transmittance. Find the percentage uncertainty in the concentration when the transmittance is (a) 10%, (b) 37%.

   Ans: (a) 2.10%, (b) 1.39%.

9. The vibration frequency of HCl corresponds to 2886 cm^{-1}. (a) What is the force constant? (b) What is the frequency for DCl? (c) Explain why there are two vibrational frequencies for HCl in a non-polar liquid medium.

   Ans: (a) 4.8×10^5 dynes/cm; (b) 2,068 cm^{-1}; (c) natural abundance of Cl is $3Cl^{35}$ to $1Cl^{37}$. A peak will be observed for each.

CHAPTER 9

Surfaces and Interfaces

Up to the present, biology and biochemistry have developed dramatically with little help from surface chemistry. One wonders how long this robust progress can continue when so little attention is paid to the fantastically large interfacial areas in living tissue. Surely the nature of the outermost layer on a cellular and subcellular membrane must be of great importance in the behavior of living systems. For example, we actually know nothing at all about the molecular structure of the arterial wall which is directly exposed to the blood. This could very well turn out to be a medical problem of first importance. The truth of the matter is that we are terribly short of suitable techniques for the study of biological films.

Surface Tension. To expand a surface, work must be done because molecules in the interior of the liquid are brought to the surface where the molecular attractive forces above the surface have been greatly diminished; the potential energy of the molecules has been increased. The work of surface expansion is expressed in ergs per square centimeter or alternatively in dynes per centimeter and the work in ergs per square centimeter is equal to the free energy change of the expansion.

Consider the expansion of a soap bubble. The surface work of expansion is

$$dW_s = 2\,\sigma dA \qquad 1$$

where σ is the surface tension and A is the area. The surface work involves both the inside and outside surfaces of the bubble, hence, the factor of 2 in eq. 1. The mechanical work of expanding the bubble is

$$dW_m = PdV \qquad 2$$

where P is the pressure in dynes per cm^2, and V is the volume of the bubble. At equilibrium these two work terms must be equal and

$$2\,\sigma dA = PdV \qquad 3$$

dA is equal to $8\,\pi r dr$ and dV is $12/3\,\pi r^2\,dr$, where r is the radius of the bubble. Substituting in eq. 3 for dA and dV and rearranging gives

$$P = \frac{4\sigma}{r} \quad 4$$

There are a variety of methods available for the measurement of surface and interfacial tensions and an investigator has to select that method which is more appropriate to his needs. A brief sketch of some of the methods will be given.

Capillary Rise. Imagine a capillary dipping into a liquid (Fig. 1). The force acting downward is $\pi r^2 hg\rho$ whereas the force upward is $2\pi r\sigma \cos \alpha$ where r is the radius of the capillary, h is the height of rise, g is the acceleration of gravity, ρ is the density of the liquid and α is the contact angle between the liquid and the wall of the capillary. At equilibrium, the upward and downward forces are equal and

$$\pi r^2 hg\rho = 2\pi\sigma r \cos \alpha \quad 5$$

or

$$\sigma = \frac{rhg\rho}{2 \cos \alpha} \quad 6$$

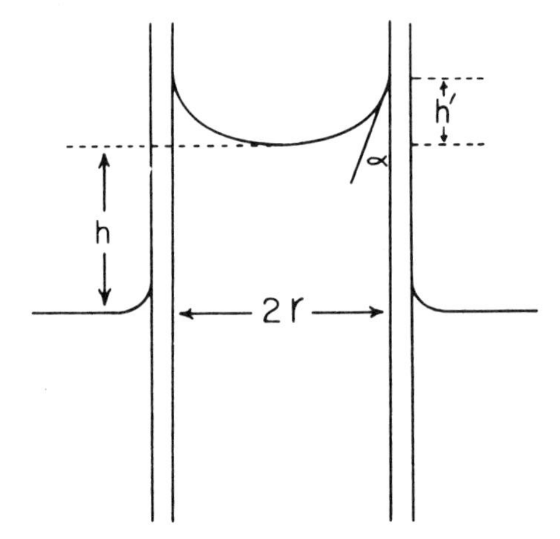

Fig. 1. Rise of a liquid in a capillary.

If the contact angle, α, is zero,

$$\sigma = \frac{rhg\rho}{2} \quad 7$$

The total volume of liquid in the capillary above the surface is not exactly equal to $\pi\ r^2 h$ because there is a small amount of liquid above the tip of the meniscus and the value of h to be used in the calculation of the surface tension is very nearly h *plus* $r/3$. Whereas the capillary rise method is capable of great accuracy, it is also experimentally very demanding; it is practically useless for the measurement of solutions containing proteins because of the tendency of the protein surface films to stick to the capillary walls.

Drop Weight Method. This method has been widely employed with biological materials and is especially useful for the measurement of interfacial tensions. The liquid forms a drop at the end of a capillary. The drop is permitted to grow slowly until it falls; simple theory would give the surface tension as

$$\sigma = \frac{gw}{2\pi r} \quad 8$$

where w is the weight of the drop. However, the drop constricts before it falls and, further, not all of the drop is detached from the capillary tip. High speed motion pictures show the situation to be further complicated by the appearance of several smaller drops following the main drop. These complications are diagramed in Figure 2C.

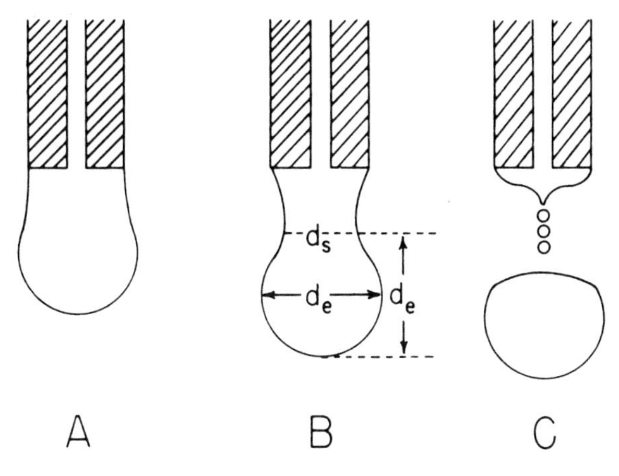

Fig. 2. Fall of a drop from a capillary tip.

Equation 8 is subject to important corrections which can increase the calculated tension by as much as 50 percent. The corrections are a function of $r/V^{1/3}$ where r is the radius of the capillary tip and V is the volume of the drop. In Figure 3 is plotted the correction factor, ϕ, by which the right side of eq. 8 must be multiplied to yield the correct tension. The drop weight method has limited utility for the measurement of the surface and interfacial tension of surfaces which age slowly as do many protein films. For such surfaces the next method to be considered is more suitable.

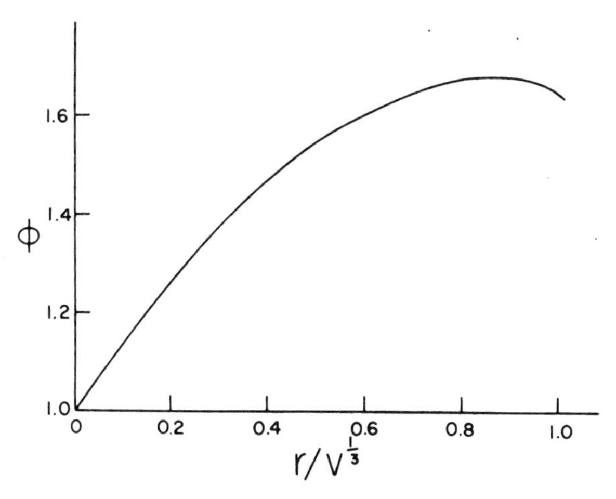

Fig. 3. Correction, ϕ, for the surface tension as measured by the dropweight method as a function of $r/v^{1/3}$. (From the data of W. D. Harkins and F. E. Brown: J. Amer. Chem. Soc. *41:* 499, 1919.)

Pendant Drop Method. The shape of a drop as it forms should be a function of the size of the drop, the density of the liquid and the surface tension of the liquid. With good optical equipment, enlarged images of a pendant or hanging drop can be made and the change of surface tension with time can be studied with minimum disturbance to the surface. Figure 2B gives the significant dimensions of the drop. d_e is the equatorial diameter of the drop and d_s the diameter measured at a distance, d_e, up from the bottom of the drop. The surface tension of the liquid is given by

$$\sigma = \rho g d_e^2 f\left(\frac{d_s}{d_e}\right) \tag{9}$$

where ρ is the density of the liquid and $f(d_s/d_e)$ is a function, dependent on the ratio d_s/d_e. A plot of this function against d_s/d_e is shown in Figure 4.

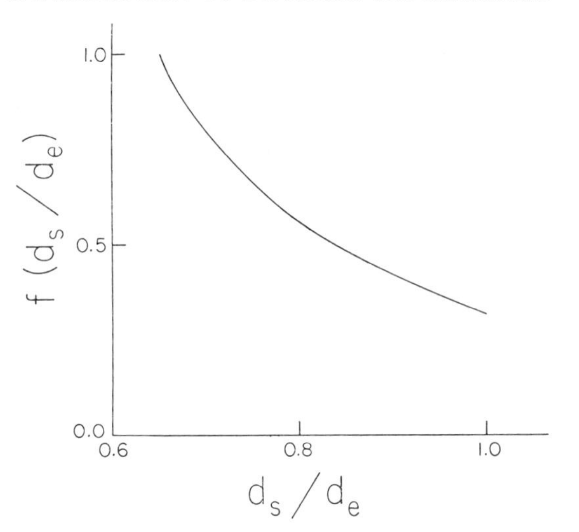

Fig. 4. The function f (d_s/d_e) plotted against d_s/d_e for use in the pendant drop method. (See eq. 5 and Fig. 2B for the meaning of the terms.)

du Nouy Ring Method. This is probably the most widely used and one is tempted to say the most abused method for surface and interfacial tension measurements. The pull required to detach a platinum ring from the surface is measured and elementary theory would relate this pull to the surface tension by

$$\sigma = \frac{gw}{4\pi R} \qquad 10$$

where gw is the upward pull and $4\pi R\sigma$ is the downward pull, R being the radius of the ring. The ring method is, however, subject to corrections which depend both on the ring diameter as well as on the diameter of the wire.

Wilhelmy Slide. The apparatus consists of a strip of glass (microscope cover glass) dipping into the liquid under investigation and suspended from the arm of an analytical balance. The weight of the dry slide in air and its weight while dipping into the liquid are determined. After the buoyancy correction of the liquid displaced by the slide is added, and the weight of the slide in air subtracted from the weight when dipping in the liquid, the surface tension can be directly calculated and

$$\sigma = \frac{gw}{2L} \qquad 11$$

where w is the net pull on the slide in grams and L is the length of the slide plus the thickness of the slide. The Wilhelmy method involves no rupture of the surface and is to be recommended. A requirement of the method is that the slide must be completely wet by the liquid (zero contact angle). It is possible to convert the commercially available du Nouy ring apparatus into a Wilhelmy slide method by the simple expedient of replacing the ring with a vertical slide.

Table 1 shows the values of the surface tension of some common liquids and Figure 5 is a plot of the surface tension of water against temperature.

In principle, any surface tension method can be adapted to the measurement of the interfacial tension between two immiscible liquids; the necessary modifications in technique must, of course, be made. The drop weight and the pendant drop methods are to be recommended.

Energy of a Surface. Since the surface tension at constant pressure and temperature is equal to the reversible work required to expand a surface by one cm^2, it is also the free energy change for the process. The entropy change in ergs per

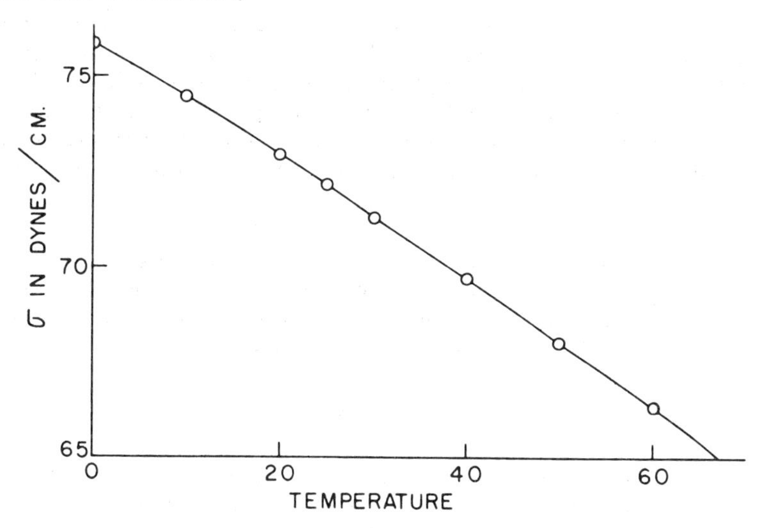

Fig. 5. Surface tension of water as a function of temperature.

cm² per degree is

$$\left(\frac{d\sigma}{dT}\right)_{T,P} = -\Delta s \qquad 12$$

and the Gibbs-Helmholtz equation for a surface at constant temperature and pressure is

$$\sigma = \Delta h + T\left(\frac{d\sigma}{dT}\right)_{T,P} \qquad 13$$

where Δh is the enthalpy of the surface in ergs per cm². Lower-case s and h have been used in eqs. 12 and 13 to indicate that these terms refer to quantities per unit area rather than quantities per mole. The term $T(d\sigma/dT)$ represents the amount of heat which must be supplied to an expanding surface to maintain a constant temperature. Since, in general, the surface tension of liquids decreases with increasing temperature, expanding liquid surfaces tend to cool. For a water surface at 20°, $T(d\sigma/dT)$ is -47.7 ergs per cm², and σ, the surface tension, is 72.75 ergs per cm² (ergs per cm² and dynes per cm are numerically and dimensionally identical) and, accordingly, Δh is 118.5 ergs per cm². Δh, unlike the surface tension, is very nearly independent of temperature.

The negative temperature coefficient for the surface tension of liquids means that the entropy of the surface must increase on expansion.

TABLE 1. SURFACE TENSION OF SOME COMMON LIQUIDS IN DYNES PER CENTIMETER AGAINST AIR

Substance	*Surface Tensions*	*Temperature*
Acetic acid	27.6	20
	24.7	50
Acetone	23.7	20
	18.6	60
Benzene	28.88	20
	25.0	50
n-Butyric acid	26.8	20
	24.0	50
Ethanol	22.3	20
	19.8	50
Methanol	22.6	20
	20.1	50
Water	71.97	25

When a surface forms, the entropy increases because a molecule is now given the opportunity of occupying a position in the surface as well as in the liquid. These two possibilities give rise to an entropy change of Rln 2 per mole or +1.4 E.U. A molecule in the surface is already partly in the vapor phase. It was noted above that the surface tension of a liquid decreases with increasing temperature. It has been found empirically that the surface tension is the following function of temperature

$$(M_v)^{2/3}\sigma = K(T_c - T - 6) \qquad 14$$

T_c is the critical temperature of the liquid and K is Eötvös constant and has a value of about 2 for non-associating or dissociating liquids. M_v is the mole volume and, accordingly, $M_v^{2/3}$ is the surface area of one face of a cube of liquid containing one mole of the liquid. The differentiation of eq. 14 leads to

$$(M_v)^{2/3}\left(\frac{d\sigma}{dT}\right) = -K \qquad 15$$

By this calculation, it turns out that the entropy change per mole is approximately 2, as compared with 1.4 obtained above. Molecules which orient at the surface such as do fatty acids pack into a unit surface area a larger number of molecules and, accordingly, the Eötvös constant is larger, about 4 for liquid fatty acids.

Cohesion and Adhesion. The work of separating a rod of liquid one sq. centimeter in cross sectional area into two parts is 2σ and is known as the work of cohesion where σ is the surface tension of the liquid. The work of cohesion of water at 25° is 143.9 ergs. The work of separating water from water would be completed over a very short distance amounting to about 2 A-units. The force which would need to be exerted would be $143.9/2 \times 10^{-8}$ or about 7×10^9 dynes and is equivalent to about 7×10^6 grams per square centimeter. The yield point for a good quality steel is about 10×10^6 grams per square centimeter.

The work of adhesion, w_a, is the work done in separating one square centimeter of two unlike substances. The work of adhesion between a solid and a liquid is

$$w_a = \sigma_L + \sigma_{SV} - \sigma_{SL} \qquad 16$$

where σ_L is the surface tension of the liquid, σ_{SV} that of the solid and σ_{SL} is the interfacial tension between the solid and the liquid. Unfortunately, there is serious danger of confusion in respect to the work of adhesion. To separate a liquid from a solid leaves the surface of the solid saturated with the liquid and only for so-called low energy solid surfaces such as solid paraffin exposed to water vapor is the surface tension of the solid equal to σ_{SV}.

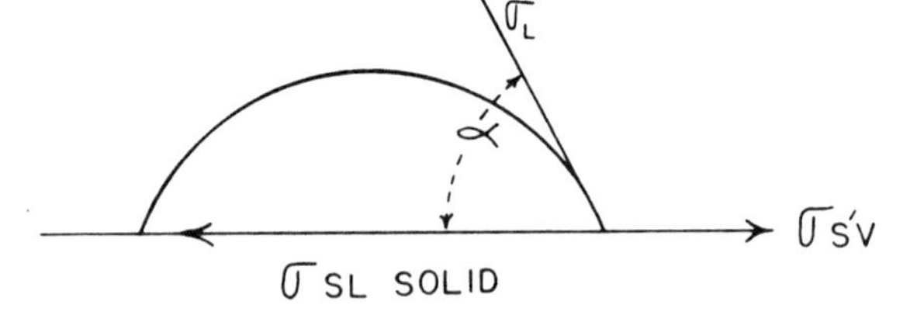

Fig. 6. Resolution of surface forces acting on a liquid drop at a solid surface.

Contact Angles. A drop of liquid placed upon a solid surface may spread over the solid and completely wet it or remain as a drop. The situation is diagramed in Figure 6. The angle α is called the contact angle and a formal resolution of forces acting at the line of contact between the liquid drop and the solid gives

$$\sigma_{S'V} = \sigma_{SL} + \sigma_L \cos\alpha \qquad 17$$

where $\sigma_{S'V}$ is the surface tension of the solid saturated with the vapor of the liquid,

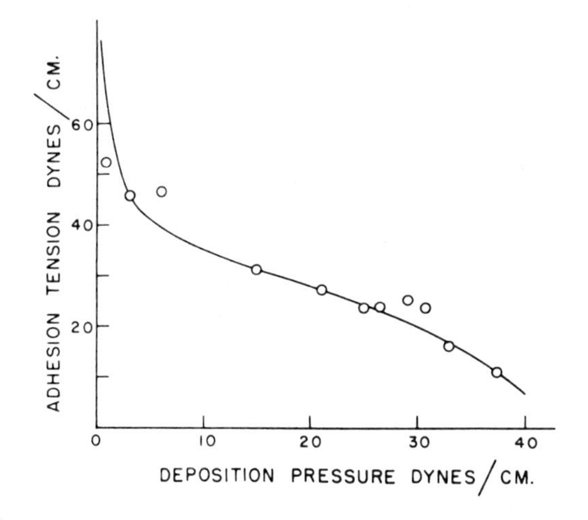

Fig. 7. Adhesion tension in dynes per centimeter between water and a deposited spread monolayer of egg albumin compressed to the indicated film pressures and deposited on a glass slide. (H. B. Bull: J. Biol. Chem. *125:* 585, 1938.)

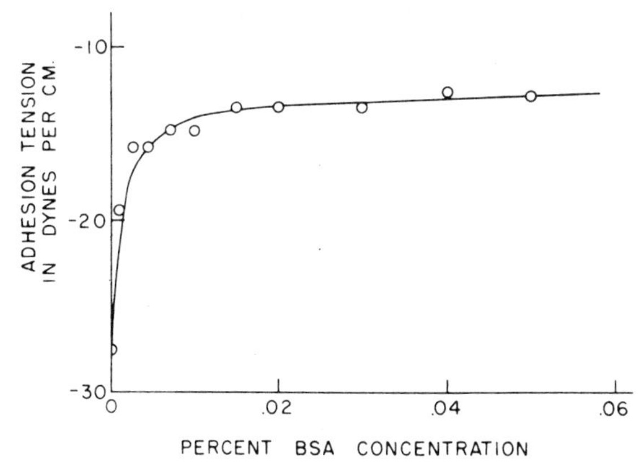

Fig. 8. Adhesion tension between bovine serum albumin solutions (BSA) and paraffin as a function of the protein concentration; pH 7.59, 0.05 ionic strength tris buffer at 25°. Advancing surfaces. (S. Ghosh, K. Breese, and H. B. Bull: Unpublished results.)

σ_L is the surface tension of the liquid and σ_{sl} is the solid-liquid interfacial tension.

Identifying σ'_{SL} of eq. 17 with σ_{SL} of eq. 16 and combining these two equations yields

$$W_a = \sigma_L + \sigma_L \cos \alpha \qquad 18$$

As expressed in eq. 18, W_a is the reversible work required per square centimeter to separate the liquid from the solid, leaving an adsorbed layer of liquid in equilibrium with its saturated vapor on the surface of the solid. When a solid slide is pulled vertically out of a liquid, a vapor-solid surface is created and a solid-liquid surface is destroyed. The term $\sigma_L \cos\alpha$ is equal to the work done per square centimeter and is known as the adhesion tension.

The contact angle may be measured in several ways; one of the simplest is to project the enlarged image of the liquid drop as it rests on the solid on a screen and measure the contact angle directly with a protractor.

Figure 7 shows the adhesion tension of water for monomolecular layers of egg albumin deposited on glass slides at the indicated film pressures; the adhesion tensions were calculated from contact angles. The adhesion tension between water and a smooth paraffin surface is about −25 dynes per square centimeter; the deposited egg albumin film becomes more hydrophobic as the film pressure is increased.

An outstanding and often disconcerting feature of contact angle measurements is the pronounced hysteresis of such determinations; the advancing liquid surface gives an angle which is often significantly greater than that of the receding surface. Usually, hysteresis is a reflection of the roughness of the surface.

Wetting Balance. Adhesion tension measurements can be made much more conveniently and more accurately with a wetting balance devised by J. Guastalla rather than with conventional contact angle technique. The experimental arrangement of the wetting balance is simple. For example, a thin cardboard strip can be saturated with paraffin above its melting point, the paraffin solidified and smoothed out. The cardboard strip is then suspended from a du Nouy tensiometer in place of the ring, i.e., used as a Wilhelmy balance. The pull in dynes per centimeter for the paraffined slide is then compared with that for a glass slide which is completely wetted. It is also possible to deposit a thin layer of solid paraffin on a microscope coverglass. The cosine of the angle of contact against the paraffin surface is evidently equal to the ratio of the apparent surface tension using the paraffined slide to the true surface tension. The adhesion tension, $\sigma_L \cos\alpha$, is simply equal to the apparent surface tension as given by the coated slide. Figure 8 shows the adhesion tension of bovine serum albumin solutions against paraffin at pH 7.59 in 0.05 ionic strength tris buffers as a function of the protein concentration; bovine serum albumin significantly increases the adhesion tension between water and paraffin.

Surface Tension of a Solution. Let A be the surface area of a solution containing one mole excess of solute. The transfer of a small amount of solute to the surface from the solution results in a change $Ad\sigma$ in the surface energy. The removal of material from solution will change its osmotic pressure and, if V is the volume of solution containing one mole of solute, the osmotic work is $-Vd\Pi_0$. At equilibrium these two work terms are equal and

$$Ad\sigma = -Vd\Pi_0 \qquad 19$$

Since the solution is dilute and obeys the ideal solution law, $V\Pi_0$ is equal to RT and eq. 19 can be written

$$Ad\sigma = -\frac{RTd\Pi_0}{\Pi_0} \qquad 20$$

Since Π_0 is equal to RTC and $d\Pi_0$ to $RTdC$, eq. 20 becomes upon rearrangement

$$\frac{d\sigma}{dC} = -\frac{RT}{AC} \qquad 21$$

The excess solute adsorbed in moles per square centimeter is $1/A$ and eq. 21, after rearrangements, gives

$$n_a = -\frac{Cd\sigma}{RTdC} \qquad 22$$

which is the Gibbs adsorption equation where n_a is expressed in moles per square centimeter. This is an expression of the behavior of an ideal solution and an ideal surface and is analogous to the expression for the osmotic pressure of an ideal solution.

It is clear that at significant concentration levels the Gibbs equation should be stated in terms of the activity of the solute in the bulk phase and

$$n_a = -\frac{ad\sigma}{RTda} \qquad 23$$

where a is the activity of the solute in bulk.

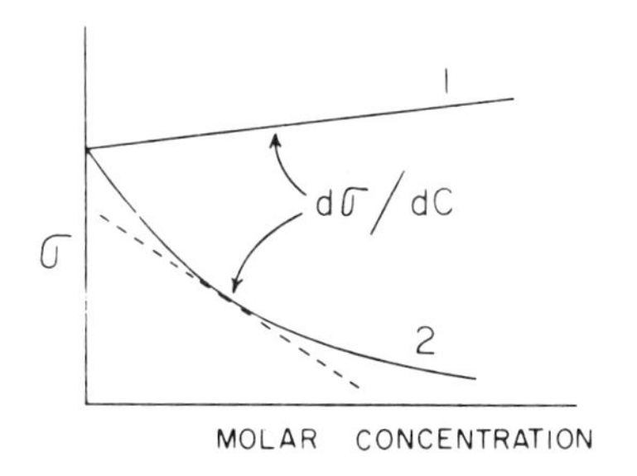

Fig. 9. Diagrammatic plot of the surface tension against concentration. Curve 1, Surface inactive solute; curve 2, surface active solute.

From the slope $d\sigma/dC$ of a plot of the surface tension of a solution against the concentration, the amount of the solute adsorbed at a given concentration can be calculated. Figure 9 is a diagrammatic representation of such a plot. Curve 1 is typical of many inorganic electrolytes such as sodium chloride and potassium chloride dissolved in water. Such salts increase the surface tension of water; they are negatively adsorbed, i.e., the concentration of the salt in the surface is less than that in bulk. Curve 2 could represent the surface tension changes of a protein solution, of the soluble fatty acids, soaps, and a vast number of surface active organic compounds dissolved in water; the differential $d\sigma/dC$ is negative indicating a positive adsorption.

The surface tension of protein solutions has a marked tendency to decrease with time. Protein molecules arriving at a surface typically undergo surface denaturation and the molecules spread out on the surface forming a very viscous film at comparatively low pressures indicating considerable interaction between the surface molecules. Figures 10 and 11 give some surface and interfacial tension results obtained with egg albumin and with bovine serum albumin as functions of time and

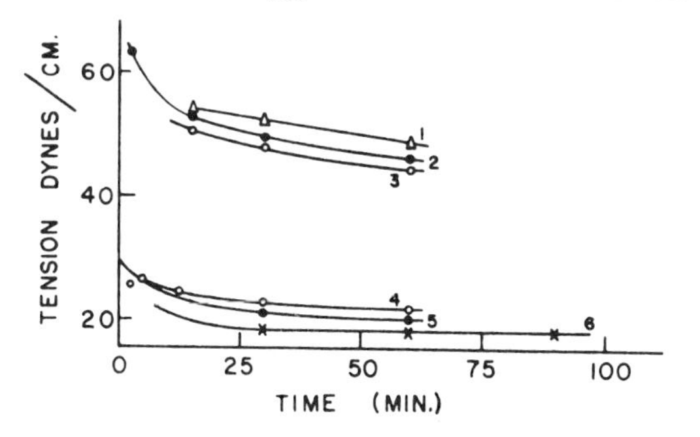

Fig. 10. Change of tensions with time. Curve 1, 0.06% egg albumin in air (pH 5.85); curve 2, 0.06% egg albumin in air (pH 4.9); curve 3, 0.06% BSA in air (pH 4.9); curve 4, 0.06% egg albumin against *n*-octadecane (pH 4.2); curve 5, 0.0018% BSA against *n*-octadecane (pH 4.17); curve 6, 0.05% BSA against *n*-octadecane (pH 4.90). Sodium acetate buffers at 30°. (S. Ghosh and H. B. Bull: Biochem. Biophys. Acta *66:* 150, 1963.)

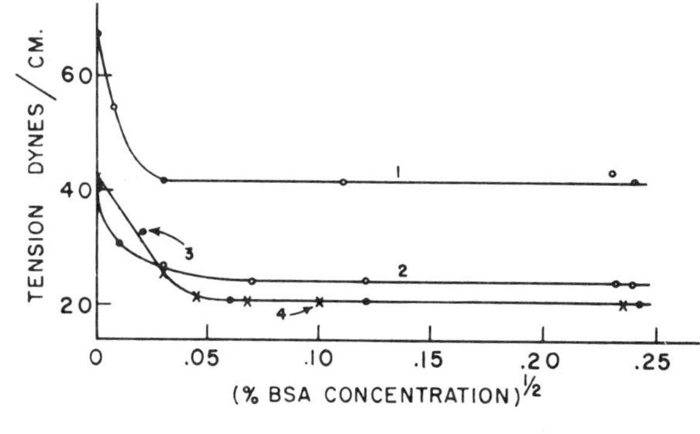

Fig. 11. Influence of BSA concentration on one hour tensions. Curve 1, Solution-air (pH 5.9) at 25°; curve 2, *n*-octadecane-solution interface (pH 5.9) at 30°; curve 3, *n*-octadecane-solution interface (pH 4.34) at 30°; curve 4, *n*-octadecane-solution interface (pH 4.20) at 30°. (S. Ghosh and H. B. Bull: Biochim. Biophys. Acta *66:* 150, 1963.)

of protein concentrations. The tensions were measured with the pendant drop method.

The relative surface tension lowering produced by a solute can be set equal to an expansion series and

$$\frac{\sigma_0 - \sigma}{\sigma_0} = a_0 + a_1 C + a_2 C^2 + \qquad 24$$

where σ_0 is the surface tension of the pure solvent. When C becomes zero $(\sigma_0 - \sigma)/\sigma_0$ must be zero so that in the above series a_0 is zero. Taking this into account and rearranging the series

$$\frac{\sigma_0 - \sigma}{C} = a_1\sigma_0 + a_2\sigma_0 C + \qquad 25$$

and a plot of $(\sigma_0 - \sigma)/C$ against C should yield a linear relation for smaller values of C. The intercept on the y-axis is $a_1\sigma_0$ and the slope of the line is $a_2\sigma_0$. Differentiation of eq. 25 gives

$$-\frac{d\sigma}{dC} = a_1\sigma_0 + 2a_2\sigma_0 C + \qquad 26$$

and at zero concentration $-d\sigma/dC$ equals $a_1\sigma_0$. Shown in Table 2 are the initial surface tension-concentration slopes for a series of fatty acids at 20°.

The average ratio of the initial surface tension lowering produced by fatty acids as the length of the carbon chain is increased by one CH_2-group (see Table 2) is about 3.94; this ratio is actually the ratio of the distribution coefficient between the surface and the bulk solution. Applying the Boltzmann distribution equation and from the ratio of 3.94, it turns out that the free energy difference for each successive CH_2-group added to a fatty acid is at 20° about 795 calories per mole of CH_2-group.

TABLE 2. INITIAL SURFACE TENSION-CONCENTRATIONS SLOPES FOR FATTY ACIDS

Fatty Acid	*Number Carbon Atoms*	*Initial Slope in Dynes/cm./moles/cc.*	*Ratio of Successive Initial Slopes*
Butyric	4	-0.162×10^6	
Valeric	5	-0.600×10^6	3.70
Caproic	6	-1.62×10^6	2.70
Heptoic	7	-6.5×10^6	4.01
Caprylic	8	-30×10^6	4.62
Pelagonic	9	-101×10^6	3.37
Capric	10	-530×10^6	5.25

Comparing the Gibbs adsorption equation with eq. 26 it is seen that a_1, the first coefficient of the expansion series, is, for dilute solutions,

$$a_1 = \frac{n_a RT}{\sigma_0 C} \qquad 27$$

Substituting the value of a_1 into eq. 25, there results

$$\sigma_0 - \sigma = \Delta\sigma = n_a RT \qquad 28$$

$\sigma_0 - \sigma$ or $\Delta\sigma$ is the so-called film pressure and is usually designated by the symbol Π. The reciprocal of n_a is evidently the total surface area, A, occupied by one mole of solute and, accordingly, eq. 28 can be written as

$$\Pi A = RT \qquad 29$$

Equation 29 is the equation of state for an ideal surface film. It will be noticed from eqs. 28 and 29 that RT is the change in free surface energy per mole of substance adsorbed on the surface for an infinitely dilute solution; the concentration of an infinitely dilute solution is constant and the osmotic term associated with adsorption from solution has been eliminated. Neglecting the interacting energies between adsorbed molecules but considering the area occupied by the adsorbed molecules, eq. 29 becomes per mole

$$\Pi \, (A - A_0) = RT \qquad 30$$

where A_0 is the limiting area of the surface film at high film pressures.

There is a close analogy between the osmotic pressure of a solution and the surface tension lowering. The same factors which are responsible for the departure of a solution from ideal osmotic behavior are also operating to cause the surface to depart from ideality; the surface phase, however, is apt to become very concentrated at low bulk concentrations and, accordingly, departure from ideality starts at far lower concentrations than does the osmotic pressure of a solution. The surface work can be separated into a part due to the heat of mixing on the surface as well as to the entropy of mixing of the solute and solvent molecules on the surface.

The entropy of surface mixing is related to the relative surface areas of the solute and solvent molecules as well as to the shape and flexibility of the solute molecules. If the solute molecule is an ion, the electrostatic work of charging the surface must be considered; there are two general approaches to the electrical effects. A Donnan equilibrium exists between the surface and the bulk solution which produces a higher film pressure than would be expected from uncharged molecules. This approach is difficult, however, because of the uncertainty regarding ion distribution between the bulk and surface phases. The other approach is to calculate the electrical potential of the surface through some modification of the Gouy equation (see Chapter 13, Electrophoresis and Electrokinetic Potentials). Equation 30 is then modified to read per molecule

$$\Pi = \frac{kT}{A - A_0} + \int_0^{\psi_0} \frac{1}{A} \, d\epsilon\psi_0 \qquad 31$$

where the integral represents the electrical work. ϵ is the charge per ion and ψ_0 is the electrostatic potential of the surface. The departure of surfaces from ideality will again be considered in the discussion of spread monolayers.

Surface Potentials. Electrostatic potentials at a surface can not only be calculated as noted above but likewise they can be measured.

There are two techniques used for the measurement of the potential of films. A small gold or platinum electrode, on which a small amount of an α-emitter such as polonium (Canadian Radium and Uranium Corp., 630 Fifth Ave., New York) has been applied, is brought to within about a millimeter of the surface of the liquid. A non-polarizable electrode is inserted in the aqueous substrate solution and the potential difference between the air electrode and the reversible electrode measured. The polonium radiation ionizes the air molecules between the electrode and the surface thus rendering the air conducting; however, the resistance is still extremely high and the measuring device has to be appropriate. The potentiometer associated with a glass electrode works quite well; good shielding and insulation are imperative.

The other method uses a vibrating plate held above the surface. The vibration of the plate produces an alternating current flow which is amplified and fed into an oscilloscope. On a separate circuit, the potential across the surface film is brought to zero by an opposite and equal potential supplied by a potentiometer. The vibrating plate method is especially suited for the measurement of potentials of films at oil-water interfaces.

A surface layer containing organic ions is accompanied by an ionic double layer in the aqueous phase and a potential exists across the double layer which is part of the potential of the spread film as measured by techniques described above. The magnitude of the double layer potential can be calculated with the Gouy equation (see Chapter 13, Electrophoresis and Electrokinetic Potentials). The potential resulting from the dipole moments of the spread film will not be responsive to changes in ionic strength but the potential of the ionic double layer decreases as the ionic strength is increased. Not only will the ionic double layer lead to a surface potential but this potential is related to the repulsive energy between the organic ions in the spread film; the film pressure is increased as a result. This represents an additional correction which must be applied to the surface tension and which was noted previously.

The surface potential is clearly analogous to the potential across a parallel plate condenser. The absolute magnitude of this potential is not of significance but the changes of the measured potential resulting from the formation of a surface film and the compression of this monolayer do reflect the molecular structure of the adsorbed molecules and the orientation of the surface molecules as well as the conditions in the electrical double layer underlying the spread film.

If the surface film is assumed to be built of n dipoles per unit area with μ being the vertical component of each dipole, then the surface potential ΔV should be given by

$$\Delta V = 4\pi n\mu + \psi_0 \qquad 32$$

where ψ_0 is the potential of the ion atmosphere and is a function of the number of charged molecules adsorbed in the film as well as of the ionic strength. For small values of the double layer potential, ψ_0 is approximately equal to $4\pi\ nZ/D\kappa$ where Z is the net charge on the adsorbed molecules, κ is the Debye-Hückle reciprocal distance in the ionic double layer and D is the dielectric constant. Substituting the value of ψ_0 in eq. 32 gives

$$\Delta V = 4\pi n\mu + \frac{4\pi nZ}{D\kappa} \qquad 33$$

Figure 12 shows the surface potentials of egg albumin surfaces as a function of the age of the surface and of protein concentrations; the potentials were measured with a rapid flow technique. Figure 13 shows the surface potentials of quiescent protein solutions as a function of the pH of the bulk solutions. The marked influence of pH on the static values of ΔV indicates an important contribution of the electrical double layer in the solution below the surface. At the same pH, the ΔV values are larger, the higher the isoelectric points of the proteins. At the respective isoelectric points of the proteins, the potential of all the protein surface films is about 350 mv; at the isoelectric point there is no double layer.

Surface Viscosity. The viscosity of surface films can be measured by flow through a canal or by the drag of the surface film on a metal ring placed in the surface. The elastic and viscous properties of the surface films can also be studied by producing standing waves on the surface; the velocity as well as the damping of the ripples reflects the film properties. There are various arrangements of the apparatus for film viscosity. In the canal method the film can be forced through a canal with a movable barrier or the underlying solution can be caused to move by a difference in solution level and the movement of the film observed. Resistance to the motion of the film is along the solid edges of the canal. Talcum powder sprinkled on the surface is used to indicate the velocity of flow. As a spread monolayer is moved, there is a viscous drag on the underlying water and this effect extends about 1×10^{-3} cm. into the liquid. A fairly complicated and substantial correction has to be applied to the measured flow rates of the film as a result of this viscous drag in order to obtain the film viscosity.

The rotational method employs a metal ring in the surface attached to a fine torsion wire and the substrate solution in a petri dish is caused to rotate; the displacement of the ring is a measure of the surface viscosity. Another version of this

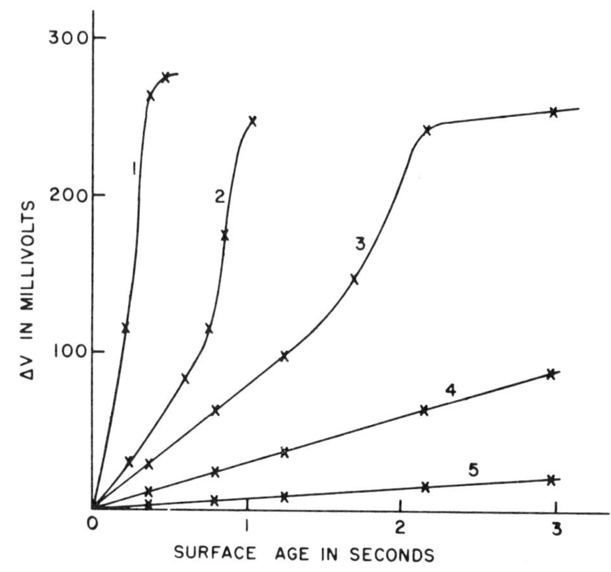

Fig. 12. Variation of ΔV as a function of the age of the air-solution surface of egg albumin solutions. Isoionic egg albumin in 0.001 M NaCl. Egg albumin concentrations: curve 1, 0.05%; curve 2, 0.016%; curve 3, 0.008%; curve 4, 0.004%; curve 5, 0.001%. (S. Ghosh and H. B. Bull; Biochem. *2:* 411, 1963.)

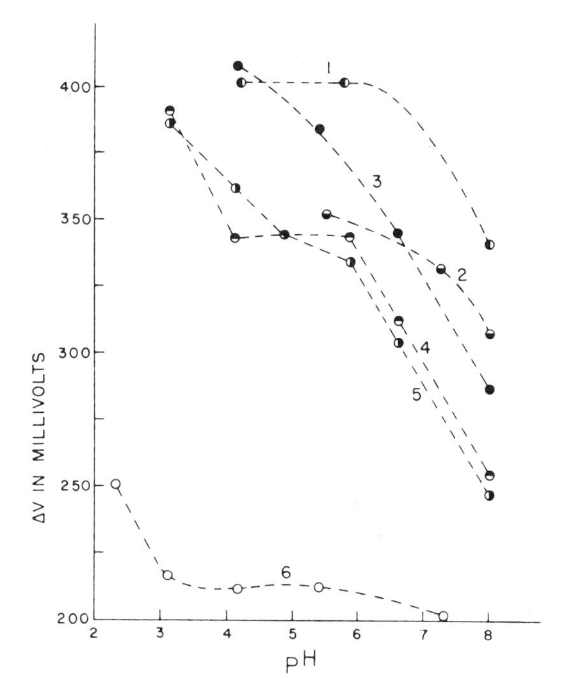

Fig. 13. Static potentials at air-solution surfaces as a function of pH. Protein, 0.06%. Ionic strength 0.05. Curve 1, Salmine; curve 2, ribonuclease; curve 3, HbO_2; curve 4, egg albumin; curve 5, bovine serum albumin; curve 6, pepsin. (S. Ghosh and H. B. Bull: Biochem. *2:* 411, 1963.)

method is to apply a torque to the ring through the torsion wire, the solution remaining stationary. The response of the ring to the torque is a measure of the film viscosity. These methods may be applied not only to spread films but also to films formed on solutions.

Some films, especially those of proteins, show elastic as well as viscous behavior; when the stress applied to a film is removed, the film tends to return to its former position. The film viscosity is responsive to the various phase changes and other structural changes of the spread monolayers. At moderate compressions the film viscosities compared to bulk viscosities as estimated with the help of the film thickness are very high; fatty acid films have a viscosity corresponding to that of butter and those of protein films to asphalt.

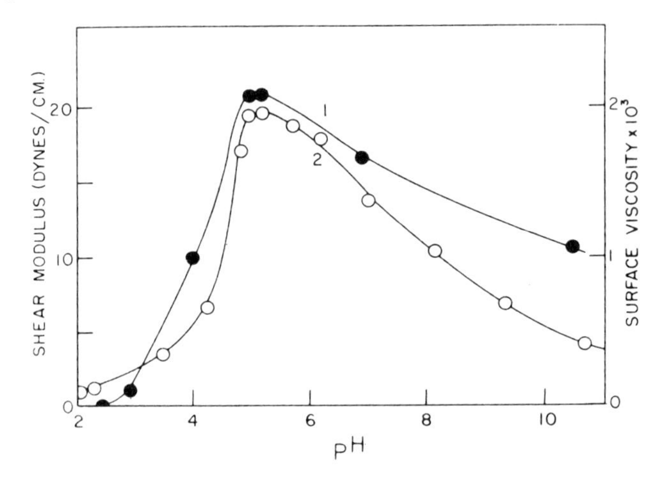

Fig. 14. Rigidity and viscosity of films on bovine serum albumin solutions as a function of pH at 20°. Curve 1, Viscosity; curve 2, rigidity. (B. Biswas and D. A. Haydon: Proc. Roy. Soc. *A271:* 296, 1963.)

Shown in Figure 14 are the viscosity and the rigidity of surface films on 0.10 percent bovine serum albumin solution as a function of pH.

Adsorption from Solution. The rate at which a substance diffuses to or forms a surface depends among other things upon whether the solution is stirred or not. For an unstirred solution, the amount diffusing is proportional to the square root of the time, whereas for a stirred solution the amount is directly proportional to time. Diffusion to surfaces is discussed in Chapter 11, Diffusion. By equating the rate of adsorption of solute molecules on a surface to the rate of departure of adsorbed molecules from a surface, Langmuir derived his well-known adsorption equation. This equation takes the form

$$a = \frac{\alpha\beta C}{1 + \alpha C} \tag{34}$$

where a is the amount of solute adsorbed, C is the molar concentration of the solute and α and β are the adsorption constants. It is to be recalled that an equation of exactly the same form was derived from the mass action law and discussed in Chapter 5, Acids and Buffers, in connection with ion binding. The various linear forms of this equation were also considered along with the determination of the constants. Inspection of eq. 34 reveals that α is an association constant and β is a measure of the number of binding sites on the surface. It should also be noted that eq. 34 is valid only if statistical binding is occurring and there is no interaction between the binding sites and only one intrinsic binding constant is involved.

The Langmuir equation assumes monolayer adsorption and, if adsorption occurs in multilayers, some form of the Brunauer, Emmett and Teller approach must be used.

A vast variety of adsorption studies have been made. The experimental techniques employed are for the most part very simple but the theoretical difficulties are usually massive and meaningful interpretations are hard to arrive at. In studying adsorption from solution, it has to be appreciated that solvent molecules must be displaced from the surface to make room for the solute molecules. It is possible to have a greater adsorption of the solvent than of the solute; the net effect is a negative adsorption of the solute.

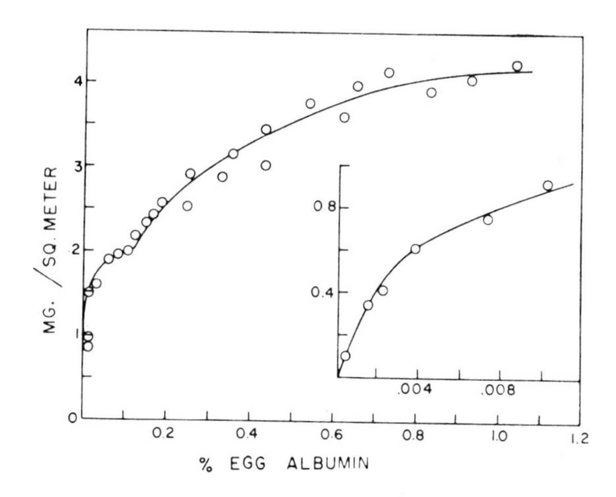

Fig. 15. Adsorption isotherm of egg albumin on Pyrex glass powder at 30°; pH 4.66 in 0.05 ionic strength sodium acetate buffer. Insert graph has expanded scales to accommodate low protein concentrations. (H. B. Bull: Arch. Biochem. Biophys. *68:* 102, 1957.)

There are many surface active materials in animals and plants and extensive interfaces at which adsorption can occur including blood vessel walls, cell walls, surfaces of mitochondria and other cellular particles. Of the surface active materials in living tissue, proteins are outstanding. In Figure 15 is shown the adsorption isotherm of egg albumin on Pyrex glass. Whereas protein adsorbs rapidly on glass and adsorption reaches its final value for a given protein concentration in a few minutes, the protein is removed only with great difficulty; in fact, part of the protein remains bound to the glass even after exhaustive washing. The fact that glass adsorbs protein so readily and irreversibly should be of interest to anyone working with very dilute protein solutions; a substantial part of the protein in solution could be adsorbed on the glassware. This remark has special pertinence for the enzymologist. It is difficult to understand how it is possible to obtain a well-defined and reproducible adsorption curve for a protein if the protein is irreversibly adsorbed. Offhand, one would expect the protein to be removed completely from solution up to a concentration sufficient to saturate the surface. It is probable that, upon adsorption, the protein molecules spread at the solid-solution interface until they come in contact with neighboring adsorbed molecules. The amount of protein actually adsorbed at a given bulk concentration then represents a balance between the rate of adsorption and the rate of spreading.

The pH of the solution influences the amount of protein adsorbed; as the pH is made progressively alkaline to the isoelectric point of the protein, higher and higher protein concentrations are required to achieve a given level of adsorption. Figure 16 shows the effect of pH on the adsorption of bovine serum albumin on Pyrex glass.

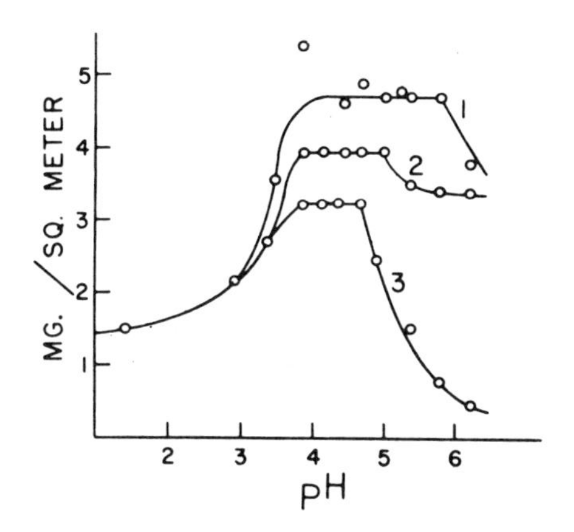

Fig. 16. Adsorption of bovine serum albumin on Pyrex glass powder as a function of pH. Ionic strength 0.05, 30°. Curve 1, 0.56 percent protein; curve 2, 0.40 percent protein; curve 3, 0.05 percent protein. (H. B. Bull: Biochem. Biophys. Acta *19:* 464, 1956.)

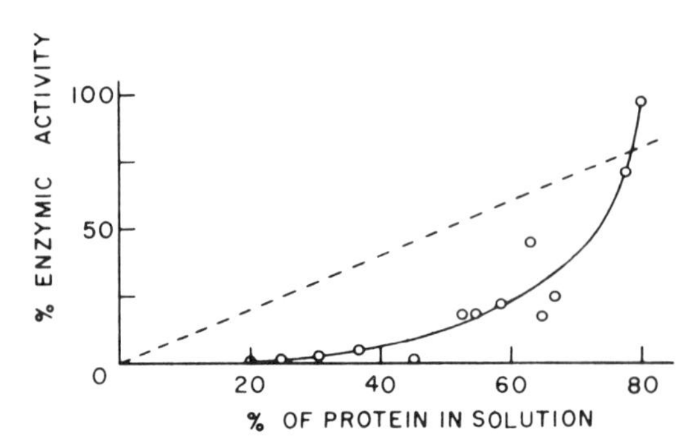

Fig. 17. Percent of added chymotrypsin activity in the presence of emulsion plotted against percent of the enzyme protein in solution. Dotted line is based on the assumption that all of the dissolved enzyme is completely active and all of the adsorbed enzyme is completely inactive; 30°. (S. Ghosh and H. B. Bull: Arch. Biochem. Biophys. *99:* 121, 1962.)

A glass surface would be expected to have many sites available for hydrogen bonding and attachment of protein. Glass also carries a negative electrostatic charge in water which becomes progressively larger as the solution becomes more alkaline. An air-water surface or an oil-water interface, however, provides no fixed binding sites; nevertheless adsorption occurs at such surfaces and interfaces. For example, the proteolytic enzyme chymotrypsin adsorbs at an oil-water interface and the extent of adsorption as well as the enzymatic activity has been studied at a water-n-octadecane surface. In order to have sufficient interface, oil-in-water emulsions of the hydrocarbon were prepared. At low concentrations of enzyme the adsorbed enzyme was inactive, but with increasing concentration the activity progressively increased, finally reaching that of the native enzyme in bulk. These results are summarized in Figure 17. Apparently, at low protein concentration, the enzyme is surface denatured at the interface and loses its activity. As the protein concentration is increased, protein molecules arrive at the interface so rapidly that neighboring adsorbed protein molecules have insufficient time and space at the interface to spread out and surface denature.

Energy of Adsorption. The free energy change associated with the adsorption of water vapor on solid surfaces (dry protein) was discussed in Chapter 7, Osmotic Pressure and Related Topics. The free energy change for adsorption of a solute from solution on a solid surface is a closely related problem. The free energy of adsorption per square centimeter according to the Gibbs adsorption equation from zero concentration to the concentration corresponding to the saturation of the surface is

$$\Delta F_{cm^2} = -RT \int_0^{C_s} n_a' \frac{dC}{C} \qquad 35$$

The free energy change for the total surface A is

$$\Delta F_G = -RT \int_0^{C_s} A\, n_a' \frac{dC}{C} \qquad 36$$

An_a' is the total number of moles of the solute adsorbed and

$$\Delta F_G = -RT \int_0^{C_s} n_a \frac{dC}{C} \qquad 37$$

where n_a is the total number of moles of solute adsorbed. The free energy of adsorption must be referred to a standard state since the surface saturation concentration will differ for each adsorbing system. A suitable standard state is the hypothetical one molal solution. The free energy of dilution to the saturation concentration is then (assuming ideal behavior)

$$\Delta F_D = n_a RT \ln C_s \qquad 38$$

The total free energy change for adsorption is obtained by adding eqs. 37 and 38 and

$$\Delta F = n_a RT \ln C_s - RT \int_0^{C_s} \frac{n_a dC}{C} \qquad 39$$

To understand the process being conducted, examine Figure 18 which is a diagrammatic representation of an adsorption curve.

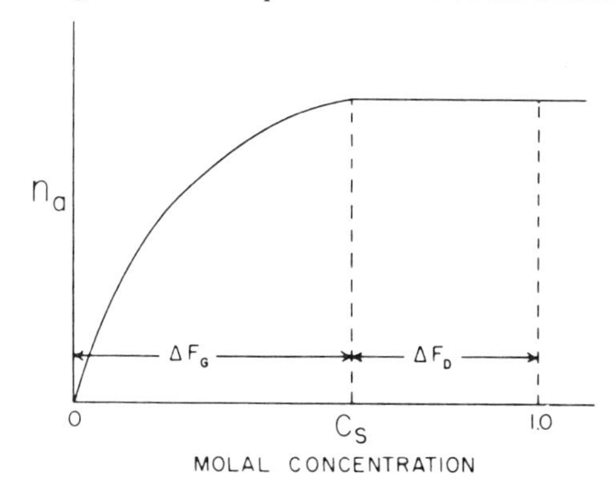

Fig. 18. Diagrammatic representation of an adsorption curve. C_s is the concentration at which the surface is saturated. See text for definitions of ΔF_G and ΔF_D.

The amount of the solute adsorbed can be expressed in milligrams and, at 30°, eq. 39 can be written

$$\Delta F = \frac{1385.5 \times 10^{-3} W}{M} \log C_s - \frac{602.38 \times 10^{-3}}{M} \int_0^{C_s} \frac{W dC}{C} \qquad 40$$

where W is the weight of the solute adsorbed expressed in milligrams and M is the molecular weight of the solute. The second term on the right side of eq. 40 is evaluated by graphical integration by plotting W/C against C and multiplying the area by $602.38 \times 10^{-3}/M$. For exact calculations the concentrations should be replaced by activities. The enthalpy and entropy changes associated with adsorption from solution can be calculated by slight modification of the methods described in Chapter 7 for the adsorption of water vapor on solid surfaces.

Binding and Adsorption. Gibbs defined adsorption as the surface excess and this is the definition used in the above discussion (see, for example, eq. 23). The measurement of the adsorption of ions or small molecules by proteins, providing the affinity of the added solute for the protein greatly exceeds that of water, presents no serious experimental or theoretical difficulty; many such studies have been reported. By definition

$$\frac{n_2}{n_1} - \frac{n_2'}{n_1'} = a \qquad 41$$

where n_1 and n_2 are the number of moles of water and solute respectively initially present and n_1' and n_2' are the number of moles of water and solute respectively in equilibrium with the protein after adsorption has occurred. a is the amount of solute adsorbed and can be positive, zero or negative. In most adsorption experiments n_1 and n_1' are in great excess and, accordingly, it is assumed that n_1 and n_1' are equal to each other. Equation 41 can be then written

$$a = \frac{n_2 - n_2'}{n_1} \qquad 42$$

and the amount adsorbed can be determined by simple analytical procedures.

Suppose, however, that dry protein is added to a solution containing n_1 moles of water and n_2 moles of solute. Δn_1 moles of water and Δn_2 moles of solute are bound by the protein and at equilibrium

$$\frac{n_2 - \Delta n_2}{n_1 - \Delta n_1} = \frac{n_2'}{n_1'} \quad 43$$

Comparing eqs. 41 and 43, it is evident that adsorption and binding are not identical. In contrast to adsorption, binding can be zero or positive but never negative. Bound solute and solvent molecules are defined as those molecules which cannot participate in the colligative properties of the protein solution.

The determination of binding is more extensively discussed in Chapter 7 in connection with the isopiestic method for the determination of the vapor pressure of the solvent.

Chromatography. Undoubtedly, the widest application of adsorption is that of chromatography. This is a kinetic adsorptive process in which an adsorbent is packed in a vertical tube and a solution of the mixture of substances to be separated into its various components is poured into the tube. The chromatogram is now developed by passing a suitable solvent into the column, whereupon each component of the mixture dissolves slightly as the solvent passes through the column but is readsorbed lower down. The net effect is to remove the less readily adsorbed material from the upper layer and carry it farther down the column. Under favorable circumstances the column will exhibit a separate band for each component. The technique of chromatography is closely related in principle to counter-current distribution; in an adsorbing column the number of transfer "tubes" is exceedingly large and corresponds to each grain of the adsorbent.

The equation for the relative position of an adsorbing band in chromatography is

$$\frac{W}{V} = \frac{V_T}{\rho V_a \alpha + V_s} \quad 44$$

where W is the volume rate of migration of a band expressed in cubic centimeters per second, V is the volume of the solvent passed into the column in cubic centimeters per second. V_T is the total volume enclosed in the tube, V_a is the volume of the adsorbent in the tube, V_s is the volume of the solvent in the tube, ρ is the density of the adsorbent, and α is the adsorption coefficient of the solute at the concentration in question, i.e., the ratio between the amount of solute adsorbed per gram of dry column material and the concentration of the solute in solution. Filter paper chromatography is, in principle, identical with column chromatography.

Spread Monolayers. Some substances, notably fatty acids, phospholipids, sterols and proteins, form insoluble monolayers when spread at an air-water surface or at an oil-water interface. In principle, such films differ in no way from those formed by adsorption from solution and discussed above. However, the situation is uncomplicated by solute in solution beneath the films and thus presents a simpler experimental approach; the amount of solute on the surface is easily defined.

Benjamin Franklin spread olive oil on a pond in Clapham Commons, London, in 1765 and, by measuring the quantity of oil required to form a complete layer on the pond, concluded that the film was about one ten millionth of an inch thick (25 A-units). This first recorded observation on spread monolayers was followed around the turn of the present century by experiments on spread monolayers by Miss Pockels and by Devaux. It was not, however, until Langmuir introduced his surface balance and the conception of molecular orientation at surfaces that meaningful experiments could be made on spread monolayers.

Langmuir devised a surface balance which consisted of a mica float separating the clean solution surface from the surface containing the spread film. By measuring the force required to maintain the float in an equilibrium position, he was able to measure the difference in surface tension between the separated surfaces; this difference in surface tension he called the film pressure. Spread films of fatty acids were compressed with a movable barrier and the surface pressure studied as a function of the area of the spread film. This early work of Langmuir was followed by the precise and detailed investigations of W. D. Harkins and of N. K. Adam as well as of many other workers. Investigation of spread monolayers of proteins was initiated by the Dutch pediatrician, E. Gorter, in the late 1920s.

The fact that fatty acids contain the hydrophilic-COOH group suggested that in compressed films the molecules were oriented with these groups sticking into the water and the hydrocarbon chain pointed into the air. Langmuir further suggested that in the uncompressed film the molecules lie flat on the surface. The films of fatty acids containing less than 14 carbon atoms per molecule dissolved in the substrate solution when the film was compressed; evidently, the "buoyancy" of the hydrocarbon chain was not sufficient to overcome the attraction of the —COOH group for water.

Spread monolayers of fatty acids on aqueous solutions have been explored in some detail and it has been shown that the films can exist in condensed, liquid-expanded, vapor-expanded and "gaseous" states. There is thus a close analogy between phase transitions in two dimensions (surface films) and in three dimensions, and a single substance, under proper conditions of temperature, surface pressure and substrate solution compositions, may exist in several states.

Force-area experiments are done with the Langmuir float balance or some modification of it or with the Wilhelmy balance. The Langmuir float balance is commercially available under the name of the Hydrophile Balance, Central Scientific Co., Chicago. The balance as supplied does not have a high degree of sensitivity but with a few simple changes can be made to respond to film pressures of the order of 0.002 dynes per centimeter. A technical problem associated with the Langmuir float for which there is no elegant solution is the danger of leakage of the spread film around the ends of the float.

The Wilhelmy balance has considerable utility and is capable of a high degree of sensitivity; it measures the surface tension of the spread film directly, the film pressure being obtained from the difference in surface tension of the solution in the absence and in the presence of the spread film. The slide must be completely wet by water, impurities can cause much trouble, and temperature changes are of more importance with the Wilhelmy balance than with the Langmuir float method. The Wilhelmy balance, of course, avoids completely the film leakage problem.

The technique for the spreading of a monolayer depends on the substance being spread. Fatty acids, steroids and other lipid-like substances are dissolved in a volatile fat solvent and the solution added directly to the surface in small drops. The volatile solvent takes more time to disappear from the surface than might be expected and can modify the film characteristics considerably. Proteins are best spread from dilute aqueous solutions on a fairly concentrated buffer substrate (0.10 to 0.5 M) or better still on a salt solution such as 5 percent sodium or ammonium sulfate. There are many small but important points of technique associated with good spreading.

In addition to force-area measurements on spread films, other properties of the spread film, such as film viscosity, electrostatic potentials of the films and the optical properties, can be examined. The chemical reactivity of spread molecules can also be studied.

Spread monolayers of numerous substances have been studied and in many instances the properties of the monolayers have been related to the chemical structure of the spread substances. Following the early and important work on fatty acid films, which led to the conception of molecular polarity in the Langmuir sense, apparently there existed an expectation that monolayer technique would develop into a useful approach to molecular structure. This expectation has not been fulfilled and spread monolayer technique turned out to be a dull and undiscriminating tool for this purpose. In other respects, however, monolayer studies have been fruitful. Among other possibilities the method is of aid in providing a background technique for the study of cellular membranes. Among substances which form spread monolayers, and are of interest as membrane components, are phospholipids, sterols and proteins.

Lecithin Films. The phospholipids are a complex group of substances containing phosphate, a nitrogen base and an alcohol and embrace a large number of different compounds; the lecithins are an important branch of the larger family. They are the choline derivatives of phosphatidic acid. In phosphatidic acid, two of the hydroxyls of glycerol are esterified with fatty acids; the individuality of the phosphatidic acids (as well as of the lecithins) arises from the nature and position of

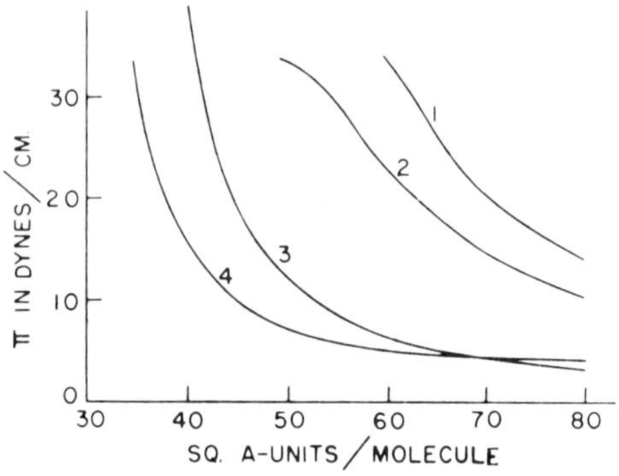

Fig. 19. Film pressure as a function of the film area. Substrate solution: phosphate buffer pH 7.4, ionic strength 0.14, temperature 21°-24°. Curve 1, γ-stearoyl-β-oleoyl-*D*-α-lecithin film. Curve 2, film of *L*-α-lecithin containing two 15 carbon saturated fatty acids. Curve 3, *L*-α-distearoyl-lecithin film. Curve 4, distearoyl 1 phosphatidic acid film. (Data of L. L. M. van Deenen, V. M. T. Houtsmuller, G. H. de Haas and E. Mulder: J. Pharmacy and Pharmacology *14:* 429, 1962.)

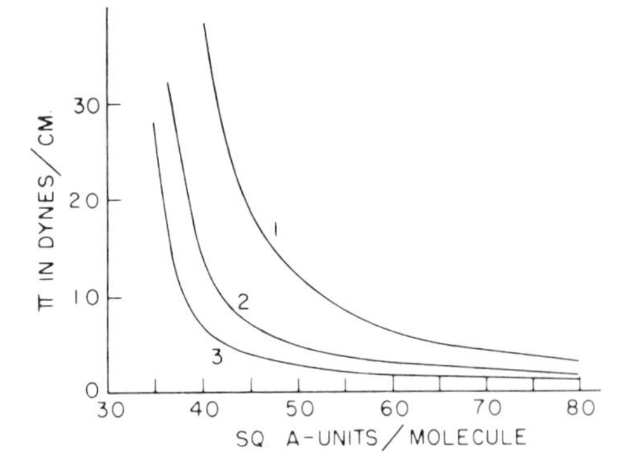

Fig. 20. Film pressure as a function of the film area. Substrate solution: phosphate buffer pH 7.4, ionic strength 0.14, temperature 21°-24°. Curve 1, *L*-α-distearoyl lecithin film. Curve 2, mixed film of equimolar quantities of lecithin and cholesterol. Curve 3, cholesterol film. (Data of L. L. M. van Deenan, V. M. T. Houtsmuller, G. H. de Haas and E. Mulder:. J. Pharmacy and Pharmacology *14:* 429, 1962.)

the fatty acids. The third glycerol hydroxyl is esterified to phosphoric acid and completes the phosphatidic acid molecule. In lecithin the phosphoric acid is further esterified with the hydroxyl of choline, leaving the quaternary ammonium group free. Since lecithin contains a free primary hydrogen of phosphoric acid as well as a strong basic group, it exists as a zwitter ion over a broad range of pH with its isoelectric point close to neutrality.

Lecithin contains strongly hydrophobic (fatty acids) and hydrophilic (phosphate and choline) centers and is very surface active. Figure 19 contrasts the properties of spread films of synthetic lecithins containing different fatty acids; also shown is the force-area curve for phosphatidic acid.

Figure 20 displays the force-area curve of a film composed of equimolecular mixtures of cholesterol and of lecithin; the mixed film seems to resemble that of the pure cholesterol more than it does the lecithin film; cholesterol is the ubiquitous sterol of animal tissue.

If lecithin is observed under the microscope in contact with water, it can be seen that growths are proceeding outwardly from the water-lecithin interface. These are myelin forms and consist of concentric tubes of bimolecular leaflets. The amount of agitation required to form an emulsion of lecithin in water is so gentle that, in effect, lecithin undergoes spontaneous emulsification; evidently the interfacial ten-

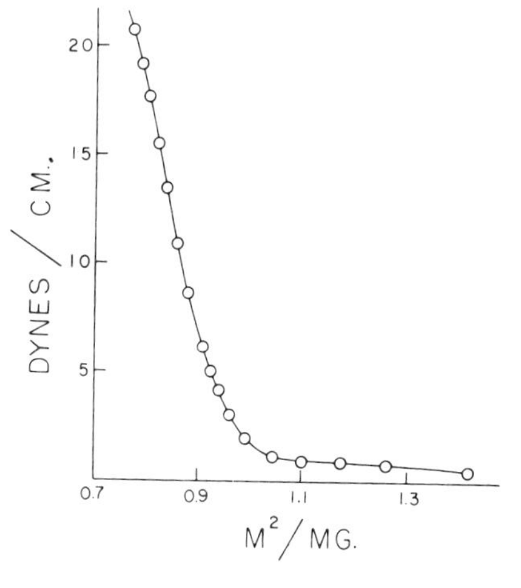

Fig. 21. Force-area curve of egg albumin spread on 35 percent ammonium sulfate. (H. B. Bull: J. Amer. Chem. Soc. *67:* 4, 1945.)

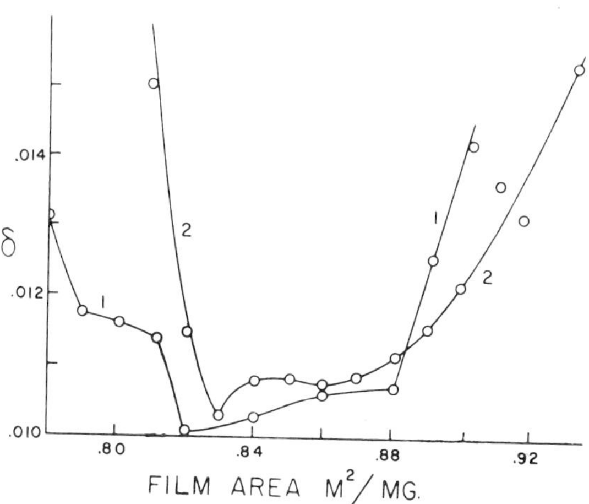

Fig. 22. Compressibilities (δ) as functions of film areas on 35 percent ammonium sulfate. Curve 1, egg albumin. Curve 2, β-lactoglobulin.

sion between lecithin and water is very low.

Protein Films. As a spread film of protein is compressed with a movable barrier, the film pressure increases and the plot of the film pressure against the area of the spread film yields a rather characteristic type of curve; there are no phase transitions such as are found with many other types of spread films; the film simply gels at higher pressures. Figure 21 shows a force area curve of egg albumin spread on 35 percent ammonium sulfate.

The coefficient of compressibility of a spread film offers a convenient characteristic especially in the higher pressure regions. The coefficient of compressibility is

$$\delta = -\frac{dA}{Ad\Pi} \qquad 45$$

where A is the area of the film in square meters per milligram and Π is the film pressure in dynes per centimeter. Figure 22 shows the compressibility curves of spread films of egg albumin and of β-lactoglobulin on 35 percent ammonium sulfate.

When the coefficient of compressibility is plotted against the area of the film (see Fig. 22) a well-defined minimum is usually observed which permits the area and pressure at minimum compressibility to be assigned without ambiguity. It is probable that the point of minimum compressibility of a protein film corresponds to the smallest area to which a film can be compressed without partial collapse of the film occurring. Calculations show that the areas corresponding to point of minimum compressibility of proteins is nearly 0.80 square meters per milligram. This value agrees well with the dimensions of a compact layer of peptide chains of average composition as determined from X-ray diffraction studies on proteins in bulk.

As mentioned earlier the viscosity of compressed protein films is considerable and, accordingly, the rate of compression of the film is significant. If the film is compressed rapidly the measured area for a given film pressure is larger than for a slower compression. Also proteins apparently differ in their rates of surface denaturation. Egg albumin and bovine serum albumin spread very rapidly and, at least on ammonium sulfate solutions, the reaction is complete in less than 5 minutes. There is evidence, however, that bovine oxyhemoglobin undergoes spreading rather slowly.

The actual structure of spread monolayers of proteins is about as obscure as the molecular structure of proteins themselves. It is one of the remarkable facts of nature that, when water soluble, highly organized, native protein molecules are placed on an aqueous surface, they promptly spread on the surface to form insoluble films whose thickness, when compressed, corresponds to one peptide chain irrespective of the dimensions of the original, native molecule.

The composition of spread monolayers undoubtedly depends on the initial spreading pressure. When protein is permitted to spread under low pressures and ample time is allowed for spreading, the protein is surface denatured and corresponds more or less to peptide chains lying flat on the surface. If the protein is an enzyme, the enzymic properties are completely lost. Curiously, however, such films, if compressed and deposited on slides, retain the ability to react specifically with antibodies. A different kind of monolayer results when the protein is spread against an appreciable pressure. The areas of such films at their point of minimum compressibility are much smaller than for the completely spread films and, further, such films retain enzymatic activity if the protein is an enzyme; the greater the initial spreading pressure the greater the retention of activity. These films contain molecules which have suffered a minimum degree of distortion; they are only partially denatured.

It is possible to transfer films of protein adsorbed from a protein solution on to a clean aqueous surface and to permit the film to expand freely. The amount of protein in the expanded film can be estimated by comparison of its force-area curve with that of a standard force-area curve of a spread film. Having the surface tension of the original protein solutions as well as the amount of adsorbed protein it is possible to construct a force-area curve for the adsorbed film. Figure 23 shows a comparison of the force-area curve for adsorbed films of lysozyme with the force-area curve for spread monolayers of this protein. As can be seen, the adsorbed film of lysozyme is much more condensed than is the spread film; the calculated thickness of a fully compressed adsorbed film is compatible with that of an adsorbed monomolecular layer of native lysozyme.

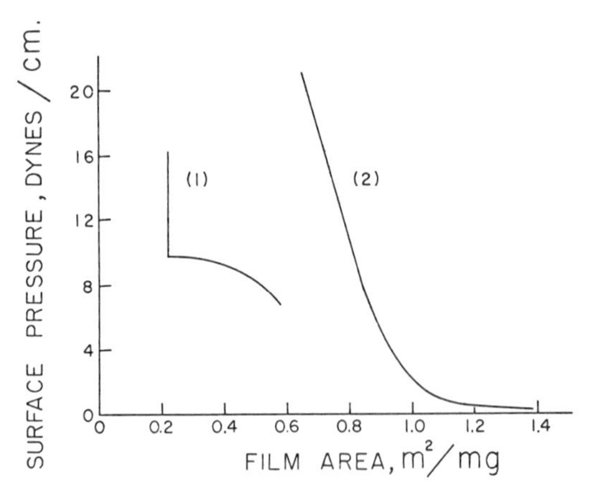

Fig. 23. Force-area curves for lysozyme. Curve 1 adsorbed on 1 M KCl; curve 2, spread monolayer on 3.5 M KCl. 23°. (Adapted from T. Yamashita and H. B. Bull: J. Coll. Interface Sci. *27:* 19, 1968.)

Gaseous Films. The term "gaseous film" is an unfortunate one. Such films are not in any sense of the word two-dimensional gases but exist rather as surface solutions. Many, if not all substances, form gaseous films at sufficiently low film pressures.

To expand a gaseous spread film, water must be transferred from the bulk phase to the surface phase. If a_o is the activity of the water in the substrate solution and a_s the activity of water in the clean surface, then the free energy required to transfer one mole of water from solution to the surface at constant temperature and pressure is

$$\Delta F = A_1\sigma_0 = RT \ln \frac{a_s}{a_o} \qquad 46$$

where A_1 is the area occupied by one mole of water and σ_0 is the surface tension of water. The corresponding free-energy change for a surface containing n_2 moles of spread substance is

$$\Delta F_1 = A_1\sigma' = RT \ln \frac{a'_s}{a_0} \qquad 47$$

where a'_s is the activity of the water in the surface containing the spread material. Subtracting eq. 47 from eq. 46, there results

$$A_1\left(\sigma_0 - \sigma'\right) = \Pi A_1 = RT \ln \frac{a_s}{a'_s} \qquad 48$$

where Π is the film pressure. Now a_s/a'_s is nearly equal to $1/N$, the mole fraction of the water in the film containing the spread substance. Equation 48 can then be written

$$\Pi A_1 = -RT \ln \left(1 - \frac{n_2}{n_1 + n_2}\right) \qquad 49$$

where n_1 and n_2 are the number of moles of water and of the spread substance respectively in the spread film. Since n_1 is very much greater than n_2, there results after expanding the logarithmic term, rearranging and neglecting higher terms of the expansion

$$n_1\Pi A_1 = n_2RT \qquad 50$$

n_1A_1 is the area occupied by the water and n_2A_2 is the area occupied by the spread material where A_2 is the area per mole of spread substance. Then A, the area of the surface, is

$$A = n_1A_1 + n_2A_2 \qquad 51$$

Substituting eq. 51 into eq. 50 and rearranging, it is found that

$$\Pi A = n_2RT + n_2A_2\Pi \qquad 52$$

and if ΠA is plotted against Π the intercept on the y-axis is n_2RT and the slope of the line is n_2A_2. When ΠA is expressed in dyne-centimeters per molecule, it is, at zero film pressure, equal to kT where k is Boltzmann's constant and T is the absolute temperature. At 25°, this product is equal to 411×10^{-16} erg and, if the area is expressed in square Angstrom units, then kT is equal to 411. Frequently, the area of spread material is expressed in square meters per milligram and at 25° RT becomes $411 \times 10^{-16} \times 6.02 \times 10^{23}/1{,}000 \times 10{,}000$ or is 2478. A study of force-area relations of a gaseous film thus offers a means for the determination of the molecular weight of the spread substance. Evidently, one mole of the spread substance will yield an intercept of 2478 at zero pressure when ΠA is plotted against Π and the area of the film is expressed in square meters per milligram; the molecular weight of the spread material is, at 25°, equal to 2478 divided by the intercept value of ΠA. The spread film technique for molecular weights of proteins and peptides has not been exploited to the full extent. The range over which the molecular weights of surface active material can be determined is from 1,500 to about 80,000. If the molecular weight exceeds this last figure, the intercept of the ΠA vs Π plot on the A-axis becomes so small that no useful estimates can be made. It is necessary to have available a film balance capable of accuracy down to about 0.01 dyne per centimeter. Ordinarily, about 25 micrograms of protein are required and, if the film balance is in order, the measurements take about 15 minutes to perform. The method gives a number average molecular weight. Figure 24 shows three typical ΠA vs Π plots for gaseous films. Figure 24A is that given by sodium dodecyl sulfate; Figure 24B that by bovine serum albumin and Figure 24C is the plot for a pepsin film.

The dodecyl sulfate (Fig. 24A) exhibits a minimum in the ΠA vs Π plot which is probably the result of the formation of micelles on the surface as the film is compressed; the micelles contain, on the average, 4 dodecyl sulfate molecules. Other substances exhibit such minima, for example, esters of dibasic acids.

The break in the ΠA vs Π curve for bovine serum albumin is attributed to a compression of the expanded gaseous molecules themselves with increasing film pressure.

From the derivation outlined above and leading to eq. 52, it is clear that the pressure exerted by a spread film is closely analogous to an osmotic pressure. According to this derivation, n_2A_2 is a direct measure of the surface occupied by the spread substance, and, indeed, experiments tend to confirm this conclusion; the gaseous areas calculated for a number of proteins are very nearly equal to those obtained by compression of the film to the point of minimum compressibility. There is something unclear about this situation, however, because according to osmotic theory the solute molecules should exhibit a co-volume and the co-volumes are significantly larger than the actual volumes.

The plot of ΠA vs Π frequently does not give a linear relation and, at sufficiently large values of Π, all such plots depart from linearity. As noted previously, departure from ideality can be ascribed to an entropy term or to the enthalpy of mixing of the molecules in the surface film. The ideal equation of state for a monolayer, i.e., ΠA is equal to RT, was modified to accommodate the area occupied by the solute molecules. This represents an entropy correction. By the methods of statistical mechanics, Singer derived an equation of state for a spread monolayer which takes into account the flexibility of the spread monolayer; his equation provides for the departure of the ΠA vs Π plots from linearity and, in fact, the departure from linearity can be used to characterize the chain flexibility; Singer's correction takes into account the entropy of mixing.

Singer considered the number of sites which a molecule can occupy on the surface; the greater the flexibility of the spread molecule the greater the number of sites. The relation which he derived can be expressed as

$$\Pi = \frac{RT}{A_0}\left[\left(\frac{x-1}{x}\right)\left(\frac{Z}{2}\right)\ln\left(1-\frac{2A_0}{ZA}\right)-\ln\left(1-\frac{A_0}{A}\right)\right] \qquad 53$$

where x is the number of flexible units in the polymer chain, Z is the coordination number or the number of available sites and A_0 is the limiting area per mole. For a rigid molecule Z has the value of 2 and if Z does have the value of 2, eq. 53 can be reduced to eq. 52. It is possible to estimate Z from experimental data and to gain an idea of the degree of flexibility of the polymer chain on the surface; the greater the departure of a plot of ΠA vs Π from linearity, the greater the chain flexibility. Protein

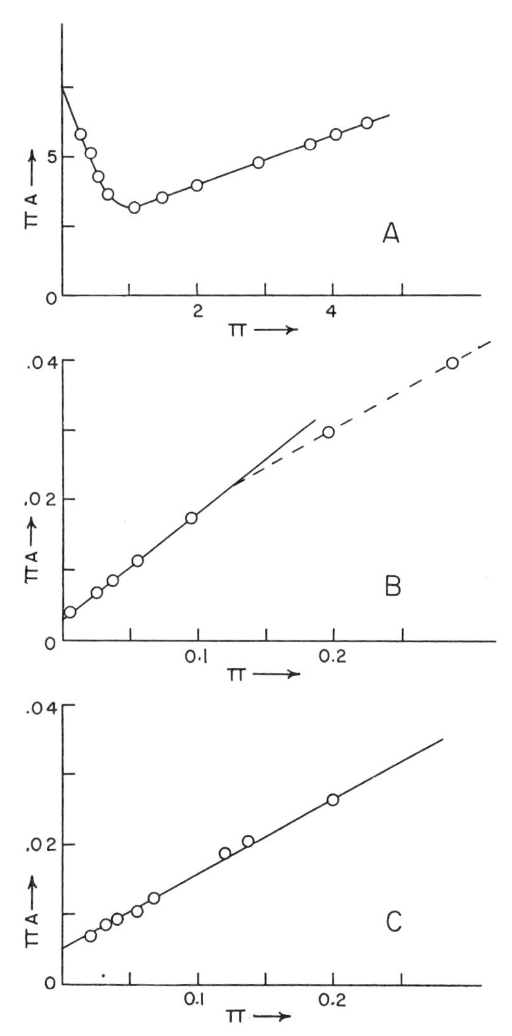

Fig. 24. Π *A* vs Π plots in low pressure region. Area expressed in square meters per milligram and the film pressure (Π) in dynes per centimeter. *A*, Sodium dodecyl sulfate on 19 percent ammonium sulfate; *B*, bovine serum albumin on 5 percent ammonium sulfate film compressed one minute after spreading; *C*, pepsin spread on 5 percent ammonium sulfate.

films spread at the air-water surface are, in general, quite rigid whereas films of synthetic polypeptides such as those of poly-D, L-alanine show a high degree of flexibility. There is a significant increase in the flexibility of proteins spread at an oil-water interface; the interaction between the peptide chains has been diminished by the presence of oil.

A number of other equations of state for surface films have been derived and some of these derivations attempt to deal with the interaction energy between the surface molecules as well as with the entropy correction. These problems could be clarified by careful studies on the influence of temperature on film pressures from which the entropy and enthalpy of the films could be calculated. The entropy in ergs per square centimeter per degree is equal to $(d\Pi/dT)$ at constant area and temperature and the enthalpy in ergs per square centimeter is equal to $T(d\Pi/dT)_{A,T} - \Pi$. Shown in Figure 25 is a plot of the calculated entropy and enthalpy of insulin monolayers spread on 0.01 NHCl. It appears that not only is the entropy term of importance but likewise the heat of mixing is worth considering. It should be mentioned that both the entropy and enthalpy of film compression are sensitive to the initial spreading area as well as to the composition of the substrate solution.

Film Penetration. If surface active molecules having hydrophilic heads and hydrophobic tails be injected under a protein film, three situations may arise. (1) If

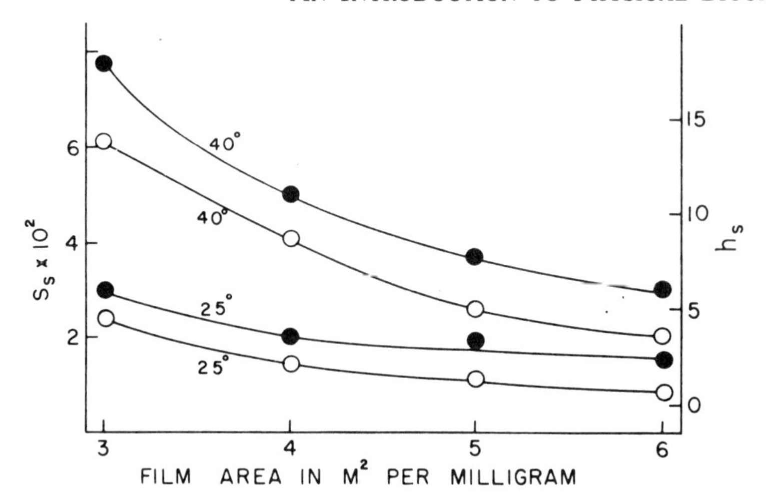

Fig. 25. Insulin monolayers spread on 0.01 N HCl at an initial spreading area of 6.8 M^2 per milligram. Open circles, S_s; closed circles, h_s. See text for further explanation. (J. Llopis and D. V. Rebollo: Arch. Biochem. Biophys. *88:* 142, 1960.)

there is no association between the polar heads of the injected molecules and the polar groups in the film, no alteration in the film characteristics is noticed. (2) If there is association between the polar groups of the injected molecules and the film but no association between the hydrophobic tails and the film, there is an adsorption of the injected molecules under the film with consequent change in the surface potential but no increase in surface pressure. (3) If there is association between both the heads and tails of the injected molecules with the film, then a polar group of the film anchors a polar group of the injected molecule. The hydrophobic portion of the injected molecule associates with the hydrophobic portion of the film, and thus penetration of the monolayer results. In this case the film pressure assumes a value that is an average one for the original monolayer and the injected molecule.

Film penetration can be completely prevented by compressing the film, prior to injection, to the equivalent pressure at which displacement of the penetrating substance would start. Film penetration, although of great interest, is very difficult to quantitate because of the unknown amount of the injected material that remains in the substrate solution.

In a mixed film containing more than one kind of surface-active molecule, an increase in film pressure will usually lead to displacement of one of the components from the film. It has been found possible to make an estimation of the molecular weight distribution of a mixture of peptides resulting from the partial hydrolysis of proteins on this basis. The partial hydrolysate is spread, and its number average molecular weight measured by extrapolating a ΠA vs Π plot to zero pressure; the slope of this line gives a measure of the amount of material in the film. The film is then compressed to higher and higher pressures and held at these successive pressures for 5 minutes. After each compression the film is expanded, and the average molecular weight determined by a compression in the low pressure region. The molecular weight and the amount of material squeezed out at each pressure can be calculated, and, accordingly, the molecular weight distribution of the peptides can be estimated.

Spread films have been studied at oil-water interfaces. The techniques which are available are, however, less convenient than those for the air-solution surface. The film balance can be modified for this purpose but there are many technical difficulties associated with its use. The film pressure can also be studied as a function of the amount of spreading material, the area of the oil-water interface being maintained constant. Protein films at an oil-water interface tend to be more expanded at a given pressure than at the air-water surface.

Bifacial Films. Of interest is a soap bubble. Such a bubble if allowed to drain in a moist and undisturbed atmosphere undergoes a series of color changes eventually becoming completely transparent to transmitted light with no ability to reflect light; such films are known as "black" films. In this state, the thickness of the film is about 50 to 90 A-units and consists of bimolecular leaflets of soap molecules with the hydrocarbon tails exposed to the air phase. Films having this character were first studied by Isaac Newton.

More recently, attention has centered on bifacial lipid films formed under water. A technique for the formation of a bifacial membrane is to paint, with an appropriate lipid solution, over a circular hole 0.5 to 5 mm^2 in the side of a polyethylene or Teflon beaker, the beaker being immersed in a solution of the same composition inside and outside at the same hydrostatic pressure. The lipid membrane thins down to a "black" film within a few minutes and is invisible to transmitted light. The film may be stable for several hours or even days. A number of lipids and lipid combinations have been used successfully such as cholesterol, lecithin, glycerol dioleate and phosphatidyl choline dissolved in a hydrocarbon or chloroform solvent. Another method for the formation of bifacial lipid membranes is to produce such a film around an aqueous drop floating in water. A sucrose density gradient containing a uniform concentration of the desired electrolyte is formed in a rectangular cell. A syringe microburette attached to a small diameter polyethylene tube is filled with sucrose solution of intermediate density containing the electrolyte of interest. The end of the tube is coated with a solution of phosphatidylcholine and n-tetradecane dissolved in a mixed solvent of chloroform-methanol and inserted in the sucrose gradient. A droplet from the microburette is discharged and simultaneously coated with the lipid solution. The detached droplet falls through the sucrose gradient until it reaches a level of equal density. The lipid phase moves to the top of the drop and forms a lens. The rest of the film thins to transparency and resembles closely the corresponding planar bilayers.

The bilayers have fair mechanical stability and are self-sealing to puncture. Unlike the soap bubble film, the polar groups of the lipid are directed outwardly with the non-polar tails oriented towards the center of the film. It is suggested that such films constitute an acceptable model for cellular membranes. The physical properties of the bifacial films have been studied by a variety of methods, and Table 3 shows some of the characteristics of such films.

TABLE 3. PHYSICAL PROPERTIES OF BIFACIAL FILMS

Property	
Thickness	40 – 80 A. units
Resistance	10^6 to 10^9 ohm · cm^2
Capacitance	0.33 to 1.3 μF per cm^2
Refractive index	1.56 to 1.66
Water permeability	0.5 – 10 × 10^{-3} cm per sec
Interfacial tension	0.2 to 6.0 dynes per cm

The water permeability of the bifacial films falls within the range of values reported for living cellular membranes. There are apparently no aqueous pores in the bifacial membranes; the water simply dissolves in the hydrocarbon phase and passes through the film. The electrical resistance of the bifacial films is, however, at least two orders of magnitude larger than that of cellular membranes.

The interfacial tensions of the films have been measured by a bubble pressure or bulging method. In this method, a pressure difference is established across the "black" film and the film bulged until its maximum extension is equal to the radius of the opening used to support the film. The interfacial tension is then calculated with the use of eq. 4. Table 4 shows the interfacial tensions of bilayers formed from dodecyl acid phosphate, cholesterol and dodecane exposed to 0.1 M NaCl as a function of the temperature. Also included are the enthalpy, Δh, and the entropy, Δs, for the expansion of the interface between the bilayer and the aqueous phase by one cm²; Δs and Δh were calculated using eqs. 12 and 13.

TABLE 4. INTERFACIAL TENSIONS, σ, AND Δh THE ENTHALPY IN ERGS PER CM², AND Δs, THE ENTROPY IN ERGS PER CM² PER DEGREE AS A FUNCTION OF THE TEMPERATURE. LIPID BILAYER IN CONTACT WITH 0.1 M NACL. (FROM THE DATA OF H. T. TIEN: J. PHYS. CHEM.: *72* 2723, 1968.)

T	σ	Δh	Δs
25	1.1	−7.3	−0.028
28	1.2	−7.3	−0.028
32	1.3	−10.9	−0.040
34	1.4	−14.5	−0.052
36	1.5	−22.2	−0.077
38	1.7	−32.5	−0.126
40	2.0	−66.3	−0.212
42	2.5	−100.2	−0.326

It is evident from Table 4 that the interfacial tensions of the bilayers are quite low. Since the interfacial tension between water and dodecane is about 50 dynes per cm., the film pressures are very large indeed, which means that the "black" films are tightly packed and condensed. Unlike the expansion of an air-liquid surface, the entropy change is large and negative and the enthalpy actually favors the expansion of the interface. The negative entropy of expansion can be understood in the following way: The expansion of the bifacial film feeds on the lipid lens around the aperture and the lipid solution is transformed into two parallel layers whose molecules are highly oriented normal to the interface with a consequent decrease of entropy.

Deposited Monolayers. Growing out of the original work of Blodgett who deposited layers of fatty acids on glass and on chromium plated metal slides, many experiments have been made on deposited films of a variety of substances. The spread monolayer is transferred from the water surface to glass or to metal slides by raising the slide slowly out of the water upon which the monolayer has been spread. It is usually possible to deposit another layer, and the up-and-down trips are continued until as many layers are deposited as wished. It is necessary to maintain a constant film pressure on the monolayer as deposition takes place. The ratio of the area of the deposited film to the decrease in the area of the spread film at constant film pressure is known as the deposition ratio.

The deposition of protein layers beyond the first up-trip requires the slide to be conditioned with a layer of calcium or barium stearate before protein deposition; otherwise the protein film deposited on the trip-out of the surface will peel off on the down-trip through the surface. The immunological, enzymic and wetting properties of deposited protein layers have been investigated.

In order to make quantitative measurements on deposited films, it is necessary to measure their thickness. If the films are deposited on polished metal slides such as those of chromium or on glass slides with sufficiently high index of refraction, beautiful interference colors may be produced. The light reflected from the surface of the slide itself may be in phase or out of phase with that reflected from the surface of the deposited film and, accordingly, lead to color production if white light is used or to maxima and minima as a function of the angle of reflection if monochromatic light is used. The condition that the two reflecting waves shall differ in phase by an integral multiple of π-radians and thus lead to cancellation for two waves of equal amplitude is

$$nNd \cos \alpha = \frac{m\lambda}{4} \qquad 54$$

where n is the refractive index of the deposited film, N is the number of monolayers deposited on top of each other, d is the thickness of each monolayer, α is the angle of reflection, m is an integer (the order of reflection) and λ is the wavelength of the light. A complicating factor in such studies is that deposited films are frequently birefringent with two indices of refraction and a decision has to be made regarding the use of the ordinary or the extraordinary ray.

The other method in use for the measurement of the thickness of deposited films, and which is capable of greater accuracy than the interference method, employs the ellipsometer. Light reflected from a surface is elliptically polarized. If the reflecting surface is covered by a thin layer of transparent material, the reflected light undergoes further changes in ellipticity due to the interaction between reflected light from the two surfaces. This change in ellipticity is a function of the film thickness and of the refractive index of the deposited film. Monochromatic light is passed through a polarizer rotated through 45° to the vertical resulting in plane polarized light transmitted at this angle. The polarized light is reflected from the slide fixed in a vertical position at an angle of incidence between 30° and 90°. The reflected light which is elliptically polarized is passed through a quarter wave plate. Light emerging from the quarter wave plate is plane polarized but rotated through a certain angle. The light is then passed through an analyzer to determine the angle of rotation and the ellipticity of the light reflected from the slide calculated. Shown in Figure 26 is a diagrammatic sketch of the arrangement of the ellipsometer.

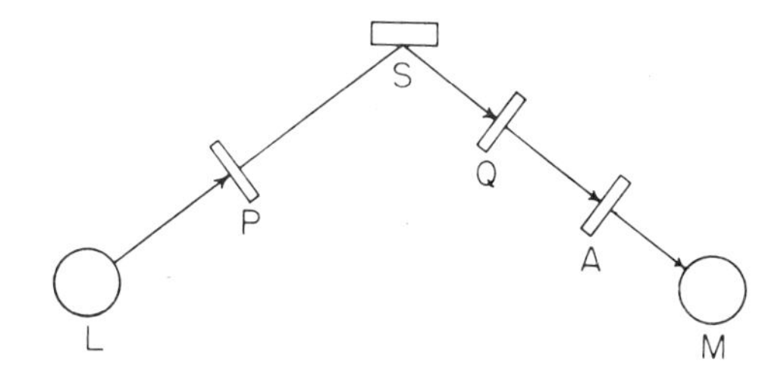

Fig. 26. Arrangement for an ellipsometer. L, Monochromatic light source; P, polarizer; S, reflecting slide with deposited monolayers; Q, quarter wave plate; A, analyzer; M, photomultiplier tube.

Not only is it possible to use the ellipsometer in air but the slide can be immersed in a solution and the rate of accumulation of adsorbed proteins on the slide measured by the change in thickness of the deposited layer. What is measured by the ellipsometer is the optical thickness of the deposited layer; a knowledge of the refractive index of the film is necessary and an exact value of the index is frequently difficult to obtain. This is a particularly troublesome question to answer for deposited layers exposed to water because an unknown amount of water is taken up by the film.

GENERAL REFERENCES

Adamson, A. W.: Physical Chemistry of Surfaces. Interscience Publishers, New York, 1960.

Danielli, J. F., H. G. A. Pankhurst, and A. C. Riddiford: Recent Progress in Surface Science, Vols. 1 and 2. Academic Press, New York, 1964.

Davies, J. T. and E. K. Rideal: Interfacial Phenomena. Academic Press, New York, 1961.

Gaines, G. L., Jr.: Insoluble Monolayers at Liquid-Gas Interfaces. Interscience Publishers, New York, 1966.

James, L. K. and L. G. Augenstein: Adsorption of enzymes at interfaces: film formation and the effect on activity. Adv. Enzymol. *28*:1, 1966.

Sobotka, H. and H. G. Trurnit: Unimolar Layers in Protein Analysis. *In* Analytical Methods in Protein Chemistry. Pergamon Press, New York, 1961.

PROBLEMS

1. Derive the relation between the surface tension and the height of rise of the liquid in a capillary. What small correction needs to be applied to obtain the true surface tension?

   Ans: See text.

2. Derive the Gibbs adsorption equation.

   Ans: See text.

3. The surface tension of a peptic digest of egg albumin was determined at 25°C as a function of the concentration of the digest with the following results:

Conc. in grams/liter	*Surface tension in dynes/cm.*
0	71.50
5×10^{-3}	70.13
10×10^{-3}	68.85
15×10^{-3}	67.48
20×10^{-3}	66.30
25×10^{-3}	65.35

   Calculate the area occupied by one mole of the peptic digest. Assume a compressed film of peptides occupies one square meter per milligram and estimate the molecular weight of the digest.

   Ans: Mol. wt. about 630.

4. The interfacial tensions of aqueous solutions of varying molar concentrations of sodium dodecyl sulfate in 0.02 M HCl and 0.03 M NaCl against purified Nujol saturated with water were determined by the drop weight method at 25°C.

Conc. NaDS $\times 10^6$	*Drop vol.* cc.	*Conc. NaDS* $\times 10^6$	*Drop vol.* cc.
0.00	0.38036	2.50	0.33941
0.50	0.37179	3.00	0.33395
1.00	0.36366	4.00	0.32097
1.50	0.35404	5.00	0.30707
2.00	0.34742	7.50	0.27815
		10.00	0.25512

   Density of aqueous solution is 0.99870, density of Nujol is 0.87553, radius of capillary tip is 0.18035 cm. Calculate the interfacial tension as a function of concentration and determine the ratio $(\sigma_0 - \sigma)/C$ as C approaches zero.

   Ans: $(\sigma_0 - \sigma)/C_{C \to 0} = 2.39 \times 10^6$

5. 2.15 gram portions of powdered Pyrex glass with a surface area of 0.335 sq. meters per gram are added to 7.00 ml. of bovine serum albumin solutions in sodium acetate buffers at pH 5.05 with an ionic strength of 0.05 at 30°. The suspensions are stirred. The suspensions are centrifuged and the optical densities of the protein solutions measured before and after adsorption and the following results obtained.

Percent conc. BSA before adsorption	*Optical densities after adsorption* OD_{278}	OD_{320}	*Percent conc. BSA before adsorption*	*Optical densities after adsorption* OD_{278}	OD_{320}
0.0104	.108	.048	0.0401	.184	.038
0.0207	.140	.056	0.0498	.234	.038
0.0311	.245	.122	0.0519	.241	.043
0.0401	.208	.053	0.1037	.535	.046
0.0498	.250	.048	0.1508	.826	.051
0.0996	.535	.060	0.0711	.171	.019
0.1659	.945	.064	0.1245	.336	.028
0.0104	.089	.034	0.2304	.668	.031
0.0207	.123	.038	0.2593	.766	.036
0.0311	.145	.035	0.2872	.853	.036

   The ratio of the optical density of a dilute suspension of powdered Pyrex glass at 278 mμ to that at 320 mμ is 2.1 and the ratio of the optical density of a solution of bovine serum albumin at 278 mμ to that at 320 mμ is 62.72. The percent concentration of the protein solution not exposed to glass is at 278 mμ equal to 0.1486 multiplied by the optical density. Assume the bovine serum albumin to be reversibly adsorbed and calculate the molar free energy of adsorption on Pyrex glass. Make a gaseous plot of your data in the low film pressure region.

   Ans: −7,920 cals/mole

6. Derive the expression for the film pressure of a spread monolayer in the gaseous condition in terms of the film area; interpret the meaning of the constant terms.

   Ans: See text.

7. 0.0134 mg. of a protein was spread on 5 percent ammonium sulfate and the surface pressure of the film in dynes per centimeter was measured as a function of the area of the film in square centimeters with the following results at 25°C.

Film Pressure	*Film Area*	*Film Pressure*	*Film Area*
0.065	322	0.234	182
0.072	280	0.292	175
0.092	252	0.352	168
0.105	224	0.494	161
0.163	196		

Calculate the molecular weight of this protein as well as its gaseous area in square meters per milligram.

Ans: Mol. wt. 34,900; area 1.03 M^2/mg.

CHAPTER 10

Viscosity and the Flow of Liquids

There is considerable ebb and flow of aqueous solutions and suspensions in living tissue with important physiological implications. The viscosity of solutions has also been related to molecular dimensions of the solute and its study can yield valuable information about the size and shape of molecules. The methods available for its measurement are remarkably simple—almost too simple, because they are apt to be abused.

Viscosity is the resistance to flow and is expressed as the force required to maintain unit velocity between two plates of unit area and unit distance apart. Examine Figure 1.

The force, F, required to maintain a velocity u between two parallel plates each of area A, d distance apart in a liquid of viscosity η is given by

$$F = \frac{Au\eta}{d} \qquad 1$$

When A and d are unity

$$\eta = \frac{F}{u} \qquad 2$$

The force in dynes required to maintain unit velocity difference between the plates is known as a poise. The dimensions of the coefficient of viscosity, η, can be obtained from eq 1 and are equal to $Mt^{-1} L^{-1}$. The term kinematic viscosity is occasionally used; it is equal to η/ρ where ρ is the density of the liquid. If the viscosity is independent of the applied stress, the liquid is said to be Newtonian.

The energy required to transfer a volume, V, of liquid from one location to another at constant pressure, P, is evidently PV and if there has been no change in potential energy of the system this energy, PV, appears as heat. The coefficient of viscosity is a measure of the rate at which it is necessary to supply energy to a system to maintain a given velocity of flow; it is also a measure of the heat production incidental to flow.

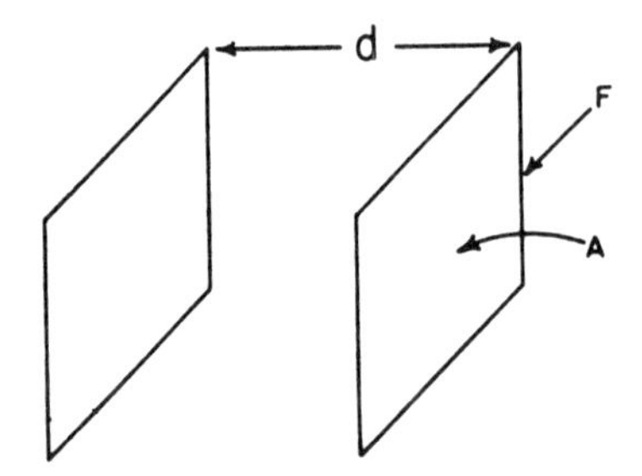

Fig. 1. Motion of two parallel plates of unit area unit distance apart at unit velocity as the result of the applied force, F.

There are a number of theories of liquid viscosity; principal among these are the hole theory and the relaxation theory. The hole theory (Erying) proposes that for a liquid to flow holes must be available for the molecules to move into. The production of such holes is akin to vaporization; accordingly, there should exist a relation between the energy of activation of the flow rate and the heat of vaporization of the liquid. The flow of a liquid is regarded as a rate process, and from its dependence on temperature the energy of activation is calculated in the usual manner (see Chapter 15, Reaction Kinetics and Enzyme Activation). The relaxation theory regards the flow of a liquid in the sense of a deformation with a very short relaxation time.

The viscosity of liquids containing molecules capable of intermolecular hydrogen bonds decreases strongly with increasing temperature with a large energy of activation. The energy of activation consists not only of the energy required to disrupt Van der Waal interactions but also that required to break hydrogen bonds. It is the presence of a network of hydrogen bonds that accounts for the very high viscosities of such compounds as glycerol and water.

Table 1 shows the coefficients of viscosity of several common liquids.

TABLE 1. VISCOSITIES OF SOME COMMON LIQUIDS EXPRESSED IN POISES.

Liquid	*Viscosity*	*Temp.* °C
Benzene	0.00763	10
	0.00654	20
Heptane	0.00416	20
	0.00341	40
Methanol	0.00596	20
	0.00456	40
Ethanol	0.01752	10
	0.01716	20
N-Butanol	0.02948	20
	0.01782	40
Glycerol	25.18	8.1
	13.87	14.3
	8.30	20.3
	4.94	26.5
Water	0.017916	0
	0.015192	5
	0.013069	10
	0.011382	15
	0.010020	20
	0.008903	25
	0.007975	30
	0.007195	35
	0.006532	40
	0.005963	45

Flow in a Capillary. A convenient point of departure for the consideration of the flow of liquids in capillaries is Poiseuille's law. It may be recalled that an equation for the flow of liquids based on dimensional analysis was presented in Chapter 1, Mathematical Review. At this time, Poiseuille's law will be derived from elementary hydrodynamics.

Imagine a capillary of radius, r, through which is flowing a liquid of viscosity, η, under a pressure head, P. It is assumed that the flow is laminar or non-turbulent and the velocity of flow at the wall of the capillary is zero. It is required to calculate the linear velocity of the liquid at any distance normal to the wall of the capillary. The first step is to calculate the velocity of flow at a distance $r - r_1$, from the wall of the capillary where r_1 is the radius of a moving cylinder of liquid. The situation is diagramed in Figure 2.

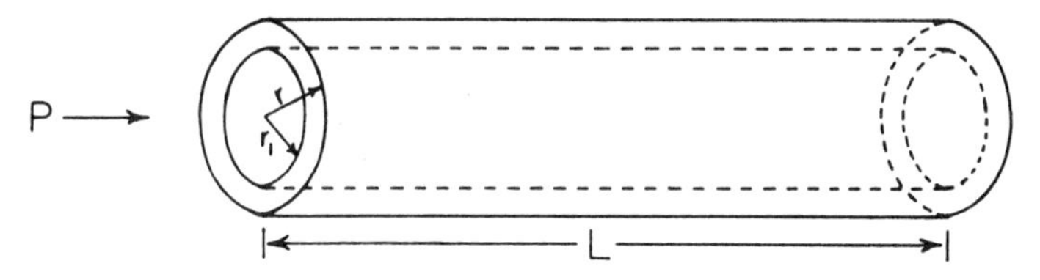

Fig. 2. Capillary showing cylinder of moving liquid.

From eq. 1, the velocity difference Δu between the wall and the layer at distance r_1 from the center of the capillary is very nearly

$$\Delta u = \frac{F\ (r - r_1)}{2\pi r_1 L\eta} \qquad 3$$

where L is the length of the capillary. F, the force, can be replaced by $\pi r^2 P$ and eq. 3 then becomes

$$\Delta u = \frac{r^2 P(r - r_1)}{2r_1 L\eta} \qquad 4$$

As the distance $r - r_1$ is made smaller and smaller eq. 4 becomes more and more exact. Finally, when $r - r_1$ is replaced by dr and Δu by du, it is exact. Under these conditions r_1 can be set equal to r and eq. 4 can be written

$$\frac{du}{dr} = \frac{rP}{2L\eta} \qquad 5$$

In the derivation of eq. 5 and purely for the sake of clarity, the velocity gradient in the immediate neighborhood of the wall of the capillary has been calculated. It is, however, possible to calculate by the same procedure the velocity gradient du/dr between any two cylindrical layers of liquid dr distance apart anywhere in the capillary. Equation 5 is, therefore, the velocity gradient anywhere in the capillary. The integration of eq. 5 between the limits r_1 and r gives

$$u = \frac{P}{4L\eta}(r^2 - r_1^2) \qquad 6$$

From eq. 6 the plot of u against r should give a parabolic velocity profile. Such a plot is shown diagrammatically in Figure 3A.

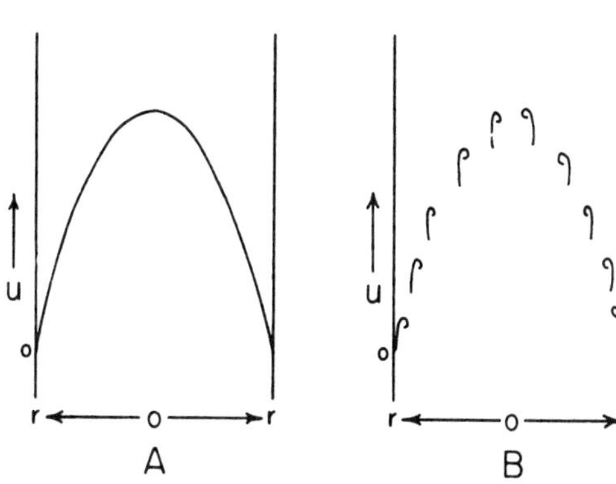

Fig. 3. Flow gradients in a capillary. *A*, Laminar, non-turbulent flow. *B*, Turbulent flow.

A more useful relationship than the linear velocity as a function of the radius is the total volume flow through a capillary. The volume of liquid flowing through any cylindrical volume element is evidently equal to the cross sectional area of the volume element, i.e., $2\pi r dr$ multiplied by the velocity of flow at r or $2\pi u r dr$. Then the rate of flow through the capillary in cubic centimeters per second, V, is

$$V = \int_0^r \frac{\pi P r^3 dr}{2L\eta} \qquad 7$$

The integration of eq. 7 gives

$$V = \frac{\pi P r^4}{8\eta L} \qquad 8$$

Equation 8 is the desired statement of Poiseuille's law.

The plot of the rate of flow of a liquid through a capillary as a function of the pressure difference between the ends of the capillary is diagramed in Figure 4.

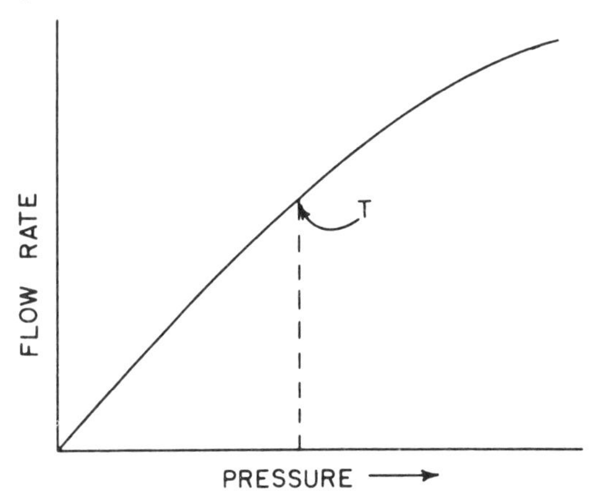

Fig. 4. Diagrammatic plot of flow rate of a liquid through a capillary as a function of the applied pressure showing the onset of turbulence at T.

At a pressure below the onset of turbulence and for long tubes, Poiseuille's law holds exactly. Above a certain critical rate of flow, the flow gradients break into eddies, the flow is no longer laminar and there is mixing between the planes of flow (see Fig. 3B) and the relation between volume flow and pressure is no longer linear (see Fig. 4). Turbulence is characterized by increased heat production for a given rate of flow and by the appearance of noise.

The critical pressure at which turbulent flow begins depends on the geometry of the system as well as on the density and viscosity of the liquid. The ratio of the inertial to the frictional force is $\rho u a/\eta$ where u is the fluid velocity and a is a characteristic linear dimension of the system. For capillaries, a is taken as the radius of the capillary. The expression $\rho u a/\eta$ is called the Reynolds number and is denoted by R. It is dimensionless and is independent of the units used. For a capillary and with the use of eq. 8 the Reynolds number is

$$R = \frac{\rho r^3 P}{8\eta^2 L} \qquad 9$$

If R for a capillary exceeds about 2,000 turbulence will be observed.

Poiseuille's law is still valid in a capillary with elastic side walls but with important complications. Under a constant applied pressure, the capillary assumes the shape of an elongated funnel due to the stretching of the walls as the result of the pressure. Since the liquid is incompressible, the rate of volume flow is uniform along the length of the capillary. Obviously, therefore, the linear velocity of the liquid increases towards the distal end of the capillary; the radius of the capillary decreases as the pressure drops along the capillary. Further, the rate of volume flow will no longer be a linear function of the applied pressure as required by eq. 8; the value of r increases as the pressure increases.

The situation becomes still more complex when the hydrostatic pressure is not constant but varies in pulses as, for example, in the great aorta because of the beating of the heart. Figure 5 shows the variation of the blood pressure in an artery with time.

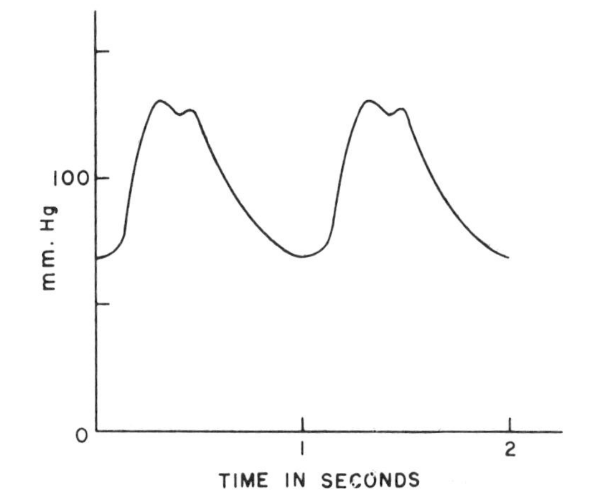

Fig. 5. Blood pressure in an artery as a function of time.

The velocity of the pulse wave in a homogeneous incompressible fluid in a cylindrical tube with elastic walls is proportional to the square root of the product of the modulus of elasticity of the wall, the wall thickness and the radius of the distended tube, and inversely proportional to the radius of the tube when the excess pressure is zero (the smaller the blood vessel, the greater the velocity of the pulse wave). In principle, therefore, the elasticity of a blood vessel can be determined by measuring the velocity of the pulse wave. Since the wall of the blood vessel is not only elastic but likewise viscous, there should be a phase lag between the pressure wave and the pulse wave. It should, therefore, be possible to calculate the frictional coefficient of the wall of the blood vessel. The actual situation is exceedingly complex —enough so to occupy the attention of many skillful and talented people for a long time.

Methods. The determination of the rate of flow through a rigid capillary is the basis for an important method for the measurement of viscosity. In this method, the time of outflow of a known volume of liquid through a capillary under the influence of gravity is determined. Such an apparatus is diagramed in Figure 6, and is known as the Ostwald viscometer.

A definite volume of liquid is placed in the viscometer, and the level of the liquid is drawn above the top mark of the bulb by suction (see Fig. 6). The liquid is allowed to flow out freely, and the time, t, required for the level to drop from the upper mark to the lower mark is measured. The relation between the viscosity and the time of outflow at constant pressure is

$$\eta = \frac{\pi P r^4 t}{8V(L + nr)} - \frac{m\rho V}{8\pi(L + nr)t} \qquad 10$$

where n is a correction for the acceleration of the liquid in the central part of the capillary as it leaves the bulb, and m is a kinetic correction related to the flow of liquid after it leaves the capillary. The other symbols have the same meaning as in

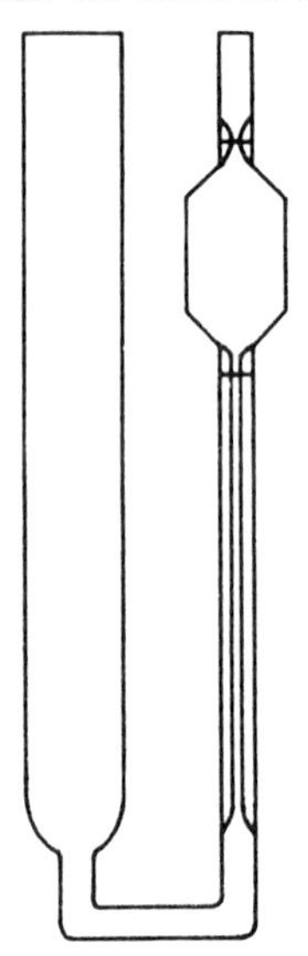

Fig. 6. Ostwald viscometer.

eq. 8. If L and t are large enough, the correction n and m can be neglected and the viscosity is then equal to $k\rho t$ where k is the viscometer constant, ρ is the density of the liquid and t is the time.

The viscometer is calibrated with water or some other appropriate liquid whose viscosity is exactly known. If the kinetic correction can be neglected, the relative viscosity of the liquid under investigation is given by the relation

$$\eta_r = \frac{\eta}{\eta_0} = \frac{\rho t}{\rho_0 t_0} \qquad 11$$

where η, ρ, and t are the coefficient of viscosity, the density and the time of outflow of the liquid under examination respectively. η_0, ρ_0 and t_0 are the corresponding quantities for the standard liquid.

The most reliable method to determine if the kinetic corrections can be neglected is to measure the viscosity of two liquids whose viscosity is known. If the two measured viscosities bear the correct ratio to each other, and if the unknown viscosity has approximately the same value as that of the two known solutions the kinetic corrections are unnecessary. Actually, kinetic flow errors begin to be appreciable at much lower pressures than does turbulence.

In addition to the kinetic correction mentioned above, the Ostwald viscometer is subject to three other errors: (1) drainage error, (2) working volume correction, and (3) surface-tensions effects. The ideal design of a viscometer strikes a compromise with all these disturbing factors. The drainage error is negligible except for very viscous liquids or for viscometers with very short times of outflow; lack of drainage amounts to decreasing the working volume. The working volume correction refers to error inherent in placing a precise volume of liquid in the viscometer. Ordinarily the error involved here is much smaller than the other types of errors. The surface-tension correction may be important for liquids where surface tension differs greatly from that of the calibrating liquid. The pull upward due to surface tension is greater per unit area the smaller the diameter of the tube. In conclusion of this discussion, it is well to emphasize that two absolute necessities for satisfactory viscosity measurements are cleanliness of the viscometer and an accurate temperature control (the viscosity of water decreases about 2 percent for each degree rise in temperature).

Many ingenious modifications of the Ostwald viscometer have been developed including viscometers with constant pressure heads.

The Couette viscometer employs a rotating cylinder containing the liquid under investigation. A smaller cylinder is suspended in the liquid by means of a fine wire. The torque produced on the suspended cylinder by the rotation of the outer cylinder is recorded by an optical lever. The type of flow in a Couette viscometer is simpler than that in an Ostwald viscometer; the flow lines are circular with velocity being zero at the surface of the stationary cylinder and equal to that of the rotating cylinder at its surface. The rate of shear between the two cylinders is given by the expression

$$G = \frac{2\omega r_i^2}{r_0^2 - r_i^2} \qquad 12$$

where r_i and r_0 are the radii of the inner and outer cylinders respectively and ω is the angular velocity in radians per second. When d, the distance between the two cylinders, is small relative to the radii of the cylinders, eq. 12 reduces to

$$G = \frac{r_i\omega}{d} \qquad 13$$

and the velocity gradient is effectively constant from the inner to the outer cylinder. Very small flow gradients can be achieved with ease. The viscosity is directly proportional to the torque (the angular displacement of the inner cylinder). Unfortunately, the construction of a Couette viscometer is a job for the highly skilled machinist.

A variation on the Couette viscometer was introduced by Zimm and Crothers and later modified by Gill and Thompson. The modified viscometer is shown diagrammatically in Figure 7.

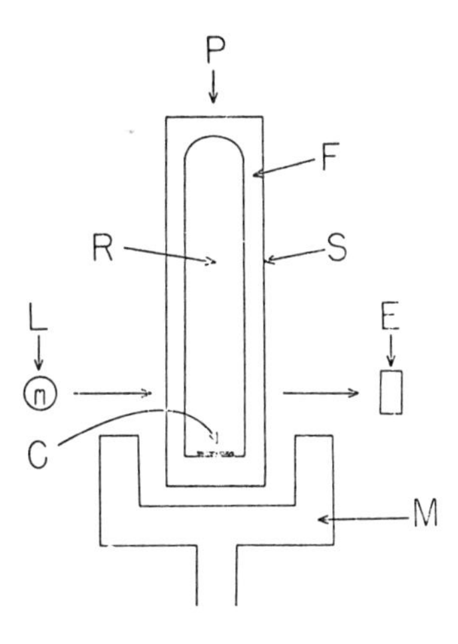

Fig. 7. Z-C-G-T viscometer. P, pressure source; S, stator; R, rotor; L, light beam; E, electric eye; C, non-ferrous conducting ring; M, rotating permanent magnet; F, liquid whose viscosity is being measured. (See S. J. Gill and D. S. Thompson: Proc. Natl. Acad. Sci. *57:* 562, 1967.)

The rotor, a glass cartesian diver, is completely submerged in the liquid whose viscosity is being measured. The rotor is maintained at a constant depth by a servomechanism actuated by an electric eye which observes the position of the rotor and applies or releases pressure to keep the cartesian diver (rotor) at the desired level. The rotor is caused to rotate relative to the outer stator by a rotating magnet which induces eddy currents in a non-ferrous conducting ring (copper or aluminium). The rate of rotation of the rotor relative to that of the rotating magnet is the information which permits the calculation of the viscosity of the liquid. The merits of the Z-C-G-T viscometer are that small volumes of solution can be used (less than one ml), and the instrument is capable of very low shear rates (as low as 10^{-2} per second; shear rates in a capillary viscometer are usually about 10^3 per second);

further the shear rate can be varied at will. The low shear rate is an especially attractive feature in investigations on solutions of elongated molecules such as DNA which tend to orient in higher flow gradients giving an erroneous notion of the true viscosity.

Suspension of Spherical Particles. It is evident that if spherical particles whose size exceeds those of the solvent by a significant amount are suspended in a liquid the streamlines of flow in the liquid will be disrupted. Such disruption will lead to a higher measured viscosity. Einstein calculated the extra energy which must be supplied to sustain the disruption of the flow lines and concluded that, for very dilute suspensions, the viscosity of the suspension would be given by the relation

$$\eta = \eta_0 (1 + 2.5\phi) \tag{14}$$

where η_0 is the viscosity of the solvent and ϕ is the volume fraction occupied by the suspended particles (the ratio of the volume of the particles to the total volume of the suspension). It has been demonstrated, principally by the measurement of the viscosity of suspensions of spherical glass particles in liquids of increased density that the coefficient 2.5 as given in eq. 14 is in fact correct.

The Einstein equation neglects the influence of the interaction of the suspended particles with each other and when this interaction is considered the equation

$$\eta = \eta_0(1 + 2.5\phi + 7.35\phi^2 + \cdots) \tag{15}$$

is obtained. The experimental equation for a suspension of glass spheres is

$$\eta = \eta_0(1 + 2.5\phi + 7.17\phi^2 + 16.2\phi^3 + \cdots) \tag{16}$$

It is a general observation that the fluidity of a solution (reciprocal of the viscosity) is a linear function of concentration and can, in general, be expressed by the relation (see, for example, Fig. 8),

$$f = \frac{1}{\eta} = f_0(1 - KC) \tag{17}$$

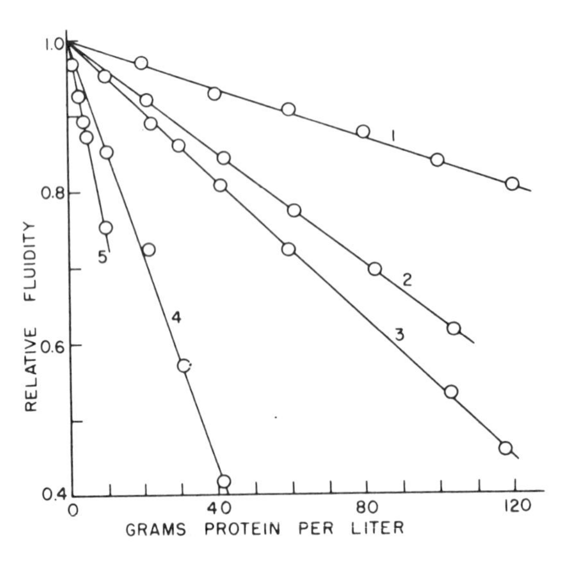

Fig. 8. Plot of relative fluidity of protein solutions against proteins concentration. Curve 1, Egg albumin; curve 2, CO-hemoglobin; curve 3, horse serum albumin; curve 4, horse serum globulin fraction; curve 5, bovine fibrinogen. (Adapted from H. P. Treffers, E. C. Bingham and R. R. Roepke: J. Amer. Chem. Soc. *62:* 1405, 1940; *64:* 1204, 1942.)

where f is the fluidity of the solution and f_0 that of the solvent; K is a constant and C the concentration of the solute. Equation 17 can be rewritten as

$$\eta = \frac{\eta_0}{(1 - KC)} \tag{18}$$

The expansion of eq. 18 leads to

$$\eta = \eta_0(1 + KC + K^2C^2 + K^3C^3 + \cdots) \tag{19}$$

If C is identified with ϕ and K is assigned the value of 2.5, eq. 19 becomes for spherical particles

$$\eta = \eta_0(1 + 2.5\phi + 6.25\phi^2 + 15.6\phi^3 + \cdots) \tag{20}$$

Compare eq. 20 with eq. 16.

Expression for the Viscosity of Suspensions. The relative viscosity is defined as

$$\eta_r = \frac{\eta}{\eta_0} \tag{21}$$

where η is the viscosity of the suspensions (solution) and η_0 is that of the pure solvent.

The specific viscosity, η_s, is defined as $\eta_r - 1$ and the reduced viscosity is η_s/ϕ or η_s/C where ϕ is as before the volume fraction occupied by the suspension, C is the concentration in grams of solute per milliliter. The intrinsic viscosity $[\eta]$ is equal to the reduced viscosity as the quantity of solute per unit volume approaches zero. There will be two intrinsic viscosities: volume intrinsic viscosity and concentration intrinsic viscosities. The two intrinsic viscosities are expressed as follows:

$$[\eta]_\phi = \left(\frac{\eta_s}{\phi}\right)_{\phi \to 0} \quad \text{and} \quad [\eta]_C = \left(\frac{\eta_s}{C}\right)_{C \to 0} \tag{22}$$

For a suspension of spheres (irrespective of their size) $[\eta]/\phi$ is equal to 2.50. Huggins relates reduced viscosity to concentration by

$$\frac{\eta_s}{C} = [\eta] + k[\eta]^2C \tag{23}$$

where k is a constant to be determined by experiment.

There is a certain element of confusion connected with the above expressions. To start with, neither the relative viscosity, the specific viscosity, the reduced viscosity nor the intrinsic viscosity has the dimensions of viscosity. Secondly, in many papers on polymer chemistry the concentration is expressed in grams per 100 ml. so that their reduced and intrinsic viscosities have to be multiplied by 100 to conform to the above definitions. These two sources of confusion are man-made; the other ambiguity, unfortunately, is much more serious and has a physical origin: except in special cases and mostly of minor interest there is no way of determining the hydrodynamic volume of the solute and of calculating the volume ratio, ϕ. The solute becomes hydrated in solution and, accordingly, occupies a larger volume in solution than would be estimated from its dry density. Not knowing the actual hydrodynamic volume of the solute, the most expedient procedure is to multiply the grams of solute in one ml. by its partial specific volume $\overline{V}_2$ to give the partial volume of the solute in solution. The partial specific volume intrinsic viscosity becomes

$$[\eta]_{\bar{V}_2} = \left(\frac{\eta_s}{C\bar{V}_2}\right)_{C \to 0} \qquad 24$$

C being expressed in grams of solute per milliliter of solution. In this expression the volume increase of the solute caused by hydration has been neglected.

Viscosity and Particle Asymmetry. It was realized rather early that a relation must exist between the asymmetry of the suspended particles and the viscosity of a suspension. The particles are turning and twisting in Brownian motion, and they thus appear to occupy a larger volume than they actually do. The twisting of the particle interrupts the streamlines of flow and requires additional expenditure of work to maintain a given velocity of flow. This work appears as an increase in the viscosity.

The theoretical relation between particle asymmetry and viscosity is quite complicated and has been approached by a number of workers. Simha has apparently succeeded in expressing this relation for rigid prolate (rotation about the major axis) and for rigid oblate (rotation about the minor axis) ellipsoids of revolution more completely than anyone else. Figure 9 shows a graph of the ratio of major to minor axes for ellipsoids of revolution against the intrinsic viscosity.

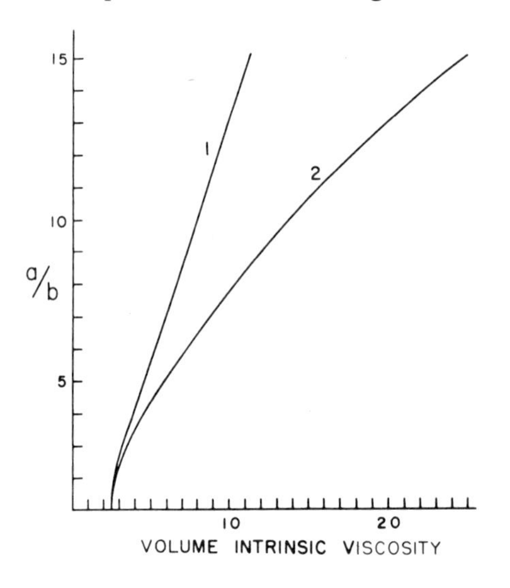

Fig. 9. Plot of Simha's viscosity equation showing relation between particle asymmetry and intrinsic viscosity. Curve 1, Oblate ellipsoids of revolution: curve 2, prolate ellipsoids of revolution.

The problem of the viscosity of long, flexible chain molecules has been treated considering the deformation produced by flow gradients acting on the flexible molecules. Most biopolymers contain so many centers capable of hydrogen bond formation that the interaction of different parts of the same molecule with each other probably produces rather rigid, stiff structures in solution. It would appear that Simha's approach to asymmetry and viscosity for this class of molecule is the more realistic. It should, however, be added that the actual shapes assumed by protein and other biopolymers in solution is a difficult matter to resolve. For molecules of low asymmetry such, for example, as bovine serum albumin, the situation is at present very nearly hopeless. However, as the molecule becomes more and more asymmetric, such, for example, as those of myosin, fibrinogen, tobacco mosaic virus, undegraded deoxyribonucleic acid (DNA), the individual irregularities of the molecules become less important in their hydrodynamic behavior; such molecules have very large intrinsic viscosities and can be treated successfully as very elongated prolate ellipsoids of revolution.

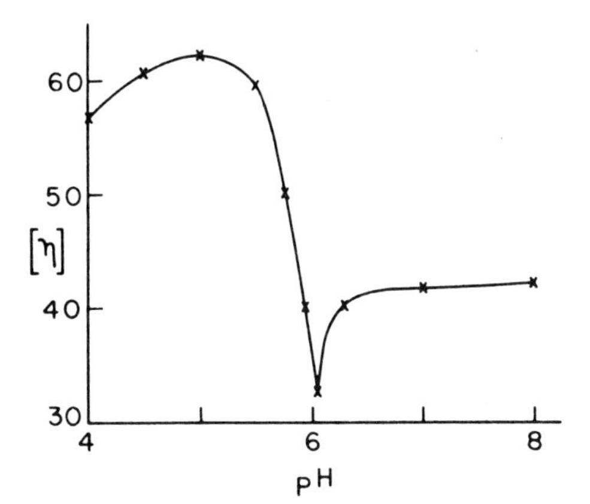

Fig. 10. Intrinsic viscosity of poly-L-glutamic acid as a function of the pH in 0.2 M NaCl-dioxane (2:1) at 25°. (Data of P. Doty, A. Wada, J. T. Yang and E. R. Blout: J. Polymer Sci. *23:* 851, 1957.)

The Simha equation has also been very helpful in investigations on the conformation of synthetic polypeptides in solution. A high molecular weight peptide in solution can exist as a disordered coil or in the form of an α-helix. The α-helix gives a stiff rod-like molecule with a much larger intrinsic viscosity than does the disordered coil for the same molecular weight. For example, shown in Figure 10 is a plot of the intrinsic viscosity of poly-L-glutamic acid with an average molecular weight of 34,000 as a function of pH. The polypeptide exists in the form of an α-helix below pH 5.5 and as a disordered coil above a pH of 6.5.

Viscosity and Molecular Weights. The intrinsic viscosity of a spherical unhydrated molecule is clearly independent of molecular weight but for linear molecules either flexible or rod-like dependence should exist. A discussion of the Simha relation for rod-shaped (very elongated prolate ellipsoids of revolution) has already been given. In general, for linear molecules the relation between viscosity and molecular weights is

$$[\eta] = KM^a \qquad 25$$

where M is the weight average molecular weight of the linear polymer and K and a are constants to be determined by experiment for a given polymer and solvent.

For a spherical molecule, a is zero, and, if the sphere is unhydrated with a density of unity, K is 2.5. For ideal random coils, a is 0.5 and K depends on the nature of the molecules, but is independent of the length of the chain if the chain is long enough. a is about 1.8 for long rigid rod-like molecules (Simha's relation). Table 2 shows a comparison of the molecular weights and the intrinsic viscosities of a series of native proteins with the corresponding values for the same proteins after treatment with 6 M guanidine·HCl in 0.5 M mercaptoethanol (mercaptoethanol ruptures the disulfide bonds in the proteins). Guanidine·HCl is a powerful denaturing agent causing the protein molecules to swell, liberating flexible peptide chains. As can be seen from Table 2, the native protein molecules have low intrinsic viscosities indicating compact globular structures, and further there is no relation between their molecular weights and intrinsic viscosities. Upon denaturation, the intrinsic viscosities are greatly increased and there is a correspondence between the molecular weights and intrinsic viscosities.

The relation between intrinsic viscosities and molecular weights of flexible chain polymers (peptide chains) and the molecular weights of molecular rods (with some flexibility) is shown more explicitly in Figure 11, where the logarithm of the intrinsic viscosity of proteins denatured by guanidine·HCl is plotted against the logarithm of the molecular weight. The slope of this line is 0.66 indicating the pep-

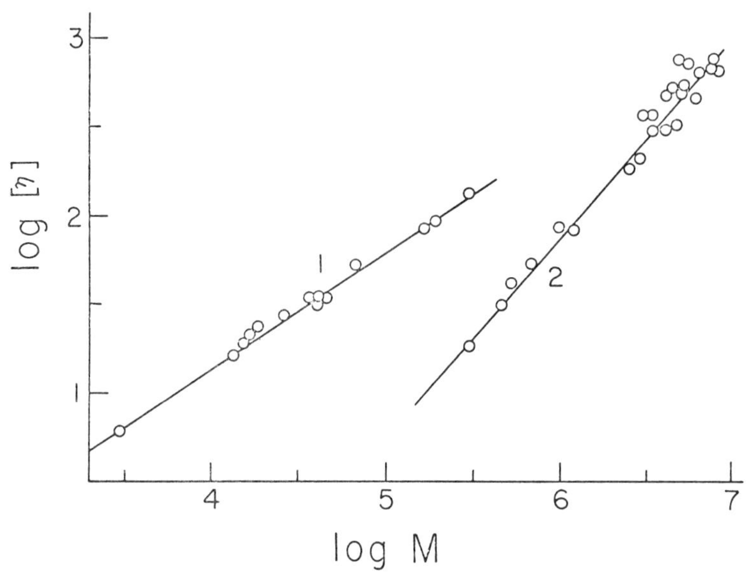

Fig. 11. Plots of logarithm [η] against logarithm molecular weights. Curve 1, proteins denatured with 6 M guanidine. HCl (refer to A. H. Reisner and J. Rowe: Nature *222:* 558, 1969). Curve 2, native DNA (refer to J. Eigner and P. Doty: J. Mol. Biol. *12:* 549, 1965).

tide chains are approaching the condition of random coils. Also shown in Figure 11 is a corresponding plot for native DNA molecules from a variety of sources; the slope of this plot is 1.13.

The above dependence of the intrinsic viscosity on molecular weight suggests that viscosity would be of use in following the kinetics of the depolymerization of polymers: their expectation has been fulfilled. For example, viscometric methods are

TABLE 2. COMPARISON OF THE MOLECULAR WEIGHTS AND THE INTRINSIC VISCOSITIES OF A SERIES OF NATIVE (N) AND OF DENATURED (D) PROTEINS. (DATA OF F. J. CASTELLINO AND R. BARKER: *Biochem. 7:* 2207, 1968.)

	*Molecular Weight**		[η]	
Protein	*N*	*D*	*N*	*D*
Serum albumin (bovine)	68,300	67,800	3.6	51.3
Egg albumin	44,600	46,500	4.4	34.6
Alcohol dehydrogenase (horse liver)	86,000	40,800	3.6	34.1
Enolase (rabbit muscle)	82,600	36,500	3.7	33.0
Methemoglobin (bovine)	63,700	15,800	3.4	19.8
Lactate dehydrogenase (bovine heart)	136,300	36,200	3.8	32.4
Aldolase (rabbit muscle)	156,500	42,400	3.4	35.4

* From osmotic pressure measurements.

used to measure the action of the enzyme hyaluronidase on hyaluronic acid. Hyaluronic acid is an important biopolymer containing equal amounts of N-acetylglucosamine and glucuronic acid residues; the molecular weight of the polymer is in the range of 200,000 to 400,000. It is highly viscous and serves as a cementing substance for body cells forming barriers preventing the spread of infectious agents. Viscosity measurements are also used as an assay method for ribonuclease, the enzyme which depolymerizes ribonucleic acid (RNA).

Viscosity has been employed to measure the rate of hydrolysis of synthetic polypeptides. Shown in Figure 12 is a plot of the reciprocal of the reduced viscosities of poly-L-glutamic acid which had been subject to hydrolysis by the enzyme papain,

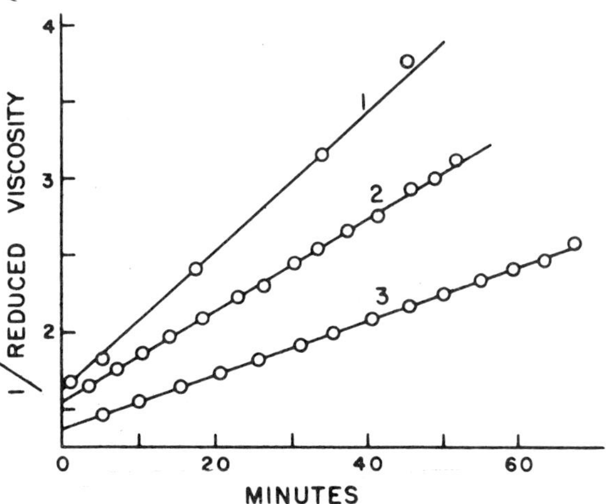

Fig. 12. Plot of the reciprocal of the reduced viscosity of poly-L-glutamic acid against time of digestion by 1.0 x 10^{-7} *M* papain at 25° and pH 5.05: Curve 1, 1.6 *g*/l substrate; curve 2, 6.6 *g*/l substrate; curve 3, 16.4 *g*/l substrate. (W. G. Miller: J. Amer. Chem. Soc. *83*: 259, 1961.)

against time. Knowing the relation between viscosity and the molecular weight of this polymer, the change of the weight average molecular weight was calculated. For random degradation of a random distribution of molecular weights, the weight average molecular weight should drop to one half of its initial value when, on the average, one bond per molecule has been broken. It is, therefore, possible to calculate the number of bonds broken during the early part of the reaction. The viscometric method provides means for kinetic studies of high molecular weight linear polymers which are most sensitive in the initial part of the reaction where chemical methods are apt to be least sensitive.

Hydrodynamic Hydration. As mentioned earlier, reduced and intrinsic viscosities, when expressed in terms of volume fractions, both require a knowledge of the volume occupied by the solute. When dissolved, a solute typically becomes hydrated and the hydrodynamic volume is larger than the volume of the anhydrous material. It is probable that any coherent, non-swollen large molecule in solution has one layer of water molecules attached to its surface. It is also likely that hydrodynamic shear begins immediately beyond this first layer of water molecules. The area occupied by one water molecule on a surface is about 10.9 sq. A-units; it is, accordingly, possible to estimate a minimum hydration for spherical protein molecules as a function of the molecular weight of the protein. If there are hills or valleys on the surface of the protein molecule thus increasing the estimated molecular surface, the hydration will, of course, be greater. Also if the protein departs from a sphere, the hydration will be greater than estimated. Figure 13 shows the estimated spherical hydration as a function of the molecular weight.

For rod-shaped molecules and neglecting end effects, the percent hydration should be independent of the molecular weight and depend only on the radius of the rod. For an α-helix of average dimensions, the hydration based on a monolayer of water would be about 0.55 gram of water per gram of peptide. If two or more peptide chains are associated together, the percent hydration would naturally be less. Hydration and molecular volumes enter into problems of viscosity, diffusion and less directly into rate sedimentation; additional comment will be made on this topic in Chapters 11 and 14 dealing with diffusion and sedimentation respectively.

Non-Newtonian Viscosity. Simha's equation for the relation between particle asymmetries and viscosity postulates complete Brownian motion of the suspended particles with no orientation of the particles in the streamlines. If the particles are sufficiently elongated and if the rate of shear is large enough, the particles spend more and more time with their long axes parallel to the streamlines as the velocity gradient of flow is increased. The oriented particles interrupt the streamlines less than do randomly oriented particles and accordingly the viscosity of such suspen-

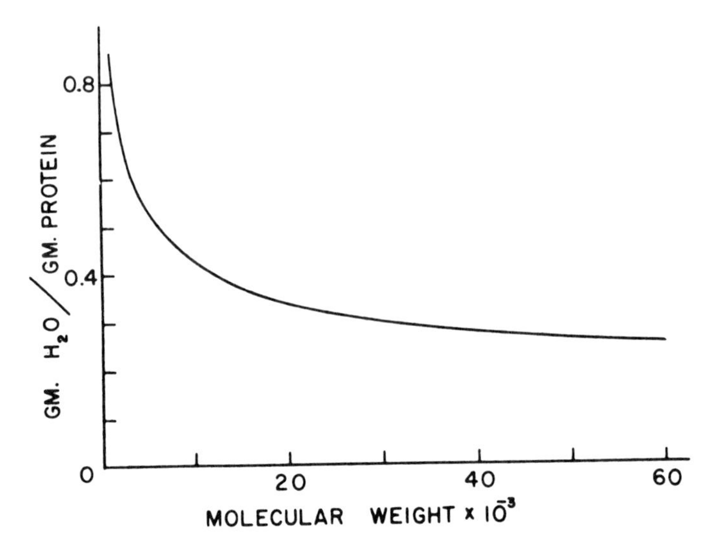

Fig. 13. Estimated hydrodynamic hydration of spherical protein molecules as a function of the molecular weight of the protein, protein density assumed to be 1.33.

sions decreases as the rate of flow of the suspension is increased. This constitutes a method for the evaluation of the rotary diffusion coefficient (see Chapter 11, Diffusion). Unfortunately, the solute particles have to be very elongated before the method becomes useful for flow gradients encountered in ordinary viscometry.

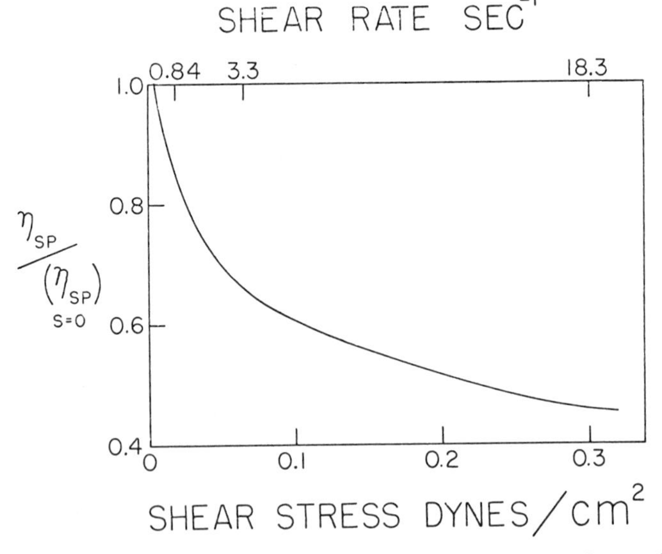

Fig. 14. Ratio of the measured specific viscosity to the specific viscosity at zero stress as a function of the shear stress. 33 μg Na salt DNA per ml. (Adapted from B. H. Zimm and D. M. Crothers: Proc. Natl. Acad. Sci. *48:* 905, 1962.)

Figure 14 shows the response of the viscosity of a solution of DNA (33 μg Na salt per ml) to the shear stress of the solution; the specific viscosity of this solution extrapolated to zero stress is 1.730.

There may be sufficient interaction between the suspended particles to lead to a gel network in the suspension. Such a network tends to be disrupted in the streamlines of flow and, accordingly, the viscosity decreases with increasing velocity of flow; this is an example of structural viscosity.

The electroviscous effect will be discussed in Chapter 13.

GENERAL REFERENCES

Reiner, Markus: Deformation, Strain and Flow: an Elementary Introduction to Rheology. H. K. Lewis and Co., London, 1960.

Tanford, Charles: Physical Chemistry of Macromolecules. John Wiley and Sons, New York, 1961.

Yang, J. T.: The viscosity of macromolecules in relation to molecular conformation. Adv. Prot. Chem. *10:* 323, 1961.

PROBLEMS

1. Define the terms relative viscosity, specific viscosity and volume intrinsic viscosity. How may each of these be determined experimentally?

   Ans: See text.

2. Egg albumin was dissolved in a series of 0.03 M phosphate buffers at pH 7.0. The times of outflow in an Ostwald viscometer as well as the relative densities of the protein solutions were measured at 25° with the following results:

Grams Protein per 100 ml.	*Outflow in Seconds*	*Relative Density*
0.00	96.17	1.0000
0.53	98.12	1.0014
1.62	102.26	1.0041
3.24	108.58	1.0080
4.85	115.57	1.0120
6.47	123.74	1.0158

Assume a hydrated density of 1.26 for the protein. Calculate the volume intrinsic viscosity and axial ratio for a prolate ellipsoid of revolution.

Ans: $[\eta]_\phi$ is 5.13 and a/b is 4.2.

CHAPTER 11

Diffusion

Diffusion of substances across membranes, in and out of living cells, is a subject which has received the assiduous attention of biologists; such attention is commensurate with the importance of the topic. Unfortunately, diffusion problems have a way of becoming exceedingly complex and their quantitative expression frequently requires the use of some fairly fancy differential equations. The diffusion of material in solution is closely related to the flow of heat from a warm to a cool region; both processes are irreversible and are accompanied by an increase in entropy.

The thermal motion of molecules in solution is a function of the kinetic energy of the molecules as well as of the frictional forces exerted on the molecules by the medium through which the dissolved molecules are moving. As time passes, the molecule or particle takes a random walk through the solution; it exhibits Brownian movement. As noted in Chapter 6, Biopolymers, the formation of a random coil by a high molecular weight polymer is a closely related problem.

Intuitively, it would be expected that the net rate of transfer of a substance from a concentrated to a more dilute solution would be proportional to the concentration gradient, to the area of the plane through which the transfer is made and to a coefficient. Expressing these intuitive conclusions in the form of an equation gives

$$\text{Flux} = J_1 = \frac{dm}{dt} = -DA\,\frac{dC}{dx} \qquad 1$$

where m is the mass of material being transferred, t is the time, x is the distance in the direction in which diffusion occurs, A is the area of the plane through which the transfer is being made and D is the proportionality coefficient known as the diffusion coefficient; the negative sign in eq. 1 indicates that the substance is moving in the direction of decreasing concentration. Substitution of dimensional quantities into eq. 1 reveals the diffusion coefficient to have the dimensions of area per unit time (D is usually expressed in sq. cm. per second). Equation 1 is an expression of Fick's first law of diffusion and is the basis for the porous disk method for the measurement of diffusion coefficients.

Porous Disk Method. This is a relatively simple technique originally proposed by Northrop and Anson and based on eq. 1. Figure 1 is a diagrammatic sketch of the apparatus in which two chambers containing solutions of a solute at different concentrations are separated by a sintered glass disk of a suitable porosity (usually a No. 4). The two solutions on the two sides of the disk are well stirred and the only concentration gradient occurs in the disk.

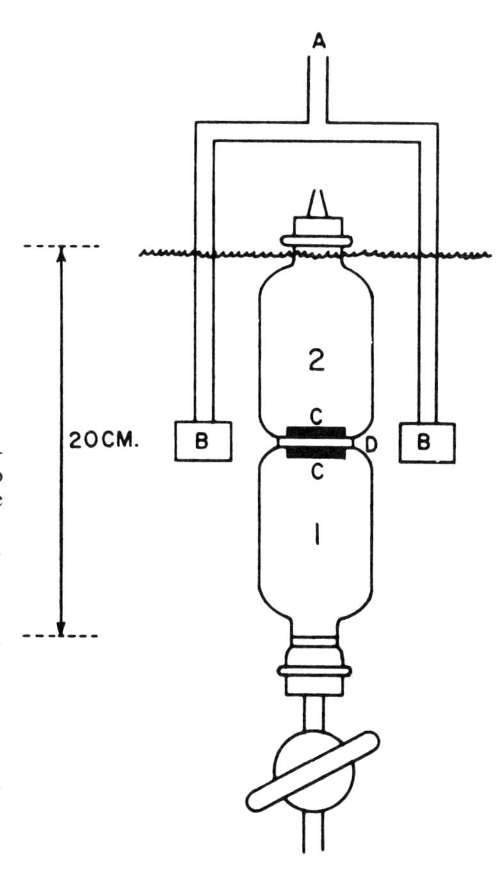

Fig. 1. Porous disk assembly for diffusion coefficients. *A*, Connection to stirring motor; *B*, two permanent magnets; *C*, two magnetic fleas, the lower one floating; *D*, sintered glass disk; 1 and 2 are solution chambers, the lower solution being more concentrated. Apparatus immersed to level shown in constant temperature water bath. (R. H. Stokes: J. Amer. Chem. Soc. *72:* 763, 1950, as modified by C. Nordschow, private communication.)

Provided the concentration of the diffusing substance in the lower compartment (Fig. 1) does not change significantly during the experiment and, further, that pure solvent is placed in the upper chamber at the start, the amount of material diffusing in time, t, is

$$m = \frac{DAtC}{h} \qquad 2$$

where A is the effective cross sectional diffusing area of the glass disk and h is the effective thickness. Neither A nor h can be directly measured. What is done is to measure the rate of diffusion of a substance whose diffusion coefficient is exactly known and obtain a cell constant for the glass disk. The ratio of A to h represents the cell constant. Since m is equal to C_2V_2 where C_2 is the concentration of the diffusing substance in the upper compartment after time, t, and V_2 is the volume of the upper compartment, eq. 2 can be written

$$D = \frac{hC_2V_2}{AtC_1} \qquad 3$$

Equations 2 and 3 suffer from several serious defects. It is best and, in fact, necessary in order to obtain reliable values for D to permit diffusion to proceed until a steady state is reached in the disk. It is also unacceptable to assume the concentration in compartment 1 remains constant during a measurement; the rate of diffusion is proportional to the concentration difference between the two compartments.

One of the defects in eq. 3 can be repaired by setting the flux proportional to the concentration difference between compartment 1 and 2; then from Fick's first law

$$-\frac{dm}{dt} = \frac{V_1 dC_1}{dt} = -\frac{DA}{h}(C_1 - C_2) = \frac{DA\Delta C}{h} \tag{4}$$

and

$$-\frac{dm}{dt} = \frac{V_2 dC_2}{dt} = -\frac{DA}{h}(C_2 - C_1) = -\frac{DA\Delta C}{h} \tag{5}$$

Rearranging eqs. 4 and 5 gives, respectively,

$$dC_1 = -\frac{DA\Delta C}{hV_1}\,dt \tag{6}$$

and

$$dC_2 = \frac{DA\Delta C}{hV_2}\,dt \tag{7}$$

Subtracting eq. 7 from eq. 6 gives

$$dC_1 - dC_2 = d\Delta C = -\frac{DA\Delta Cdt}{hV_1} - \frac{DA\Delta Cdt}{hV_2}$$

$$= -\left(\frac{1}{V_1} + \frac{1}{V_2}\right)\frac{DA\Delta Cdt}{h} \tag{8}$$

Equation 8 can be rearranged to give

$$\frac{d\Delta C}{\Delta C} = -\beta Ddt \tag{9}$$

where β is a constant and is equal to $\left(\frac{1}{V_1} + \frac{1}{V_2}\right)\frac{A}{h}$. The integration of eq. 9 yields

$$\ln\frac{\Delta C_2}{\Delta C_1} = -\beta Dt \tag{10}$$

where t is the elapsed time and ΔC_1 is the concentration difference between compartment 1 and 2 at the start of the experiment and ΔC_2 is the difference at the end of the run.

Equation 10 still suffers from an awkward disability; the diffusion coefficient is not independent of the concentration of the solute and, accordingly, what is obtained by the use of eq. 10 is an average or apparent diffusion coefficient; the concentrations in compartments 1 and 2 change as the experiment proceeds. The true diffusion coefficient can be calculated by methods described in the general reference by Gosting. The diffusion coefficient of non-electrolytes as a function of concentration does not depart greatly from a linear function but the relation for electrolytes is more complex and will receive additional comment in Chapter 12, Ion Transport. Figure 2 shows the ratio of the diffusion coefficient, D, at a finite concentration to the diffusion coefficient extrapolated to infinite dilution, D^0, as a function of the molar concentration of glucose, of urea and of KCl.

It is evident from eq. 10 that knowledge of the actual concentrations is unnecessary; what is needed is the ratio of the concentrations in the two compartments

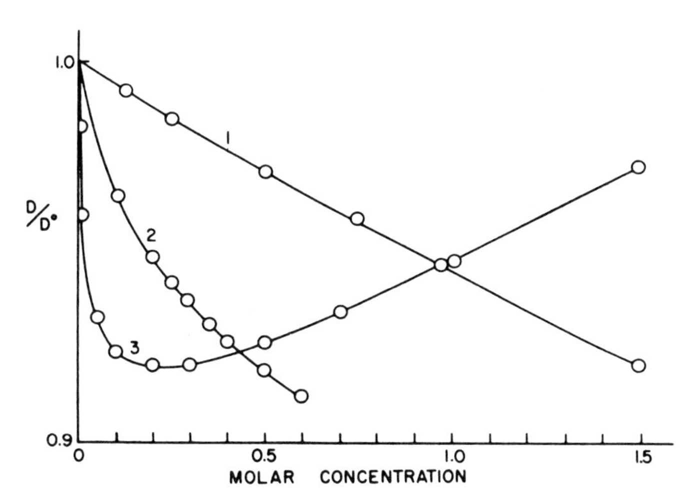

Fig. 2. Ratio D/D° as a function of the molar concentration. D is the diffusion coefficient at the indicated concentration and D° is the diffusion coefficient at infinite dilution. Curve 1, urea, D° is 1.383 x 10^{-5}, cm.^{2}/sec. Curve 2, glucose, D° is 0.678 x 10^{-5} cm.^{2}/sec. Curve 3, KCl, D° is 1.996 x 10^{-5} cm.^{2}/sec. All at 25°.

after a given time. Thus the diffusion coefficient of a biologically active material which is grossly impure can be measured.

An experimental difficulty sometimes encountered with the porous disk method is bacterial contamination, resulting in the destruction of the substrate; diffusion experiments are frequently continued for as long as two or three days.

Fick's Second Law. Equation 1 contains four variables, m, t, C and x, and, accordingly, requires more information than is sometimes available. It is possible to reduce these variables to three and to obtain a more general and useful equation. Consider a volume element in a diffusing boundary whose area is A and whose thickness is δx, x again being the distance in the direction in which diffusion occurs (see Fig. 3).

The rate of entry of the diffusing substance into the volume element is given by eq. 1. The distance δx is assumed to be so small that the concentration gradient remains constant through the volume element in the x-direction. The concentration at the distance $x + \delta x$ will be $C - \delta x dc/dx$ because $-dc/dx$ is the rate of change of the concentration with distance and the gradient multiplied by the distance, δx, gives the change in concentration from position x to $x + \delta x$. Hence the rate of exit of the diffusing substance from the right hand face of the volume element (Fig. 3) is

$$\frac{dm'}{dt} = -DA \frac{d}{dx}\left(C - \delta x \frac{dC}{dx}\right) \qquad 11$$

or

$$\frac{dm'}{dt} = -DA \left(\frac{dC}{dx} - \delta x \frac{d^2C}{dx^2}\right) \qquad 12$$

The rate of increase of the diffusing substance in the volume element is evidently equal to the difference between the rate of entrance and exit and subtracting eq. 12 from eq. 1 gives

$$\frac{dm''}{dt} = DA\, \delta x \frac{d^2C}{dx^2} \qquad 13$$

The rate of increase of the amount of the diffusing substance in the volume element is also dC/dt multiplied by the volume of the element ($A\delta x$) or is $A\delta x dC/dt$.

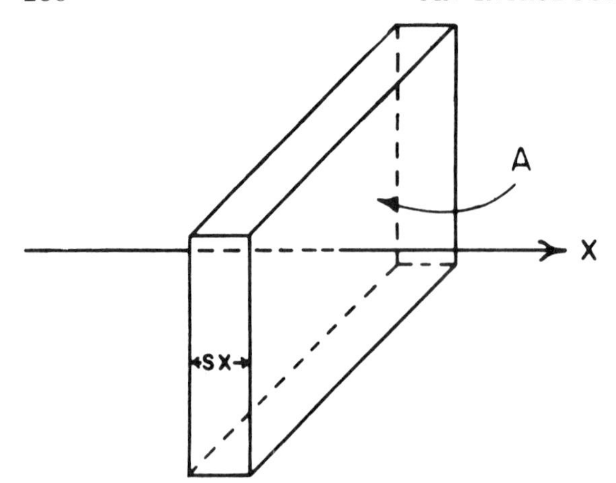

Fig. 3. Volume element of area A and thickness δx through which diffusion is occuring in the x-direction.

Setting the two expressions for the rate of increase equal to each other and rearranging gives

$$\frac{dC}{dt} = D\frac{d^2C}{dx^2} \qquad 14$$

Equation 14 is an expression of Fick's second law of diffusion. The corresponding equation in three dimensions instead of one is

$$\frac{dC}{dt} = D\left(\frac{\partial^2 C}{\partial x^2} + \frac{\partial^2 C}{\partial y^2} + \frac{\partial^2 C}{\partial z^2}\right) \qquad 15$$

Equation 14 says that the rate of change of concentration at any plane perpendicular to the direction of diffusion is proportional to the rate of change of the concentration gradient in the plane. The general solution of eq. 14 leads to a Fourier series involving the trigonometric functions; however, if diffusion occurs in an infinite system (diffusion gradients not limited by the dimensions of the container) the solution becomes

$$\frac{dC}{dx} = \frac{\Delta C}{2(\pi Dt)^{1/2}} e^{-\frac{x^2}{4Dt}} \qquad 16$$

where ΔC is the concentration difference between the two solutions in contact. That eq. 16 is, in fact, a solution of eq. 14 can be verified by taking the differential of eq. 16. It is to be noted that eq. 16 is an expression of a Gaussian distribution and is a probability equation. Equation 127, Chapter 1, Mathematical Review, is

$$y = \frac{1}{(2\pi)^{1/2}\sigma} e^{-\frac{x^2}{2\sigma^2}} \qquad 17$$

where y represents the probability, x is the deviation from the mean and σ is the standard deviation. Comparison of eq. 16 with eq. 17 reveals that σ^2 is equal to $2Dt$. Equation 16 can be integrated to give

$$C = \overline{C} - \frac{\Delta C}{\pi^{1/2}} \int_0^y e^{-y^2}\, dy \qquad 18$$

where y is an integration factor and is equal to $x/2(Dt)^{1/2}$. $\overline{C}$ is the mean solute concentration of the two solutions in contact. The integral in eq. 18 is a probability integral, numerical values for which can be found in handbooks of mathematics and statistics.

The displacement of a particle in solution in the x-direction is related to the separate displacements following each collision with another particle S_1, S_2, S_3 etc. by

$$\Delta x = S_1 + S_2 + S_3 + \qquad 19$$

It is evident that for a great many collisions an average value of x will be zero since S_1, S_2, S_3, etc., may be either positive or negative. However, the square of the average will not be zero because the squares of the negative displacements will be positive. The sum of the cross terms, $\overline{S_1 S_2}$, disappears because these terms can be positive and negative. Then the square of the average displacement after n collisions is given by

$$\overline{\Delta x^2} = n\overline{S^2} \qquad 20$$

a result which was noted in Chapter 6, Biopolymers, in the discussion of random coils. It should be realized that $\overline{\Delta x^2}$ is, in fact, the square of the standard deviations of the displacement in the x-direction. It can be seen also from eq. 16 that the plot of dC/dx against x leads to a bell-shaped probability curve as noted above and comparison of eq. 16 with the expression for the probability curve (Chapter 1, Mathematical Review) reveals that the square of the standard deviation ($\overline{\Delta x^2}$) is equal to $2Dt$.

Concentration Gradients. A sharp boundary formed between a solvent and solution becomes blurred and diffuse as time passes; the solute diffuses into the solvent. Figure 4 illustrates the formation of a boundary stabilized by gravity in which the denser solution is at the bottom of the diffusion cell.

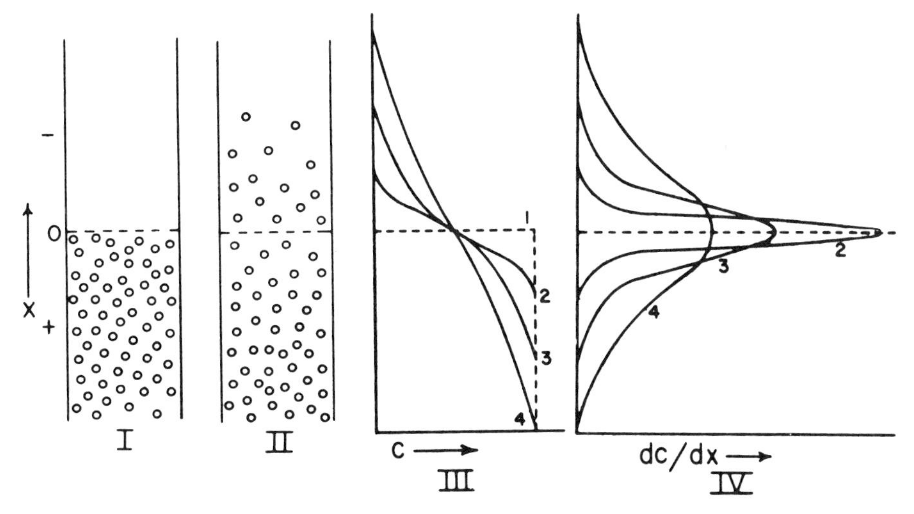

Fig. 4. *I*, Initial boundary of solute diffusion into pure solvent. *II*, Boundary after a certain time. *III*, Variation of concentration with distance. *IV*, Concentration gradients after successive times. Curve 1, Initial boundary. Curves 2, 3 and 4, Boundary after successive times.

A conventional Tiselius electrophoresis cell can be employed as a diffusion cell; the initial boundary, which is somewhat broadened due to shifting in the process of formation, can be sharpened by removing the disturbed volume between the two solutions with a fine capillary immersed to the level of the boundary and connected to a vacuum line.

According to Fick's second law (eq. 16), the plot of the concentration gradient, dC/dx, against the positive and negative values of x where x is zero at the center of the diffusing boundary will yield a bell-shaped probability (Gaussian distribution) curve (see Fig. 4, IV). The numerical value of D may be obtained with the use of eq. 16 providing dC/dx is known as a function of x, t and C. Of the several ways of solving eq. 16 numerically, perhaps the simplest is the maximum ordinate method. It will be noted from eq. 16 that at x equal zero (position at which the boundary was formed) dC/dx is equal to $C/2(\pi Dt)^{1/2}$. Since the area under the dC/dx vs x curve is proportional to the concentration difference between the upper and lower solutions, the area along with the proper proportionality factor can be substituted for C. It is clear that if the time in seconds is plotted against the reciprocal of $(dC/dx)^2$, a linear relation should result whose slope is $A^2/4\pi D$ where A is the area under the curve and includes the proportionality factor between area and concentration. D can then be evaluated.

Optical Methods. Concentration gradients at a boundary are almost always measured by an optical method. Some of these methods are widely used in biochemistry and are employed not only in diffusion studies but likewise in electrophoresis, chromatography, ultracentrifuge methods, and in other areas of modern biochemistry. It seems appropriate to present a summary of the most frequently used methods at this point.

Light absorption measures the concentration of the solute directly. Unlike the refractometric methods to be discussed below there is an element of specificity since the examination can be made at the maximum absorption of the solute. Proteins and nucleic acids both absorb in the ultraviolet, and nucleic acids especially have large extinction coefficients around 260 mμ; very dilute solutions are suitable. Originally, photographs through the boundaries were taken and, in the ultraviolet, quartz cells were used. The degree of blackening of the photographic plates was then measured by some type of recording microphotodensitometer. To take full advantage of this method, suitable filters for the wavelength of light at maximum absorption of the solute are necessary; this is difficult to manage in the ultraviolet. More recently, the boundary is scanned with a photomultiplier tube coupled to a monochromator and the optical densities through the solvent-solution boundary as a function of distance are read directly. Naturally, this modification requires sophisticated instrumentation. Absorption optics have good potentialities in the ultracentrifuge and on occasion can be of help in moving boundary electrophoresis; this is the method of choice in the study of the electrophoresis of hemoglobin solutions. It is not suitable for the measurement of diffusion.

Schlieren Optics. It was noted in Chapter 8, Solution Optics, that a linear relation exists between the concentration of a solute and the index of refraction of the solution. It is evident that a gradient in the refractive index is equal to the concentration gradient multiplied by the refractive increment or

$$\frac{dn}{dx} = \delta \frac{dC}{dx} \qquad 21$$

where n is the refractive index, x is the distance, δ is the refractive increment defined by $\Delta n/\Delta C$ where C is the concentration of the solute. Thus, the gradient in the index of refraction is simply related to the gradient in concentration and several methods exist for the determination of refractive gradients; these are based on schlieren or shadow optics.

Consider Figure 5 which describes a ray of light passing through a refractive gradient in a diffusion cell, electrophoresis cell or a cell in the ultracentrifuge. The ray passing through the solution-solvent boundary is bent downward due to the gradient in the refractive index in the boundary; the greater the gradient the greater the downward displacement. If the refractive ray is interrupted by a knife edge before it can be brought to focus by the camera lens, a dark band will appear in the image of the cell and the band will correspond to the position of the diffusion boundary in the cell. As used in this form the schlieren method can only detect and locate a diffusion boundary. The schlieren method, however, has been modified to permit the determination of concentration gradients in the boundary.

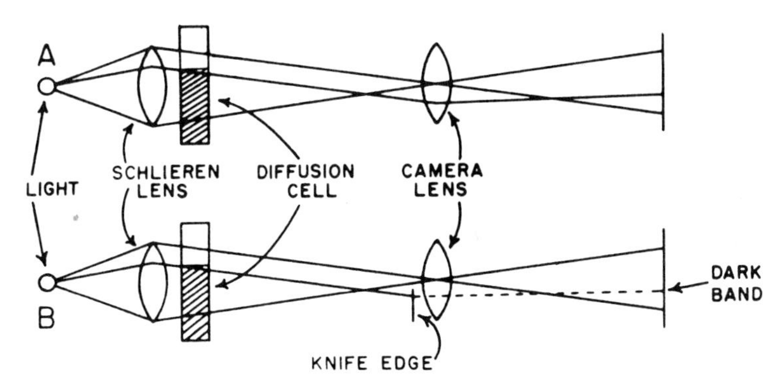

Fig. 5. Schlieren method for visualizing the position of a boundary in a diffusion cell. *A*, Without knife edge; *B*, with knife edge in place.

The Lamm scale displacement method has been extensively employed in diffusion measurements. A transparent scale ruled with horizontal lines is placed in front of the diffusion cell (see Fig. 5A), and photographed with monochromatic light through the diffusion cell. The image of the scale is distorted by the index of refraction gradients and the resulting displacements of the scale lines from their true positions are proportional to the refractive gradients. The scale-line displacements are plotted against the positive and negative distances from the original boundary to yield a curve of the type shown in Figure 4, IV, and described by eq. 16. This method, while capable of good accuracy, is exceedingly tedious and other and more convenient methods have been devised.

In the Longsworth method, a photographic plate is driven at a constant rate past a narrow vertical slit (see Fig. 5). A system of gears activated by the same mechanism moves a knife edge upward in front of the camera lens at a rate proportional to the velocity of the moving plate; the knife edge progressively interrupts the refracted ray. A contour is thus obtained on the photographic plate which is a graph of the refractive index gradient against the position in the cell.

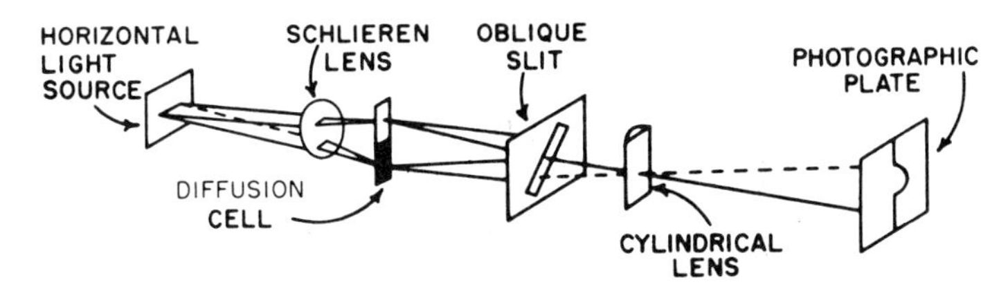

Fig. 6. Diagrammatic sketch of the cylindrical lens system to obtain the contour of the index of refraction gradients in a diffusion cell.

The cylindrical lens system is diagramed in Figure 6. The cylindrical lens has the property of focusing light horizontally but not vertically and, if there be no concentration gradient in the diffusion cell, a sharp vertical straight line will be observed on the photographic (or frosted glass) plate. In the presence of a concentration gradient, the line on the plate will be shifted horizontally to produce a hump at this point. Since the deviated beam has undergone a horizontal displacement as a result of the blocking by the oblique slit and the deviation is proportional to the refractive gradient, the area under the hump is proportional to the concentration difference between the solution and the solvent.

In the Gouy interference fringe method, light from a horizontal slit is passed through the diffusion boundary. Above and below the center of the boundary there will be positions at which the refractive index gradients will be identical. When the

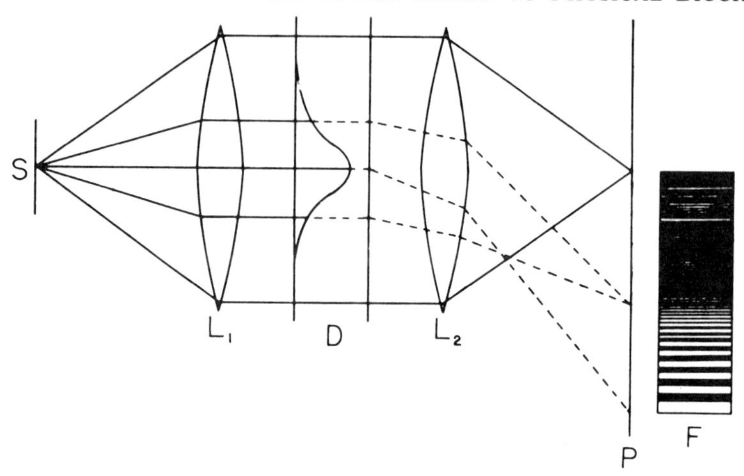

Fig. 7. Gouy interference arrangement. S is a horizontal illuminated slit. L_1 and L_2 are the collinating and focusing lens respectively. D is the diffusion cell with concentration gradients. P is the photographic plate. F is a diagrammatic representation of the interference fringes.

downward displaced light is brought to focus, the beams from above and below the boundary will suffer destructive interference or reinforcements depending on the distances traveled. The result will be a series of horizontal fringes (see Fig. 7). Light through the center of the boundary where the refractive index gradient is greatest will suffer the greatest displacement downward; as diffusion proceeds, the fringes are compressed in an upward direction.

The essential information required to calculate the diffusion coefficient is the number of fringes, J, and the maximum deflection of the light, R_m. The number of fringes, J, is equal to $a\Delta n/\lambda$ where a is the cell thickness, Δn is the refractive index difference across the initial boundary and λ is the wavelength of the monochromatic light. The maximum deflection, R_m, is proportional to the refractive index gradient in the center of the diffusion boundary. The diffusion coefficient is then calculated by the maximum ordinate method. The relation is (there are second order corrections to be applied to R_m)

$$D = \frac{(J\lambda b)^2}{4\pi t R_m^2} \qquad 22$$

where b is the optical distance from the center of the diffusion cell to the photographic plate. The Gouy interference optics have the merits of speed and accuracy in the measurement of diffusion coefficients. It has no obvious utility in electrophoretic and sedimentation work.

Rayleigh Interference Optics. There are several types of interferometers all based on the destructive interference suffered by split beams of coherent light which are forced to travel unequal optical distances. The optical distance is defined as equal to $\Sigma n_i d_i$ where n_i and d_i refer to refractive indices and thicknesses of the various media through which the light ray passes. Figure 8 shows a sketch of a conventional Rayleigh interferometer. The condition for zero light intensity using monochromatic light at a point in the focal plane of the converging lens (L_2), i.e., at the photographic plate is

$$n_2 d - n_1 d = \lambda \Delta N \qquad 23$$

where $n_1 d$ and $n_2 d$ are the optical distances through C_1 and C_2 respectively, λ is the wavelength of the monochromatic light and ΔN is the shift in the number of interference fringes as a result of replacing the solvent by the solution in C_2 (see Fig. 8).

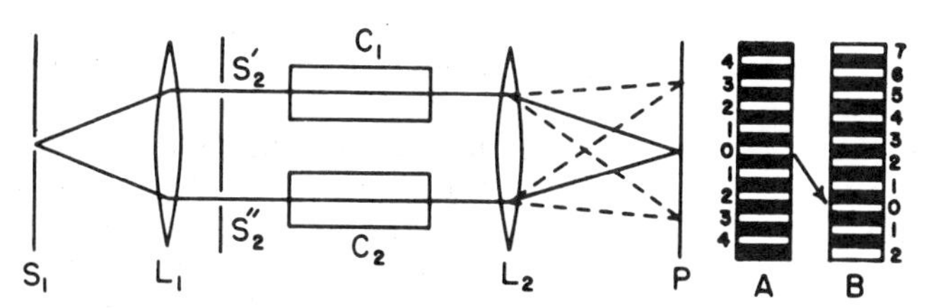

Fig. 8. Rayleigh interferometer. S_1, Slit illuminated with monochromatic light; L_1, collinating lens; S_2' and S_2'', secondary slits; C_1 and C_2, cells containing solvent and solution; L_2, converging lens; P, photographic plate. A, Interference pattern when C_1 and C_2 are both filled with solvent. B, Pattern when C_2 contains solution; note shift in the position of the zero order band.

If white light is used instead of monochromatic light, the zero order fringe will be white and all other fringes will be colored; this aids in identification of the zero order fringe. When used as a refractometer in the measurement of differences in refractive indices, the Rayleigh interferometer is capable of extreme accuracy; a difference in the order of 1×10^{-7} can be measured.

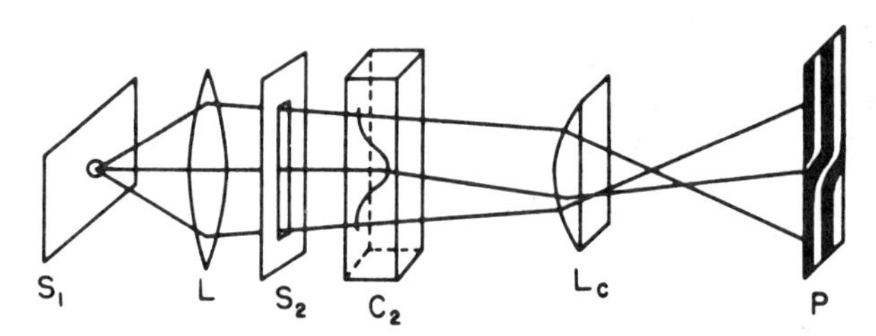

Fig. 9. Rayleigh diffusiometer. Only the split ray passing through the diffusion boundary shown. S_1, Slit illuminated with monochromatic light; L, lens of long focal length to focus image of S_1 on P; S_2, vertical slit (another vertical slit for split ray not shown). C_2 is diffusion cell with solution in bottom and solvent in top half. L_C is the horizontal cylindrical lens. P is the photographic plate to record interference fringes.

The arrangement of the Rayleigh interferometer for the purpose of scanning a diffusion boundary is somewhat other than that sketched in Figure 8. Shown in Figure 9 is a simplified diagram of the apparatus used to obtain interference fringes through a concentration boundary. Only the course of the ray through the boundary is included; the other split beam passing through the solvent is not shown. It is the interaction between the two rays which gives rise to the observed interference fringes. Note the shift of the position of the interference fringes of the rays passing through the solution relative to those passing through the solvent; the rays passing through the diffusion boundary trace out fringes which are contour lines of the solute concentrations across the boundary. The concentration of the solute at any point in the cell can be calculated with the use of eq. 23 providing the relation between the index of refraction and concentration is known. The Rayleigh interference method gives information about concentration rather than concentration gradients as does the schlieren optics. The method, thus, is directly of use in the equilibrium ultracentrifuge.

The calculation of the diffusion coefficient from the Rayleigh interference pattern is somewhat more complex. The total number of fringes is needed as well as the fringe position to obtain the variation of n as a function of the position in the cell. Details of the method will not be given although the calculations resemble those for the Gouy interference optics.

Concentration and Heterogeneity. Equation 14 has been integrated assuming D to be constant; in general, D varies with concentration. For non-electrolytes the dependence of D on concentration is nearly linear over a fair concentration range; however, for electrolytes, such, for example, as potassium chloride, the diffusion coefficient is a much more complex function of concentration and exhibits a pronounced minimum as the concentration is increased. The dependence of D on concentration produces a skewed distribution curve when the concentration gradient is plotted against the distance x (see Fig. 10). The technique for accommodating the variation of the diffusion coefficient with concentration is to make the initial difference in concentrations across the diffusing boundary small. It is customary to express the diffusion coefficients for proteins at a given temperature and at zero concentration; the measured diffusion coefficients are plotted against the mean concentration of the two solutions in contact and extrapolated to zero concentration. It is also possible but more complicated to integrate eq. 14 assuming a relationship between D and the concentration.

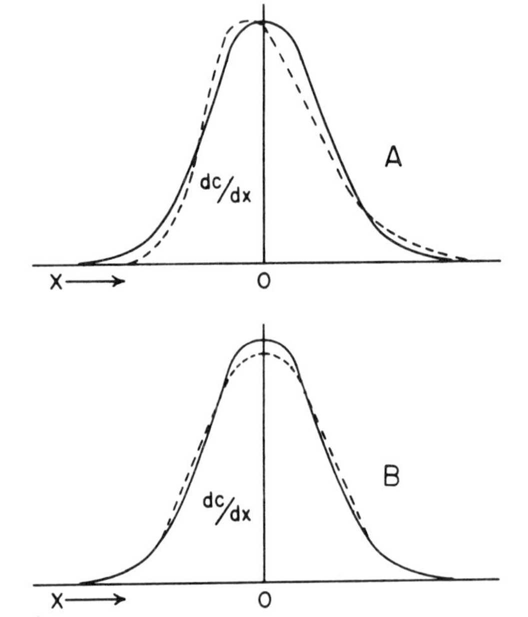

Fig. 10. Plot of concentration gradient dC/dx vs. the distance x. *A*, Contrast of ideal Gaussian distribution (solid line) with experimental showing concentration dependence (broken line); *B*, same as *A* except that broken line indicates polydispersity.

The diffusion coefficient obtained at one temperature in a given solvent can be approximately adjusted to another temperature by the use of the equation

$$D_2 = \frac{D_1 T_2 \eta_1}{T_1 \eta_2} \qquad 24$$

where the subscripts refer to the two different temperatures (absolute), η and D are the viscosity and the diffusion coefficients respectively.

Diffusion is not a discriminating tool for the detection of impurities in the solute. If more than one solute is present, the distribution of dC/dx plotted against x departs from a normal distribution curve (see Fig. 10B). However, it is difficult to express a given departure from ideality in terms of heterogeneity of the sample. It is also true that serious theoretical difficulties are encountered in a diffusing system consisting of more than two components; multiple diffusion will be considered in a later and brief discussion.

Diffusion in Gels. Gels consist of a three-dimensional network which extends throughout the liquid, leading to rigidity. Substances such as gelatin and agar form gels in water which are quite dilute and there is little obstruction to the movement of molecules even of the size of proteins through such gels. For example, the electrical conductance of a dilute solution of electrolyte in a gelatin solution before it sets is little changed when the gelatin sets as a gel.

A gel could obstruct diffusion in three ways: (1) The diffusing substance might interact with the gel substance, (2) the gel might be so concentrated that the liquid pores through the gel would be smaller than the diameters of the diffusing molecules and the gel would behave as a filter and (3) the solute molecules have to detour around the cross-links in the gel.

Lauffer has discussed the theory of diffusion in gels and has described a simple and ingenious apparatus for the measurement of diffusion in gels. Suppose the gel is dilute and does not interact with the diffusing substance, then it is only necessary to consider the detours the solute molecules make as they diffuse through the gel. Lauffer considers that diffusion under this condition is analogous to the conduction of electrical current through a suspension of randomly oriented extremely asymmetric non-conducting prolate ellipsoids of revolution (see Chapter 12, Ion Transport). He then wrote a modified Fricke's equation

$$D_g = \frac{D}{1 + (\alpha - 1)\phi} \tag{25}$$

where D_g is the diffusion coefficient in the gel, D is the coefficient of free diffusion in the solvent in the absence of gel, ϕ is the ratio of the volume occupied by the gel substance to the total volume of the gel, and α is a form factor which depends on the shape and orientation of the obstructing particles. For randomly oriented rods whose lengths are much greater than their diameter, α has the value of 5/3.

The diffusion cell consists of a 50 ml. hypodermic syringe with the needle end of the barrel cut off squarely. About a 1.5 percent hot agar solution is poured into the syringe and allowed to solidify. The syringe with the plunger in place is then placed in contact with a well-stirred solution of the solute and diffusion into the gel is permitted to occur. At the end of the diffusion experiment the column of agar gel is pushed out of the barrel of the syringe with a mechanical device and sections of known and desired thickness are cut with a fine wire pulled across the end of the barrel. The slices are then analyzed for the concentration of solute.

Since the solution is stirred, the concentration of the solute remains uniform up to the face of the gel and the concentration gradient exists in the gel; the plot of dC/dx against x would yield a half of a Gaussian distribution curve with the largest concentration gradient at x equal zero, i.e., at the surface of the gel. D_g is then obtained by solving eq. 16. The result obtained by this method compares favorably with accepted values of D in the literature.

Diffusion in gels is the basis of various laboratory techniques. For example, one method for the assay of antibiotics is to inoculate an agar plate with some test organism and to permit the antibiotic to diffuse outwardly from a circular well; the area of the clear circle around the well in which no growth of the bacteria has occurred is taken as a measure of the concentration of the antibiotic.

Diffusion in gels has also been of help to the immunologist who allows the antigen and antibody to diffuse from separate locations in an agar gel; a precipitin reaction in the gel is observed where the two diffusion zones overlap. A variation on this arrangement is immunoelectrophoresis which is a powerful tool for the investigation of protein mixtures. The protein mixture is first subject to electrophoresis in an agar gel; this separates the protein mixture more or less into the individual protein components which spread linearly along the agar block. Antigens in solution are then applied in a channel in the agar parallel to the direction of electrophoretic movement; the channel is displaced a short distance from the line of electrophoresis: Antibodies diffuse outwardly from their electrophoretic positions to meet the antigens diffusing from the channel to produce precipitin zones; these precipitin zones can be frequently identified as protein components of the original mixture.

Diffusion in a gel such as described above for the antibiotic and for the antibody occurs in two directions and Fick's second law reduces to

$$\frac{dC}{dt} = D\left(\frac{\partial^2 C}{\partial x^2} + \frac{\partial^2 C}{\partial y^2}\right) \tag{26}$$

Integration of eq. 26 requires the use of Bessel functions.

Molecular Sieves. Sausage casing (seamless cellulose) tubing has long been used to separate proteins from low molecular weight solutes by dialysis. The pores in the cellulose membrane are not large enough for the protein molecules to pass, whereas inorganic ions and small organic molecules can be accommodated. The pores of such a membrane must have a very heterogeneous character and perhaps a fair physical picture, much enlarged, of such a membrane would be a wad of cotton. A measure of porosity is provided by the rate of flow through the membrane, along with a knowledge of the free space. Thus, for a single cylindrical capillary the volume of liquid flowing per second (Poiseuille's law) is

$$v = \frac{\pi r^4 \, \Delta P}{8\eta l} \tag{27}$$

where r is the radius of the capillary, ΔP is the difference in hydrostatic pressure between the ends of the capillary in dynes per square centimeter, η is the coefficient of viscosity and l is the length of the capillary. For n pores of average radius r_a

$$v = \frac{n\pi r_a^4 \, \Delta P}{8\eta l} \tag{28}$$

The total volume of the pores is evidently

$$\phi = n\pi r_a^2 \, l \tag{29}$$

Combination of eqs. 28 and 29 gives

$$r_a = \frac{(8\eta l^2 \, v)^{1/2}}{\phi \, \Delta P} \tag{30}$$

The expression for the average pore radius, r_a, simply serves to compare membranes; the physical meaning is ambiguous since the pores are far from being the cylindrical capillaries which the application of Poiseuille's law requires.

Efforts have been made with some success to develop membranes of graded porosities and to separate molecules in solution on the basis of their rate of penetration through such membranes; the membranes act as molecular sieves.

A more fruitful approach to a molecular sieve is the use of cross-linked dextran gels (Sephadex). Other such materials can likewise prove suitable. The porous gels are packed into a column and the solution of protein poured into the column and eluted with a proper aqueous solvent. The elution is monitored with a recording spectrophotometer at a wavelength of about 280 mμ (maximum absorption of the aromatic amino acid residues). The position of the protein whose molecular weight is unknown permits an estimate of the molecular weight of the protein to be made. If the protein is of such molecular weight (size) as to be completely excluded from the pores of the gel, it passes through the column unhindered, whereas smaller protein molecules are able to penetrate into the gel and tend to be retained in the column. It is found that the volume of the liquid required for elution is inversely proportional to the logarithm of the molecular weight of the protein providing the protein molecules are not markedly asymmetric or do not interact strongly with the gel. Figure 11 shows a plot of the ratio Ve/Vo, where Ve is the elution volume of the

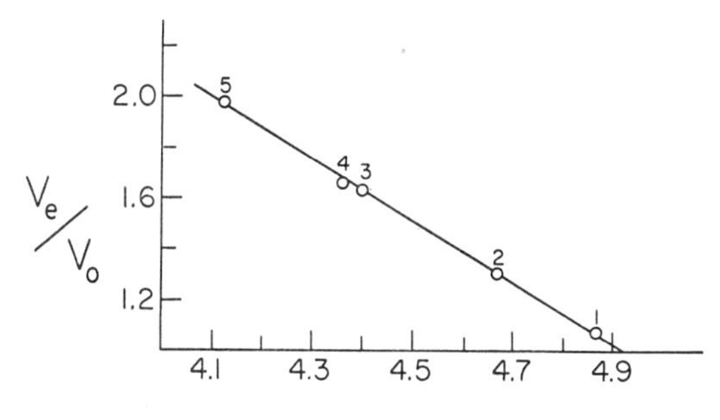

Fig. 11. Standard curve for the molecular weight determination with a porous gel using a polyacrylamide column. Plot of the ratio *Ve/Vo* against the logarithm of the molecular weight. 1, Bovine serum albumin; 2, egg albumin; 3, trypsin; 4, α-chymotrypsin; 5, ribonuclease. (Adapted from C. Schwabe and G. Kalnitsky: Biochem. *5:* 158, 1966.)

solvent and *Vo* is a constant for the column employed, against the logarithm of the molecular weights of a series of calibrating proteins (proteins whose molecular weights are known by other methods).

Diffusion to a Flat Surface. The amount of solute diffusing to a flat surface in a given time depends among other things on whether or not the solution is stirred. For an unstirred solution the concentration gradient at the surface is given by eq. 16 and at the surface where x is equal to zero

$$\frac{dC}{dx} = \frac{C}{2(\pi Dt)^{1/2}} \tag{31}$$

Substituting Fick's first law into eq. 31, there results (neglecting the negative sign)

$$\frac{dm}{dt} = \frac{AC}{2}\left(\frac{D}{\pi t}\right)^{1/2} \tag{32}$$

The integration of eq. 32 yields

$$m = AC\left(\frac{Dt}{\pi}\right)^{1/2} \tag{33}$$

In unrestricted diffusion when a molecule reaches the plane at which x is zero, the chances that the molecule will go forward or backward are each one-half. However, if all the molecules arriving at the plane pass through it or are irreversibly adsorbed on it, there is no opportunity for the molecules to go backward and, accordingly, under these conditions the right side of eq. 33 should be multiplied by 2 and

$$m = 2AC\left(\frac{Dt}{\pi}\right)^{1/2} \tag{34}$$

If the solution is stirred, the concentration of the solute is uniform to within a short distance of the surface with water as the solvent, and with vigorous stirring at room temperature this distance is about 3×10^{-3} cm. It can be assumed that over this short distance the concentration gradient dC/dx across the unstirred layer is constant and is equal to C/δ where δ is the thickness of the unstirred layer. Then from Fick's first law (neglecting the negative sign)

$$\frac{dm}{dt} = \frac{DAC}{\delta} \tag{35}$$

which upon integration gives

$$m = \frac{DACt}{\delta} \tag{36}$$

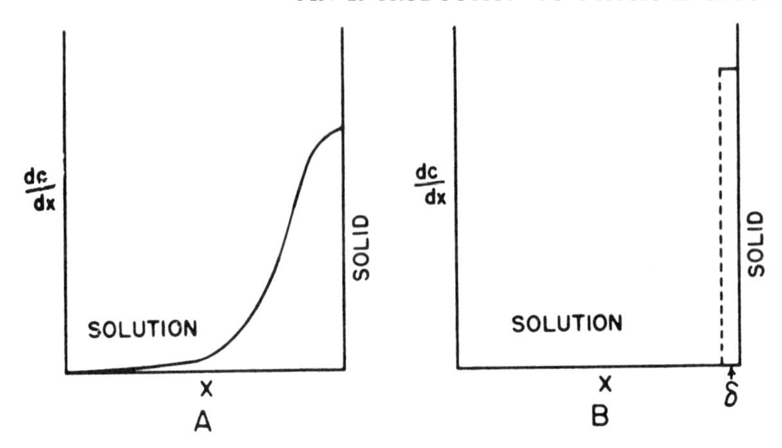

Fig. 12. Concentration gradients in a solution with adsorption occurring at a solid as a function of the distance x from the surface. *A*, Unstirred solution; *B*, stirred solution showing unstirred liquid layer of thickness δ.

Comparing eqs. 34 and 36, it is noted that the quantity of material adsorbed at a plane surface from an unstirred solution is proportional to the square root of the time, whereas, for a stirred solution, the amount is directly proportional to the time. Both eqs. 34 and 36 are dependent on the establishment of a steady state and neither is valid after a substantial part of the surface is covered by adsorbed molecules. Figure 12 contrasts the change of concentration gradients with distance from a solid flat surface for an unstirred and stirred solution.

Diffusion into and out of spherical and cylindrically shaped particles has been formulated on numerous occasions. For spherical particles, eq. 15 is expressed in polar coordinates and integrated under the conditions of the experiment. For cylindrical structures, eq. 15 is integrated with the necessary boundary conditions. References to Jacobs and to Crank listed at the end of the chapter should be consulted.

Diffusion through Membranes. If the pores of the membrane are much larger than the diameters of the diffusing molecules, the only role the membrane plays is to decrease the cross-sectional area available for diffusion and all molecules diffuse through the membrane at a rate proportional to their diffusion coefficient and to their concentration gradient across the membrane. This situation has been discussed in connection with the porous disk method and with diffusion through gels. As the pores of the membrane become smaller and approach in size that of the diffusing molecules, the interaction between these molecules and the membrane becomes important; the diffusing molecules are adsorbed on the surface of the membrane and penetrate by a process akin to solution in the membrane.

It is assumed that the rate of diffusion in the liquid phase bathing the membrane is much greater than the rate of penetration through the membrane and can be neglected in formulating the kinetics of penetration.

The processes occurring at the membrane interfaces are: (1) adsorption at the interface, (2) desorption back into solution, and (3) diffusion into the membrane. The process of penetration bears a close resemblance to the formation and decomposition of an enzyme-substrate complex of the Michaelis-Menten type, described in Chapter 15, Reaction Kinetics and Enzyme Activation. At each of the two membrane interfaces the following reactions take place:

$$X_1 + M_s \underset{k_2}{\overset{k_1}{\rightleftharpoons}} XM_s \overset{k_3}{\rightarrow} X_2 + M_s \qquad 37$$

where X_1 is the solute in solution and X_2 is the same substance in the membrane. M_s is the surface of the membrane exposed to the solution. In Chapter 15 it is shown that the process described in eq. 37 can be formulated for steady state kinetics

$$V_1 = \frac{VC_1}{K + C_1} \tag{38}$$

where V_1 is rate of penetration of X through the membrane, V is the maximum rate of penetration with the membrane saturated with X, C is the concentration of X in the liquid phase, K is the so-called dissociation constant but is actually equal to $(k_2 + k_3)/k_1$. For a completely symmetrical membrane with both sides identical, an equation exactly analogous to eq. 38 can be written for the opposite side of the membrane; the net rate of penetration of the substance through the membrane is equal to the difference in rates of penetration in the two directions, and the net rate is

$$V_1 - V_2 = V_n = \frac{KV(C_1 - C_2)}{(K + C_1)(K + C_2)} \tag{39}$$

where C_2 is the concentration on the dilute side of the membrane. When K is very large in comparison with C_1 and C_2, there results

$$V_n = \frac{V}{K}(C_1 - C_2) \tag{40}$$

Fick's first law of diffusion for membranes can be written as

$$V_n = \frac{DA}{\delta}(C_1 - C_2) \tag{41}$$

where A is the area of the membrane and δ is its thickness. If the thickness of the membrane is constant, it may be incorporated in the diffusion coefficient to give a permeability coefficient P. Equation 41 then becomes

$$V_n = PA(C_1 - C_2) \tag{42}$$

The permeability coefficient, P, is thus the number of moles of the substance passing unit cross-sectional area of the membrane in unit time under unit concentration difference. The unit of area customarily chosen by biologists in the study of cell penetration is the square micron; the concentration difference is expressed in moles per liter, and the time in seconds.

Comparison of eqs. 40 and 42 shows that if K is much larger than C_1 and C_2

$$PA = \frac{V}{K} \tag{43}$$

It was noted above that K is equal to $(k_2 + k_3)/k_1$. If k_3 is very much smaller than k_2, K becomes equal to k_2/k_1 which is equal to the reciprocal of the adsorption coefficient on the membrane surface. It is also proportional to the reciprocal of the distribution coefficient of the diffusing molecules between the membrane and the solvent bathing the membrane. Thus, when k_2 is much larger than k_3, the rate of penetration becomes directly proportional to the distribution coefficient of the diffusing molecules between the membrane and the solvent.

Of considerable interest is the observation by Scholander that the rate of oxygen transport through a millipore filter is greatly enhanced by the presence of hemoglobin in the millipore filter and this enhancement was greater the lower the oxygen pressure. This enhancement is probably related to the much greater solubility of oxygen in a hemoglobin solution as compared with serum or with water.

The penetration of material through a membrane encounters energy barriers. It is possible to apply the theory of the activated state to the penetration of membranes by diffusing molecules and under favorable conditions to calculate the free energy, the heat energy, and the entropy of activation for the process.

In our previous discussion, the flow due to a hydrostatic pressure or osmotic pressure difference has been regarded as completely distinct from the diffusion flow. However, when the two flows occur in the same system and simultaneously, they are apt to be related. This relationship can be expressed as

$$J_v = L_p \Delta P + L_{pD} RT \Delta C \qquad 44$$

and

$$J_D = L_{Dp} \Delta P + L_D RT \Delta C \qquad 45$$

where J_v is the volume flow and J_D is the diffusion flow. ΔP is the pressure difference across the membrane arising from either a hydrostatic pressure or from an osmotic pressure or from both. $RT\Delta C$ is the diffusion force where ΔC is the concentration difference of the solute across the membrane. The L-coefficients are of the nature of reciprocal frictional coefficients and are known as phenomenological coefficients; L_p is the filtration coefficient, L_D is a diffusion coefficient and L_{pD} and L_{Dp} are the cross coefficients A consequence of Onsager's law of irreversible thermodynamics is that the cross coefficients are equal to each other. The number of L-coefficients required is thereby reduced to three.

The reflection coefficient of the membrane has been defined as the ratio of the observed to the calculated osmotic pressure exerted by a solute. In terms of the L-coefficients, the reflection coefficient is

$$\sigma = -\frac{L_{pD}}{L_p} \qquad 46$$

Another coefficient is that of the mobility of the solute, ω

$$\omega = (L_D - L_p \sigma^2)\, C_m \qquad 47$$

where C_m is the mean concentration of the solute on the two sides of the membrane. To summarize; in general, three coefficients are necessary to describe the penetration of a single solute and a solvent through a membrane and these are: (1) L_p is the filtration coefficient and is the permeability coefficient through the membrane provided no permeable solute is present, (2) $RT\omega$ is the coefficient of solute flow, i.e., P of eq. 42, if there is no volume flow, and (3) σ is the reflection coefficient as defined above. The coefficients can be measured separately provided the proper experiments have been done.

In the free diffusion of a single solute through a single solvent (water), the only resistance to flow is the friction between the solute and solvent alone and the process of diffusion can be described by a single diffusion coefficient. In the presence of a third component, however, not only is the interaction between both solutes and water to be considered but likewise the interaction between the two solutes. This leads to a more complex situation involving cross terms with coupled diffusion rates; coupled diffusion rates are especially important in the case of electrolytes. In a sense, a membrane represents a third component in the diffusion of a single solute through the membrane. There is a friction between the solute and solvent, between the solute and the membrane and between the solvent and the membrane. How important the interaction of the solute and solvent with the membrane depends, among other things, upon how big the pores of the membrane might be; in the porous disk method for diffusion constants the interaction with the membrane is minimal because the pores are large as compared with molecular dimensions. Dense collodion membranes and especially living membranes, however, present an entirely different situation and strong interaction with the membrane can be expected.

Penetration through Cellular Membranes. Investigations on cellular membrane structure are numerous; this is a difficult research area. It is not always certain that materials extracted for study have, in fact, been obtained only from the membrane and not from some other area in the cell. Further, the isolation of membrane constituents automatically disrupts the membrane and destroys its unique transport properties, although there is general agreement that the essential components of a cellular membrane are water, protein and lipid.

It was proposed a number of years ago by Davson and Danielli, on rather slender experimental evidence furnished by Gorter and Grendel regarding the lipids in red cell membranes, that a living membrane consisted of a highly ordered and oriented leaflet of lipid (phospholipid) sandwiched between an inner and outer layer of proteins. Electron microscopic pictures of a variety of isolated cellular membranes do indeed reveal two parallel dense lines whose combined thickness is of the order of 50 to 150 A-units. However, after the membranes have experienced the treatment necessary for electron microscopy, it is likely that their structure has been vastly altered from their native state. X-ray diffraction studies on more or less intact membranes reveal a greatly disordered hydrocarbon layer 20 to 30 A-units thick, and the water content required for structural integrity of the membrane is about 20 to 30 percent of the dry weight. There is no evidence for proteins existing as stretched peptide chains (β-keratin structures) and spectroscopic measurements indicate the membrane proteins to have 25 to 35 percent α-helical content.

Transport across a membrane can be active, facilitated or passive. In active transport, the substrate accumulates in a cell against a concentration gradient. Facilitated transport means that the substrate passes into the cell faster than would be expected on the basis of its diffusion coefficient. Passive transport occurs when the substrate passes into the cell approximately as expected on the basis of the concentration gradient and its diffusion coefficient.

Transport of a substrate into a cell is often highly specific and, further, the transport is conducted against a concentration gradient; the transport is active. It is evident that the only membrane constituents which could provide specificity are the proteins. Needed are recognition proteins located directly on the cell surface exposed to the bathing fluid; the recognition protein binds the substrate specifically. The next step in the transport process is to couple an energy supply to the movement of the substrate from outside to the inside of the cell against a concentration gradient. It has been claimed that both the recognition as well as the energy coupling proteins have been isolated from cellular membranes.

Experimental results seem to indicate that many non-electrolytes penetrate living cells by simple diffusion in which the only significant concentration gradient is across the membrane itself.

Many years ago Overton found that a general parallelism existed between the rate of penetration of various organic substances into living cells and their lipid-water distribution coefficients, and he concluded that the membrane of living cells was fat-like in nature. A byproduct of this penetration theory was the lipid theory of narcosis, narcosis being defined as the loss of irritability by a living organism. It was found that there also existed a parallel between the oil-water distribution of narcotizing substances and the concentration of these substances required to narcotize such animals as tadpoles. Narcosis can be produced by a variety of diverse substances, for example, by many simple organic molecules such as hydrocarbons, ethers, alcohols, chlorine derivatives of hydrocarbons, by xenon and to lesser extent by krypton. Magnesium ions will also produce narcosis. It appears improbable that many of these agents can be considered as enzymatic inhibitors; emphasis has, therefore, been placed on the behavior of cell walls. Narcotizing and anesthetizing agents appear to have the ability to disorganize living structure. It seems probable that ion transport across a membrane is dependent on highly organized structural units and the function of an anesthetizing agent is to dilute this structure and to separate elements of the ion carrier system.

Collander has reported a study of the penetration of non-electrolytes into the

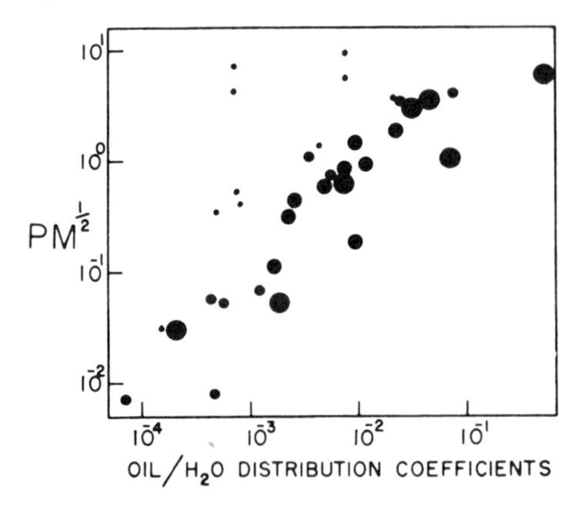

Fig. 13. Permeability of cells of Chara ceratophylla to various non-electrolytes. Ordinate indicates the permeability coefficient (cm./hr.) multiplied by the square root of the molecular weight. The abscissa gives the oil-water distribution of the substances. The relative sizes of the circles indicate the relative sizes of the molecules. (R. Collander: Physiol. Plantarum *2:* 300, 1949.)

alga Chara ceratophylla. He also determined the distribution of these non-electrolytes between olive oil and water. Shown in Figure 13 is a plot of the permeability coefficients of these compounds multiplied by the square roots of their respective molecular weights against their oil-water distribution coefficients. Clearly, there is a close correlation between oil-solubility of a substance and its permeability coefficient. However, the smallest molecules penetrate faster than would be expected on account of their oil-solubility alone.

The passage of water across cell membranes apparently involves no active mechanism, and water is transferred into and out of cells in response to differences in osmotic pressure. The osmotic flow of water into and out of cells has been discussed in Chapter 7, Osmotic Pressure and Related Topics. The rate of uptake of water by living cells in response to osmotic changes varies from one type of cell to another, but it is usually surprisingly rapid, complete adjustment frequently requiring only a few minutes. The rate of entry of water into mammalian red cells is especially fast and the reaction is complete in a matter of seconds.

If a solute penetrates the cell as easily as water, it would exert no osmotic pressure across the cell membrane irrespective of its concentration difference. However, some solutes are leaky; the membrane does not completely exclude them but they do not penetrate as easily as water and while they exert an osmotic pressure the pressure is not as large as would be calculated from the concentration. The ratio of the osmotic pressure observed to that calculated for a given solute is called the reflection coefficient and can vary from zero to unity.

The nature of the cell membrane is far from clear but it is believed by some to have in it pores or channels through which water can pass. The rate of diffusion through membrane pores should be proportional to the total area of the pores whereas the rate of flow of water in response to hydrostatic pressure is proportional not only to the area of the pores but likewise to the shape and sizes of the individual pores. Diffusion permeability can be measured with tritiated water and the flow of water through the membrane in response to an osmotic pressure difference can be easily accomplished. Expressing these two permeabilities in the same units, it is found that the osmotic permeability is always greater than is the diffusion permeability and, in fact, the two permeabilities can differ by an order of magnitude.

The attempt has been made to estimate the average pore diameter of living cells by comparing the osmotic and diffusion permeabilities, that is, by comparing Poiseuille's law with eq. 44. By this method it is estimated that the average pore diameter of the membrane of mammalian red cells is about 7 A-units. This would appear to be a highly dubious calculation. It is probable that Poiseuille's law and the constant terms in eq. 44 for capillaries of molecular dimensions no longer have the meaning assigned to them. It is also possible that osmotic flow creates openings in cell walls which are repaired as soon as the flow ceases.

Thermodynamics of Diffusion. The force causing diffusion of a solute is the gradient of chemical potential of that substance just as the flow of heat is proportional to the temperature gradient. It is thus possible to write

$$\text{Flux} = J_1' = -L_1\left(\frac{\partial \mu_1}{\partial x}\right) \qquad 48$$

where J_1', the flux, differs from J_1 in eq. 1 only in terms of the frame of reference. J_1 refers to the flux across some fixed plane in the diffusion cell where J_1' refers to the flux of the solute relative to the local center of mass. When a solute diffuses into water, there has to be a corresponding diffusion of water in the opposite direction. If there is little volume change when the solute is mixed with water, then J_1 and J_1' would be nearly equal. L_1 in eq. 48 is the diffusional mobility and is equal to c_1/Nf' where c_1 is the concentration of the solute in moles per milliliter, f' is the frictional coefficient per molecule and N, Avogadro's number, is introduced to give the frictional coefficient per mole. J_1', then has the units of moles per square centimeter per second.

Multiplying the right side of eq. 48 by dC_1/dC_1 and rearranging and replacing $f'N$ by f the frictional coefficient per mole, there results

$$J_1' = -\frac{c_1}{f}\left(\frac{d\mu_1}{dC_1}\right)\frac{\partial C_1}{\partial x} = -\frac{C_1}{f}\left(\frac{d\mu_1}{dC_1}\right)\frac{dc_1}{dx} \qquad 49$$

In eq. 49 the lower case c refers to the concentration in moles per milliliter, whereas the capital C refers to moles per liter.

It is interesting to compare eq. 49 with eq. 1, Fick's first law. If, on mixing the solute and solvent, there is no change in total volume, then the relation between J_1 and J_1' becomes

$$J_1 = \overline{V}_0 \rho J_1' \qquad 50$$

where $\overline{V}_0$ is the partial specific volume of the solvent and ρ is the density of the solution in grams per milliliter. Combining eqs. 49 and 50 there results

$$D = \frac{C_1 \overline{V}_0 \rho}{f}\left(\frac{d\mu_1}{dC_1}\right) \qquad 51$$

Since the chemical potentials of all the components of a system are related (Chapter 2, Energetics) it is clear that the diffusion coefficient of all the components at any finite concentration will be interrelated. It will be recalled that the relation between the chemical potential and the concentration is (Chapter 2, Energetics)

$$\mu = \mu_0 + RT \ln \gamma C \qquad 52$$

where μ_0 is the chemical potential in the standard state of a given component and γ is its activity coefficient. Differentiation of eq. 52 in respect to concentration and substitution in eq. 51 gives

$$D = \frac{RT}{f}\left(1 + C_1 \frac{d \ln \gamma}{dC_1}\right) \overline{V}_0 \rho \qquad 53$$

In dilute solutions $\overline{V}_0 \rho$ is very nearly unity and as C approaches zero eq. 53 reduces to

$$D^0 = \frac{RT}{f} \qquad 54$$

Equation 53 is not entirely suitable for the prediction of the variation of the diffusion coefficient with concentration because the frictional factor, f, varies with the concentration. Assuming that the frictional resistance is proportional to the relative viscosity (η_r) of the solution and combining eqs. 53 and 54 and neglecting $\overline{V}_0\rho$ gives

$$D = \frac{D^0}{\eta_r}\left(1 + C_1 \frac{d \ln \gamma}{dC_1}\right) \tag{55}$$

Diffusion and Molecule Size. According to eq. 54, the diffusion coefficient (at infinite dilution of the solute) is inversely proportional to the frictional coefficient between the solute and solvent. The derivation of eq. 54 can perhaps be more clearly seen from the following argument. The force causing diffusion of a dissolved substance is the gradient of the chemical potential for that substance, and, for dilute solutions where the activity and concentration are very nearly equal, the force per mole is $RT\ dC/dx$. Actually, this expression has the dimensions of a pressure—the diffusion pressure. In a steady state the resisting forces must equal the force causing diffusion and

$$RT \frac{dC}{dx} = fu = f \frac{dm}{dt} \tag{56}$$

where f is the frictional coefficient per mole and u is the rate of diffusion or is dm/dt. Substituting Fick's first law of diffusion in eq. 56 and rearranging, eq. 54 is obtained. As the size of the solute molecules increases, the frictional coefficient should increase; accordingly, there should exist a relation between the diffusion coefficient and the molecular size of the solute. Solute molecules whose size is roughly that of the solvent molecules would not experience a constant resistance to motion but the resistance would vary depending on the positions of the solute molecules relative to the solvent molecules. On the other hand, solute molecules whose size is much greater than that of the solvent molecules would be subject to a constant frictional resistance as they moved through the solvent; variations in the structure of the solvent would average out.

Longsworth has measured the diffusion coefficients of a number of smaller molecules including amino acids, peptides and sugars in water at 25°. Some of his results are shown in Figure 14 where the measured diffusion constant has been multiplied by the cube root of the partial molal volume V of the solute and plotted against the diffusion coefficient.

The empirical equation corresponding to the best straight line drawn through the points in Figure 14 is

$$D = \frac{24.182 \times 10^{-6}}{\overline{V}_1^{1/3} - 1.280} \tag{57}$$

If it is assumed that the partial molal volume of the solute is approximately equal to the molecular weight divided by the density, then the molecular weight of the solute (small molecules) can be estimated from eq. 57.

If the solute molecules are large in comparison with the molecules of the solvent and furthermore if the solute molecules are spherical, the frictional force per molecule is simply Stokes factor $6\pi r\eta$ where r is the radius of the solute molecule and η is the coefficient of viscosity. To obtain the resisting forces per mole instead of per molecule, the Stokes factor has to be multiplied by N, Avogadro's number. Substituting this information in eq. 54 there results for large spherical solute molecules

$$D_s = \frac{RT}{6\pi r\eta N} \tag{58}$$

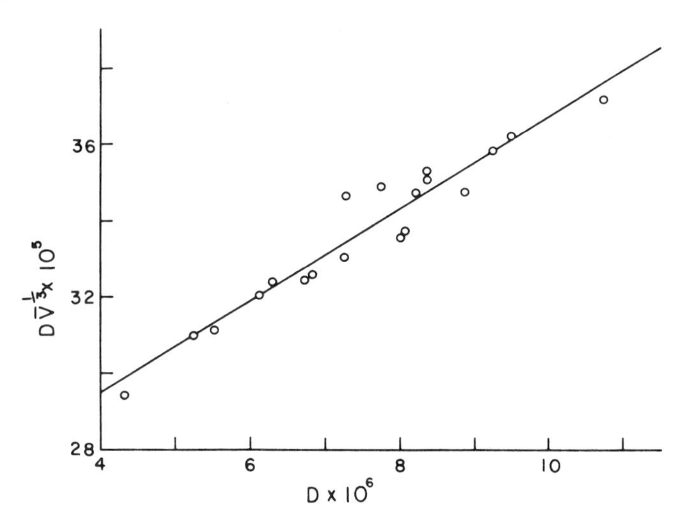

Fig. 14. Product of the diffusion coefficient and the cube root of the partial molar volume, $\bar{V}$, of amino acids, sugars and peptides plotted against the corresponding diffusion coefficient. (L. G. Longsworth: J. Amer. Chem. Soc. *75:* 5705, 1953.)

Since eq. 58 applies to spherical molecules, it can be written

$$D_s = \frac{RT}{6\pi\eta N\left(\frac{3}{4}\frac{V_m}{\pi N}\right)^{1/3}} \quad 59$$

where V_m is the molar volume of the solute expressed in cubic centimeters. Substituting numerical values for the constants in eq. 59 and for spherical molecules in water

$$D_s^{20°} = \frac{28.82 \times 10^{-6}}{V_m^{1/3}} \quad 60$$

and

$$D_s^{25°} = \frac{33.06 \times 10^{-6}}{V_m^{1/3}} \quad 61$$

The superscripts attached to the diffusion coefficients in eqs. 60 and 61 indicate the temperature at which the diffusion is conducted. The diffusion of a spherical molecule is thus inversely proportional to the cube root of the volume of the molecule in solution. If the molecule is hydrated, the diffusion coefficient will be reduced and the ratio of the anhydrous diffusion coefficient to the hydrated diffusion coefficient will be

$$\left(\frac{D_a}{D_h}\right)_s = \left(\frac{V_{mh}}{V_{ma}}\right)^{1/3} \quad 62$$

where D_a is the diffusion coefficient of the anhydrous spherical molecule, D_h that of the hydrated spherical molecule; V_{mh} is the hydrated molar volume and V_{ma} is the anhydrous molar volume. The subscript s indicates that both the anhydrous and hydrated molecules are spherical. It should be noted that, whereas diffusion varies inversely with the cube root of the molecular volume, intrinsic viscosity varies directly with the molecular volume; viscosity is a more critical function of hydration than is diffusion.

Diffusion and the Shape of Molecules. A molecule which departs from a spherical shape will have a smaller diffusion coefficient than does a spherical molecule of the same volume. Qualitatively, it is easy to see why asymmetry should decrease the diffusion coefficient of a molecule; the amount of surface exposed to the frictional forces of the medium by an asymmetric particle is larger than that exposed by a spherical particle of the same volume. The frictional coefficient, f, in eq. 54 has been calculated for prolate (cigar-shaped) and for oblate (disk-shaped) ellipsoids of revolution. Figure 15 shows the results of the numerical solution of the published equations. In this figure, the ratio D_s/D, where D_s is the calculated diffusion coefficient of a spherical molecule (eq. 58) with the same molar volume as the actual molecule and D is the experimentally measured diffusion coefficient, is plotted against the calculated axial ratios of the ellipsoids of revolution. The equations upon which Figure 15 is based are usually referred to as Perrin's equations although they were actually derived earlier by Herzog, Illig and Kudar.

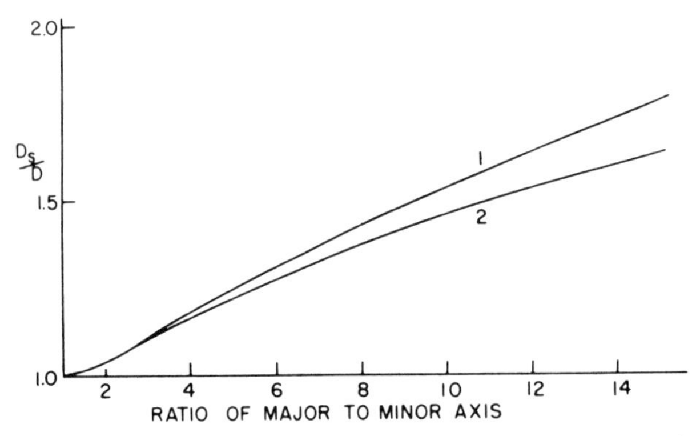

Fig. 15. Relation between asymmetry of prolate (Curve 1) and of oblate (Curve 2) ellipsoids and their diffusion coefficients.

It is unlikely that any biopolymeric molecule has a regular or easy shape such as a sphere, or ellipsoid, although some, no doubt, can be described as elongated rods. Only approximate results can, therefore, be expected when molecular asymmetries are calculated from diffusion measurements. The effect of hydration, which is a complicating factor, is also unresolved. This is the same situation which has been discussed in relation to viscosity of biopolymeric molecules and will be considered again in Chapter 14, Sedimentation.

Values for the ratios D_s/D as well as the intrinsic viscosities for several proteins have been gathered from the literature and the plot of D_s/D against the corresponding intrinsic viscosities is shown in Figure 16. To obtain the intrinsic viscosities, concentration has been expressed in grams per cubic centimeter multiplied by the partial specific volume of the protein. Examination of Figure 16 does indeed reveal a general correspondence between the diffusion ratio, D_s/D, and the intrinsic viscosities. There is, however, considerable scatter of the points in Figure 16 and

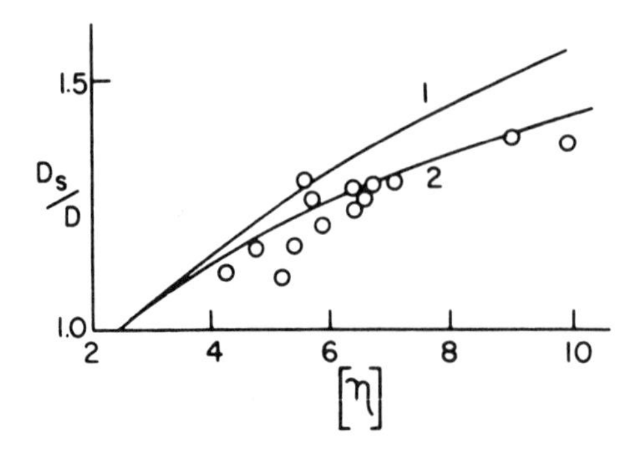

Fig. 16. Plot of the ratios D_s/D against the corresponding intrinsic viscosities of proteins (data from the literature). Solid lines are theoretical curves. Curve 1, Oblate ellipsoids of revolution. Curve 2, Prolate ellipsoids of revolution.

altogether the relationship is not very impressive. Theoretical curves for prolate and oblate ellipsoids of revolution are also included.

Rotary Diffusion Coefficient. Molecules in solution are undergoing constant random rotation due to Brownian motion. Imagine that all solute molecules in solution are oriented by the application of some external force so that their principal axes are parallel and suddenly the external force is removed. The position of any molecule after an interval of time may be characterized by the cosine of the angle α between its axis and the original direction of orientation. When the mean value of the cosine of α for all the solute molecules has fallen to $1/e$ where e is the base of the natural logarithms, the elapsed time is defined as the relaxation time, τ; the angle whose cosine corresponds to $1/e$ is 68° 25′.

As with the translational diffusion coefficient, the rotary diffusion coefficient, θ, is directly proportional to the kinetic energy of rotation, kT, and is inversely proportional to the rotary frictional coefficient f_r' and

$$\theta = \frac{kT}{f_r'} \qquad 63$$

The frictional coefficient, f_r', is dependent on the size and shape of the solute molecules as well as on the viscosity of the medium.

If the three axes of the solute molecule are all unequal, three rotary diffusion coefficients are needed and each would have a different frictional coefficient, but the kinetic energy would be the same for rotation about any of the three axes. The relations between the three relaxation times τ_a, τ_b and τ_c for rotation about each of the three axes and the corresponding rotary diffusion coefficients are

$$\tau_a = \frac{1}{\theta_b + \theta_c};\ \tau_b = \frac{1}{\theta_a + \theta_c};\ \tau_c = \frac{1}{\theta_a + \theta_b} \qquad 64$$

For an ellipsoid of revolution with b and c axes equal, the relations shown in eq. 64 become

$$\tau_a = \frac{1}{2\theta_b};\ \tau_b = \tau_c = \frac{1}{\theta_a + \theta_b} \qquad 65$$

If the solute molecule is spherical, there is only one rotary diffusion coefficient and only one relaxation time. For a spherical molecule f_r' is equal to $8\pi\eta r^3$ where η is the coefficient of viscosity and r is the radius of the molecule. For a sphere, eq. 65 becomes

$$\theta = \frac{1}{2\tau} = \frac{kT}{8\pi\eta r^3} \qquad 66$$

For a prolate ellipsoid of revolution rotated about the b-axis and whose major axis, $2a$, is at least 5 times larger than its minor axis, $2b$, the relation between the rotary diffusion coefficient and the dimension of the ellipsoid is

$$\theta_b = \frac{1}{2\tau_a} = -\frac{3kT}{16\pi\eta a^3} + \frac{3kT}{8\pi\eta a^3} \ln \frac{2a}{b} \qquad 67$$

The rotary diffusion coefficient given by eq. 67 is relatively insensitive to variation in the logarithmic term; the approximate value of a/b can be found from viscosity or translational diffusion measurements and the length of the elongated ellipsoid (or rod) calculated with the help of 67.

Measurement of Rotary Diffusion Coefficients. Several methods exist for the determination of rotary diffusion coefficients and these will be summarized briefly.

1. In flow birefringence, two concentric cylinders are employed, one of which is stationary and the other rotated; the solution of elongated molecules occupies the

space between the cylinders. It is seen that the experimental arrangement is somewhat similar to that described for the Couette viscometer but the flow gradients employed are usually much greater than for those used in viscosity studies. The solution between the two cylinders is viewed with crossed nicols or polaroids using monochromatic light. The liquid is isotropic and dark when at rest. Upon rotation of one of the cylinders, stream double refraction is observed; dark brushes or arms of a cross appear. The position of these brushes relative to the optic axis and to the flow gradient is related to the rotary diffusion coefficient of the solute molecules. The utility of flow birefringence is limited by the requirement of high flow gradients and is applicable only to fairly elongated molecules.

2. Dielectric dispersion can be used to estimate the rotary diffusion coefficient. Evidently, if the frequency of alternations of the charge on the condenser plates, between which is a solution, is progressively increased, a frequency will be reached at which neither the solute nor the solvent dipoles will be able to rotate with sufficient rapidity to respond to the alterations in the electric field and the measured dielectric constant will drop to low levels. If the molecules of the solute are enough larger than those of the solvent and the dielectric constant of the solution is plotted against the logarithm of the frequency, distinct steps in the decrease of the dielectric constant will be observed. The frequency corresponding to the half point of these steps is the critical frequency and is related in a simple way to the relaxation time; the rotary diffusion coefficient of the solute may then be obtained. Complicating factors can enter into the determination of the rotary diffusion coefficient by this method; the electrical double layer of the large solute molecule can provide an electrical vector and the electrolyte concentration has to be maintained at a low level.

3. The electrical birefringence method resembles that of dielectric dispersion; it is also a variation on the method used to determine the Kerr effect. A rectangular electrical pulse of considerable potential but of millisecond duration is applied across two electrodes about 2 mm. apart. The solution between the electrodes is illuminated between crossed nicols. The resulting light transmitted by the analyzing nicol is displayed as a photomultiplier signal on an oscilloscope screen. The rate of the decay of the birefringence after the electrical pulse has passed is directly proportional to the rotary diffusion coefficient. For relatively elongated molecules such as those of fibrinogen and of tobacco mosaic virus, this method yields satisfactory results. The technique is, however, limited by the requirement that the electrolyte concentration be small; otherwise it will be impossible to establish a sufficiently high potential between the electrodes without excessive heating and other effects. It was mentioned above that the method is related to the Kerr effect. Many years ago it was discovered that a transparent isotropic substance becomes doubly refracting if placed in a stationary electric field. The electric field tends to bring about an orientation of the molecules, partly because of the moments induced in them and partly because of the permanent dipoles they may possess. The Kerr constant for a given substance is given by

$$K = \frac{n_p - n_s}{n} \frac{1}{E^2} \tag{68}$$

where n is the refractive index of the incident light and n_p and n_s are the values for the emergent polarized light whose electric vectors vibrate parallel and perpendicular to the direction of the applied electric field of strength E.

4. Fluorescent polarization is the more interesting of the methods available for the study of phenomena related to rotary diffusion. Fluorescence has been discussed to some extent in Chapter 8, Solution Optics. Light is absorbed by a molecule and an electron is promoted to a higher energy level and on returning to the ground state emits radiation. Some energy is, however, lost prior to re-radiation and the return to the ground state is then accompanied by the emission of light of a longer wave-

length than that absorbed. The emission of the light is delayed by a time interval τ^* which represents the lifetime of the excited state. τ^* is usually in the range 10^{-9} to 10^{-7} seconds. Light is absorbed by the solute molecule only if the plane of its oscillator coincides or nearly so with the plane of vibration of the incident radiation and, on emission, each molecule emits a plane polarized wave parallel to its emission oscillator. Evidently, if the relaxation time of the solute molecule, τ, is less than the emission time of fluorescence, τ^*, the emitted light will be unpolarized. However, if in time τ^* the molecules whose oscillators were by chance oriented properly to receive the incident light have rotated to only a limited extent, the emitted light will be partially polarized.

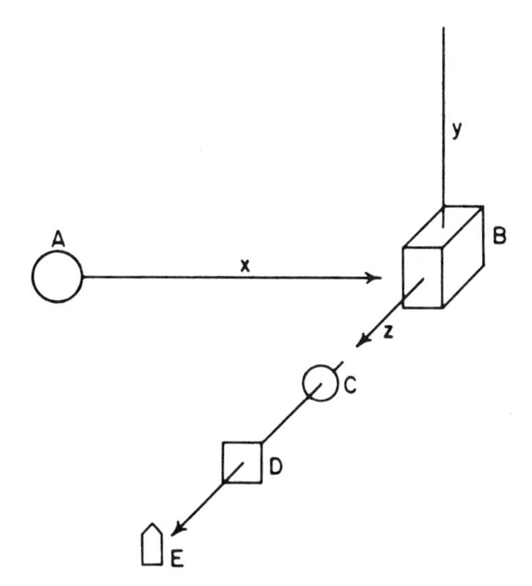

Fig. 17. Schematic diagram of the experimental arrangement for polarization fluorescence. *A*, Light source; *B*, cell containing solution of fluorescent solute; *C*, quartz depolarizer; *D*, polaroid filter; *E*, photomultiplier tube. (Adapted from P. Johnson and E. G. Richards: Arch. Biochem. Biophys. *97:* 250, 1962.)

The experimental arrangement for polarization fluorescence is schematically diagramed in Figure 17.

The fluorescent solution is placed at the origin of X, Y and Z axes. The X-axis is the direction of the incident light and the observation is made at 90° to this direction along the Z axis. The intensities of the horizontal, i_x, and of the vertical, i_y, components of the light are measured and the polarization, P, calculated as

$$P = \frac{i_y - i_x}{i_y + i_x} \qquad 69$$

The relaxation time, τ, is related to the polarization by

$$\frac{1}{P} + \frac{1}{3} = \left(\frac{1}{P_0} + \frac{1}{3}\right)\left(1 + \frac{3\tau^*}{\tau}\right) \qquad 70$$

where P_0 is a constant. Since τ is proportional to η/T where η is the coefficient of viscosity and T is the absolute temperature, a plot of $1/P$ vs T should yield a straight line whose intercept on the $1/P$ axis gives P_0 and the slope of the line will give τ provided τ^* is known. In case the solute molecule has two relaxation times corresponding to rotation about two axes, then τ, the relaxation time computed from experiment, is the harmonic mean of the two relaxation times and is given by

$$\frac{1}{\tau} = \frac{1}{2}\left(\frac{1}{\tau_a} + \frac{1}{\tau_b}\right) \qquad 71$$

The natural fluorescence of proteins, which is due to the aromatic residues in the proteins, is of low order and, accordingly, it is customary to conjugate the protein under mild conditions to a strongly fluorescent molecule such as l-dimethyl-amino-naphthalene 5 sulphonyl chloride. If the protein molecule is not rigid but shows flexibility, the study of the polarization fluorescence will reveal this because the calculated relaxation time will no longer be that of the molecule as a whole but will be shorter, corresponding to that of the flexible unit. Polarization fluorescence is not limited by electrolyte content or pH and, by increasing the viscosity of the medium with glycerol, measurement can be extended to smaller molecules.

GENERAL REFERENCES

Crank, J.: The Mathematics of Diffusion. Oxford University Press, London, 1956.

Fish, W. W., K. G. Mann, and C. Tanford: The estimation of polypeptide chain molecular weights by gel filtration in 6 M guanidine hydrochloride. J. Biol. Chem. *244:* 4989, 1969.

Gosting, L. G.: Measurement and interpretation of diffusion coefficients of proteins. Adv. Prot. Chem. *11*: 429, 1956.

Griffith, O. M. and C. R. McEwen: Determination of diffusion coefficients from Gouy fringes. Anal. Biochem. *18*: 397, 1967.

Jacobs, M. H.; Diffusion processes. Ergeb d. Biologie *12*: 1, 1935. (Reprinted Springer-Verlag, New York, 1967.)

Katchalsky, A. and P. F. Curran: Non-Equilibrium Thermodynamics in Biophysics. Harvard University Press, Cambridge, 1965.

Weber, G.: Rotational Brownian motion and polarization of fluorescence of solutions. Adv. Prot. Chem. *8*: 415, 1953.

PROBLEMS

1. The volume intrinsic viscosity of egg albumin was found to be 5.13 (assuming a hydrated density of the protein of 1.26). The diffusion coefficient of egg albumin in water at 25° is 8.8×10^{-7} sq. cm. per second. Estimate the ratio D_s/D for egg albumin as well as the molecular weight. (The anhydrous protein density is 1.35.)

   Ans: D_s/D equals 1.20, the mol. wt. 41,400

2. The measured diffusion coefficient of a certain protein is 9.3×10^{-7} sq. cm. per second at 25°. The molecular weight of this protein is 16,500 and its specific volume is 0.75. Estimate the ratio of the major to the minor axis of this protein, assuming the molecule to be an anhydrous prolate ellipsoid of revolution. Calculate the rotary diffusion coefficient about the minor axis.

   Ans: a/b is 9.6 and θ is 3.52×10^6

3. What are the dimensions of a rotary diffusion coefficient?

   Ans: Reciprocal time.

CHAPTER 12

Ion Transport

The transport of ions across cellular and subcellular membranes is a biological problem of the first magnitude. An ion in solution will move in response to gradients in electrical and chemical potentials. There are, however, other ways of transporting ions such as by the flow of an ionic solution or by combining the ion with a carrier molecule for which a gradient in chemical potential exists. Since ions move in response to electrical potential gradients, the movement of ions is apt to generate potential gradients. Such potentials arise at liquid junctions and across membranes.

Conductance in Solution. Specific resistance is the resistance in ohms of a column of a substance one centimeter long and one square centimeter in cross section; an ohm is the resistance which requires the application of one volt to produce a current flow of one ampere and Ohm's law is expressed by the relation

$$i=\frac{E}{R} \qquad 1$$

where i is the current flow in amperes, E is the voltage and R is the resistance in ohms.

Specific conductance (or conductivity) is the reciprocal of the specific resistance and is expressed in reciprocal ohm-centimeters or in mhos per centimeter. The specific conductance, K_C, is also equal to

$$K_C=\frac{iL}{EA} \qquad 2$$

where A is the area of each of two identical electrodes L distance apart. K_C can, therefore, be expressed in amperes per volt-centimeter. Equivalent conductance is the conductance in mhos of a solution containing one equivalent of the electrolyte enclosed between two electrodes one centimeter apart. The more dilute the solution, the greater the area of the electrodes required to accommodate the volume of the solution and, accordingly, the equivalent conductance increases with decreasing electrolyte concentration whereas the specific conductance decreases as the concentration is decreased. The relation between the specific conductance and the equivalent conductance is expressed as

$$\Lambda = K_c V_e \tag{3}$$

where Λ is the equivalent conductance, K_c is the specific conductance and V_e is the volume of solution which contains one gram equivalent weight of the solute, then

$$V_e = \frac{1{,}000}{C} \tag{4}$$

where C is the equivalent concentration of the solute. Combining eqs. 3 and 4, there results

$$\Lambda = \frac{1{,}000\ K_c}{C} \tag{5}$$

Equivalent conductance, therefore, has the dimensions of iL^2/E and can be expressed in ampere-centimeters per volt per centimeter.

Figure 1 shows the plot of the specific and equivalent conductance of potassium chloride as a function of the molar concentration.

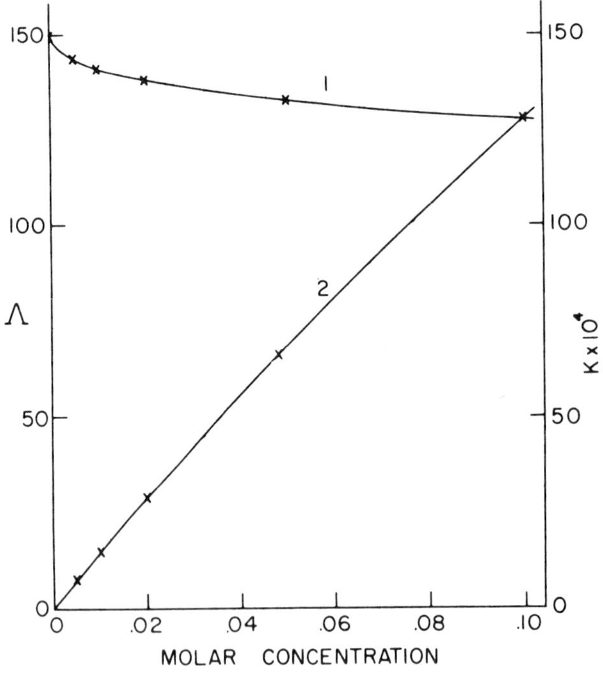

Fig. 1. Conductance of potassium chloride solutions as a function of concentration at 25°. Curve 1, Equivalent conductance; curve 2, specific conductance.

Conductance Measurement. The principle of the method is to provide two pathways for flow of the electrical current; one of these paths includes a known resistance and the other the unknown, and the two resistances are balanced against each other (see Fig. 2). The apparatus includes a source of alternating current, a detecting device for current flow and a variable condenser to compensate for the capacity of the conductivity cell. At the null point, the ratio R_1/R_2 is equal to a/b from which R_1 can be calculated. The specific conductance of the solution is determined by comparing the resistance of a known solution of electrolyte (usually

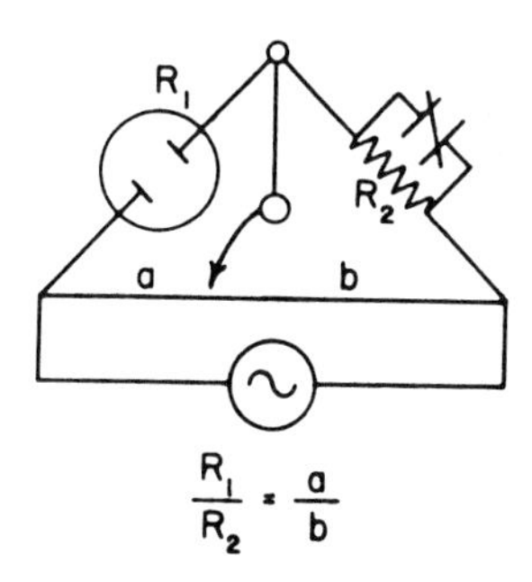

Fig. 2. Conductivity circuit. R_1 and R_2 are the resistances of the conductivity cell and the known resistance respectively. a and b are the arms of the Wheatstone bridge (fractions of the slide wire).

solutions of potassium chloride) with that of the unknown using the same conductivity cell. The cell constant is equal to the resistance of the potassium chloride solution multiplied by the specific conductance of the potassium chloride solution. Finally, the specific conductance of the unknown is equal to the cell constant divided by the resistance of the unknown. To obtain the specific conductance due to the dissolved electrolyte, the specific conductance of water is subtracted from the measured specific conductance. Water which has been distilled and preserved with reasonable care has a specific conductance of about 1×10^{-6} mho. To reduce the conductance significantly below this value requires an elaborate experimental procedure. Two common forms of conductivity cells are shown in Figure 3.

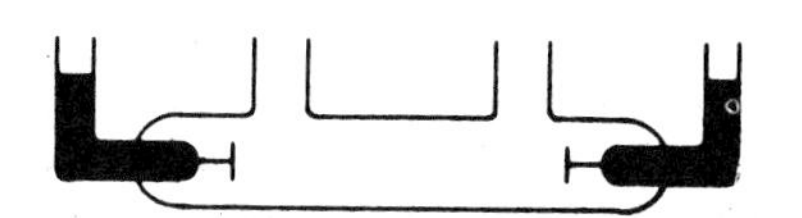
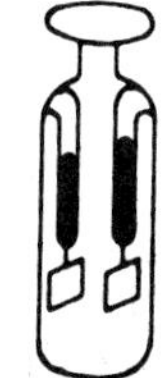

Fig. 3. Two common forms of conductivity cells.

Table 1 shows the values of the specific conductance of KCl solutions which are suitable for the calibration of conductivity cells. Concentrations are expressed as the grams of KCl per kilogram of solution in vacuum (weights must be corrected for the buoyancy of air at the atmospheric pressure in the laboratory).

TABLE 1. STANDARD VALUES FOR THE SPECIFIC CONDUCTANCE OF KCl SOLUTIONS AT 25° ACCORDING TO G. JONES AND B. C. BRADSHAW. (J. AMER. CHEM. SOC. 55: 1780, 1933.)

Grams KCl per Kilogram Solution	*Specific Conductance*
0.745263	0.0014088
7.41913	0.012856
71.1352	0.11134

Independent Migration of Ions. The equivalent conductance of an electrolyte can be set equal to the sum of the equivalent conductance of the cations and anions and

$$\Lambda = \lambda_+ + \lambda_- \qquad 6$$

where λ_+ and λ_- are the equivalent conductances of the cation and anion respectively. The fraction of the total current carried by an ion is its transference number defined as

$$t_+ = \frac{U}{U+V} = \frac{\lambda_+}{\lambda_+ + \lambda_-};\ t_- = \frac{V}{U+V} = \frac{\lambda_-}{\lambda_+ + \lambda_-} \qquad 7$$

where U and V are the mobilities of the cation and anion respectively. The ion conductances λ_+ and λ_- are expressed in coulomb-centimeters per second per volt per centimeter. Since there are 96,493 coulombs carried by one equivalent of ions, dividing λ_+ or λ_- by 96,493 gives the mobilities of the cation or anion in centimeters per second per volt centimeter.

Table 2 shows the individual mobilities of some ions at infinite dilutions at 25°.

TABLE 2. MOBILITIES OF IONS AT INFINITE DILUTION AND AT 25° EXPRESSED IN CENTIMETERS PER SECOND PER VOLT PER CENTIMETER

Ion	*Mobility*	*Ion*	*Mobility*
H^+	36.2×10^{-4}	OH^-	20.5×10^{-4}
Cs^+	8.0×10^{-4}	Br^-	8.13×10^{-4}
K^+	7.61×10^{-4}	I^-	7.97×10^{-4}
NH_4^+	7.60×10^{-4}	Cl^-	7.92×10^{-4}
Na^+	5.20×10^{-4}	NO_3^-	7.40×10^{-4}
Li^+	4.01×10^{-4}	CH_3COO^-	4.25×10^{-4}
Ca^{++}	6.16×10^{-4}	SO_4^{--}	8.28×10^{-4}

At infinite dilution, the force exerted on an ion by an electrical potential is ϵZE where ϵ is unit ionic charge, Z is the valence of the ion and E is the potential gradient in electrostatic volts. For a unit potential gradient and for a univalent ion this force is 1.60×10^{-12} dyne. The velocity of any particle moving through a viscous medium is equal to the ratio of the force acting divided by the resisting forces. For a spherical particle, the resisting force is Stokes factor $6\pi\eta r$ where η is the coefficient of viscosity and r is the radius of the particle. Accordingly, the ideal velocity of an ion should be

$$u = \frac{\epsilon ZE}{6\pi\eta r} \qquad 8$$

The calculated mobility of a sodium ion with a radius of 0.95×10^{-8} cm. at 25° is 10×10^{-4} m. per second per volt per centimeter. The observed velocity is 5.2×10^{-4} (see Table 2). The discrepancy between the observed and calculated mobility is due principally to two causes: (1) the effective diameter of the sodium ion is larger than the crystal lattice radius because the ion is hydrated and (2) the Stokes factor for the resisting forces experienced by a spherical particle is no longer valid because the ionic radius is comparable with the radius of the water molecules.

As noted previously, the equivalent conductance of an electrolyte decreases with increasing concentration; the electrophoretic and the relaxation effects are primarily responsible for the influence of concentration on conductance. The electrophoretic effect arises because an ion moving in a viscous medium drags solution along with it. The neighboring ions, therefore, move against or with the motion of the solvent. This effect will be clearly dependent on the distance between the ions and, accordingly, concentration dependent.

The relaxation effect exists because the symmetrical distribution of ions in the neighborhood of a given ion will be disturbed by the externally applied field. This disturbance in the ionic atmosphere produces an electrical force acting on the ion which opposes the applied external field and, accordingly, leads to a decrease in the mobility of the ion—a decrease which is concentration dependent. Taking both the electrophoretic and the relaxation effects into consideration gives an intricate equation for the conductance which, however, reduces to Onsager's limiting law for dilute solutions and has the form

$$\Lambda = \Lambda_0 - A\left(\frac{\Gamma}{2}\right)^{1/2} \qquad 9$$

where Λ_0 is the equivalent conductance at infinite dilution and $\Gamma/2$ is the ionic strength. The coefficient A is fairly complex; for a uni-univalent electrolyte and after inserting the values for the physical constants, A reduces to a simple expression, that is

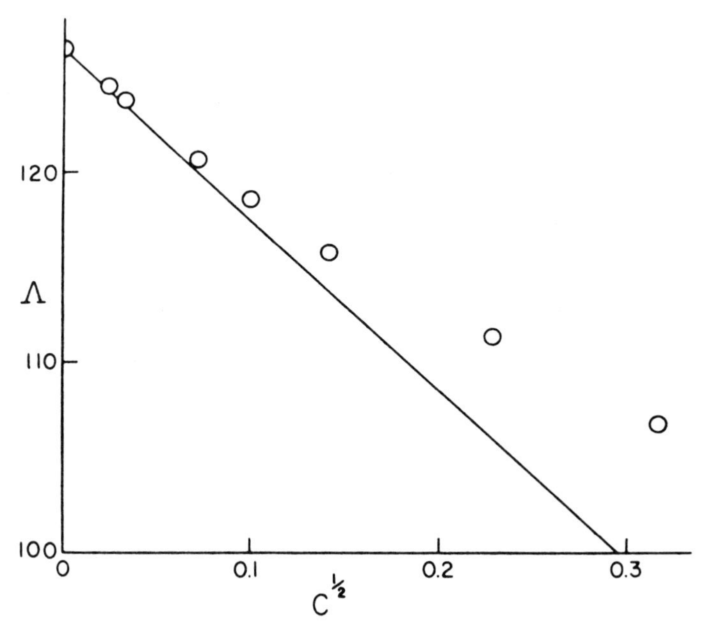

Fig. 4. Equivalent conductance of sodium chloride solutions as a function of the square root of the concentration at 25°. Open circles are experimental points. The line is a plot of Onsager's limiting law (eq. 9).

$$A = \frac{0.9945 \times 10^4}{(DT)^{3/2}} + \frac{1}{\eta(DT)^{1/2}} \tag{10}$$

where D is the dielectric constant, T the absolute temperature and η is the coefficient of viscosity. Figure 4 shows a plot of eq. 9 along with the experimental points for solutions of sodium chloride.

The Onsager limiting law is of use in the extrapolation of equivalent conductance data to infinite dilution. It is also a substantial adjunct in the calculation of the ionization constant of weak electrolytes from conductance measurements.

Ionization Constants from Conductance. Conductance measurements have long been used to obtain the ionization constants of weak electrolytes and the method still has utility for this purpose; under some conditions it may be the only method which is experimentally feasible.

The degree of dissociation of a weak electrolyte, α, is the fraction of the electrolyte existing in the ionic forms. It is equal to Λ/Λ_e where Λ is the measured equivalent conductance and Λ_e is the equivalent conductance of the ions at the concentration in question. For a very dilute solution of a pure weak electrolyte, the ionization constant is given by Ostwald's dilution law which is

$$K = \frac{\alpha^2 C}{1 - \alpha} = \frac{\Lambda^2 C}{\Lambda_e(\Lambda_e - \Lambda)} \tag{11}$$

where C is the total concentration of the electrolyte—both ionized and un-ionized. To calculate the ionization constant of a weak electrolyte from conductance measurements, three pieces of information are needed: (1) The value of the equivalent conductance at infinite dilution, Λ_0, must be known, (2) the value of Λ_e at the experimental concentrations must be calculated, and (3) the specific conductance of dilute solutions of the pure weak electrolyte must be determined by experiment.

It is not practical to extrapolate the measured conductance of the weak electrolyte to infinite dilution to evaluate Λ_0 directly because the conductance is too critical a function of concentration. Instead, the following procedure is adopted:

Suppose, for example, it is necessary to estimate the ionization constant of acetic acid. Since each ion has its own equivalent conductance and the sum of the ionic conductance is equal to the conductance of the completely ionized electrolyte, the following relation is valid

$$\Lambda_{HAc} = \Lambda_{HCl} + \Lambda_{NaAc} - \Lambda_{NaCl} = \lambda_{H^+} + \lambda_{Cl^-} + \lambda_{Na^+} + \lambda_{Ac^-} - \lambda_{Na^+} - \lambda_{Cl^-} \quad 12$$

To obtain Λ_0 for acetic acid, it is, therefore, necessary to measure the conductance of solutions of HCl, of NaAc and of NaCl as a function of concentration and extrapolate each to zero concentration. The specific conductance of the ionized part of acetic acid is then calculated as a function of concentration by the use of Onsager's limiting law and a knowledge of Λ_0. The calculated values of the specific conductance are then compared with the experimental from which Λ_e can be obtained, and the ionization constant calculated with the help of eq. 11.

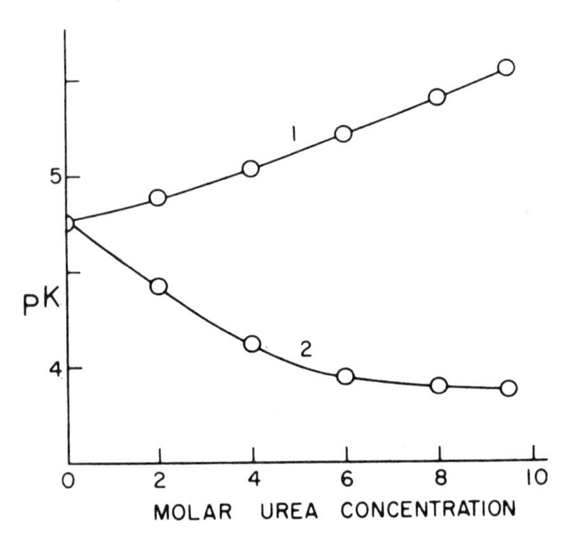

Fig. 5. Variation of the pK of acetic acid with urea concentration at 30°. Curve 1, From pH measurements; curve 2, from conductance measurements. (H. B. Bull, K. Breese, G. L. Ferguson and C. A. Swenson: Arch. Biochem. Biophys. 104:297, 1964.)

It has been known for some time that the pH of a buffer increases in the presence of urea. Shown in Figure 5, curve 1 is the pH of a 0.05 molar sodium acetate-acetic acid buffer as a function of the molar concentration of urea; the ratio of salt to acid was fixed at unity. Under these conditions the measured pH should very nearly equal the pK of acetic acid. It appears that the pK of acetic acid increases with increasing urea concentration; the acid becomes weaker. However, curve 2 shows that the pK as measured from conductance decreases with increasing urea concentration; the acid becomes stronger in the presence of urea. For reasons which are still not completely clear, the glass electrode (and also the hydrogen electrode) does not give a true measure when urea is present. In any event, Figure 5 serves to illustrate the importance of conductometric methods for the evaluation of ionization constants of weak electrolytes.

Conductance of Suspensions. The equation describing the conductance of a suspension of conducting particles is

$$\frac{K_c - K_{c_1}}{K_c + fK_{c_1}} = \phi \frac{K_{c_2} - K_{c_1}}{K_{c_2} + 2K_{c_1}} \quad 13$$

where K_c is the specific conductance of the suspension, K_{c_1} that of the suspending liquid, K_{c_2} that of the suspended particles, and ϕ is the ratio of the volume of the particles to the total volume of the suspension. For spherical particles f, the form factor is 2.0 and equation 13 reduces to the well-known Maxwell equation. Fricke developed the dependence of f on the ratio of the axes of suspension of ellipsoids of revolution. As the ratio of the major to the minor axis increases, f approaches 1.5 as the limiting value for prolate ellipsoids and zero for oblate ellipsoids. The Maxwell-Fricke equation has been widely used in the analysis of conductance experiments on suspensions of living cells.

It is found experimentally for isoelectric protein solutions that the function $(1 - R_1/R)/w$ is independent of w, the weight protein per ml., and of R_1/R where R is the specific resistance of the protein solution and R_1 is the specific resistance of the dialysate of the protein-salt solution. In analogy to viscosity, the function $(1 - R_1/R)/w$ can be called the intrinsic resistivity and denoted by $[R]$. Table 3 shows the intrinsic resistivities of a series of proteins at their isoelectric points.

The fibrinogen molecule is rod-shaped (actually, three modules on a peptide thread) with an asymmetry of about 8 to 1 and yet its intrinsic resistivity does not differ greatly from those of the globular proteins (see Table 3). It seems likely that the intrinsic resistivities reflect fairly directly the volume occupied by the protein in solution.

TABLE 3. INTRINSIC RESISTIVITIES [R] OF SOME PROTEINS. 25° (H. B. Bull and K. Breese: J. Coll. Interface Sci. *29*: 492, 1969.)

Salt	*Protein*	$[R]$
0.20M NaCl	Egg albumin	2.04
0.20M NaCl	Methemoglobin (bovine)	1.96
0.30M KCl	Fibrinogen (bovine)	2.23
0.20M NaCl	Myoglobin (whale)	1.95
0.20M NaCl	Serum albumin (bovine)	2.11

The Coulter Counter is an important instrument for the analysis of the size distribution of suspended particles and covers the range from about 0.25μ radius up to fairly large particles. A measured quantity of the dilute suspension is drawn through a fine aperture; as each particle passes through the aperture, it momentarily interrupts an electrical circuit which is completed through the aperture. This fluctuation in the current is recorded on an electronic counter. The sensitivity of the counter can be varied and, accordingly, the particle size distribution of the suspension can be scanned.

Diffusion of Ions. The conditions leading to the transport of ions in an electrical potential gradient have been briefly reviewed. Ions will also move in response to a concentration gradient; the heat motion of the ion (Brownian motion) results in diffusion. The force acting on an ion is the gradient in the chemical potential and is equal to $-\frac{1}{N}\frac{d\mu_1}{dx}$ where N is Avogadro's number. However, the anions and cations must move with the same velocity to maintain electrical neutrality so that the faster moving ion is slowed down and the slower moving ion is speeded up. The effect arising from unequal mobilities of the ions may be represented as an electrical field intensity, E, which exerts on each ion an additional force equal to $Z_1\epsilon E$ *and* $Z_2\epsilon E$ for the cations and anions respectively. The total forces are, therefore,

$$F_1 = -\frac{1}{N}\frac{d\mu_1}{dx} + Z_1\epsilon E \qquad 14$$

and

$$F_2 = -\frac{1}{N}\frac{d\mu_2}{dx} + Z_2\epsilon E \qquad 15$$

The force acting on the ions of mobilities U_1 and U_2 to produce the same velocity V is then

$$V = U_1\left(-\frac{1}{N}\frac{d\mu_1}{dx} + Z_1\epsilon E\right) = U_2\left(-\frac{1}{N}\frac{d\mu_2}{dx} + Z_2\epsilon E\right) \tag{16}$$

ϵE is the same for both ions and, accordingly,

$$\frac{1}{Z_1}\left(\frac{V}{U_1} + \frac{1}{N}\frac{d\mu_1}{dx}\right) = \frac{1}{Z_2}\left(\frac{V}{U_2} + \frac{1}{N}\frac{d\mu_2}{dx}\right) \tag{17}$$

since to maintain electrical neutrality,

$$\nu_1 Z_1 + \nu_2 Z_2 = 0 \tag{18}$$

where ν_1 and ν_2 are the number of mole ions produced per mole of electrolyte. Combining eqs. 17 and 18, there results

$$V = -\frac{U_1 U_2}{N\,(\nu_1 U_2 + \nu_2 U_1)}\frac{d\mu}{dx} \tag{19}$$

where μ is equal to the sum of μ_1 and μ_2. The flux of the electrolyte is

$$J = CV = -\frac{U_1 U_2 C}{N\,(\nu_1 U_2 + \nu_2 U_1)}\frac{d\mu}{dC}\frac{dC}{dx} \tag{20}$$

The flux, J, is related to the diffusion coefficient by

$$J = -D\frac{dC}{dx} \tag{21}$$

Then, D, the diffusion coefficient of the electrolyte, is

$$D = \frac{U_1 U_2}{\nu_1 U_2 + \nu_2 U_1}\frac{1}{N}\frac{d\mu}{d\ln C} \tag{22}$$

But

$$\frac{d\mu}{d\ln C} = RT\,(\nu_1 + \nu_2)\left(1 + \frac{d\ln\gamma_\pm}{d\ln C}\right) \tag{23}$$

then

$$D = \frac{U_1 U_2}{\nu_1 U_2 + \nu_2 U_1}\frac{RT\,(\nu_1 + \nu_2)}{N}\left(1 + \frac{d\ln\gamma_\pm}{d\ln C}\right) \tag{24}$$

where $\gamma_\pm$ is the mean activity coefficient of the electrolyte.

The relaxation effect is absent in diffusion but there is a small electrophoretic effect and, in more concentrated solutions, the motion of the solvent must be considered. At zero concentration, the factor $1 + \frac{d\ln\gamma_\pm}{d\ln C}$ becomes unity and for a uni-univalent salt the diffusion coefficient is

$$D^0 = \frac{2U_1 U_2}{U_1 + U_2}\frac{RT}{N} \tag{25}$$

Dividing eq. 24 by eq. 25 gives

$$\frac{D}{D^0} = \left(1 + \frac{d\ln\gamma_\pm}{d\ln C}\right) \tag{26}$$

It should be remarked immediately that eq. 26 is valid only at very low electrolyte concentrations. In a mixture of electrolytes there is strong coupling between the diffusion of the various ions present and, in general, it is not possible to predict diffusion rates in electrolyte mixtures. The theory of ionic diffusion is complex and will not be further considered. Figure 6 shows the experimental values of D/D^0 as a function of salt concentration for three uni-univalent electrolytes.

The measurement of diffusion coefficients has been discussed in the previous chapter. Potassium chloride solutions are frequently used as a standard for the calibration of a porous disk. At 25°, D for a 0.1 molar KCl solution is 1.844×10^{-5} sq. cm. per second.

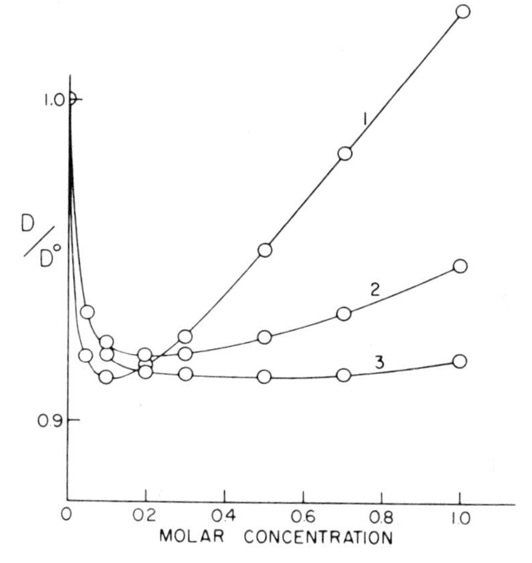

Fig. 6. Experimental variation of the diffusion coefficient with concentration. D° is the diffusion coefficient at zero concentration, 25°. Curve 1, HCl; curve 2, KCl; curve 3, NaCl.

Liquid Junction Potentials. If a boundary is formed between an electrolyte solution and the pure solvent or between a concentrated and more dilute solution, diffusion of the electrolyte immediately begins. The more mobile ion will tend to diffuse ahead; this will lead to the establishment of a potential difference which will retard its motion and a steady state is rapidly established in which both kinds of ions diffuse at the same rate. The free energy change for a transfer of an infinitesimal of electricity is

$$dF = -dqdE \tag{27}$$

where q is the quantity of electricity and E is the difference in potential. Then per Faraday transferred

$$-FdE = \Sigma \frac{t_i}{z_i} d\mu_i \tag{28}$$

where t_i is the transport number of the ion i, Z_i is its valency and the summation includes all ions.

Equation 28 can be written

$$-dE = \frac{RT}{F} \Sigma \frac{t_i}{z_i} d \ln a_i \tag{29}$$

The transport numbers vary with concentration so that t_i is not constant but varies through the diffusion boundary and, accordingly, the total free energy change per Faraday transferred must be found by integration across the diffusion boundary and eq. 29 should be stated as

$$-E = \frac{RT}{F} \int_1^2 \frac{t_i}{z_i} d \ln a_i \tag{30}$$

Equation 30 is to include not only the transport of all ions but likewise the transport of water and as such is a valid expression for liquid junction potentials.

Unfortunately, it is not possible to integrate this equation without some simplifying assumptions.

Suppose a boundary is formed between solutions of a uni-univalent salt such as sodium chloride at two different concentrations. If it is assumed that the transport numbers of sodium and chloride do not change significantly through the boundary and that there is no water transfer, eq. 30 can be written

$$-E_L = \frac{RTt_{Na^+}}{F} \int_1^2 d \ln a_{Na^+} - \frac{RTt_{Cl^-}}{F} \int_1^2 d \ln a_{Cl^-} \qquad 31$$

The next assumption to be made is that the activities of the sodium ions and the chloride ions, at any given location, are equal. If this be true, then eq. 31 becomes

$$-E_L = (t_{Na^+} - t_{Cl^-}) \frac{RT}{F} \int_1^2 d \ln a_{NaCl} \qquad 32$$

which can now be integrated to give

$$-E_L = \frac{U - V}{U + V} \frac{RT}{F} \ln \frac{a_2}{a_1} \qquad 33$$

The above discusses the boundary formed between solutions of the same electrolyte at two different concentrations. The general problem of liquid junction potentials was considered a number of years ago by M. Planck and by P. Henderson who derived expressions for potentials for several different situations. Suppose all ions are univalent but the ions on one side of the boundary are all different from those on the other side, for example, solutions of KCl and of NaCNS. The equation which was obtained is

$$E_L = \frac{RT}{F} \frac{(U_1 - V_1) - (U_2 - V_2)}{(U_1 + V_1) - (U_2 + V_2)} \ln \frac{U_1 + V_1}{U_2 + V_2} \qquad 34$$

where U_1, U_2, V_1 and V_2 are defined as follows:

$$\mathrm{U}_1 = C_1^{+\prime} U_1^{+\prime} + C_2^{+\prime} U_2^{+\prime} \qquad 35$$

$$U_2 = C_1^{+\prime\prime} U_1^{-\prime\prime} + C_2^{+\prime\prime} U_2^{+\prime\prime} \qquad 36$$

$$V_1 = C_1^{-\prime} U_1^{-\prime} + C_2^{-\prime} U_2^{-\prime} \qquad 37$$

$$V_2 = C_1^{-\prime\prime} U_1^{-\prime\prime} + C_2^{-\prime\prime} U_2^{-\prime\prime} \qquad 38$$

where C_1', C_2' are the equivalent concentrations of the ions in solution 1 and C_1'', C_2'' are the corresponding concentrations in solution 2. $U_1^{+\prime}$, $U_2^{+\prime}$, $U_1^{+\prime\prime}$, $U_2^{+\prime\prime}$ and $V_1^{-\prime}$, $V_2^{-\prime}$, $V_1^{-\prime\prime}$, $V_2^{-\prime\prime}$ are the mobilities of the cations and anions in the two solutions respectively.

Suppose both ionic solutions contain uni-univalent electrolytes but they have a common ion between them such as between solutions of KCl and NaCl, and further suppose both solutions are at the same concentration, then the expression for the liquid junction potential becomes

$$E_L = \frac{RT}{F} \ln \frac{U^{+\prime} + U^{-\prime}}{U^{+\prime\prime} + U^{-\prime\prime}} \qquad 39$$

For strong electrolytes (completely ionized) and for two solutions at the same concentration, eq. 39 reduces to

$$E_L = \frac{RT}{F} \ln \frac{\Lambda'}{\Lambda''} \qquad 40$$

where Λ' and Λ'' are the equivalent conductances of solutions 1 and 2 respectively.

Liquid junction potentials can be of considerable magnitude. An EMF cell in the form

$$\text{Ag} \mid \text{AgCl, M'Cl} \parallel \text{M''Cl, AgCl} \mid \text{Ag} \quad 41$$

can be used to measure liquid junction potentials (assuming the chloride ion activities in the two compartments to be identical). M′Cl and M″Cl are two solutions of two different electrolytes separated by a liquid junction. Both solutions are at the same concentration and both contain chloride ions. The only potential is at the liquid junction and eq. 40 should apply. Table 4 shows a comparison of the observed and calculated (eq. 40) potentials.

TABLE 4. LIQUID JUNCTION POTENTIALS OBSERVED AND CALCULATED (eq. 40) FROM THE CELL 42. BOTH SOLUTIONS ARE 0.1 M. (DATA OF D. A. MACINNES AND Y. L. YEH: J. AMER. CHEM. SOC. 43: 2563, 1921.)

	EMF IN MV	
Junction	*Observed*	*Calculated*
HCl:KCl	+ 26.78	+ 28.52
KCl:NaCl	+ 6.42	+ 4.86
KCl:LiCl	+ 8.76	+ 7.62
NaCl:LiCl	+ 2.62	+ 2.76
LiCl:NH_4CL	− 6.93	− 7.57

There are several modern pH meters commercially available which can be read to 0.001 pH unit. At 25°, 0.001 pH unit corresponds to 0.059 millivolts. It is clear that if the liquid junction potential in the electrode arrangement changes significantly when the standard buffer is replaced by the test solution, a claim of an accuracy of 0.001 pH unit is fallacious. Liquid junction potentials are apt to be especially troublesome in the determination of the pH of protein solutions because of the presence of the slow moving protein ion.

Membrane Potentials. There are, in general, three sources of potentials at and across membranes, i.e., (1) distribution potentials, (2) diffusion potentials and (3) dipolar potentials. The dipolar potentials arise from the orientation of polar molecules at the membrane interface along with the potential of the ionic double layer. The potential across a non-conducting membrane is the difference in potential between two interfacial potentials; this is the situation of a plane plate condenser with a dielectric between the plates. Interfacial potentials have been described in Chapter 9 and it is to be recalled that such potentials are measured with a vibrating plate electrode or with an electrode upon which polonium has been deposited. Interfacial potentials cannot be measured with reversible electrodes because there is no mechanism for ion flow. Interfacial potentials can, however, modify the membrane surface in respect to the rate of ion transport across the membrane and in this sense make a contribution to the diffusion potential.

A distribution potential results from the unequal solubilities of the cation and anion in an oil phase. However, when a solution containing an organic ion and an inorganic ion is placed in contact with an oil phase, the system comes to equilibrium fairly quickly and the distribution potential disappears and a diffusion potential remains. These potentials can be measured with reversible electrodes such as silver-silver chloride.

A great many measurements have been made on systems represented in Figure 7 and the membrane which separates the two aqueous salt solutions at concentrations C_1 and C_2 has been composed of a vast variety of substances. The materials which have attracted the greatest interest are collodion membranes, ion exchange resin membranes and various oil or lipid membranes.

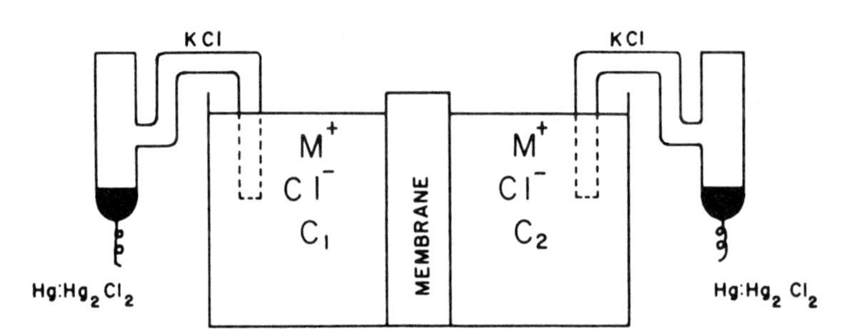

Fig. 7. Diagrammatic arrangement for the measurement of membrane potentials.

Aside from the filtering properties of collodion membranes, much interest is attached to their electrochemical properties. By casting collodion membranes and permitting them to dry rather thoroughly before water is added, membranes with very small pores are obtained. If these membranes have ionogenic groups, they show pronounced electrochemical character.

Sollner and co-workers have studied the electrical behavior of collodion membranes in much detail. Collodion membranes oxidized by treatment with sodium hydroxide produce carboxyl groups in the membrane. Such membranes are negatively charged and restrict the movement of anions. Positively charged collodion membranes can be prepared by immersing the membranes in solutions of protamine chloride or sulfate buffered to a pH of about 10.5. The protamine which is a basic protein with a relatively small molecular weight is adsorbed on the membrane and, after several days' exposure to the protamine solution, the membranes are washed and ready for use.

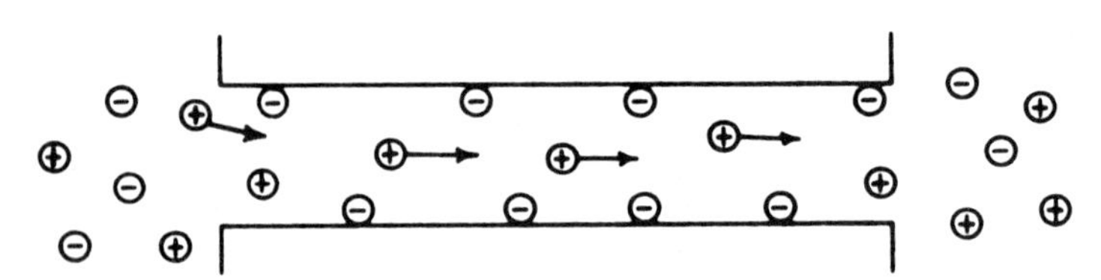

Fig. 8. Charged pore with fixed negative sites showing exclusion of soluble negative ions.

The reason for the partial and, sometimes, the almost complete exclusion of cations or anions from the pores of a membrane can perhaps be understood by an inspection of Figure 8 which is a diagrammatic representation of a membrane containing fixed negative sites (ionized carboxyl or sulfonic acid groups).

Provided that the mobility of either the cation or of the anion through the membrane has been effectively reduced to zero, eq. 33 becomes (the contribution of water movement is neglected)

$$E = \frac{RT}{nF} \ln \frac{a_1^+}{a_2^+} \tag{42}$$

or

$$E = -\frac{RT}{nF} \ln \frac{a_1^-}{a_2^-} \tag{43}$$

for negative and positive membranes respectively.

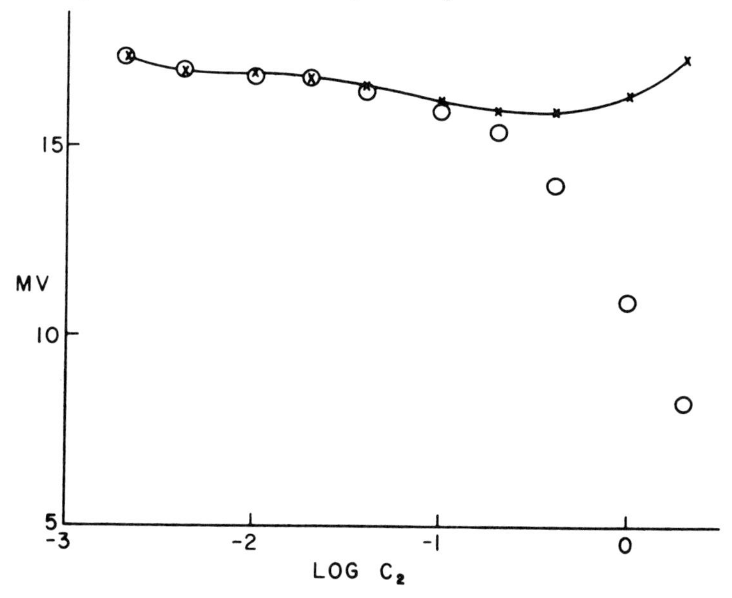

Fig. 9. Membrane potentials in millivolts as a function of log C_2, C_2/C_1 being maintained constant at a ratio of 2:1, potassium chloride solutions, oxidized collodion, 25°. Circles are experimental and solid line is calculated (eq. 42). (Data of K. Sollner, S. Dray, E. Grim and R. Neihof: Ion Transport Across Membranes. Academic Press, New York, 1954.)

Figure 9 shows a comparison of the theoretical potentials across an oxidized collodion membrane with those determined experimentally for solutions of potassium chloride with the ratio of the salt concentration in one compartment to that in others maintained at 2 to 1 and the total salt concentration varied and expressed in terms of the log of the larger concentration of potassium chloride.

Anomalous Osmosis. The establishment of a potential gradient across a charged membrane results in an electroosmotic flow of water through the membrane (see Chapter 13 for a description of electroosmosis). This flow is known as anomalous osmosis. If the flow takes place from the dilute to the concentrated side, the process is said to be positive anomalous osmosis and if water flows from the concentrated to the more dilute side, it is called negative anomalous osmosis.

The conditions for negative anomalous osmosis may be outlined as follows: If the ion whose sign of charge is the same as that of the membrane has the greater mobility through the membrane, negative anomalous osmosis will occur. This situation is apt to arise with lithium (slow ion) salts and with a negatively charged membrane with fairly large pores.

Experimentally, it is much easier to produce positive anomalous osmosis than negative anomalous osmosis. The conditions to be fulfilled are small charged membrane pores which exclude the ion of opposite sign. The water then flows from the dilute into the concentrated salt solution. The conditions for positive and negative

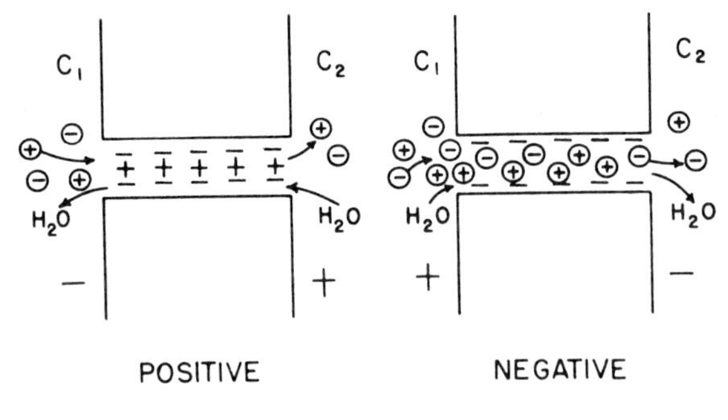

Fig. 10. Conditions for positive and negative anomalous osmosis through a charged membrane. The electrolyte concentration C_1 is greater than C_2. In negative anomalous osmosis the mobility of the anion is much greater than that of the cation.

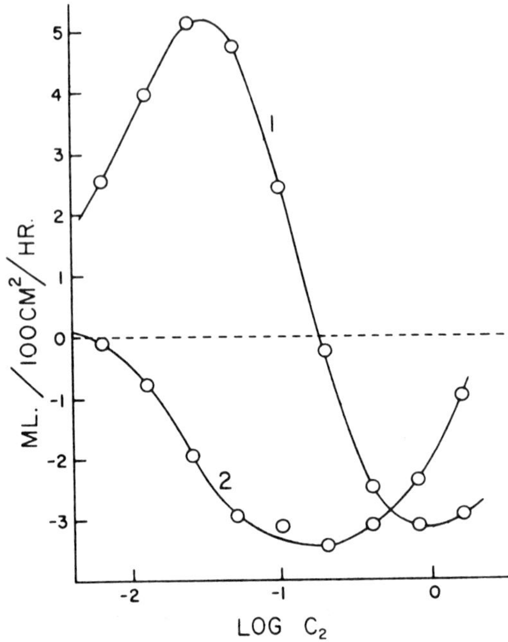

Fig. 11. Water flow rates across a charged oxyhemoglobin collodion membrane, C_2/C_1, is 2:1. Curve 1, LiCl. Curve 2, $MgCl_2$. 25°. (Data of K. Sollner, S. Dray, E. Grim and R. Neihof: Ion Transport across Membranes. Academic Press, New York, 1954.)

anomalous osmosis are diagramed in Figure 10 where C_1 is greater than C_2.

To decide which part of the water flow is due to ordinary osmosis and which part to positive anomalous osmosis either a non-electrolyte of equal osmolar concentration is compared with the flow produced by the electrolyte or a membrane is prepared which can be made isoelectric. Such a membrane is produced by permitting the collodion membrane to adsorb a protein such as hemoglobin and adjusting the membrane solution to the isoelectric point of hemoglobin.

Figure 11 shows the flow rates across a charged oxyhemoglobin-collodion membrane as a function of the electrolyte concentration. Notable is the fact that lithium chloride solutions exhibit positive anomalous osmosis at low electrolyte concentration which becomes negative as the concentration is increased.

It should be noted that anomalous osmosis is a transient phenomenon and disappears as an equilibrium state is approached. Anomalous osmosis is of considerable interest in its own right but how far model systems of this kind can serve as an analogy for living membranes is questionable.

The ion exchange resin membranes are useful for the determination of ionic activities. For example, Scatchard et al. have employed them to determine the anion binding of serum albumin, the potential being established lengthwise through a strip of the membrane thus providing a path of about 3 cm. through which ion discrimination can occur.

Lipid membranes are easily constructed by placing a millipore filter in a solution of the lipid in a volatile fat solvent. After the membrane has been saturated, it is withdrawn and the volatile solvent is permitted to evaporate. The membrane is then clamped in a suitable container. The electrical resistance varies greatly depending on the nature of the lipid used. The resistance of cholesterol membranes is very great whereas that of phospholipids is more moderate. The resistance of phospholipid membranes is responsive to electrolyte enviroment; calcium salts tend to increase the resistance whereas potassium salts make the membrane more permeable. Lipid membranes are of interest because they provide a more realistic model for living membranes.

Cellular Membranes. Living membranes always exhibit a potential difference, the inside of the cell being, in general, negative to the outside. If a living cell or tissue is injured so that the membrane is destroyed or the metabolism is interrupted, the electrolytes diffuse into or out of the cell depending on the direction of the concentration gradients and this diffusion of ions gives rise to injury potentials in which the injured tissue is negative to the uninjured tissue. Action currents are closely related to, if not identical in origin with, injury potentials, and action currents are associated with any stimulation of protoplasm. The most dramatic example of an action current is that produced by an electric eel. The electric organ of the electric eel is formed by compartments each containing one cell, the electric plate, and arranged in columns. The action potential developed by a single plate is about 0.15 volts. Electrophorus electricus has 5000 to 6000 plates and the voltage discharge is about 500 volts whereas in the Torpedo marmorta the number of elements is about 500 and the voltage is about 40.

Conductance and Frequency. If a condenser is connected in series to a direct current source, current will flow momentarily until the condenser is charged and the build-up of the capacitor voltage, E_c, is given by

$$E_c = E\ (1 - e^{-t/RC}) \qquad 44$$

where E is the applied voltage, t is the time in seconds, R is the resistance in ohms and C is the capacity of the condenser in farads; RC is the time constant for the condenser. However, with an alternating current the condenser is charged and discharged and charged again as the polarity of the alternating current changes and the current appears to flow through the condenser. In a pure resistance, the voltage and current are always in phase but in a pure capacitance the current leads the voltage by 90°. In a mixed circuit containing both resistance and capacitance, the difference in phase between the voltage and current can be anything between 0° and 90° depending on the relative magnitudes of the resistance and capacitance. This difference in phase is known as the phase angle. For a sinusoidal alternating current the ratio of the average, the maximum or the root mean square of the potential difference to the corresponding measure of the current flow is known as the capacitive reactance and, in a sense, is equivalent to a resistance. (There is also an inductive reactance for a circuit containing an inductance.) The capacitive reactance is given by

$$X_c = \frac{1}{2\pi\nu C} \qquad 45$$

Where ν is the frequency of the alternating current. It can be seen from eq. 45 that as the frequency increases the reactance decreases. A combination of resistance and reactance is known as an impedance.

Biological systems (living cells), in general, constitute mixed circuits containing both resistance and capacitance and can be represented by an equivalent circuit (see Fig. 12). The paths of current flow as a function of the frequency is

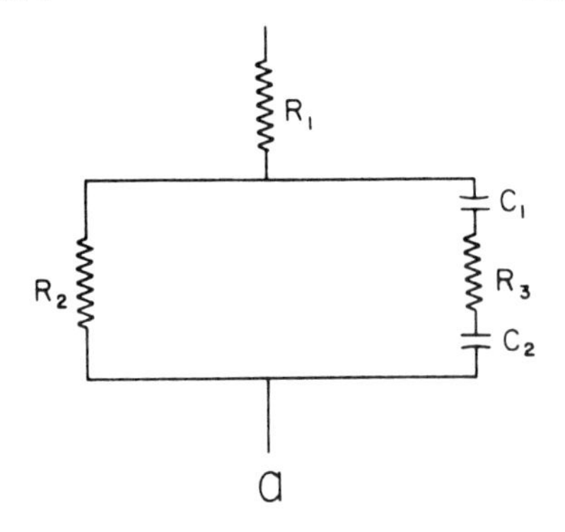

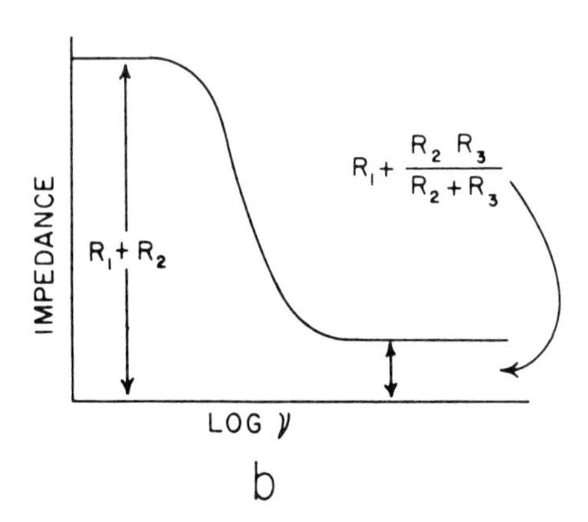

Fig. 12. *a*, Circuit equivalent to a living cell. R_1 is the resistance through the medium to the cell, R_2 resistance around the cell, R_3 resistance inside the cell; C_1 and C_2 are the capacities of the cell walls. *b*, The change of the impedance of the equivalent circuit with frequency.

further clarified in Figure 13.

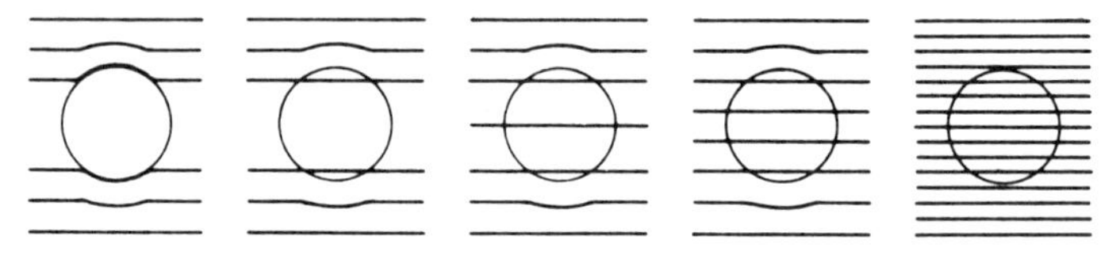

Fig. 13. Paths of current flow around a spherical cell. Frequency progressively increased from left to right. (Adapted from K. S. Cole.)

The capacity of cellular membranes is, in general, about one microfarad per square centimeter and does not vary under a variety of conditions. The observed changes range from a few percent during excitation (of nerve), current flow and variation of the external environment up to about two-fold at fertilization, under narcotics and anesthetics, in rigor and even close to death. Table 5 shows the capacities of some membranes which have been measured.

TABLE 5. CAPACITIES OF SOME CELLULAR MEMBRANES (DATA OF K. S. COLE, FOUR LECTURES ON BIOPHYSICS, UNIVERSITY OF BRAZIL, 1947)

Cell	*Capacities in Microfarads per sq. cm.*
Plant cells	0.33-1.0
Marine eggs	0.70-2.7
Red blood cells	0.80-1.0
Nerve fibers	0.55-1.1
Muscle fibers	1.5

The capacitance of one farad requires one coulomb of electricity to raise its potential one volt and for a plane plate condenser (one farad is equal to 9×10^{11}

electrostatic units)

$$C = \frac{D}{4\pi d} \tag{46}$$

where d is the distance between the plates and D is the dielectric constant. The distance corresponding to one microfarad per square centimeter is, therefore, $8.8 \times 10^{-8}D$ cm. It is clear that if the dielectric constant has a value of about 10 in the membrane, the thickness of cellular membranes is in the neighborhood of 100 A-units.

The specific resistance of cellular membranes is extremely high. On the assumption that the living membrane is about 100 A-units, its specific resistance is in the order of 10^9 ohm-cm. or about 1×10^7 times that of protoplasm. However, this resistance is sensitive to the state of the cell and can change by several orders of magnitude.

Curiously enough, some cell membranes act as rectifiers; the ratio of the outward current to the potential difference increases with the potential difference whereas the ratio for the inward current decreases with the potential difference. In the squid axone, the ratio of the maximum conductance for outward flow to the minimum conductance for inward flow is about one hundred to one. The conductance of the interior of a cell is high and is equivalent to about that of a tenth molar potassium chloride solution. The specific conductance inside of a red blood cell is somewhat less being about 5×10^{-3}; the decreased conductance of the red cell interior may be associated with the high protein content of the cell.

Accumulation of Ions by Cells. There is ample evidence that living cells can and do accumulate ions against a concentration gradient and further that this accumulation is linked to the cell metabolism. If the cell is poisoned or if there is a deficiency of material to be metabolized, the ability to accumulate ions is lost. Upon death, the cell becomes leaky and ions tend to become distributed across the cell wall in accord with the Donnan equilibrium. The accumulation of ions through metabolic reactions has lead to the conception of active transport.

The analysis of the situation in terms of measurable quantities is far from simple. Active ion transport can be defined as transfer that takes place from a lower to a higher electrochemical potential. The electrochemical potential difference for an ionic species between two dilute solutions at the same pressure and temperature is per mole $RT \ln a_1/a_2 + ZF(\psi_1 - \psi_2)$, where a_1 and a_2 are the activities, $\psi_1 - \psi_2$ is the electrical potential difference between the solutions and Z is the valence of the ion. The condition that

$$-ZF(\psi_1 - \psi_2) = RT \ln \frac{a_1}{a_2} \tag{47}$$

is a statement sufficient for a Donnan equilibrium and no work is required to maintain the ion concentration inside the cell. The free energy associated with the accumulation of a mole of a given ion can be written

$$\Delta F = \Delta F_1 + \Delta F_2 + \Delta F_3 + \Delta F_4 \tag{48}$$

where ΔF_1 is identified with $RT \ln a_1/a_2$, ΔF_2 with $ZF(\psi_1 - \psi_2)$, ΔF_3 with the transport of solvent and ΔF_4 with the free energy of the "active process" which in one way or another involves a metabolic reaction which is coupled with the ion transport. ΔF itself must be negative otherwise the accumulation could not proceed. If ΔF_4 is zero, there is no "active" contribution and the accumulation is passive and, correspondingly, if ΔF_4 is negative the accumulation is active. Unfortunately, there is no independent way in which ΔF_4 can be determined. Under favorable conditions ΔF_1, ΔF_2 and ΔF_3 can be measured; if their sum is zero or negative the accumulation is said to be passive and if their sum is positive active transport is said to have occurred. There is, however, still no estimation of the contribution of the coupled reaction, i.e., ΔF_4, and no actual decision as to whether or not the accumulation is linked metabolically.

The magnitudes of ΔF_1, ΔF_2 and ΔF_3 insofar as they can be estimated for a particular case are usually not large; neither ΔF_1 or ΔF_2 ordinarily exceed 2,500 calories per mole and they are frequently much smaller. The most conspicuous exception is the energy required to concentrate the hydrogen ions in blood plasma to that of the parietal secretion of the stomach which amounts to about 9,400 calories per mole.

Ussing has preferred to state the problem in terms of ion fluxes and in this connection he finds it useful to employ the double tracer method. For example, in studying the accumulation of sodium by a cell or tissue, at zero time, Na^{24} is added to the inside solution and simultaneously Na^{22} is added to the outside solution and the inward and outward flux of sodium across the cell membrane can thus be established. The flux is the amount of a substance which passes unit area per unit time in a given direction and the flux of an ion is proportional to the ratio of the force acting on the ion to the resisting force. The forces acting are derived from the chemical potential, the electrical potential and the movement of the solvent. The resisting forces arise from the frictional forces acting on the ion and on the solvent. Thus both the uni-directional flux and the net flux are complex functions of several parameters; however, many of these unknown factors are found to cancel out in an expression derived to give the ratio of the influx to the outflux. The equation for the flux ratios in passive transport is comparatively simple and the ratios can be predicted; if the calculated ratios do not agree with the observed ratios, active transport can be suspected but not demonstrated.

To derive the expression for the flux ratios, Ussing proceeds in the following manner: The force causing diffusion of an ion is the gradient in the electrochemical potential and, for the inward flux, M_i (see Chapter 11, eq. 48)

$$M_i = -\frac{C_i}{f}\frac{d\bar{\mu}_i}{dx} \tag{49}$$

and for the outward flux, M_0

$$M_0 = \frac{C_0}{f}\frac{d\bar{\mu}_0}{dx} \tag{50}$$

where C_i, C_0, $\bar{\mu}_i$ and $\bar{\mu}_0$ are the ion concentrations and electrochemical potentials on the two sides of the membrane respectively. f is the frictional coefficient per mole of ions. Dividing eq. 49 by 50 yields

$$\frac{M_i}{M_0} = -\frac{C_i}{C_0}\frac{d\bar{\mu}_i}{d\bar{\mu}_0} \tag{51}$$

Ussing integrates eq. 51, substitutes the values for the electrochemical potentials and rearranges to obtain the flux ratios in the following form

$$\ln\frac{M_i}{M_0} = \ln\frac{a_0}{a_i} + \frac{ZF}{RT}(\psi_0 - \psi_i) + \frac{f'}{RT}\int_0^x \frac{1}{f''}\left(\frac{d\mu_{H_2O}}{dx}\right) dx \tag{52}$$

where a_o, a_i, ψ_o and ψ_i are the activities of the ion and the electrical potentials on the two sides of the membrane respectively. The last term on the right side of eq. 52 refers to the movement of water across the membrane, μ_{H_2O} being the chemical potential of the water. f' is the friction between one mole of solute and water, f'' is the friction encountered by one mole of water passing through the membrane. All terms in eq. 52 can be easily determined with the exception of the water term. In many cases there is no net transfer of water across the membrane and the flux ratio for water is unity.

The success of the application of Ussing's flux ratio equation has been varied. The equation appears to be a valid description for chloride ions on the two sides of an isolated surviving frog skin. On the other hand, the passive flux ratio for potassium through a metabolically poisoned nerve membrane (neglecting the water movement) departed considerably from that predicted and the entire right hand side of equation 52 had to be multiplied by 2.5 to bring it in agreement with the experiment.

The above discussion of active and passive transport is far from satisfactory. In a sense, the situation is inherently ambiguous. It is impossible to transport a single ion and the transport of any ion is apt to influence and effect the transport of several other ions. However, in many cases the concentration differences between the inside and the outside of the cell are so large that there can be no doubt that an active ion-transport, i.e., coupled reaction, is involved. For example, the concentration of the potassium ion in the cell sap of the marine alga Valonia macrophysa is about 500 mEq. per liter against 12 mEq. per liter in the surrounding sea water and mention has already been made of the high energy requirements for the secretion of HCl in the stomach. It appears that cell membranes can discriminate between cations more effectively than between anions and, in general, the membrane is much less permeable to cations than to anions.

Much has been done with radioactive isotopes of the ions in following the movement of ions across cellular membranes. For example, it has been established with the use of radioactive potassium and sodium that the ratio of the potassium influx to the sodium influx in the human red cell at 37° is 21.1 whereas the corresponding outflux ratio is 0.047. In other animals, however, such as the cat and the dog these flux ratios favoring the accumulation of potassium in red blood cells are reversed. It is a general observation, however, that potassium tends to be the principal intracellular cation and sodium the extracellular cation in body cells.

It was realized that in many cases it was only necessary to postulate an active transport in terms of a single cation and such active transport would lead automatically to the transport of other cations and the necessary anions. For various reasons, emphasis has been laid on the active transport of sodium and it has become conventional to speak of the sodium pump of the cell without, however, specifying the anatomy of this pump.

Ussing's Short-Circuit. It has been long known that a frog skin is able to pump sodium ions from the outside to the inside against a concentration gradient and also that the skin maintains a potential difference between the outside and the inside. If the outside solution is not too dilute, the inside solution is positive to the outside and the potential difference across the skin can amount to as much as 100 mv. Evidently, if the electrical potential across the skin be brought to zero by the application of an external potential, the only forces acting on the sodium ions are the gradients in chemical potential from concentration differences and from metabolic reaction. If the concentration difference is abolished by employing identical solutions on the two sides of the skin, the only force acting to produce diffusion is that arising from metabolic reactions.

Ussing's apparatus is diagramed in Figure 14. The frog skin is clamped between two Lucite chambers containing Ringer's solution. The potential across the skin is brought to zero by an external potential as judged by the two sensing electrodes close to the skin and connected to a vacuum tube voltmeter. The current flow in the external circuit is read on a microammeter. Shown in Figure 15 is an electrical circuit which is possibly equivalent to the Ussing short-circuited preparation. Conditions in the skin may, however, be considerably more complex than the equivalent circuit would indicate.

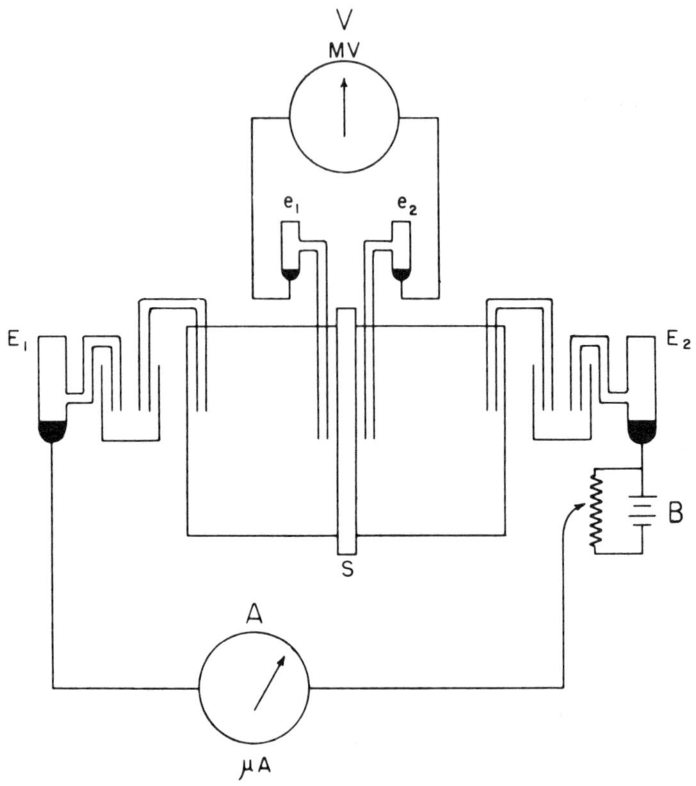

Fig. 14. Diagram of the Ussing short-circuit. *A* is a microammeter; *B*, a battery connected through a potentiometer circuit; E_1 and E_2 are two reversible electrodes through which the external potential is applied; e_1 and e_2 are two reversible sensing electrodes close to the frog skin *S* and *V* is a millivolt meter. (Adapted from H. H. Ussing and K. Zerahn: Acta Physiol. Scand. *23:* 110, 1951.)

Simultaneously with the current measurement, the flux through the skin of one or more ionic species was measured by adding radioactive tracers to one side of the skin and making activity measurements on samples taken on the other side at suitable intervals. Figure 16 shows the net sodium ion transported expressed in microamperes per square centimeter plotted against the measured current in microamperes per square centimeter. Evidently, there is a close correspondence. Whereas a very pronounced reduction of the sodium transport is brought about in the absence of K^+, there is no current carried by K^+ and only Li^+ can partially substitute for the sodium ions.

It is possible to modify the short-circuit arrangement (Fig. 14) and to measure

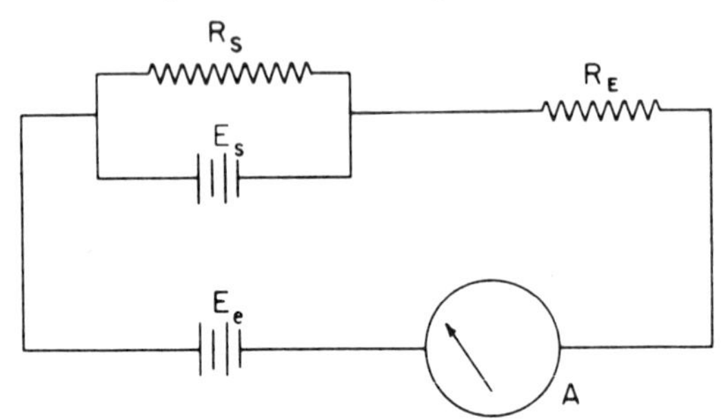

Fig. 15. Equivalent electrical circuit for the Ussing short-circuited preparation. E_e is the external potential and E_s is the skin potential. R_s is the resistance of the frog skin and R_E that of the external circuit. *A* is a microammeter.

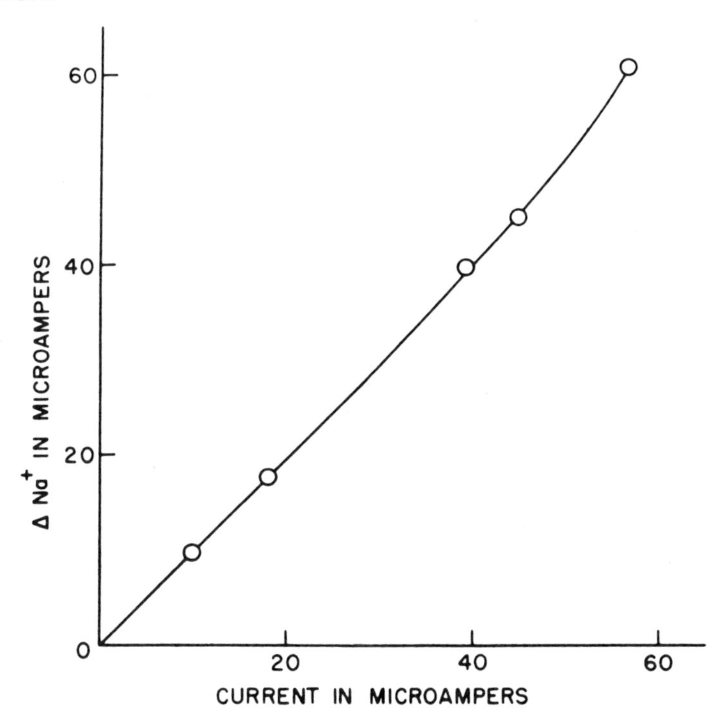

Fig. 16. Comparison between Na^+ transport across frog skin and current flow. (Data of H. H. Ussing: Ion Transport Across Membranes. Academic Press, New York, 1954.)

the oxygen uptake by the frog skin along with the other measurements. It has been shown that the number of sodium ions transported per mole of oxygen consumed ranges from 2 to 13 with a mean of 6.82. These results are incompatible with the idea that O_2 is the final electron acceptor since, if O_2 were the electron acceptor for the ion transport, the maximum ratio of sodium to oxygen should be not more than 4.

Several ingenious schemes have been proposed to link cellular metabolism to ion transport and probably the most concrete of these is the red-ox pump of Conway which connected ion transport to oxidation-reduction of the cytochrome system. A close couple to the cytochrome system, however, would permit no more than 4 univalent ions to be transported per mole of oxygen comsumed and this number is certainly exceeded by the sodium ion transport in the frog skin (see above).

A number of carrier models for ion transport have been proposed and in some instances elaborate systems of differential equations have been derived to describe the kinetics based on such models. Without, however, more biochemical knowledge about the process than is now available, mathematical model building of this kind appears unrealistic. To present the carrier idea in its simplest terms, consider an ion M^+ on both sides of a membrane. Within the membrane is a substance A^- which can combine with the ion and the complex MA is free to diffuse across the membrane in either direction. In this form the carrier system would provide simply for ion exchange across the membrane in response to electrochemical gradients. Suppose, however, the carrier, A^-, is continuously produced at the surface of one side of the membrane and continuously destroyed at the surface on the opposite side. Evidently, the ion would be transported across the membrane unidirectionally and the transport would be coupled to the reaction involving the creation and the destruction of the carrier.

There is considerable evidence from several sources that adenosinetriphosphate (ATP) is rather closely connected to K^+ and Na^+ transport and that the enzyme is

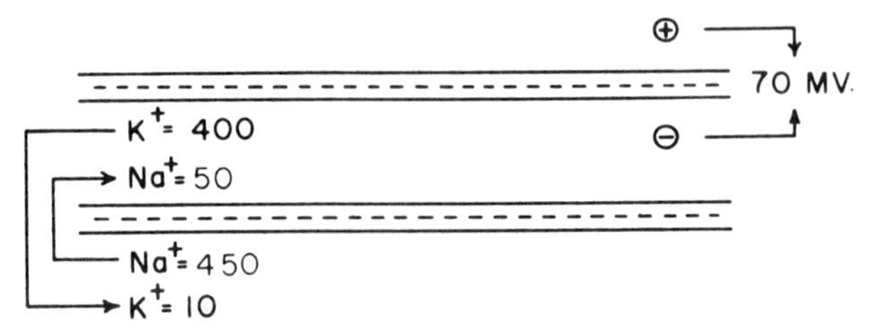

Fig. 17. Resting giant axon of the squid. Ion concentrations in milliequivalents per liter.

part of the carrier system. For example, a membrane-bound (brain tissue) adenosinetriphosphatase which is activated by Na^+ and K^+ has been isolated. The terminal phosphate of ATP is transferred to the enzyme in the presence of Na^+ and the enzyme is dephosphorylated upon the subsequent addition of K^+. Furthermore, the enzyme is inhibited by ouabain, a cardiac glycoside; ouabain is a potent inhibitor of Na^+ and K^+ transport across membranes.

Nerve Conduction. The transmission of a nerve impulse is dependent on ion transport across the nerve membrane. Considerable information and understanding of this process have been obtained by studies on the giant axon of the squid. This is a cylindrical rod of clear protoplasm about 0.5 mm. in diameter with the consistency of a jelly enclosed in a tubular sheath. In the squid's body the axon is surrounded by fluid which differs little from the composition of sea water but the electrolyte concentration inside the axon is very different. The sodium and potas-

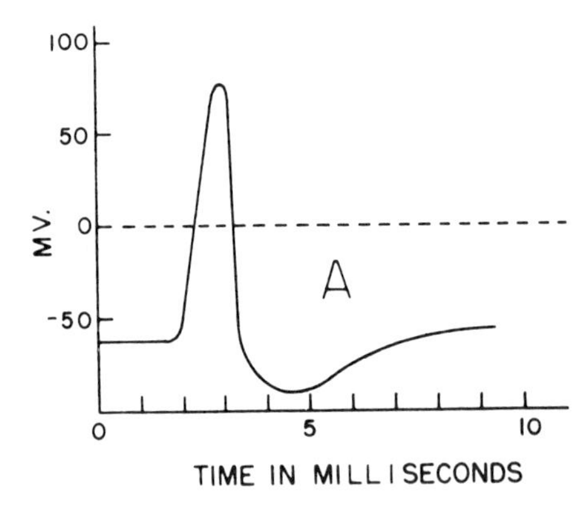

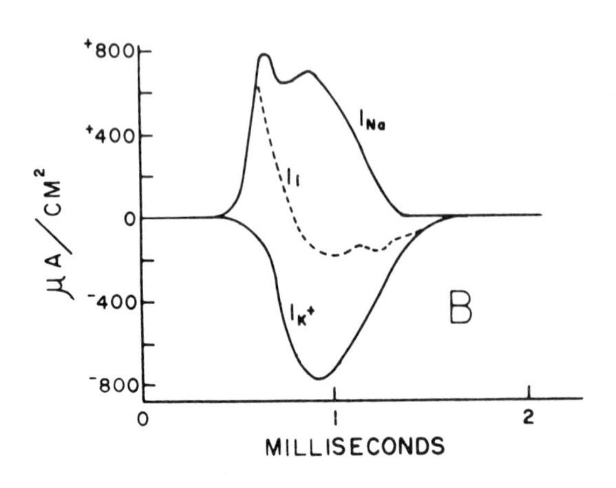

Fig. 18. Potential and current flow in the giant axon of the squid. *A*, Action potential as a function of time in milliseconds. (K. S. Cole and H. J. Curtis.) *B*, Current flow as a function of time. (A. L. Hodgkin and A. F. Huxley: J. Physiol. *117:* 500, 1952.)

sium ion concentrations in milliequivalents per liter are indicated in Figure 17 in the resting nerve.

Most of the sheath around the axon is fibrous connective tissue to give strength and the sheath is permeated by the external fluid. Somewhere in this sheath, however, is a very thin membrane which is perhaps 50 A-units in thickness and which shows a high degree of impermeability with a very large electrical resistance.

The potential difference across the membrane can be measured by inserting a micro KC1 salt bridge into the axon. The interior is found to be about 50 mv negative to the outside and, when in the animal, this resting potential is probably about 70 mv.

The nerve impulse is a transient change in the potential difference across the membrane; the inside, initially at about −50 mv relative to the exterior, goes to about +50 mv for a period of about one millisecond and this alteration travels along the fiber as a wave without change of shape or size at a speed of about 25 meters per second.

The expenditure of energy is required to sustain a nerve impulse. The immediate source of this energy is the exchange of cations across the nerve membrane. In the resting state, the membrane is slightly permeable to K^+ and to a lesser extent to Na^+; these conductances decrease if the inside of the axon is made more negative. If, however, the inside of the axon is made positive there is a prompt and large increase in conductance relative to Na^+ followed by a conductance relative to K^+. The potential difference established by the ion movements thus creates a wave of excitation (the membrane becoming ion permeable) which moves down the fiber. As the K^+ moves out of the interior of the fiber the potential of the interior of the fiber becomes less negative until the resting potential of −50 mv is again reached. The fiber is not, however, in its resting state; for about a millisecond the fiber is refractory.

Figures 18A and 18B show the action potential and the ion current flow in the giant axon as a function of time.

An important contribution to the understanding of nerve conduction was made by Hodgkin and Huxley in their well-known paper in which they set forth explicit equations involving the ion fluxes in the giant axon of the squid. Their analysis was based upon a network analogous to conditions at the nerve membrane. Their representation is shown in Figure 19.

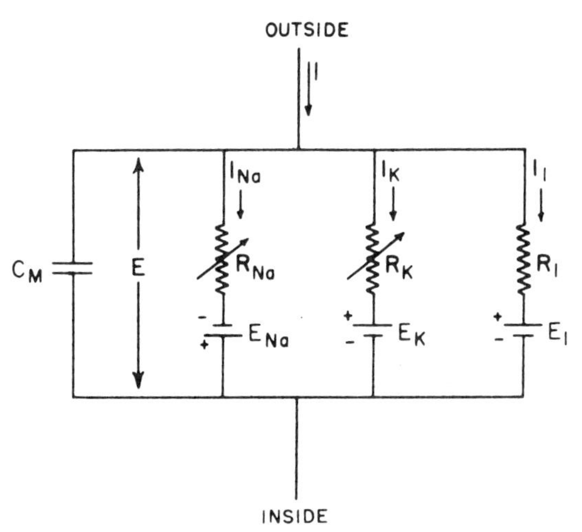

Fig. 19. Electrical circuit representing membrane of the giant axon of the squid. (A. L. Hodgkin and A. F. Huxley: J. Physiol. *117:* 500, 1952.)

Current can be carried through the membrane either by charging the membrane capacity or by movement of ions through the resistance in parallel with the capacity. The ionic currents are carried by the sodium ions (i_{Na}), by potassium ions (i_k) and a small leakage current (i_l) carried by chloride and other ions.

Baker, Hodgkin and Shaw have found it possible to squeeze the axoplasm out of the giant axon, to replace the axoplasm with artificial salt mixtures, and to obtain action potentials which were indistinguishable from the action potentials of the original nerve. This report emphasizes the importance of the nerve membrane in the conduction process.

GENERAL REFERENCES

Cole, K. S.: Membranes, Ions and Impulses. University of California Press, Berkeley, 1968.
Katz, B.: Nerve, Muscle and Synapse. McGraw-Hill Book Co., New York, 1966.
MacInnes, D. A.: The Principles of Electrochemistry. Dover Publications, New York, 1961.
Robinson, R. A. and R. H. Stokes: Electrolyte Solutions, 2nd ed. Butterworth, London, 1968.

PROBLEMS

1. A conductivity cell was filled with 0.01 M KCl whose specific conductivity at 25°C is 0.0014088. The measured resistance was 198.0 ohms. Calculate the cell constant. The cell was then filled with 0.0625 molar NaCl solution. The resistance was 46.5 ohms. What is the specific conductance of the NaCl solution? What is the equivalent conductance of the NaCl solution?

   Ans: 6.00×10^{-3}, 95.97

2. Assume Stokes' law and calculate the equivalent conductance at 25°C of an electrolyte at infinite dilution whose cation and anion are univalent and whose diameters are 3 A-units each.

   Ans: 122.16

3. The following table shows the specific conductances at 30° of acetic acid solutions of the indicated molar concentrations. The equivalent conductance at infinite dilution is 412.8. Calculate the ionization constant of acetic acid.

$C \times 10^4$	*Specific Conductance* $\times 10^6$	$K \times 10^5$
3.10	26.67	1.70
6.21	39.30	1.74
12.40	56.94	1.73
24.90	82.63	1.78
49.70	118.75	1.78
99.50	169.51	1.80

   Ans: See third column above.

4. (a) Outline clearly how you would decide if a given accumulation of ions by a living cell requires an active process or is spontaneous.
   (b) Illustrate (a) above with a specific example along with the necessary calculations.

   Ans: See text and outside reading.

5. (a) Describe the ionic movements involved in the conductance of a nerve impulse and relate these movements to the action potential.
   (b) What would happen in respect to the conduction of the nerve impulse if we replace the sodium ions outside of a resting nerve with an equal concentration of potassium ions?

   Ans: See text and outside reading.

6. Suppose you have a dense, negatively charged collodion membrane separating 0.01 M KCl from 0.10 M KCl and the potassium ion shows 5 times the mobility of the chloride ion through the membrane. The system is at 30°.
   (a) What would be the potential difference between two identical calomel electrodes connected by KCl salt bridges to the two compartments on the two sides of the membrane?
   (b) What would be the potential difference between two silver:silver chloride electrodes placed directly in the two compartments?

   Ans: (a) 37.5 mv. (b) 93.7 mv.

CHAPTER 13

Electrophoresis and Electrokinetic Potentials

Phase potentials have been discussed previously but only in terms of potentials which exist normal to the interface and measured from the interior of one phase to the interior of the boundary phase. In this chapter, potentials are to be considered which are still normal to the interface but which manifest themselves as the result of the tangential motion of one phase relative to the other. Figure 1 is a diagrammatic sketch of the situation. On the solid are fixed ion sites and extending into the liquid is an ion atmosphere in which the ions of opposite sign predominate. At greater distances from the solid, the ion atmosphere becomes less dense and the net charge per unit volume tends to zero as the distance is extended into the bulk of the liquid. This conception has already been introduced in the discussion of the Debye-Hückel theory of strong electrolytes (Chapter 3). The distance $1/\kappa$ is considered to be the thickness of the ionic double layer. A complication to this picture is the existence of an immobile layer of uncertain thickness (of the order of molecular distances) which exists in the liquid immediately adjacent to the solid. The potential from the interior of the liquid phase up to the immobile liquid layer is known as the zeta potential.

Electrophoretic Mobility. Figure 2 shows a solid charged particle suspended in an ionic aqueous solution in an externally applied potential gradient. It will be observed that the negatively charged particle migrates towards the positive electrode. The resistance to motion of the particle is $\eta du/dx$, where η is the coefficient of viscosity of the liquid and du/dx is the velocity gradient normal to the particle surface. It is assumed that u varies directly with x, the distance normal to the particle, and, accordingly, du/dx is equal to u/x. The electrical force acting and tending to produce motion is $E\sigma$, where E is the potential gradient of the external field and σ is the electrostatic charge per unit area of the particle. After the external field is applied, the particle reaches its terminal velocity in a short interval—in the order of a millionth of a second for microscopically visible particles suspended in water. At constant velocity the resisting force is equal to the applied force and

$$E\sigma = \eta \frac{u}{x} \qquad 1$$

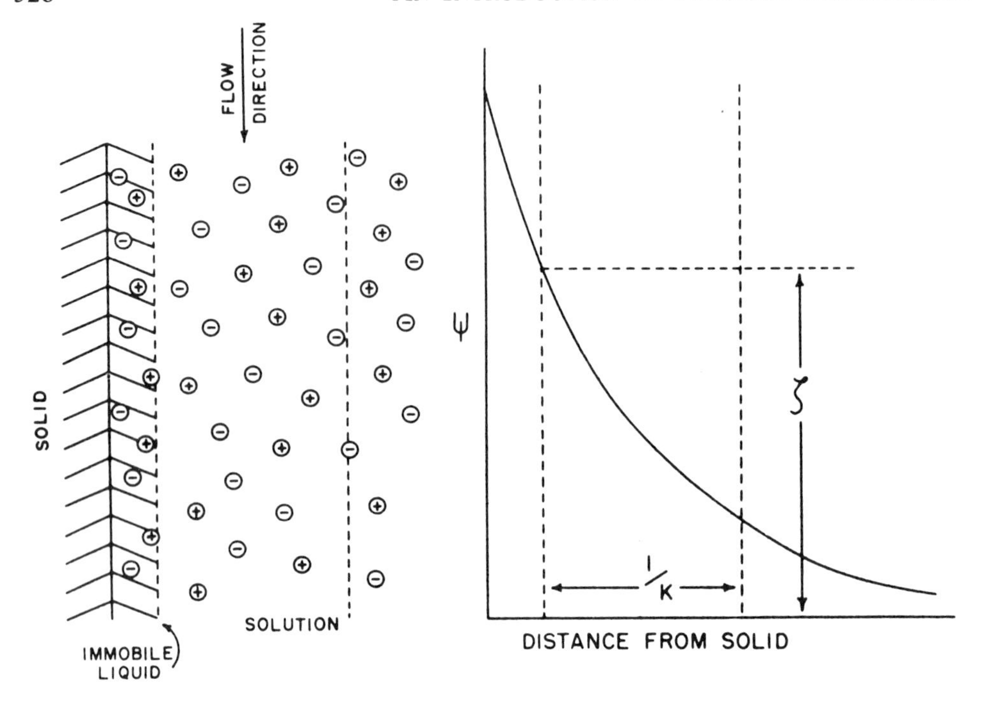

Fig. 1. Schematic representation of the charge and potential distribution at a solid-water interface.

For larger particles in which the linear dimension of the particle is much larger than the thickness of the electrical double layer, the electrical double layer can be treated as a plane condenser and ζ, the potential across the electrical double layer, is given by

$$\zeta = \frac{4\pi\sigma}{D\kappa} \tag{2}$$

where $1/\kappa$ is the thickness of the electrical double layer and D is the dielectric constant. Combining eqs. 1 and 2 and identifying the thickness of the electrical double layer with its hydrodynamic thickness, there results

$$u = \frac{\zeta D E}{4\pi\eta} \tag{3}$$

where u/E is the electrophoretic mobility of the particle and will be designated by U.

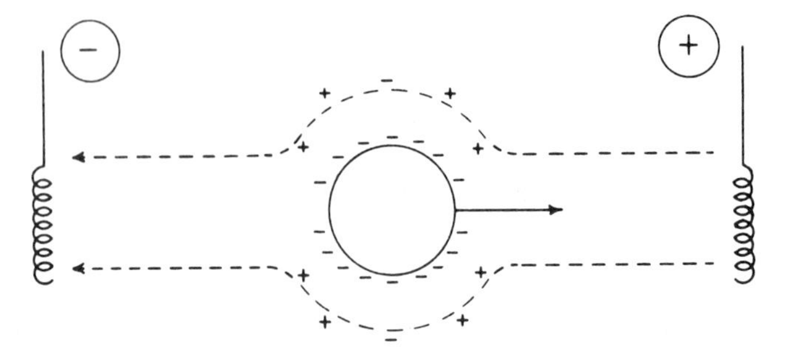

Fig. 2. Solid, negatively charged particle suspended in an aqueous ionic solution and subject to an external electric field. Arrows indicate direction of motion.

Actually, electrophoresis is only one of four electrical-hydrodynamic relations at an interface which can be observed. The general subject of the motion of two phases in respect to each other and the electrical phenomena associated with such motion were called electrokinetics by Freundlich and embrace the following:

1. *Electrophoresis,* which is, as noted above, the motion of suspended or dissolved particles in response to an applied electrical field.
2. *Electroosmosis,* which is the motion of a liquid relative to a fixed solid phase as the result of the application of an external electrical field.
3. *Streaming potential,* which refers to the potential produced as the result of the motion of a liquid relative to a stationary phase resulting from a hydrostatic pressure.
4. *Sedimentation potential (Dorn effect),* which is the potential produced by the sedimentation of suspended particles as the result of gravity or of a centrifugal force.

The electroosmotic flow of fluid through a capillary is closely related to the electrophoretic motion. If the solid phase is stationary, the mobility of the liquid phase is equal to the electrophoretic mobility U and the volume of the liquid flowing through the capillary of radius, r, is

$$V = \pi r^2 U \qquad 4$$

Combining eqs. 3 and 4, the electroosmotic flow is seen to be

$$V = \frac{\zeta D E r^2}{4\eta} \qquad 5$$

E, the potential gradient, is equal to iR where i is the current flowing and R is the resistance. R per unit length of capillary is $1/\pi r^2 K_c$ where K_c is the specific conductance of the liquid. Substituting this information in eq. 5, eq. 6 is obtained

$$V = \frac{\zeta D i}{4\pi K_c \eta} \qquad 6$$

The same assumptions and mathematical treatment as used in the derivation of the equations for electrophoresis and electroosmosis are employed in the derivation for the streaming potential and for the sedimentation potential. These derivations are associated principally with the name of Smoluchowski. The derivation of the equation for the streaming potential can be outlined as follows:

Imagine a liquid being forced through a capillary under pressure, P. This liquid moves with a velocity $x du/dx$ where x is the distance normal to the capillary wall and u is the linear velocity of the liquid. The amount of charge (q) transported by the liquid is

$$q = \sigma x \frac{du}{dx} \qquad 7$$

where σ is the charge density per unit volme at a distance x from the capillary wall. If x is identified with the thickness of the electrical double layer and the radius of the capillary is very much greater than this thickness, the double layer can be treated as a plane plate condenser and ζ, the potential across the double layer, is

$$\zeta = \frac{4\pi\sigma}{D\kappa} \qquad 8$$

where D is the dielectric constant in the double layer. From eqs. 7 and 8 there results

$$q = \frac{\zeta D}{4\pi} \frac{du}{dx} \qquad 9$$

It had been noted above in the derivation for the expression for electrophoresis that the force resisting motion per unit length of the capillary is $\eta du/dx$ where η is the coefficient of viscosity and, accordingly, the total force is $L\eta du/dx$ where L is the length of the capillary. At constant velocity the acting force must equal the resisting force and the pressure exerted on the liquid is

$$P = \frac{L\eta}{A}\frac{du}{dx} \qquad 10$$

where A is the cross sectional area of the capillary.

Combining eqs. 9 and 10 gives

$$P = \frac{L\eta 4\pi q}{A\zeta D} \qquad 11$$

The transport of surface charge by the flow of the liquid produces a potential H across the two ends of the capillary and the current flowing back through the capillary must equal q and

$$q = \frac{K_c AH}{L} \qquad 12$$

where K_c is the specific conductance of the liquid. Substituting eq. 12 into eq. 11, there results

$$\zeta = \frac{4\pi\eta K_c H}{DP} \qquad 13$$

Cognizance has to be taken of the relation between electrostatic units and practical electrical units (see Chapter 3). Substituting numerical factors and converting eqs. 3, 6, and 13 to practical electrical units, ζ is expressed in ordinary millivolts by

$$\zeta = \underset{\text{Electrophoresis}}{\frac{1.129 \times 10^9 \eta u}{DE}} = \underset{\text{Electroosmosis}}{\frac{1.129 \times 10^6 K_c \eta V}{Di}} = \underset{\text{Streaming Potential}}{\frac{1.129 \times 10^{16} K_c \eta H}{DP}} \qquad 14$$

where u, the electrophoretic mobility, is expressed in centimeters per second; E, the potential gradient, in ordinary volts per centimeter; K_c, the specific conductance, in ordinary reciprocal ohm-centimeters; V, the rate of electroosmotic flow, in cubic centimeters per second; i in amperes; H in ordinary volts, and P in dynes per square centimeter.

The above is a very simple and incomplete exposition of electrokinetic theory which is used as an introduction to the experimental methods. After a description of these methods, an attempt will be made to present a more complete theoretical discussion.

Electrophoresis has attracted by far the greatest interest and has proved to be the most rewarding of study and, accordingly, that to follow will deal mostly with electrophoresis. It will be noted that the electrophoresis of a particle is closely related to the motion of a simple ion in an electrical field; ion conductance and electrophoresis are two aspects of the same phenomen.

Electrophoretic Measurements. There are two general methods available for the determination of electrophoretic mobilities—the micro and moving boundary methods.

The micro method measures directly the mobility of particles as observed with a microscope. The particles are exposed to an externally applied electrical field; conditions must be carefully controlled to obtain meaningful mobility data. The

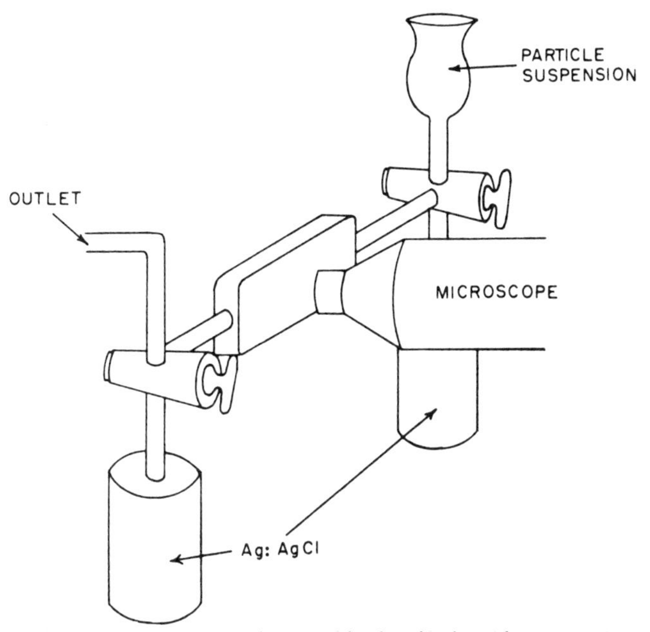

Fig. 3. Microelectrophoresis assembly showing lateral arrangement introduced by R. S. Hartman, J. B. Bateman and M. A. Lauffer. (Arch. Biochem. Biophys. *39*: 56, 1952.)

microelectrophoretic cell is either cylindrical or is rectangular in shape. A cylindrical cell can be constructed from a 10 cm. length of Pyrex tubing, 0.5 mm. internal diameter. It is important to correct for the lens effect as the outer wall is ground flat where electrophoresis is to be observed and the correction is non-linear.

The rectangular cell is constructed of parallel glass plates fused together with an open channel between, the channel being connected to reversible electrodes through appropriate salt bridges. In earlier work, the electrophoresis cell rested on a horizontal microscope stage with the barrel of the microscope in a vertical position, and the mobilities of the particles were measured at the proper levels in the cell. Unfortunately, the suspended particles either sank or rose out of focus. An important modification of the arrangement of the electrophoretic cell was introduced by Hartman, Bateman and Lauffer who adopted what they called a lateral orientation of the cell; this is shown in Figure 3. The stage of the microscope is brought in a vertical position and the barrel of the microscope in a horizontal position. As the particles sediment in response to gravity, they do not fall out of focus but instead they fall in the same plane of focus and the mobility of the particles is determined by measuring the lateral displacement of the particles as they fall.

Since the electrophoretic cell is a closed system, water which has undergone electroosmotic flow along the faces of the cell must return through the center of the cell; obviously there must be two levels in the cell at which the electroosmotic flow is zero. It is at these two levels that electrophoretic observations on the particles must be made. The situation is diagramed in Figure 4.

If the observed mobilities are plotted against the depth of focus in the cell, a parabola which can be described by the equation

$$V_{obs} = a(x - x^2) + c \qquad 15$$

is obtained, where a and c are constants and x is the fraction of the total depth of the cell (see Fig. 5A). The plot of V_{obs} against $x - x^2$ yields a linear relation (Fig. 5B); such plots are of considerable value in judging the performance of electrophoretic equipment.

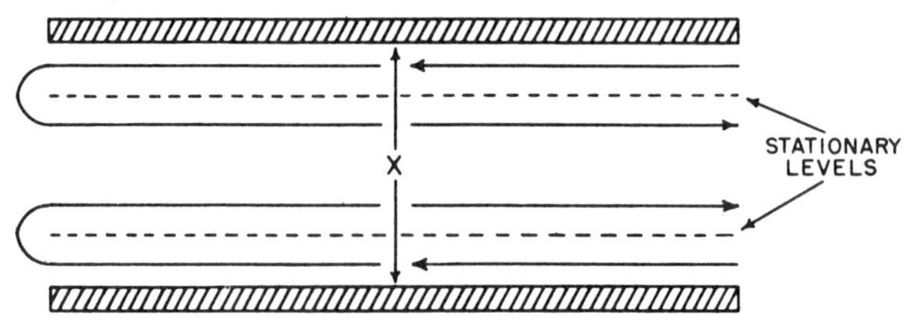

Fig. 4. Stationary levels in a microelectrophoresis cell showing lines of electroosmotic flow.

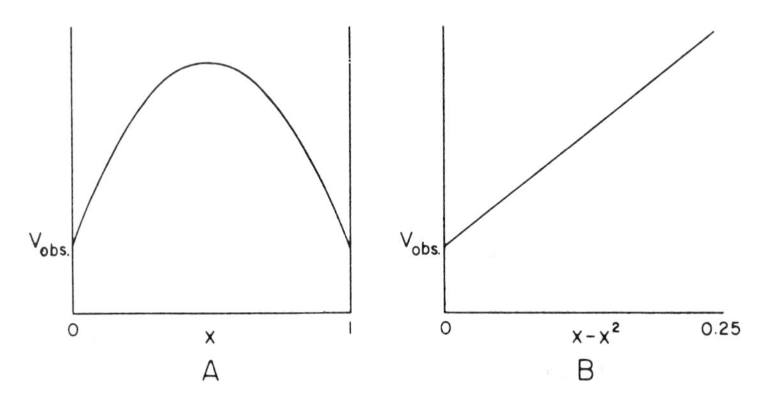

Fig. 5. Plots of eq. 15 showing electrophoretic mobilities as a function of depth in the electrophoretic cell.

The positions of the stationary levels can be calculated in the following simple manner. Since the electrophoretic cell is a closed system, the net flow of liquid due to electroosmosis must be zero. This fact can be expressed by the equation

$$\int_0^1 u dx = 0 \quad 16$$

where u is the velocity of motion of the liquid at any level x. Substituting eq. 15 into eq. 16 there results

$$\int_0^1 axdx - ax^2dx + cdx = 0 \quad 17$$

The integration of eq. 17 between the limits 0 and 1 shows that a is equal to $-6c$. Substituting the value of a in eq. 15 and setting u equal zero (stationary levels) and solving for x with the quadratic formula gives

$$x = 0.5 \pm (1/12)^{1/2} \quad 18$$

and the stationary levels at x are equal to 0.2113 and 0.7887. These values are valid only if the width of the cell is large in comparison with its thickness (about 40 to 1). If this is not true, then Komagata's correction must be applied, and the positions of the stationary levels become

$$x = 0.5 \pm \left(\frac{1}{12} + \frac{0.1046}{r}\right)^{1/2} \quad 19$$

where r is the ratio of cell width to cell thickness. In a cell of circular cross-section, the velocity of flow of liquid is zero at 0.147 of the cell's diameter measured from the perimeter of the cell.

Returning to the consideration of a cell with a rectangular cross-section, it is evident that the observed velocity of a particle at the walls of the cell (x is 0 or 1) is equal to the algebraic sum of the electroosmotic and the electrophoretic velocities. Now if the surfaces of the suspended particles and of the walls of the cell are identical (layers of adsorbed protein), the electroosmotic and the electrophoretic mobilities must be equal and of opposite sign. The observed mobility of the particle at the walls will then be zero and the constant c in eq. 15 is zero and, when x is set equal to 0.5 (midpoint of cell), the constant a is equal to 4 times the observed velocity, and, accordingly, at any level in the cell

$$V_{obs} = 4V_{obs_{1/2}}(x - x^2) \qquad 20$$

At the stationary levels (x equals 0.2113 and 0.7887 for a ratio of cell width to thickness greater than 40), the observed velocity is equal to the electrophoretic velocity which is evidently equal to $2/3V_{obs_{1/2}}$.

Equation 20 is sometimes useful since the midpoint of the cell is inherently an easier level to make observations; the observed mobility of suspended particles changes least with distance through the cell at this level (see Fig. 4). The relation between the electrophoretic mobility and $V_{obs_{1/2}}$ is, of course, subject to Komataga's correction if such a correction is necessary: eq. 20 can be used to calculate this relation.

The potential gradient (E) in the microelectrophoretic cell is given by

$$E = \frac{i}{AK_c} \qquad 21$$

where A is the measured cross-sectional area of the cell, K_c is the specific conductance of the solution and i is the current flowing in amperes. The cross-sectional area of the cell is most easily estimated by measuring the distance of focus through the cell when filled with water at regular and numerous points across the width of the cell. The thickness of the cell so determined is multiplied by the index of refraction of water and plotted against the corresponding cell width. The area under the curve is then measured with a planimeter to yield the cross-sectional area of the cell.

The microelectrophoretic technique is well adapted to the study of the electrophoretic properties of living cells and much information on such cells is scattered through the literature; bacterial cells and red blood cells in particular have received attention. Of interest is the finding by Cook, Heard and Seaman that the dominant ionogenic group at the surface of the human erythrocyte is the carboxyl group of sialic acid. Table 1 gives the mobilities of some mammalian erythrocytes in isotonic 0.067 M phosphate buffers at pH 7.4.

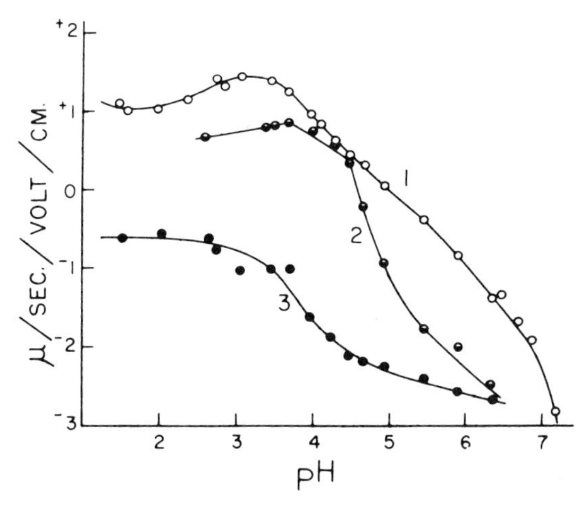

Fig. 6. Mobilities of Pyrex glass particles as functions of pH. Ionic strength 0.05. Above pH 3.4, sodium acetate buffers; below pH 3.4, hydrochloric acid-sodium chloride solutions; 25°. Curve 1, 0.173% egg albumin; curve 2, egg albumin adsorbed and washed; curve 3, bare Pyrex glass particles. (D. K. Chattoraj and H. B. Bull: J. Amer. Chem. Soc. *81*: 5128, 1959.)

The microelectrophoresis method can also be employed for the study of proteins. Non-conducting particles such as those of glass, hydrocarbons, etc., are placed in a protein solution and the particles become covered with protein by adsorption and yield mobilities characteristic of proteins. Figure 6 shows the electrophoretic mobilities as a function of pH at constant ionic strength for bare Pyrex glass particles, for glass particles in an egg albumin solution and for particles covered with egg albumin and subsequently thoroughly washed all at 25°C.

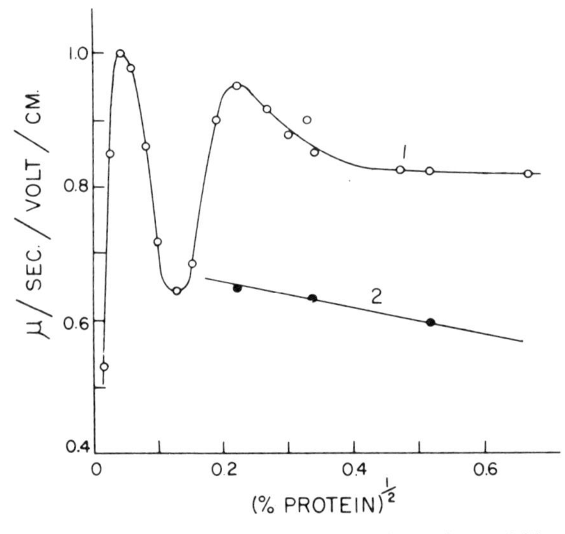

Fig. 7. Mobilities as functions of square root of percent bovine serum albumin concentration. Sodium acetate buffer ionic strength 0.05, pH 4.22 at 25°. Curve 1, Nujol particles; curve 2, dissolved protein with moving boundary. (D. K. Chattoraj and H. B. Bull: J. Amer. Chem. Soc. *81*: 5128, 1959.)

Figure 7 shows the electrophoretic mobility of bovine serum albumin as a function of the square root of the protein concentration both for protein adsorbed on Nujol droplets (microelectrophoresis) as well as for dissolved protein (moving boundary).

TABLE 1. MOBILITIES OF MAMMALIAN RED BLOOD CELLS IN 0.067 M PHOSPHATE BUFFER, pH 7.4 AND 25°. (DATA OF H. A. ABRAMSON: J. GEN. PHYSIOL. 12: 711, 1929.)

Animal	*μ/sec/volt/cm.*
Rabbit	−0.55
Sloth	−0.97
Pig	−0.98
Opossum	−1.07
Guinea pig	−1.11
Man	−1.31
Rhesus monkey	−1.33
Cat	−1.39
Mouse	−1.40
Rat	−1.45
Dog	−1.65

The first maximum in the mobility-concentration curve (Fig. 7) probably represents the completion of a layer of surface denatured protein, whereas the second maximum is probably the mobility of a layer of native protein. Probably other proteins in addition to bovine serum albumin and egg albumin would exhibit similar electrophoretic behavior.

The microelectrophoretic method can also be used to study the electrophoresis of adsorbed nucleic acids. Figure 8 shows the electrophoretic mobilities of DNA, RNA, nucleohistone and histone adsorbed on alumina as a function of pH.

Anyone planning to use the microelectrophoretic technique would do well to read the paper by A. P. Black and A. L. Smith: J. Amer. Water Works Assoc. *58*: 445, 1966.

Moving Boundary. Through the years, many workers have used variations of the moving boundary method for the measurement of electrophoretic mobilities

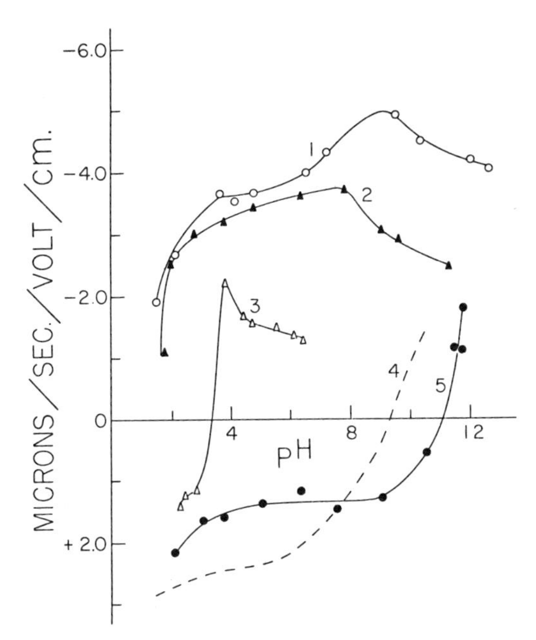

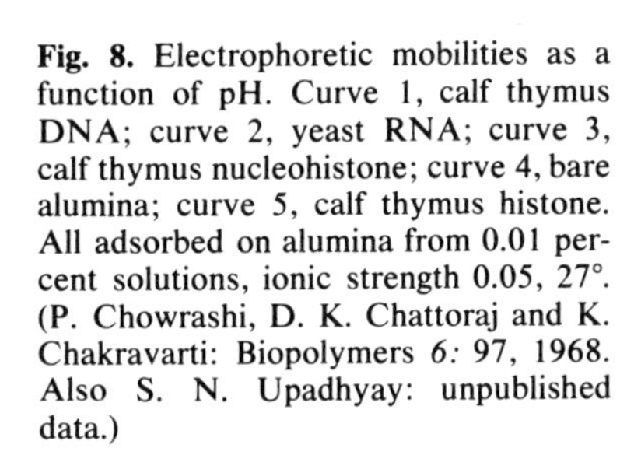

Fig. 8. Electrophoretic mobilities as a function of pH. Curve 1, calf thymus DNA; curve 2, yeast RNA; curve 3, calf thymus nucleohistone; curve 4, bare alumina; curve 5, calf thymus histone. All adsorbed on alumina from 0.01 percent solutions, ionic strength 0.05, 27°. (P. Chowrashi, D. K. Chattoraj and K. Chakravarti: Biopolymers *6:* 97, 1968. Also S. N. Upadhyay: unpublished data.)

in which U-shaped tubes were employed. A number of difficulties were experienced but the greatest ambiguity was associated with the conductance of the solutions. Finally, Tiselius in 1930 showed that the condition to be fulfilled was a uniform conductivity throughout the U-tube during an experiment. This condition could never be met exactly but could be approached.

The experimental procedure with the U-tube is to place the protein in the bottom of the tube, and an ultrafiltrate or a dialyzate of the protein in the form of a buffer at the desired pH is layered over the protein solution. The U-tube is then connected to reversible electrodes and the motion of the protein boundaries observed. The drawback to this original method was the difficulty of locating the protein boundaries and dealing with mixtures of proteins.

Later Tiselius proposed extensive modifications of the moving boundary method. He introduced a three-piece cell which increased the flexibility of the method, allowing the formation of sharper boundaries and, under favorable conditions, the isolation of pure protein fractions (see Fig. 9).

The most important modification which Tiselius introduced, however, had to do with the optical system used to observe the protein boundaries. He employed the Schlieren or shadow method to visualize the protein boundaries in the electrophoretic cell. The optical systems employed in diffusion, electrophoresis and ultracentrifugation are described in Chapter 11, Diffusion, and this material will not be repeated here. The two optical methods which are most appropriate in electrophoretic work are the Longsworth scanning method and the cylindrical lens system.

The kind of information that can be obtained by use of the Longsworth scanning method is identical with that which may be had with the cylindrical lens system. The measured area under the index of refraction gradient curve represents the integral

$$\text{area} = \int \frac{dn}{dx}\, dx \qquad 22$$

where n is the refractive index of the solution and x is the distance along the cell. Since the refractive index of the solution is proportional to the protein concentration, the area under the segment of the curve for a given electrophoretic com-

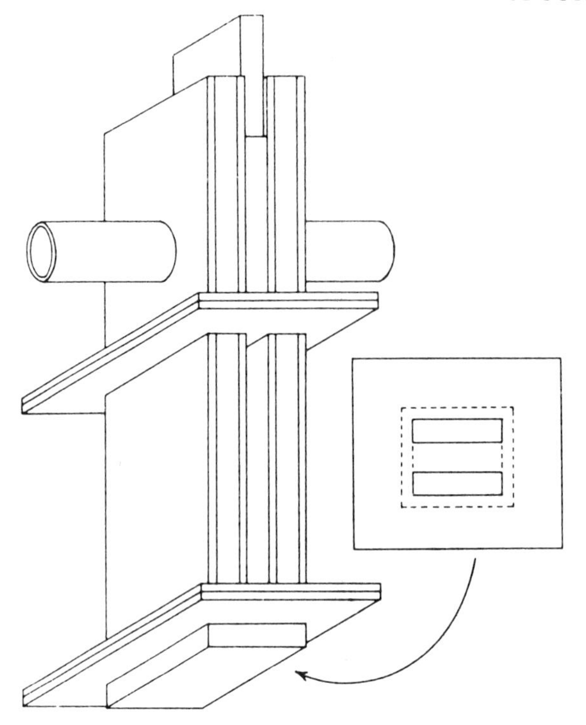

Fig. 9. Three-piece moving boundary electrophoresis cell introduced by Tiselius.

ponent is proportional to the concentration of this particular component. Two methods are available for the estimation of the areas due to the various components: (1) An ordinate is drawn from the lowest point between two adjacent peaks or (2) the pattern is resolved into a series of symmetrical curves.

To understand the origin of the boundaries and the conditions at the boundaries, examine Figure 10, which is a diagrammatic representation of a system containing two electrophoretic components distinguished by open and closed circles.

The electrophoretic mobility is calculated by

$$U = \frac{h}{tE} \qquad 23$$

where h is the distance in centimeters moved by the boundary in t seconds and E is the potential gradient in volts per centimeter. Since E is equal to iR, where R is the resistance in ohms and i is the current in amperes, and, further, since R is equal to $1/AK_c$, where A is the cross-sectional area of the channel of the electrophoretic cell and K_c is the specific conductance in reciprocal ohms, eq. 23 becomes

$$U = \frac{hAK_c}{it} \qquad 24$$

U can be expressed in centimeters per second per volt per centimeter or in square centimeters per second per volt; U has the dimensions of a diffusion coefficient per volt. U is also expressed in microns per second per volt per centimeter as is frequently the practice when microelectrophoresis is employed. Attention has not been paid to the details of the experimental procedures for moving boundary electrophoresis. For these and for a valuable discussion of moving boundary electrophoresis the chapters on electrophoresis by Longsworth in "Electrophoresis" edited by M. Bier should be consulted.

Boundary Conditions. Unfortunately, there is always a lack of symmetry between the ascending and the descending boundaries both in respect to concen-

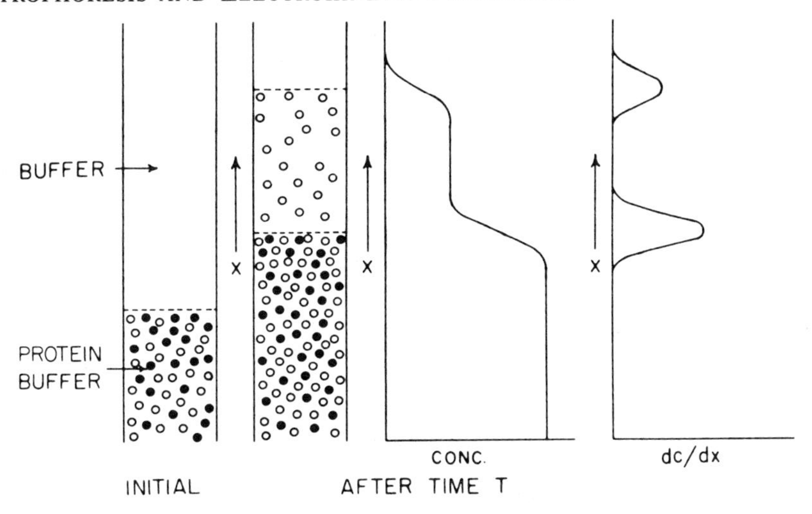

Fig. 10. Electrophoresis of a two-component system (open and closed circles). x is the vertical distance along the cell and C is the concentration.

tration gradients as well as the distance moved in a given time. This lack of symmetry in the two limbs of the electrophoretic cell arises from the different ionic environments of the two limbs; there is conductivity as well as pH gradients in the cell. The ascending boundary is moving into the buffer solution while the descending boundary is moving into the protein solution. In spite of the fact that the protein solution has been exhaustively dialyzed against the buffer solution before electrophoresis, the ionic composition and the specific conductance of the two solutions are not identical; a Donnan equilibrium was established between the protein-buffer and the buffer alone. The protein ion mobility is almost always less than that of any of the smaller buffer ions and, accordingly, the protein-buffer solution has a smaller conductance than does the buffer solution; the pH differences between the buffer and the buffer-protein solution will depend on the sign and magnitude of the charge on the protein ion and on the ionic strength.

In general, the ascending boundary is sharper and less diffuse than is the descending boundary. The ascending boundary moves into a region of higher conductance and lower potential gradient and, accordingly, those particles which by chance move ahead of the boundary and into a region of lower potential gradient slow down and are overtaken by the rising boundary; the ascending boundary is continuously sharpened. Conditions which lead to sharpening do not obtain with the descending boundary and, accordingly, this boundary tends to become diffuse.

The un-ionized part of the buffer acid being uncharged, its concentration is uniform throughout the electrophoresis cell. The concentration of the buffer salt, however, will not be uniform but will change at each boundary and, accordingly,

$$\mathrm{pH}_1 - \mathrm{pH}_2 = \log \frac{\mathrm{salt}_1}{\mathrm{salt}_2} \qquad 25$$

where the subscripts 1 and 2 refer to the pH and the buffer salt concentration in the buffer and in the protein-buffer respectively. Since the mobility of the protein is a function of the pH, the pH effect can either enhance or counter the conductivity effect and under appropriate conditions the descending boundary can be actually the sharper boundary due to the predominance of the pH effect. In general, errors in mobility are smaller on the descending side.

It is possible to decrease very considerably the lack of symmetry between the ascending and descending boundaries by the proper choice of experimental conditions. It has been found that, the lower the protein concentration and the higher the ionic strength of the buffer, the more closely the ascending and descending boundaries resemble each other; these are likewise the conditions that lead to more accurate estimates of the concentrations and mobilities of mixtures containing two or more proteins. Obviously, if the protein concentration be made too low and the ionic strength too high, no exact measure of protein concentration or of mobility can be made.

Boundary Anomalies. In addition to the protein boundaries there is always at least one salt boundary or false boundary in each limb of the electrophoresis cell which arises as a result of the difference in buffer concentration above and below the original boundaries. The false boundary in the descending limb is called the ϵ-boundary and that in the ascending side is the δ-boundary and both are either stationary during electrophoresis or move only slightly. Their magnitude depends on the size of the charge on the protein and on the ionic strength. Actually, the total number of salt boundaries is given by the expression $n-v-1$, where $n-1$ is the total number of the ionic species less one and v is the number of ionic forms absent on one side of the boundary.

To treat the boundary conditions with sufficient completeness to understand the various ionic changes which take place upon the passage of current would require a more elaborate discussion than can be afforded. Moving boundary electrophoresis in its basic aspects belongs in the field of electrochemistry.

It is, however, useful to point out that the moving boundary method is one of the standard methods for the measurement of the transport number of ions. The basis for this method is simple. In Figure 11 are shown two salt solutions with a common anion A^- which forms a boundary at *ab*. After the passage of current the cations move up the cell and the anions down the cell. After a time the boundary will have arrived at *cd* and in this interval all the cations M^+ in the volume between *ab* and *cd* must have crossed the boundary at *cd*. The current transported by M^+ is then $t_M Q$ where Q is the total amount of current passed and t_M is the transport number of M^+. If V is the volume between *ab* and *cd* and the concentration of $M^+ A^-$ is C ion equivalents of M^+ per unit volume, the amount of electricity con-

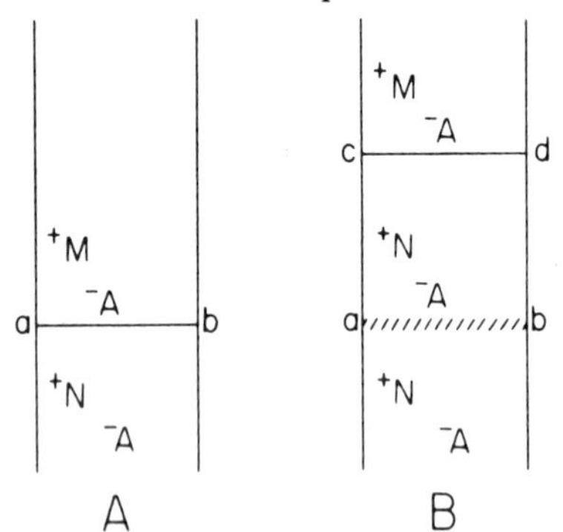

Fig. 11. Determination of the mobilities and transport number of ions by the moving boundary method. *A*, Initial formation of salt boundary; *B*, boundary after passage of current.

ducted by M^+ must be VCF, where F is the Faraday. Then

$$t_{M^+} = \frac{VCF}{Q} \tag{26}$$

and in terms of the mobility of M^+ (see eq. 24)

$$U_{M^+} = \frac{hAK_c}{it} \tag{27}$$

Diffuse Boundaries. Convection arising from heating produced by the passage of an electric current can lead to diffuse boundaries. The amount of heat produced by the current is equal to Ri^2 joules per second, where R is the resistance in ohms and i is the current in amperes. Heat production causes density gradients in the cross-sectional area of the cell, the outer layers of the liquid in the cell being cooler than the liquid in the center of the cell. The reason for the customary low temperature (near 0°C) used in electrophoretic work is that the density of water changes least with temperature in this range. By employing lower current densities, it is possible to work at higher temperatures, for example, at 25°C. Since the viscosity of water at 25°C is about half that at zero degrees, measurable electrophoretic mobilities can usually be observed at this temperature with current densities which avoid thermal gradients.

Diffusion also tends to produce diffuse boundaries; in fact, the electrophoresis cell is a diffusion cell through which a potential gradient has been established; diffusion has been discussed in Chapter 11. Neither the heat convection effect nor diffusion can be reversed by reversing the direction of the passage of the current.

The boundaries may become diffuse due to electroosmotic flow of liquid in the electrophoretic cell; this is, in principle, the same problem as encountered in the microelectrophoresis method. Moving boundary electrophoresis may or may not involve a closed hydrodynamic system but, even if the system is not closed, a hydrostatic pressure will be built up due to electroosmotic flow; this pressure will in time reduce the net flow of liquid to zero. The boundary across the shortest cross-sectional dimension of the cell will assume a parabolic shape, the acuteness of which will depend only on the ratio of the width to the thickness of the cell and to the velocity of electroosmotic flow. The parabolic shape of the boundary will appear in the optical system as diffuseness in the boundary. This apparent diffuseness due to electroosmotic movement should be reversible upon reversal of the direction of the current flow during the time it takes to reestablish a parabolic boundary in the reverse direction.

Heterogeneity. Diffuse boundaries can also be produced by electrophoretic heterogeneity of a protein component and effort has been made to develop electrophoresis into a sensitive test for protein purity.

It is noted that the peaks obtained in an electrophoretic run are constantly broadening as the electrophoresis proceeds and it is considered that this diffuseness of the boundary is due to two causes: (1) to diffusion of the protein molecules at the electrophoretic boundary, and (2) to electrophoretic inhomogeneity. The concentration gradient at a boundary usually follows a Gaussian distribution curve and, accordingly, a determination of the standard deviation of the curve should yield a measure of the heterogeneity when appropriate correction is made for the broadening (increase of the standard deviation of the Gaussian curve) due to diffusion alone.

The heterogeneity coefficient, h, may be calculated from the equation

$$D^* = D + \frac{E^2h^2t}{2} \qquad 28$$

where D^* is the apparent diffusion coefficient calculated for the protein boundary by methods outlined in Chapter 11. Thus a plot of the apparent diffusion coefficient against time of electrophoresis should yield a straight line with a slope $E^2h^2/2$ extrapolating to the normal diffusion coefficient at zero time. The heterogeneity coefficient has the units of mobility and is the standard deviation for the mobility distribution. Electrophoresis is conducted at the average isoelectric point of the protein which diminishes the influence of conductance gradients, pH gradients and electroosmosis. Figure 12 shows the refractive index gradient curves for γ_2-globulin at pH 7.2, the current being reversed after 240 minutes. Note that the curves sharpen up after reversal of the current; clearly there is electrophoretic heterogeneity present. Apparently, the Alberty test for electrophoretic heterogeneity is a very critical one as even the most highly purified and crystalline proteins show some heterogeneity.

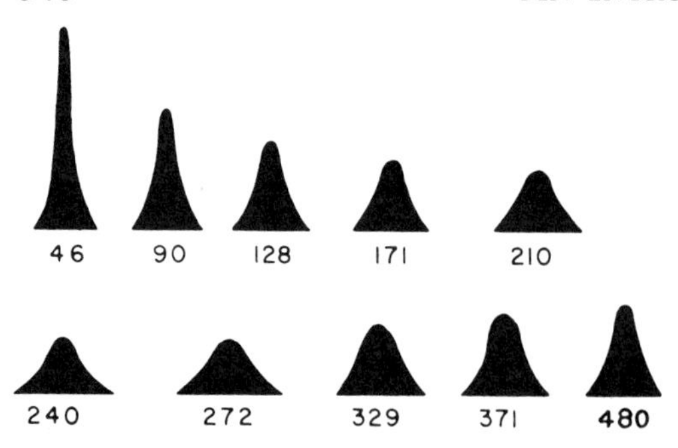

Fig. 12. Refractive index gradient curves for human γ_2-globulin at pH 7.2 during electrophoresis. E = 2.40 volts/cm. Current reversed at 240 min. Elapsed time in minutes indicated under each curve. (R. A. Alberty, E. A. Anderson and J. W. Williams: J. Phys. Coll. Chem. *52:* 217, 1948; *53:* 114, 1949.)

Interacting Systems. A protein ion exists in a number of ionic forms; these forms with their distribution of net charge are in equilibrium with each other and are continually changing into each other. If the rate of interconversion is high in respect to the separation produced by electrophoresis, this family of ions will travel in the field as a single component the mobility of which is known as the constituent mobility. The constituent mobility is given by the relation

$$\overline{U} = a_1U_1 + a_2U_2 + \cdots a_nU_n \qquad 29$$

where a_1 is the fraction of the total number of protein ions with a mobility U_1; it is also the fraction of time the molecules of the constituent exist in the ionic form with the mobility U_1. The meaning of a_2, $a_3 \cdots$ is similar but refers to the particular ionic form with mobilities $U_2, \cdots U_n$ respectively.

Evidently, the rate of reaction between the ion and the protein, or between protein molecules, as well as the rate of dissociation of the complex will determine the electrophoretic behavior of such a system and in the reaction

$$A + B \underset{k_2}{\overset{k_1}{\rightleftharpoons}} C \qquad 30$$

the magnitudes and the relative magnitudes of the kinetic constants k_1 and k_2 will be important in the electrophoretic behavior of the system. Five classes of systems can be distinguished as follows:

Class 1. k_1 and k_2 are both small and k_1/k_2 is approximately unity. A, B and C will be present in solution and each will be represented by a separate boundary.

Class 2. $k_2 >> k_1$ and k_1/k_2 is very much smaller than unity. Only A and B will exhibit electrophoretic boundaries. This situation can lead to a large heterogeneity coefficient and is usually difficult to identify.

Class 3. $k_1 >> k_2$ and k_1/k_2 is very much larger than unity. The complex C and the reactant in excess will produce electrophoretic boundaries. Trypsin in the presence of the soybean tryptic inhibitor and protein in the presence of dodecyl sulfate are examples of this class.

Class 4. k_1 and k_2 are very large and k_1/k_2 is approximately unity. This is a complex situation, and electrophoretic pattern can show a bimodal curve. Chymotrypsin undergoing association provides an example.

Class 5. k_1 and k_2 are intermediate in magnitude and k_1/k_2 is approximately unity. In this situation bimodal curves are possible. Buffer acids interacting with proteins sometimes fall into this class.

It will be gathered from the above remarks that, whereas it is possible to obtain much valuable information from electrophoresis of interacting systems, caution must be applied in their interpretation. It is indeed possible to obtain several boundaries in an electrophoretic experiment which bear no relation to the number of pure components present.

Electrophoresis of Complex Mixtures. One of the more useful ends to which boundary electrophoresis has been put has been the analysis of complex mixtures of proteins, and, indeed, the detection of the components of blood serum proteins was the purpose to which Tiselius first applied his improved electrophoretic technique. The literature dealing with electrophoresis of blood plasma and sera has grown to massive size and the mere listing of papers would occupy many pages. Shown in Figure 13 is a typical electrophoretic pattern of normal human plasma.

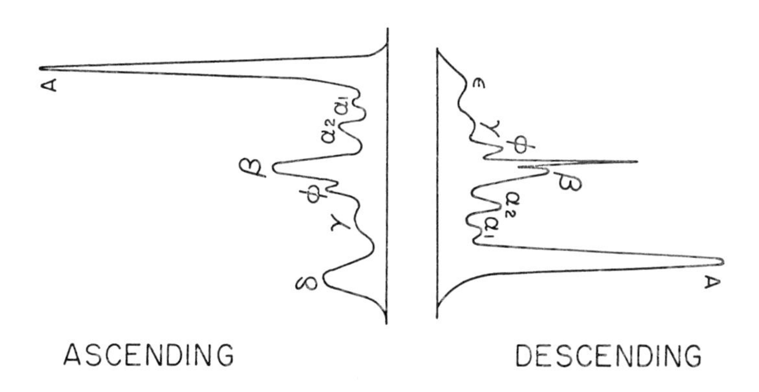

Fig. 13. Electrophoretic pattern of a 1.87 percent solution of normal human plasma in sodium diethylbarbiturate buffer pH 8.6 and at 0.10 ionic strength after electrophoresis for 13.020 seconds at 5.21 volts/cm. A represents albumin, α_1, β and γ, the globulins, ϕ fibrinogen. δ and ϵ are salt boundaries. (G. E. Perlmann and D. Kaufman: J. Am. Chem. Soc. *67:* 638, 1945.)

Any mixture of proteins is apt to have a degree of interaction and it is generally found that the lower the protein concentration and the higher the ionic strength, the more accurate electrophoretic analysis tends to become. In fact, in careful work it is well to study the electrophoretic behavior of a protein mixture as a function of protein concentration and extrapolate the mobilities to zero protein concentration.

Whereas the moving boundary method (Tiselius) has been very successful for the measurement of electrophoretic mobilities and for the analysis of complex mixtures, it has limited utility in the isolation of pure protein fractions and for this purpose study has been directed to other and more suitable electrophoretic techniques.

Zone Electrophoresis. Many ingenious devices have been developed for the isolation of pure fractions by means of electrophoresis. Typically, these devices use an inert supporting medium through which electrophoresis takes place. The advantages over moving boundary electrophoresis can be summarized as follows: (1) Complete separation of all electrophoretically active components can be achieved; (2) small ions, amino acids as well as proteins, can be separated; (3) the apparatus is usually simple and relatively inexpensive; (4) the amount of material required is small.

Zone electrophoresis is, however, poorly adapted to physicochemical studies on protein systems; the stabilizing medium can introduce several complications. For example, there is electroosmotic flow through the medium and adsorption on the medium can occur.

Isoelectric focusing has considerable promise as a means of separating protein fractions. Actually, the basic idea of isoelectric focusing has a long history and goes back many years. A pH gradient is established in a column. At a point acid to the isoelectric point of the protein, the protein exists as a cation and, conversely, basic to the isoelectric point as an anion; accordingly, if the electric field is arranged so that the positive electrode is placed towards the more acid end of the gradient cell, the protein will move towards the level in the cell which is isoelectric with the protein; the protein fractions collect in layers in the electrophoresis cell corresponding to their various isoelectric points. Unfortunately, the pH gradient itself moves when the electric field is applied and proper fractionation will not take place unless the substances to be separated migrate faster than does the pH gradient.

A more recent development is the production of low molecular weight (300 to 600) ampholyte buffers (peptide mixtures) covering the ranges of isoelectric points of interest. The ampholyte buffers under the influence of the electric field arrange themselves in the column according to their isoelectric points thereby creating and maintaining a pH gradient along the column. By this means it is possible to separate proteins whose isoelectric points differ by only 0.01 pH unit.

Electroosmosis and Streaming Potential. Our principal concern must be with electrophoresis because of its greater utility. Nevertheless, both electroosmosis and streaming potential present interesting features and, if some defects in their theory could be repaired, they could become highly effective experimental tools.

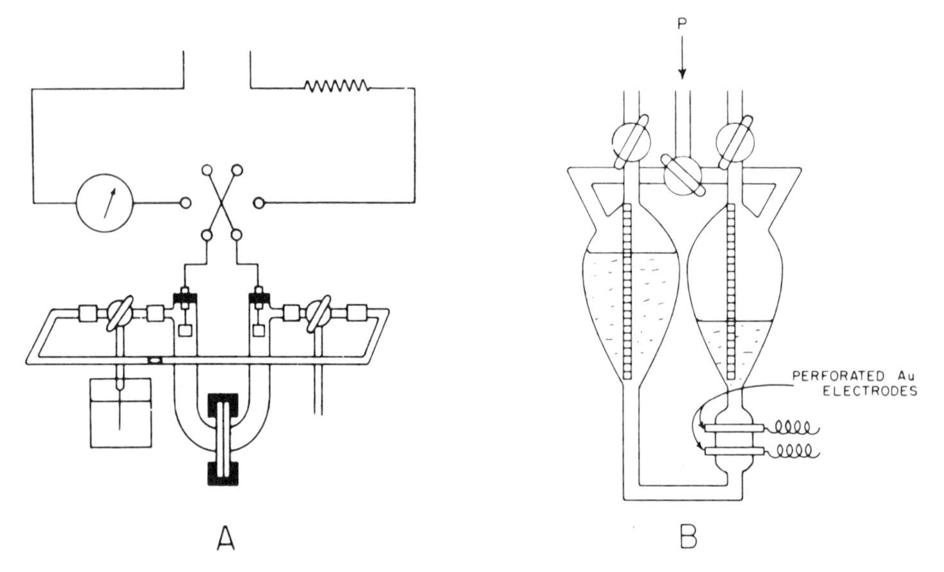

Fig. 14. *A*, Electroosmotic assembly. (B. N. Ghosh and S. Ghosh: Bull. Cent. Leather Inst. [India] *4:* 259, 1958.) B, Streaming potential assembly. (W. M. Martin and R. A. Gortner: J. Phys. Chem. *34:* 1509, 1929.)

The experimental arrangements for the measurement of electroosmotic flow and for streaming potentials are simple and are diagramed in Figures 14A and B which are almost self-explanatory. The electroosmotic flow is measured by the movement of an air bubble in an outside capillary. The measurement of the streaming potential requires the use of a potentiometer in connection with a very high resistance and sensitive indicating device because the current flow produced by reasonable pressures is very small. In earlier work, quadrant electrometers were employed but properly designed vacuum tube galvanometers should be suitable for this purpose. Figures 14A and B show both pieces of equipment with liquid motion taking place through membranes or diaphragms; membranes could be substituted by single capillaries of glass or other non-conducting material.

Simple derivations for the expressions for both electroosmosis and for streaming potential were presented earlier in this chapter and are valid for single large capillaries. It has, however, been noticed that the electroosmotic flow and the streaming potentials progressively decrease as the capillary radius is made smaller. Shown in Figure 15 is a plot of the function K_cH/P for the streaming of 2×10^{-4} M NaCl solution through powdered quartz diaphragms against the logarithm of the average diameter of the quartz particles in microns, where H is the streaming potential in millivolts, K_c is the specific conductance and P is the hydrostatic pressure in centimeters of mercury.

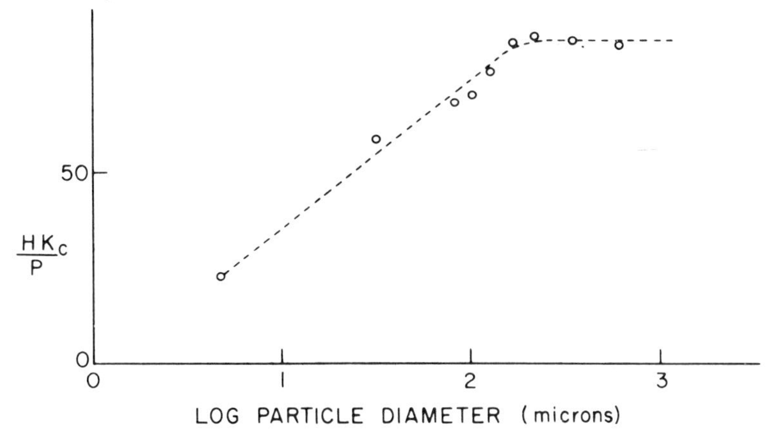

Fig. 15. Streaming potential expressed as HK_c/P produced by 2 x 10^{-4} M NaCl streaming through a diaphragm of packed quartz powder as a function of the logarithm of the average diameter of quartz particles. (H. B. Bull and R. A. Gortner: J. Phys. Chem. *36:* 111, 1932.)

There are several possible causes for the decrease of the streaming potential with capillary radius; of importance is surface conductance. In the expression for electroosmosis as well as for streaming potential, the specific conductance appears explicitly. It was first realized by Briggs that the value of the specific conductance to be used was that of the solution in the capillary rather than that of the solution in bulk; this is so because of surface conductance. Surface conductance arises not only from the movement of the net charge at the interface but also from the total accumulation of ions (both positive and negative) in the double layer. At low electrolyte concentration the contribution due to the net interfacial charge predominates but, as the ionic concentration increases, the ionic conductance in the double layer becomes more important.

The technique of Briggs to deal with the influence of surface conductance on the streaming potential was to fill the capillary (actually a packed diaphragm of cellulose) with 0.10 M KC1 and measure the cell constant of the system of capillaries. Having the cell constant as well as the resistance of the membrane in the presence of the far more dilute solution used in the streaming potential measurement, the specific conductance of the liquid in the capillaries in the diaphragm can be calculated and is the sum of the specific bulk and surface conductance. In a tightly packed diaphragm of cellulose or of powdered glass, the surface conductance with dilute salt solutions can be many times that due to bulk conductance and, therefore, represents a very important correction. The Briggs correction assumes that the surface conductance in the presence of 0.10 M KCl is trivial in respect to the bulk conductance and this is undoubtedly true, and when the correction is applied to a single capillary of uniform cross-section the correction is suitable, but when dealing with a system of capillaries of different sizes such as found in a real membrane or diaphragm the situation is more complex; surface conductance plays a far greater role in the smaller capillaries than it does in the larger ones and no really satisfactory way has been found to deal with the situation which involves a broad distribution of capillary radii. The influence of surface conductance has been further considered by B. N. Ghosh and co-workers who plot the reciprocal of the apparent ζ-potential against the reciprocal of the pore radius and extrapolate to infinite pore size to obtain the ζ-potential which is independent of the radius of the capillaries.

There are other complications to be expected with small capillaries. For example, the streaming potential itself produces a counter pressure as the result of electroosmosis so that the effective pressure producing motion of the liquid is smaller than that which is measured. Also to be noted is the possible interference of the electrical double layers from opposite sides of the capillary as the radius of the capillary approaches the thickness of the electrical double layer.

Thus, in spite of attractive material which awaits investigation by the methods of streaming potential or of electroosmosis such as membranes, etc., because of the smallness of the capillaries involved these techniques will not yield unambiguous results.

It should, however, be emphasized that the simple equations of electrophoresis, of electroosmosis and of streaming potential as given at the beginning of this chapter are consistent with each other when applied to those systems which fulfill the proper theoretical conditions, i.e., large capillaries and large particles. Indeed, the agreement between the electrokinetic techniques constitutes part of the experimental verification of the Onsager reciprocal relations of the thermodynamics of irreversible processes.

Electrophoresis and Particle Size and Shape. Earlier, the Smoluchowski derivation of the electrophoretic mobility of suspended particles was derived (eq. 3). This derivation assumes: (1) The usual hydrodynamic equations for the motion of a viscous fluid are valid in the double layer. (2) The particle distorts the electrical field so that the current passes tangentially along the surface of the particle. (3) The thickness of the electrical double layer is so much less than the particle size that the electrical field is effectively parallel to the double layer over the entire surface of the particle. (4) The double layer is not deformed by the electrical field.

There is general agreement that for large, non-conducting particles of any shape, Smoluchowski's expression (eq. 3) is valid. For smaller particles, however, complications develop and the electrophoretic mobility becomes a function of particle size, shape, and the ζ-potential.

It is easy to show from elementary considerations that the numerical factor of 4π in the Smoluchowski equation should approach 6π as the particle becomes very small. For example, the force acting to produce motion is QE where Q is the total charge on the particle and E is the potential gradient. The resisting force is the Stokes factor $6\pi r\eta V$ and at a steady velocity these two forces are equal and

$$QE = 6\pi r\eta V \tag{31}$$

The potential of an isolated charged sphere is

$$\zeta = \frac{Q}{Dr} \tag{32}$$

Substituting eq. 31 into eq. 32, there results

$$U = \frac{\zeta D}{6\pi\eta} \tag{33}$$

Fig. 16. f_1, f_2, f_3 and f_4 are forces acting during electrophoresis of a particle. Dotted line represents the ionic atmosphere. r is the radius of the spherical, non-conducting particle.

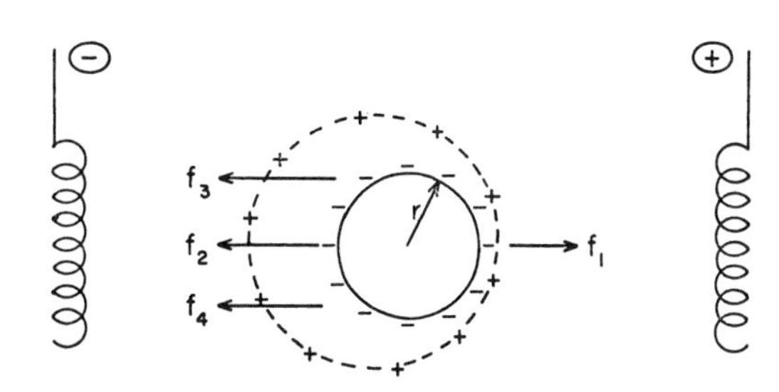

Figure 16, which is a modification of Figure 2, illustrates the effects which need to be considered in the treatment of the electrophoresis of smaller particles.

The force exerted by the applied external voltage is represented by f_1, which is equal to QE where Q is the charge on the particle and E is the potential gradient. f_2 is the frictional force and is again Stokes factor $6\pi\eta rU$ where η is the viscosity coefficient of the liquid medium and U is the electrophoretic velocity, f_3 is a force produced by a flow of the solvent as a result of movement of the ions in response to the applied external field; this is the electrophoretic retardation effect. Finally, f_4 is a force which results from a distortion of the ionic atmosphere around the particle. This gives rise to an additional electric field acting on the particle which is directly opposed to the applied field; the particle and its atmosphere become polarized and retard the electrophoretic motion. This is known as the relaxation effect.

At constant velocity, the sum of all the forces acting must be zero, and

$$f_1 + f_2 + f_3 + f_4 = 0 \qquad 34$$

Substitution for f_1 and f_2 and rearranging gives

$$U = \frac{1}{6\pi\eta r}(QE + f_3 + f_4) \qquad 35$$

The forces f_3 and f_4 are functions of the ζ-potential, the radius of the sphere as well as of the valences, concentrations and mobilities of the ions in the double layer. This is clearly a complex situation. The needed differential equations have been assembled by Wiersema, Loeb and Overbeek, and solved numerically with the help of a computer. Some of their results are shown graphically in Figure 17. The y-axis in Figure 17 refers to the numerical factor to be used in the electrophoretic equation relating mobility and the ζ-potential (see eq. 3). The results shown in Figure 17 are based on limiting mobilities of the small ions in solution of 70 $ohm^{-1}cm^2$ eq. $^{-1}$. Whereas electrophoretic mobilities are dependent on the mobilities of the small ions, the dependence is not critical. Surface conductance which is a complicating factor has been implicitly included in the solution of eq. 35 and the values shown in Figure 17 take surface conductance into account. It is clear that if the particles are conducting, the electrical field will be disturbed due to conduction through the particle. Fortunately, most particles with which the biochemist is apt to deal have much lower conductances than that of the surrounding medium and particle conduction is not ordinarily a difficulty. For example, the alternating current conductance of ion exchange resins is rather high but, as the frequency of the current decreases,

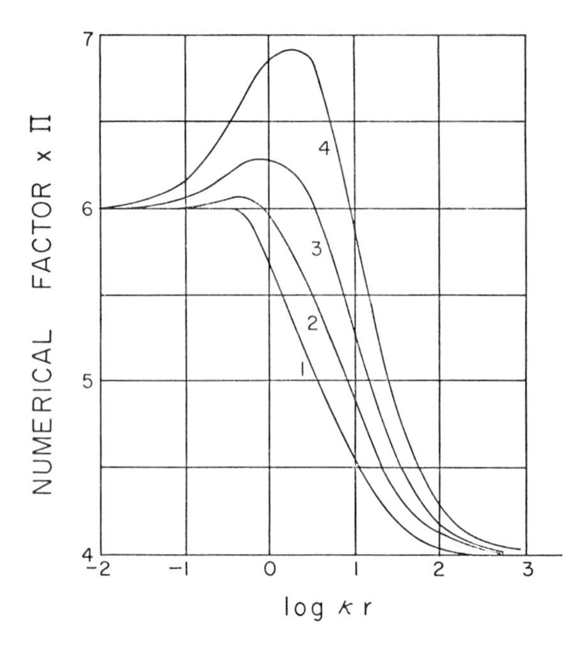

Fig. 17. Plot of the numerical coefficient for the electrophoretic mobility of spherical, non-conducting particles as a function of log κr. Curve 1, $\zeta = 0$; curve 2, $\zeta = 25.69\ mV$; *curve 3,* $\zeta = 51.38 mV$; curve 4$\zeta = 77.07 mV$. 25°. (Uni-univalent electrolyte calculated from Wiersema, Loeb and Overbeek: J. Coll. Intf. Sci. *22:* 78, 1966.)

the conductance decreases; the particles become polarized. Resins show practically no direct current conductance.

The above discussion concerns itself only with uni-univalent electrolytes. Higher valence types introduce additional complications and the publication of Wiersema, Loeb and Overbeek should be consulted.

It can be said in summary that for larger non-conducting particles, particles whose dimensions are much greater than the thickness of the double layer, Smoluchowski's equation is valid and the electrophoretic mobility is independent of the size and shape of the particle. The particle size for the validity of Smoluchowski relation is reached when κr is greater than about 1,000. For systems in which κr is less than about 1,000, the mobility becomes a function of both the size and shape of the particle. If the particle departs from a spherical shape, and $\kappa r < 1{,}000$, there is no simple relation between the ζ-potential and particle mobility, although there does exist a derivation for the mobility of small cylinders.

The Zeta Potential. Throughout this discussion, the zeta potential has been used without an explicit definition. There have been a number of workers who have doubted the utility of the zeta potential, preferring to express their electrophoretic results in terms of mobilities, and some have been outspoken in opposition to the entire concept of the zeta potential. It has to be admitted immediately that there is a measure of ambiguity associated with meaning of the zeta potential. This potential refers to the potential difference across the plane of motion between two phases and has no necessary relation to the total potential from the interior of one phase to the interior of the other phase. On the surface of proteins and other biopolymers, exposed to water, would be a fixed layer of water molecules embedded in which are the ion sites, and some of these sites would be beneath the fixed layer of water. All motion would take place beyond the fixed layer of water. Some of the ion sites might be located in cracks and crevices on the surface of a protein molecule or particle undergoing electrophoresis, and part of the ion double layer associated with the hidden sites would move with the particle. It is to be anticipated that at higher ionic strengths (decreased thickness of the double layer) the more apt are the electrophoretically hidden sites to be present.

The real excuse for the use of the zeta potential lies in the help which this approach affords in the discussion of electrophoretic theory and in the theory of streaming potential and of electroosmosis as well. In addition, the zeta potential can be related rather directly to a number of physical phenomena such as stability of suspensions, the electroviscous effect and ion binding.

Stern has presented a fairly elaborate theory of the potential across the fixed double layer as well as that of the diffuse double layer. Whereas the Stern theory is undoubtedly conceptionally correct, with few exceptions it is impossible to calculate the potential of the fixed ion layer and, accordingly, the theory has limited utility.

Interfacial Charge. The relation between electrostatic charge and potential was noted in Chapter 3, Electrolytes and Water, and the intention is to extend this discussion to interfaces. For plane solid surfaces in contact with water or for larger particles for which κr is greater than about 300, the relation between the charge and the potential is straightforward and fairly simple. The appropriate equation was first derived by Gouy and by Chapman.

Poisson's equation for the potential near a flat surface is

$$\frac{d^2\psi}{dx^2} = -\frac{4\pi\rho}{D} \qquad 36$$

where ψ is the potential at a distance x from the solid surface and ρ is the volume density of electrostatic charge; D is the dielectric constant. The expression for the distribution of charges in the liquid phase is given by Boltzmann's equation and

$$\rho = n_1 Z_1 \epsilon e^{-\frac{Z_1\psi\epsilon}{kT}} - n_2 Z_2 \epsilon e^{+\frac{Z_2\psi\epsilon}{kT}} \qquad 37$$

where n_1 and n_2 are the number of cations and anions per unit volume and Z_1 and Z_2 are their valences respectively. ϵ is the elementary charge, k is Boltzmann constant and T is the absolute temperature.

Combining the Poisson and the Boltzmann equations gives

$$\frac{d^2\psi}{dx^2} = \frac{4\pi\epsilon}{D}\left[\sum n_1 Z_1 e^{-\frac{Z_1\epsilon\psi}{kT}} - \sum n_2 Z_2 e^{+\frac{Z_2\epsilon\psi}{kT}}\right] \qquad 38$$

Integration of eq. 38 for conditions at the interface yields

$$\sigma = \left(\frac{NDkT}{2{,}000\pi}\right)^{1/2}\left[\sum C_1\left(e^{-\frac{Z_1\epsilon\zeta}{kT}} - 1\right) + \sum C_2\left(e^{+\frac{Z_2\epsilon\zeta}{kT}} - 1\right)\right]^{1/2} \qquad 39$$

where σ is the surface charge per unit area of interface, N is Avogadro's number and C is the ion concentration in moles per liter.

For ions of single valence type such that Z_1 is equal to Z_2, eq. 39 may be expressed as

$$\sigma = \left[\frac{RDTC}{2{,}000\pi}\left(e^{x} - 1 + e^{-x} - 1\right)\right]^{1/2} \qquad 40$$

where x is equal to $Z\zeta\epsilon/kT$. Since

$$\left(e^{\frac{x}{2}} - e^{-\frac{x}{2}}\right)^2 = e^{x} - 1 + e^{-x} - 1 \qquad 41$$

eq. 39 becomes

$$\sigma = 2\left(\frac{DRTC}{2{,}000\pi}\right)^{1/2}\frac{\left(e^{\frac{Z\epsilon\zeta}{2kT}} - e^{-\frac{Z\epsilon\zeta}{2kT}}\right)}{2} \qquad 42$$

and since by definition

$$\sinh x = \frac{e^{x} - e^{-x}}{2} \qquad 43$$

eq. 42 can be written

$$\sigma = \left(\frac{2DRTC}{1{,}000\pi}\right)^{1/2} \sinh \frac{Z\epsilon\zeta}{2kT} \qquad 44$$

Substituting the value of κ in eq. 44.

$$\sigma = \frac{DRT\kappa}{2\pi N\epsilon} \sinh \frac{Z\epsilon\zeta}{2kT} \qquad 45$$

It will be noticed that if ζ is small enough such that $\sinh \frac{Z\zeta\epsilon}{2kT}$ can be replaced by $\frac{Z\zeta\epsilon}{2kT}$, and provided Z is unity, eq. 45 becomes

$$\sigma = \frac{D\zeta\kappa}{4\pi} \qquad 46$$

which is the expression for the relation between the potential and the charge for a plane plate condenser where $1/\kappa$ is the distance between the plates.

Substituting numerical values at 25° in eq. 45 and converting ζ to ordinary millivolts

$$\sigma = 1.072 \times 10^{-3}\kappa \sinh 0.0194\zeta \qquad 47$$

Converting ζ into mobility, U, in microns per second per volt per centimeter (in water at 25°C) and for uni-univalent electrolytes and for larger particles ($\kappa r > 300$)

$$\sigma = 0.3514 \times 10^{5} C^{1/2} \sinh 0.250U \qquad 48$$

where C is the molar concentration of the uni-univalent electrolyte and σ is expressed in electrostatic charge per square centimeter of interface. To express the interfacial charge in terms of the number of ions or the number of mole ions per unit area, the right side of eq. 48 has to be multiplied by the appropriate conversion factors.

Charge on Spherical Particles ($\kappa r < 300$). For a spherical particle of radius r, the charge Q can be expressed as

$$Q = -\int_{r}^{\infty} 4\pi\rho a^2 da = -\int_{r}^{\infty} \frac{D}{4\pi} \frac{1}{a^2} \frac{d}{da}\left(a^2 \frac{d\psi}{da}\right) 4\pi a^2 da \qquad 49$$

which gives on integration

$$Q = -Dr^2 \left(\frac{d\psi}{da}\right)_{a=r} \qquad 50$$

where a is the radial distance from the center of the particle, extending into the solution.

The relation between the potential ψ and ζ is given by

$$\psi = \frac{\zeta r e^{k(r-a)}}{a} \qquad 51$$

and, therefore, $d\psi/da$ becomes

$$\left(\frac{d\psi}{da}\right)_{a=r} = -\frac{\zeta}{r}(1 + kr) \qquad 52$$

and substituting eq. 52 into eq. 50 gives

$$Q = D\zeta r(1 + \kappa r) \tag{53}$$

and the charge per unit area (σ) is

$$\sigma = \frac{D\zeta}{4\pi r}(1 + \kappa r) \tag{54}$$

Substituting the expression for the electrophoretic mobility into eq. 53 gives

$$Q = f(\kappa r)\pi\eta r(1 + \kappa r)\,U \tag{55}$$

An important limitation to eq. 50 must be considered. The Debye-Hückel expression for the relation between the charge and potential cannot be integrated for spherical symmetry without an approximation which involves the neglect of all higher terms of the expansion of the Boltzmann equation into a power series; this approximation has been discussed in Chapter 3 and is only valid for potentials of less than 25 millivolts.

It will be recalled from Chapter 3, Electrolytes and Water, that the relation between charge and potential depends upon the shape of the particle; the above derivation is valid only for spheres and for other shapes other relations would be needed.

A small ion correction has been proposed for eq. 53. That is, the small ions are not point charges and the apparent radius of the particle is larger than its real radius because the closest approach of the small ion to the particle is equal to the radius of the small ion. In view of the uncertainties inherent in the situation, however, it appears best to omit this correction. For example, the small ions on the surface of the particle may be hidden in the surface roughness of the particle and the small ion correction would overcompensate for the effect.

It is evident from eq. 55 that if κ is maintained constant by employing buffers of constant ionic strength but at different pH, U, for a polyelectrolyte, becomes directly proportional to Q, the charge on the particle at constant radius.

Protein Charge and Electrophoresis. The essentials of the argument presented in the last paragraph were first pointed out by Abramson who concluded that, if the thickness of the ionic double layer was held constant by the use of a series of buffers at the same ionic strength but at different pH values, the electrophoretic mobilities of protein molecules in solution should be directly proportional to the charge on the protein molecules. This should be true irrespective of the size and shape of the particle provided these factors are independent of pH. The electrophoretic mobilities of a number of proteins have been compared with the charge calculated from titration data. It has been found that there is a close proportionality between these two factors although the points of zero charge as obtained from the titration curve do not coincide with the pH of zero mobility and the proportionalities between the mobilities and titration do not agree exactly with those to be expected according to eq. 55 above.

Figure 18 illustrates the relation between the electrophoretic mobilities and the number of hydrogen ions bound by bovine serum albumin (from titration curve).

A decrease of the isoelectric points with increasing ionic strength in solutions of monovalent cations has been observed for all proteins which have been studied in this respect and, furthermore, this decrease is proportional to the square root of the ionic strength as is shown in Figure 19. The decrease of the isoelectric point with increasing ionic strength has been attributed to increased binding of anions. Whereas anion binding is certainly partly responsible, a quantitative examination of the problem leads to the conclusion that additional factors must be considered.

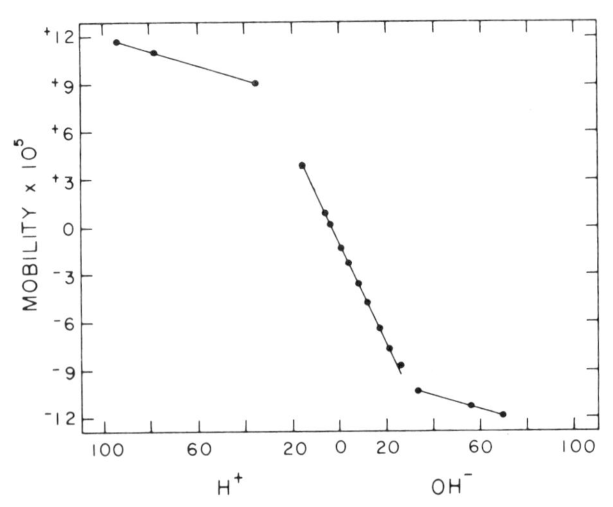

Fig. 18. Electrophoretic mobilities of bovine serum albumin compared with equivalents of H+ bound per mole of BSA. 0° ionic strength, 0.01 (NaCl + HCl or NaOH). (B. S. Schlessinger: J. Phys. Chem. *62:* 916, 1958.)

For example, correcting the charge on the protein molecules from proton binding with the values for anion binding shows a discrepancy with the electrophoretic results. Thus a plot of the net charge on the protein against the electrophoretic

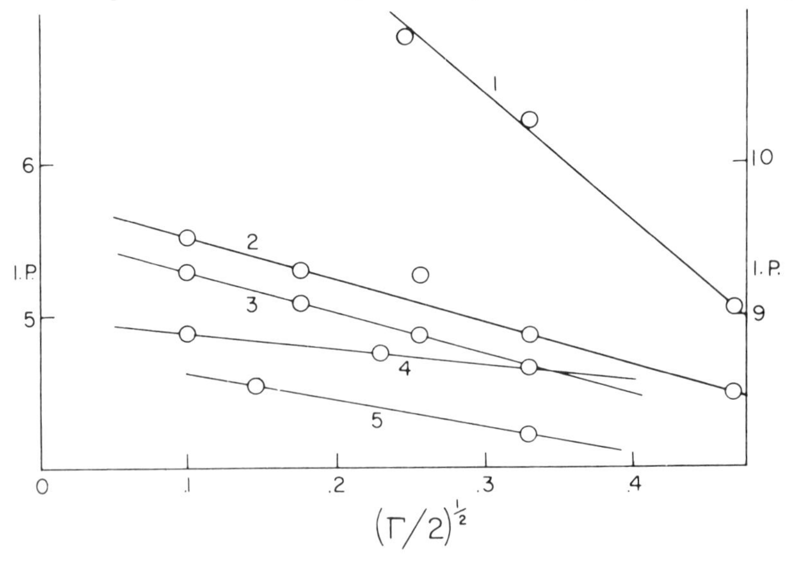

Fig. 19. Variation of the isoelectric point of proteins with the square root of the ionic strength. Moving boundary method, 0°. Curve 1, Aldolase in phosphate buffer; curve 2, ribonuclease in KCl (scale on right); curve 3, β-lactoglobulin in KCl; curve 4, bovine serum albumin in acetate; curve 5, bovine serum albumin in NaCl. (L. B. Barnett: Ph.D. Thesis, University of Iowa, 1959.)

mobility shows that the points of zero mobilities still do not agree with the points of zero net charge and furthermore the difference becomes larger as the ionic strength increases. Table 2 shows the net charge per mole at zero electrophoretic mobility for three proteins at several ionic strengths.

TABLE 2. NET CHARGE PER MOLE OF PROTEIN AT THE ISOELECTROPHORETIC POINTS. (DATA OF L. B. BARNETT AND H. B. BULL: ARCH. BIOCHEM. BIOPHYS. 89: 167, 1960.)

Protein	Γ/2	*Net Charge*	
Ribonuclease	0.01	+1.4	
	0.03	+1.6	
	0.10	+2.0	
β-Lactoglobulin	0.01	+3.0	
	0.03	+3.6	
	0.10	+12.2	
		Fast component	*Slow component*
Bovine serum albumin	0.02	+3.0	0.0
	0.10	+8.5	+4.0

The divergence between the isoelectrophoretic point and the isoelectric points may be explained on the following basis: It is assumed that the protein molecules have rough surfaces with numerous hills and valleys. At low ionic strength, the electrical double layer extends far out into the liquid and the particle acts electrophoretically as though it had a smooth surface. As the ionic strength is increased, the electrical double layer moves in towards the particle and the charges in the surface valleys fail to exert their full electrophoretic effect—part of the ions in the double layer which are counter to the hidden fixed charges now travel with the protein molecule. The hidden or valley charges would be part of the net charge on the protein and would be detectable by ion-binding studies.

In Chapter 5 the isoionic point of a protein was defined as the pH resulting after the removal of all small ions other than the hydrogen and hydroxyl ions from a protein solution. The isoelectric point of a protein has been defined as the pH at which the net charge on the protein is zero. The isoelectrophoretic point is here defined as the pH at which the electrophoretic mobility of the protein is zero; there is no necessity that the isoelectric point and the isoelectrophoretic point should coincide.

In principle, eq. 55 can be solved for the radius of the particle provided the net charge on the particle and its electrophoretic mobility are known. Such a radius would be the effective electrophoretic radius. Its actual physical significance and meaning would be dubious since in the size range considered the mobility is also dependent on the shape of the particle and also, as noted above, not all the charges on a protein molecule contribute to the electrophoretic mobility. In spite of the above qualifications, however, protein molecular radii calculated from carefully performed electrophoretic measurements, and with due regard to the estimation of the net charge as represented by the binding of cations and anions, yield values for the radii not greatly different from those derived from other physical measurements. More importantly, electrophoresis is a very sharp tool for the detection of changes in shape and size of protein molecules as a function of pH. Shown in Figure 20 are the calculated radii of the fast-moving modification of bovine serum albumin as a function of pH at an ionic strength of 0.02, chloride at 0° as reported by K. Aoki and J. F. Foster. Evidently, the molecule undergoes significant expansion below a pH of about 3.5—an expansion confirmed by other means.

Electrophoresis of Adsorbed Protein. It was noted in the discussion of the microelectrophoresis that inert particles such as those of paraffin, glass, etc., adsorb protein from solutions and acquire an electrophoretic mobility which resembles more nearly the mobility of the protein than of the inert particle. In earlier and less exact work it was believed that the mobility of the adsorbed protein was the same as that of the dissolved protein as measured by the moving boundary method under the same conditions of pH and of ionic strength. If this observation had been correct, it would have indicated that the effective electrophoretic radii of the large (microscopically visible) inert particles covered with protein were identical with the radii of the protein molecules. In the size range of many protein molecules, the electrophoretic mobility should be a pronounced function of the radius of the molecules

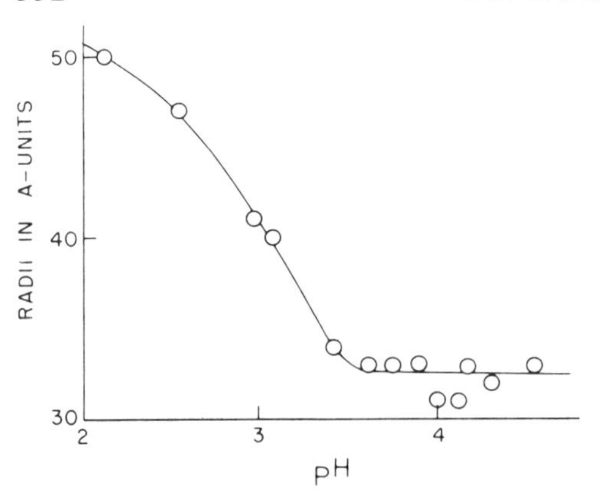

Fig. 20. Radius of fast-moving component of bovine serum albumin calculated from electrophoresis as a function of pH. 0.02 ionic strength, HCl and NaCl, 0°. (K. Aoki and J. F. Foster: J. Amer. Chem. Soc. *79:* 3385, 1957.)

since κr is, in general, much less than 300. For example, at constant charge, a plot of the electrophoretic mobility against the reciprocal of the square root of the ionic strength should be linear for κr values greater than 300, whereas for values of κr less than 300 the plot will depart greatly from linearity. Accordingly, by a study of the dependence of the electrophoretic mobilities of adsorbed protein on ionic strength at constant charge, it should be possible to estimate the effective electrophoretic radii of the adsorbed protein. Such studies have been made and the results indicate rather strongly that the effective electrophoretic radii of inert particles covered with protein are much larger than the radii of the dissolved protein molecules and these particles are well into the region where $\kappa r >> 300$.

By the use of eqs. 48 and 54 it is possible to calculate the charge per unit area for a dissolved protein and for the same protein adsorbed on an inert particle at the same pH and ionic strength, provided the appropriate mobility data are available. The electrophoretic results obtained on dissolved bovine serum albumin at an ionic strength of 0.05 by Schlessinger, using the moving boundary method, have been compared with the charge calculated from the mobility of octadecane particles covered with bovine serum albumin under the same conditions of pH, buffers and ionic strength as measured by microelectrophoresis on the visible particles. This comparison is shown in Figure 21. Evidently, when corrections are applied for the effective electrophoretic radii of the protein molecules, there is excellent proportionality between the mobilities of the adsorbed and dissolved protein and furthermore this proportionality is very nearly unity.

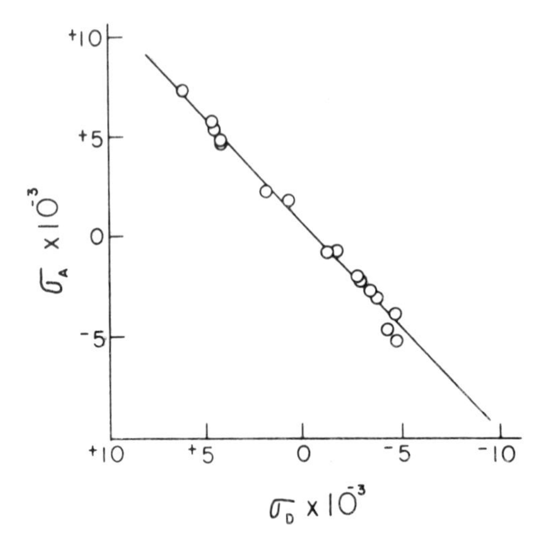

Fig. 21. Comparison of electrostatic charge per unit area for adsorbed (σ_A) and for dissolved (σ_D) bovine serum albumin (data of Schlessinger for dissolved) calculated from electrophoretic mobilities.

It will be noted in Figure 21 that zero charge for the dissolved protein does not coincide with the point of zero charge for the adsorbed protein; the isoelectrophoretic point of the bovine serum albumin has shifted to a higher pH upon adsorption. It is generally observed that the isoelectrophoretic points of adsorbed proteins are different from those of the same protein in solution and such studies as have been made indicate that the shift upon adsorption is towards a more neutral pH (see Table 3).

TABLE 3. COMPARISON OF THE ISOELECTROPHORETIC POINTS OF ADSORBED AND DISSOLVED PROTEINS; IONIC STRENGTH OF 0.05

Protein	*Adsorbed*	*In Solution*
Egg albumin	4.8	4.5
Bovine serum albumin	4.75	4.5
Trypsin	6.2	10.8
Trypsinogen	8.8	9.3
Chymotrypsin	6.2	9.5
Ribonuclease	7.8	9.2
Cytochrome C	8.6	10.7
Lysozyme	9.5	11.4

Surface Potential and Ionization. Hartley and Roe proposed an equation relating the apparent pK of a weak acid at an interface to the electrophoretic mobility. This equation has been used and discussed on several occasions and the following simple derivation serves to illustrate the effect being considered.

The total free energy of ionization of an acid can be separated into two parts such that

$$\Delta F = \Delta F_i + \Delta F_e \qquad 56$$

where ΔF_i is the intrinsic free energy change and ΔF_e is the contribution made by the electrostatic interaction. The electrostatic contribution to the free energy change is simply

$$\Delta F_e = \epsilon\zeta \qquad 57$$

where ϵ is the elementary charge. Combining eqs. 56 and 57 the result per molecule is

$$-kT \ln K = -kT \ln K_i + \epsilon\zeta \qquad 58$$

or

$$\text{pK} = \text{pK}_i + \frac{0.4343\ \epsilon\zeta}{kT} \qquad 59$$

In analogy to the expression for the pH of a weak acid, we can write

$$\text{pH} = \text{pK} + \log \frac{N}{N_0 - N} \qquad 60$$

where N_0 is the number of mole ions per unit area of the surface of the particle at surface saturation and N is the number of mole ions for intermediate stages of ionization corresponding to the existing pH in solution.

Substituting eq. 60 into eq. 59, it is found that

$$\text{pH} - \log \frac{N}{N_0 - N} = \text{pK}_i + \frac{0.4343\ \epsilon\zeta}{kT} \qquad 61$$

If the expression for the relation between the charge and the potential (eq. 53) be substituted in eq. 61, an expression will be obtained which is identical with that given in Chapter 5, eq. 88 taking account of the shift of the apparent pK determined from the titration curve of a protein and which is due to the electrostatic work of charging a spherical protein molecule. This method of evaluating W, the electrostatic factor in the acid-base titration of proteins, is more satisfactory and less ambiguous than attempting to assign a value to W from the titration curve itself; the dispersion of the intrinsic ionization constants of the protein does not complicate the calculation of W from electrophoresis.

Considering larger particles in which $\kappa r >> 300$, the ζ potential can be expressed in terms of the Smoluchowski equation and the value of ζ substituted by the electrophoretic mobility. Carrying out this substitution and introducing numerical values at 25°, eq. 61 becomes

$$\text{pH} - \log \frac{N}{N_0 - N} = \text{pK}_i + 0.217\ U \qquad 62$$

where U is the electrophoretic mobility in microns per second per volt per centimeter. The sign of the last term of eq. 62 will depend on the sign of the electrostatic charge on the particle. For a negative particle, the sign is positive whereas if the particle is positive the sign is negative (for the ionization of protons). Equation 62 has been applied to the electrophoresis of paraffin particles to which had been added small quantities of octadecyl amine on one hand and stearic acid on the other; the results are shown plotted in Figure 22.

In Figure 22 the slope for the stearic acid line is +0.292 and −0.224 for octadecyl amine. The theoretical slope as indicated by eq. 62 should be ±0.217. The intrinsic pK for stearic acid is 4.65 and that for octadecyl amine is 9.41. It is clear that the equation of Hartley and Roe is in fairly good agreement with experiment.

Closely related to the shift of the apparent ionization constant of an acid at a surface and discussed above is the accumulation of ions in the diffuse double layer under the influence of the zeta potential. The free energy change for the transfer of a mole of ions from an activity a_0 to activity a is

$$\Delta F = RT \ln \frac{a}{a_0} \qquad 63$$

The electrical free energy per mole of univalent ions for such a transfer is $N\epsilon\zeta$ where N is Avogadro's number and ϵ is the elementary charge. Equation 63 can now be written

$$a = a_0 e^{\frac{N\epsilon\zeta}{RT}} \qquad 64$$

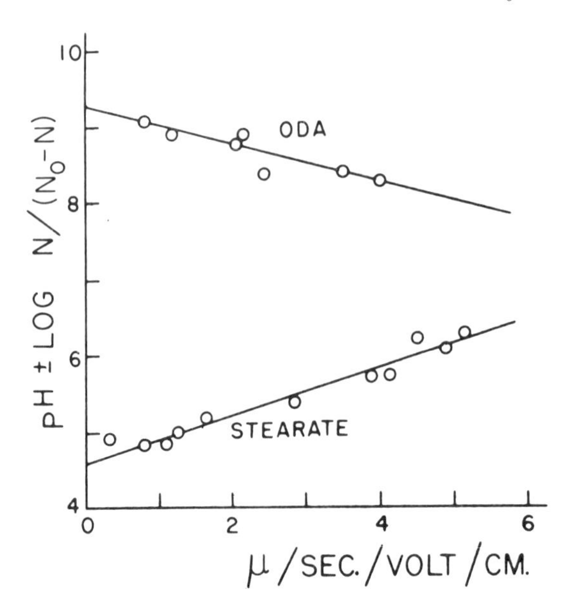

Fig. 22. Apparent pK as a function of the electrophoretic mobilities of solid paraffin particles with octadecyl amine (ODA) and with stearate; ionic strength 0.05 and at 25°. (D. K. Chattoraj and H. B. Bull: J. Phys. Chem. *63:* 1809, 1959.)

where a_0 is the activity of the ions in the absence of the surface potential. Expressing zeta in ordinary millivolts and converting eq. 64 to logarithm base 10 and at 25°

$$\zeta = 59.1 \log \frac{a}{a_0} \qquad 65$$

and for hydrogen ions

$$\zeta = 59.1 \text{ pH}_0 - 59.1 \text{ pH}_s \qquad 66$$

where pH_s refers to the pH at the surface. According to eq. 66, the pH in the immediate neighborhood of a surface is not the same as that in the bulk of the liquid. There is, however, considerable ambiguity associated with the calculation of the pH shift produced by a charged surface. The calculation using eq. 66 refers to the surface pH at a distance $1/\kappa$ from the surface and, at all other distances, the activities of the ions will be different from what they are at a distance $1/\kappa$ because the gradient of the surface potential rises steeply as the surface is approached, and the influence of individual ion sites whose sign of charge may be opposite to that of the surface as a whole makes their influence felt as the distance to the surface becomes very small.

Electroviscous Effect. A polyelectrolyte consisting of flexible chains and having a predominance of positive or negative groups along the chain will swell because the charges of like sign repel each other. This swelling is sensitive to the ionic strength of the medium; a Donnan equilibrium is established between the inside and outside of the molecule. As the molecule swells, its intrinsic viscosity increases and empirically it is found that

$$[\eta] = A\left(1 + \frac{B}{(\Gamma/2)^{1/2}}\right) \qquad 67$$

where A and B are constants and $\Gamma/2$ is the ionic strength. In the absence of salt the reduced viscosity of a flexible polyelectrolyte appears to increase without limit as the concentration of the polyelectrolyte approaches zero. In the presence of a moderate amount of added salt the reduced viscosity first increases and then decreases as the concentration of the polyelectrolyte is decreased. The behavior of the reduced viscosity as a function of the polyelectrolyte concentration is diagrammatically represented in Figure 23.

Most biopolymers do not behave as flexible chain polyelectrolytes and do not swell over a broad range of pH. Nevertheless, the viscosities of their solutions are dependent on the electrical charge they carry; this is known as the electroviscous effect. Investigation on the electroviscous effect has had a long history. Smoluchowski published an equation in 1916 describing the effect. As the suspended particle

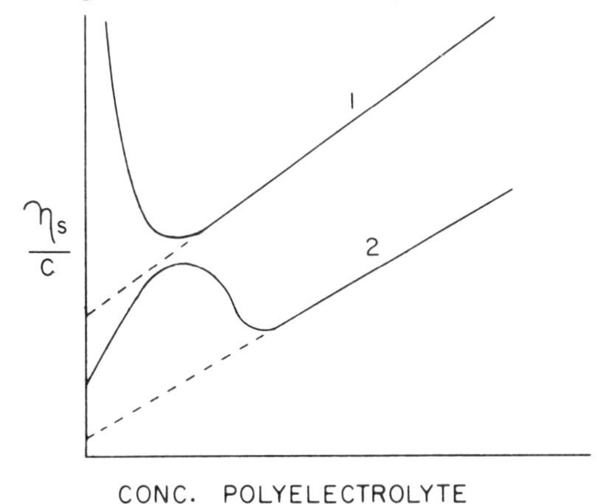

Fig. 23. Diagrammatic representation to the reduced viscosity, η_s/C, plotted against the concentration of a flexible (swelling) polyelectrolyte. Curve 1, No added salt; curve 2, salt added.

moves relative to the flow gradient, an electrical potential is established tangential to the surface particle; this is a streaming potential. This potential results in an electroosmotic force acting in the reverse direction to the flow and, accordingly, decreases the flow gradient in the immediate vicinity of the particle; the particle appears to occupy a larger hydrodynamic volume than it actually does and the viscosity of the suspension increases.

Booth has brought this argument up to date and has considered the screening effect of the small ions. If it is assumed that the mobilities of the small ions are all identical, his equation for the viscosity of dilute suppression of spherical particles reduces to

$$\eta = \eta_0 \left[1 + 2.5\,\phi \left[1 + \frac{1}{K_c \eta_0 r^2}\left(\frac{\zeta D}{2\pi}\right)^2 \pi(\kappa r)^2\,(1+\kappa r)^2 f(\kappa r)\right]\right] \qquad 68$$

η_0 is the viscosity of the solvent, ϕ is the volume fraction of the solution occupied by spherical particles of radius r. K_c is the specific conductance due to the small ions and κ is the reciprocal thickness of the electrical double layer. It will be noticed that Booth's equation is identical with that of Smoluchowski's from left to right through the term $(\zeta D/2\pi)^2$; the right hand side of Smoluchowski's equation is to be multiplied by $\pi(\kappa r)^2\,(1 + \kappa r)^2 f(\kappa r)$. The function of $f(\kappa r)$ is complex and its numerical evaluation is tedious.

The Booth equation has been subject to several experimental tests and general agreement between theory and experiment has been reported. The most impressive test is that of Rutgers and Nagels. These workers measured the viscosities of dilute suspensions of silver iodide in potassium nitrate solutions along with other necessary measurements. Comparison of their results with those predicted by the Booth equation are shown in Figure 24.

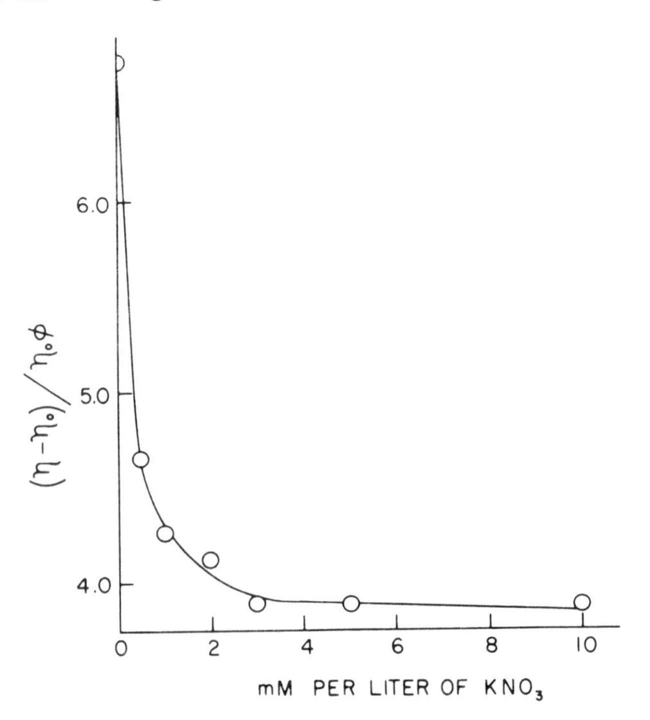

Fig. 24. Viscosities of dilute suspensions of silver bromide in solutions of KNO_3 of the indicated concentrations. Circles, experimental points; solid line, viscosities predicted from Booth's equation. (Data of A. J. Rutgers and P. Nagels: J. Coll. Sci. *13:* 148, 1958.)

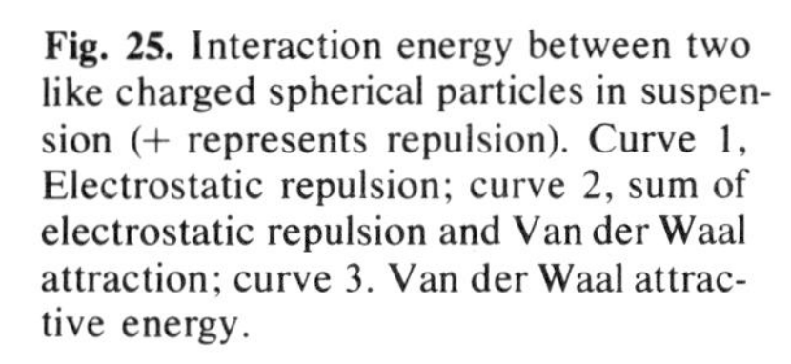

Fig. 25. Interaction energy between two like charged spherical particles in suspension (+ represents repulsion). Curve 1, Electrostatic repulsion; curve 2, sum of electrostatic repulsion and Van der Waal attraction; curve 3. Van der Waal attractive energy.

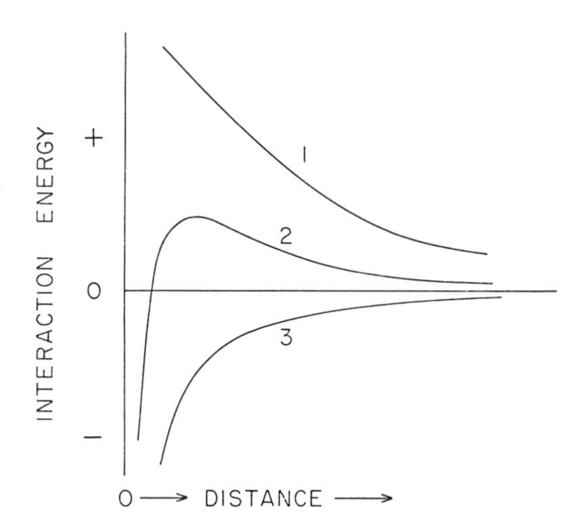

Electrostatic Forces. The stability of a hydrophobic suspension is very dependent on the concentration of the electrolytes present and, further, the valency of the ion whose charge is opposite to that of the sol plays a key role. For example, the concentration of electrolytes required to flocculate a negatively charged As_2S_3 sol stand in the ratio of about 1 to 50 to 500 for tri-, di- and univalent cations respectively. The relative importance of the valence of the ion of opposite sign from that of the sol is known as the Hardy-Schulze rule. The concept is durable and supported by evidence that the ζ-potential is one of the principal factors responsible for the stability of hydrophobic colloidal suspensions such as represented by a gold sol. That is, in order to bring two gold particles together, the electrostatic repulsive forces arising from the ζ-potential must be overcome. Electrostatic or coulombic forces diminish as the reciprocal of the square of the distance between the particles with an additional complication of screening produced by the small ions. In dilute electrolyte solutions the electrostatic forces can extend hundreds of A-units into the solution from a particle.

On closer approach of particles, the Van der Waal or London forces become of importance; these are attractive and arise as a result of mutually induced dipoles (see Chapter 6 for a discussion of Van der Waal forces). If the particles have sufficient kinetic energy to hurdle the barrier represented by the ζ-potential, the Van der Waal interaction assures the coalescence of the particles; the sol flocculates. Figure 25 is a diagrammatic representation of the variation of the interaction energy between two approaching spherical particles as a function of the distance of separation.

The situation is somewhat otherwise with solutions of most biopolymers. A protein can be adjusted to its isoelectric point and the ζ-potential is zero or nearly so but an isoelectric solution of a protein can be infinitely stable; upon collision, the isoelectric protein molecules do not stick but move apart. In a qualitative way it is easy to see why this is true. The protein molecules undoubtedly have a layer of firmly attached water at their surface and there is no more attraction between the layers of bound water on two molecules than there is for the bound water and free water. Further, the hydration energy of the protein is sufficiently large so that the water:protein interface cannot be replaced by a protein:protein interface.

The creation of a net charge on protein molecules leads to electrostatic forces of repulsion and the forces are a complex function of the ζ-potential as measured by electrophoresis. There are many examples of specific interaction between protein molecules, for example, between antigens and antibodies and between enzymes and substrates. Specific forces are characteristically short-range and hardly extend beyond a few A-units. If the non-specific repulsive electrostatic forces become large enough, specific interaction is prevented.

A direct measure of the forces existing between protein ions can be had from light scatter and from osmotic pressure measurements. The electrostatic part of the interaction constant, B, can be evaluated (Chapter 7) and related to the forces of repulsion. The forces are, however, not between just two particles but instead a vast number of particles are interacting simultaneously.

Kirkwood has considered a different type of electrostatic interaction leading to a long-range attractive force. A protein at or near its isoionic point has many possible configurations in respect to its charge distribution differing little in free energy; the configurations are constantly fluctuating. When two protein molecules approach, each alters the charge distribution on the other such as to induce interacting dipoles resulting in a long range attractive force diminishing asymptotically as $1/d^2$ where d is the distance between the protein molecules. There is screening by the small ions and the attractive force decreases as the ionic strength is increased. The dependence of the Kirkwood force on electrolyte concentration could provide an explanation as to why some proteins, for example β-lactoglobulin, are insoluble in pure water but dissolve on the addition of small quantities of electrolyte.

GENERAL REFERENCES

Abramson, H. A., L. S. Moyer and M. H. Gorin: Electrophoresis of Proteins. Reinhold Publishing Corp., New York, 1942.
Ambrose, E. J.: Cell Electrophoresis. Little, Brown and Co., Boston, 1965.
Bier, M.: Electrophoresis, Vols. 1 and 2. Academic Press, New York, 1959, 1967.
Haglund, H.: Isoelectric focusing in natural pH gradients. Science Tools *14:* 17, 1967.
Nichols, L. W., J. L. Bethune, G. Kegeles and E. L. Hess: The Proteins, Vol. 2. Academic Press, New York, 1964.
Shaw, D. J.: Electrophoresis. Academic Press, New York, 1969.
Wiersema, P. H., A. L. Loeb and J. Th. G. Overbeek: Calculation of the electrophoretic mobility of a spherical colloid particle. J. Coll. Interf. Sci. *22:* 78, 1966.

PROBLEMS

1. Derive the Smoluchowski equation for the electrophoretic mobility of a particle for which κr is larger than 300. Consider and indicate the role of the following factors in the electrophoresis of particles of any size:
   (a) Surface conductance
   (b) Relaxation
   (c) Surface charge
   (d) Ionic strength

   Ans: See text.

2. (a) Outline the micro-method for the determination of the electrophoretic mobility; include in your description how mobilities of particles in μ/sec/volt/cm are obtained from a measurement of the current flow through the cell and the rate of migration of the particles.
   (b) The mobility of a human red cell at 25°C in M/15 phosphate buffer at pH 7.4 is 1.31 μ/sec/volt/cm. Calculate the zeta potential and the thickness of the electrical double layer.

   Ans: (a) See text. (b) 16.85 mv., 7.6 *A*

3. Describe the Schlieren lens system and show how it is used to locate the positions of the concentration boundaries in the electrophoretic cell. How can the optical system be modified to permit the determination of protein concentrations of the several electrophoretic components?

   Ans: See text.

4. The specific conductance of a buffered protein solution is 6.5×10^{-4} reciprocal ohms, the cross sectional area of the electrophoresis cell is 35 sq. mm., the current is 12 ma. and the protein boundary moves 1.50 cm. in 2 hours. Calculate the electrophoretic mobility of the moving protein boundary in cm.^{2}/sec./volt.

   Ans: 3.95×10^{-6}

5. An aqueous solution of 2×10^{-4} M NaCl is forced through a capillary 20 microns in diameter at 25°C under a pressure of 3.50 cm. of mercury. The observed streaming potential is 90.0 millivolts and the specific conductance is 28.90×10^{-6} reciprocal ohms. (a) Calculate the zeta potential in millivolts. (b) what is the electrical efficiency of this process in terms of the total energy required to force the solution through the capillary?

   Ans: (a) 71.99 (b) 7.75×10^{-3}%

6. A current of 7.60 ma was passed through a microelectrophoresis cell. The cell had a cross-sectional area of 0.074 sq. cm. and contained 0.10 M NaCl whose specific conductance was 1.103×10^{-2} reciprocal ohms. Calculate the number of calories liberated per second per centimeter length of the cell.

   Ans: 1.69×10^{-2} cals/sec.

7. The following are the electrophoretic mobilities of Pyrex glass powder covered with adsorbed egg albumin as a function of pH in 0.05 acetate buffer and at 25°C.

pH	*μ/sec./volt/cm.*	*pH*	*μ/sec./volt/cm.*
3.44	+1.42	4.67	+0.32
3.68	+1.24	4.94	+0.05
3.94	+0.95	5.41	−0.33
4.23	+0.66	5.85	−0.88
4.44	+0.49	6.31	−1.41

   Calculate and plot the number of mole ions per square centimeter of protein surface as a function of the pH of the solution.

   Answer is too extensive to give.

8. Electrophoretic measurements on bovine serum albumin with the moving boundary method at 25°C as a function of pH in solutions containing NaCl and HCl with a total chloride concentration of 0.02 M were conducted and the results are shown in the table below. The cross-sectional area of the electrophoresis cell is 0.30 sq. cm. and the cell constant of the conductivity cell is 0.8392.

   (a) Calculate the mobility of B.S.A. at each pH and compare with the titration curve for B.S.A. (Tanford et al.: J. Amer. Chem. Soc. *77:* 6814, 1955.)

   (b) Calculate the effective electrophoresis radius of B.S.A. molecules as a function of pH.

pH	$i \times 10^3$	*Ohms*	*Seconds*	*Distance (cm.)*	*Moles Cl⁻ bound/mole BSA*
2.10	5.5	180	2655	2.78	35.0
2.42	5.5	205	2930	3.45	26.3
2.97	5.5	220	2725	3.11	20.1
3.48	5.5	243	2805	2.77	14.9
3.65	5.5	260	2900	2.67	13.0
3.71	5.5	260	2865	2.49	12.6
3.91	5.5	270	2595	1.81	11.6
4.01	5.5	278	2660	1.66	11.1
4.05	5.5	280	2810	1.57	11.0
4.12	5.5	285	2865	1.43	10.8
4.25	5.5	290	2775	1.06	10.4
4.39	5.5	288	2700	0.72	9.2

   Ans: (a) Results too extensive to give. (b) About 35*A* at the isoelectric point.

CHAPTER 14

Sedimentation

Movement of Particles. Suspended particles subject to an external force will move through a viscous medium in response to the force and, in a brief interval, attain a constant velocity. At constant velocity, the forces acting to produce motion are exactly equal to the resisting forces. The resisting force is equal to the product of a frictional coefficient and the velocity and

$$fu = \text{acting forces} \qquad 1$$

where f is the frictional coefficient and u is the velocity of the particles. The force causing diffusion of a dissolved substance is the gradient of the chemical potential for the substance; for dilute solutions the force per mole is $RTdC/dx$, where C is the concentration. Accordingly, for diffusion eq. 1 can be written

$$fu = \frac{RTdC}{dx}$$

where f is now the frictional coefficient per mole and u is the rate of diffusion. By comparing eq. 2 with Fick's first law (Chapter 11), it is found that

$$f = \frac{RT}{D} \qquad 3$$

where D is the diffusion coefficient.

The electrophoretic mobility per mole at constant velocity is

$$fu = SQE = \frac{RT}{D}u \qquad 4$$

where Q is the charge on the particles, E is the potential gradient and S is a screening factor resulting from the ion atmosphere and is proportional to $\kappa r/(1 + \kappa r)$ where κ is the reciprocal Debye-Hückel distance, r is the radius of the particles and u is the rate of electrophoresis.

In a gravitational field, the force acting is ma where a is the acceleration of the field and m is the mass. Setting f equal to RT/D

$$\frac{RT}{D}u = ma \tag{5}$$

In the earth gravitational field, a is very nearly equal to 981 dynes and in a centrifugal field a is equal to $\omega^2 X$ where ω is the angular velocity and X is the distance from the center of rotation.

In 1856, Stokes, interested in the rate of fall of raindrops, derived an expression for the viscous drag exerted on spherical particles. He found that f, the frictional coefficient of a spherical particle, is

$$f = 6\pi\eta r \frac{dx}{dt} \tag{6}$$

where η is the coefficient of viscosity of the medium, r the radius of the spherical particle and dx/dt is the velocity of the particle. Stokes' law holds accurately up to a Reynolds number of 0.5 beyond which second order terms become significant. There is no simple derivation of Stokes' law; the problem is not simple. It was nearly 80 years later that expressions for the viscous drag exerted on prolate and on oblate ellipsoids of revolution were derived; these expressions have been discussed in Chapter 11, Diffusion. The viscous drag exerted on ellipsoids was compared with that exerted on spherical particles of the same volume; the ratios were plotted against the axial ratios of the ellipsoids (see Fig. 15, Chapter 11). The rate of sedimentation of particles serves as a basis for the estimation of particle sizes. A short section will be devoted to the determination of the sizes of microscopic particles followed by a summary of the ultracentrifuge.

Microscopic Particle Sizes. It is a remarkable fact that it is frequently easier to measure the sizes of protein molecules than it is to determine the size of particles that are visible in the microscope. Often the need arises to measure the sizes of living cells, of mitochondria and of other cellular elements, of emulsion particles, etc. The related problem of counting the number of such particles in a given volume is also encountered. A variety of methods have been developed and some of these such as light scatter and Coulter counter have already been mentioned; here the concern is with sedimentation methods.

By setting the force of gravity equal to Stokes' factor, the rate of fall of spherical particles can be calculated readily, that is

$$(\rho_2 - \rho_1)\, gV = 6\pi\eta r \frac{dx}{dt} \tag{7}$$

where V is the volume of the particle, ρ_1 and ρ_2 are the densities of the medium and the particle respectively, g is the acceleration of gravity, η is the coefficient of viscosity, r is the radius of the particle and dx/dt is the rate of fall of the particle. Substituting the expression for the volume of a sphere and rearranging, eq. 7 becomes

$$\frac{dx}{dt} = \frac{2}{9}\frac{r^2 g\,(\rho_2 - \rho_1)}{\eta} \tag{8}$$

Since the rate of fall is constant, the radius of the particle is given by

$$r = \left(\frac{9\eta x}{2g\,(\rho_2 - \rho_1)\, t}\right)^{1/2} \tag{9}$$

where x is the distance traveled in time, t.

Suppose the suspension consists of two sizes of spherical particles of radii r_1 and r_2. After a given time of sedimentation, the situation can be diagramed as shown in Figure 1.

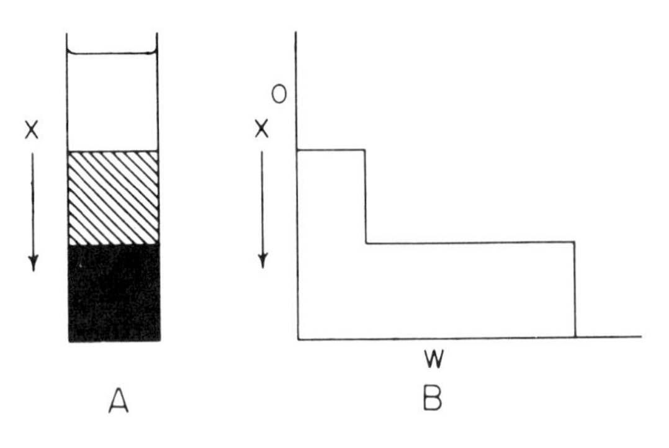

Fig. 1. Sedimentation in a suspension consisting of two sizes of spherical particles. *A*, Sedimentation column. *B*, Plot of distance, x, against the weight of particles per unit volume.

For a large number of different particle sizes, Figure 1 B would have the appearance shown in Figure 2A. Figure 2B shows the distance sedimented, x, plotted against the increment in the weight of particles per unit volume of the suspension at equal intervals of x; evidently, Figure 2B is the particle size distribution of the

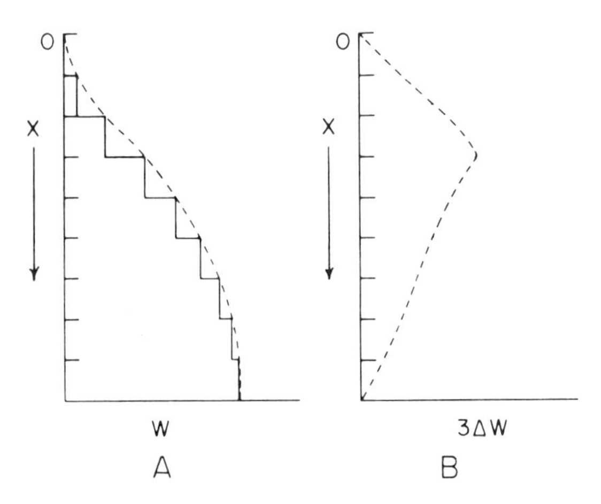

Fig. 2. Sedimentation in a suspension consisting of many sizes of spherical particles. *A*, Plot of distance, x, against the weight of particles, w, in unit volume. *B*. Plot of x against the increment of w, (ΔW), shown in *A*. ΔW has been multiplied by 3 to expand the scale of the abscissa.

suspension. The sizes of the particles could be calculated from x and the elapsed time and the area under the distribution curve for a given interval of size would give the total weight of the particles of this size interval. Several different methods have been developed for the determination of the rate of settling of a suspension. One of these is to immerse a pan of an analytical balance into the suspension and measure the weight of the particles which have sedimented. The weight is then plotted against time; the intercepts of tangents drawn to the curve on the weight axis yield the relative weights of the particles whose sizes correspond to the time required for the particles to sediment. This is the method of intercepting tangents developed by Oden in 1916.

The determination of particle size by rate of sedimentation requires that the suspension remains completely undisturbed by gradients of flow produced either by temperature or movement of the suspension. The suspensions must also be dilute without particle interaction. Experiments of the kind described above led to the development of the ultracentrifuge by Svedberg in 1923.

The Ultracentrifuge. In principle, an analytical ultracentrifuge operates in the same manner as an ordinary laboratory centrifuge. It is capable of higher speeds of rotation but, more importantly, it is provided with accurate control of speed, of temperature and optical systems similar to those previously described in the discussion of diffusion and of electrophoresis which permit an estimation of the position of the sedimenting boundary as well as the concentration of the component. Figure 3 is a diagrammatic sketch of an ultracentrifuge. The ultracentrifuge has played and continues to play an important role in investigations on biopolymers. In addition to the analytical ultracentrifuge there is the preparative ultracentrifuge used primarily for the isolation of cellular fragments such as ribosomes and mitochondria. The preparative ultracentrifuge has good speed and temperature control but no optical system; it is able to handle larger amounts of material.

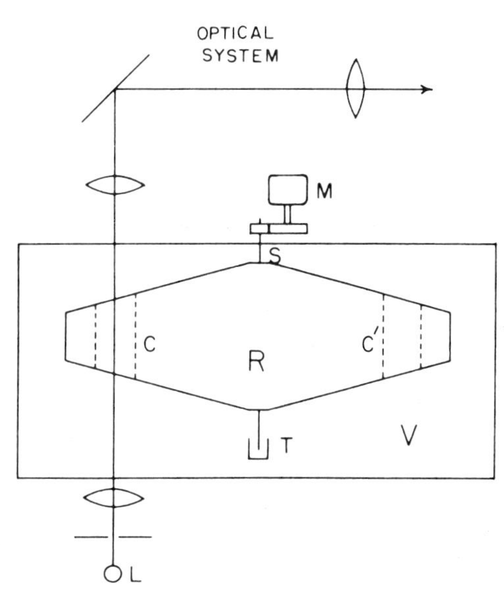

Fig. 3. Diagrammatic representation of the ultracentrifuge. C is cell with windows containing solution. C' is matched cell. L is light source. M is electric motor. R is rotor. S is a wire suspending the rotor and connected to motor by gears. T is thermistor connected by mercury cup to electrical circuit. V is a high vacuum chamber in which the rotor spins.

The action of a gravitational field and of a centrifugal field on matter are essentially equivalent. The centrifugal potential is proportional to N^2x^2 and the centrifugal field of force is proportional to N^2x where N is the number of revolutions per second and x is the radius of the rotor. A body which moves at a uniform speed in a circular path is continually accelerated towards the center of that path and the acceleration is ω^2x where ω is the velocity of rotation expressed in radians per second (see Chapter 1, Mathematical Review). The product of the acceleration and the mass gives the force, F, acting and

$$F = V(\rho_2 - \rho_1)\,\omega^2x = V\rho_2\,\omega^2x - V\rho_1\,\omega^2x \tag{10}$$

where the symbols have the meaning previously defined. Equation 10 is valid irrespective of the size of the particles and for one mole of solute eq. 10 can be written

$$F = M\omega^2x - M\overline{V}_2\rho_1\omega^2x = M\omega^2x\,(1 - \overline{V}_2\rho_1) \tag{11}$$

where M is the molecular weight of the solute and $\overline{V}_2$ is its partial specific volume. ρ_1 is the density of the solvent. The force resisting sedimentation in the centrifuge is $f dx/dt$ where f is the frictional coefficient per mole and dx/dt is the rate of sedimentation. Setting the force acting equal to the resisting force gives

$$M\omega^2x\,(1 - \overline{V}_2\rho_1) = f\frac{dx}{dt} \tag{12}$$

and rearranging

$$M = \frac{f}{(1 - \overline{V}_2\rho_1)}\,\frac{1}{\omega^2x}\,\frac{dx}{dt} \tag{13}$$

The factor $(dx/dt)/\omega^2 x$ is the rate of sedimentation per unit field of force and is denoted by S, the sedimentation coefficient. Equation 13 then becomes

$$M = \frac{fS}{(1 - \overline{V}_2\rho_1)} \quad 14$$

In Chapter 11, Diffusion, it was shown that

$$D = \frac{RT}{f} \quad 15$$

where D is the diffusion coefficient, R is the gas constant, T the absolute temperature and f is as above the frictional coefficient. Substituting eq. 15 into eq. 14 gives

$$M = \frac{RTS}{D\,(1 - \rho_1\overline{V}_2)} \quad 16$$

Evidently, if the particles are spherical, Stokes' law could have been substituted in eq. 14 and the molecular weight calculated without a determination of the diffusion coefficient. In general, it is necessary to have a separate measurement of the diffusion coefficient and this requirement is the primary reason for the considerable amount of work which has been done on the diffusion coefficients of proteins.

S, the sedimentation coefficient, is usually expressed in Svedberg units. Since ω has the dimension of reciprocal time, S itself has the dimensions of time. In rate sedimentation it is usual practice to employ schlieren optics in connection with a cylindrical lens (see Chapter 11, Diffusion, for a discussion of optical systems). From the definition of S, a plot of $\ln x$ against t, where x is the distance of the boundary from the axis of rotation, should be linear with a slope of $\omega^2 S$. The value of S observed for proteins in solution is generally of the order of 10^{-13} to 10^{-12} seconds and the value of S equal to 10^{-13} seconds is customarily referred to as one Svedberg unit (denoted by the letter S). Sedimentation data are usually reduced to 20° in pure water by the appropriate corrections and are expressed as S_{20}.

Electrophoresis is diffusion under an electrochemical gradient; diffusion itself occurs because of a gradient in chemical potential and rate sedimentation is produced by a gradient in gravitational potential. Rate sedimentation is subject to all of the qualifications of diffusion. The influence of electrical charges on the sedimentation must be minimized by the addition of electrolyte. There are also coupled sedimentation rates between two or more solutes just as there are in diffusion and there is a dependence of the sedimentation rate on solute concentration. For example, a more accurate expression for rate sedimentation than is eq. 16, and which considers the activity coefficient of the solute in a two component system, is given by

$$S = \frac{DM(1 - \rho_1\overline{V}_2)}{RT\left(1 + C\dfrac{d\ln\gamma}{dc}\right)} \quad 17$$

where C is the concentration of the solute in grams per milliliter and γ is the activity coefficient of the solute (compare with eq. 53, Chapter 11, Diffusion).

Whereas rate sedimentation has utility in the study of mixtures of biopolymers, it is not nearly so sharp a tool for this purpose as, say, electrophoresis. Consider, for example, an elongated rod-shaped molecule. If this elongated rod-like molecule is depolymerized to form a series of shorter rods of the same diameter, it will be discovered that the original rod as well as all the shorter rods of varying lengths sediment at exactly the same rate; the ratio of the mass of the various particle sizes to their frictional resistance remains constant.

The relative failure of the ultracentrifuge in the analysis of mixtures is strikingly shown in sedimentation studies on whole blood plasma or serum. All human sera show two main components, one of which has an S_{20} of about 4.5 and the other about 7. In addition, most plasmas show significant amounts of a much more rapidly sedimenting component with a S_{20} value between 17 and 20. The contrast with the rich and detailed separations found in careful electrophoretic work is considerable.

Gofman and co-workers have, however, been able to improve the situation to some extent by studying blood plasma in flotation gradients, that is, the density of the plasma has been increased by the addition of a solute and the protein fractions containing substantial amounts of lipid tend to cream in the centrifuge. By this means at least five components containing varying amounts of lipid were observed. Particular significance was attached to the fraction with S_f of 10-20 corresponding to molecular weights of about a million; this fraction was believed to be related in some way with the developement of atherosclerosis.

Density gradients have become of considerable aid in separating components in the ultracentrifuge. For example, samples of deoxyribonucleic acid (DNA) are dissolved in cesium chloride solutions and centrifuged. The cesium chloride establishes a concentration gradient and, accordingly, a density gradient, the density becoming greater outwardly from the axis of rotation as a result of the centrifugal field. The DNA then sediments to a level in the cell at which its density is equal to that of the cesium chloride solution. DNA derived from living cells which have been given the opportunity of incorporating deuterium can then be separated into deuterium containing DNA from DNA which has not assimilated the deuterium.

Hydrodynamic Hydration. The hydration of biopolymers in solution has been discussed in Chapter 10, Viscosity and the Flow of Liquids, as well as in Chapter 11, Diffusion; it is possible to approach this problem with the ultracentrifuge. In this method, which has received attention on several occasions, the density of the solvent in which the biopolymer is dissolved is varied by the addition of a third component such as sucrose, salts, etc. The rate sedimentation, S, is multiplied by the coefficient of viscosity, η, and the product plotted against the density of the solution. The plot which should be linear is extrapolated to zero ηS; at the density of the solvent corresponding to zero rate of sedimentation, the density of the protein in solution is equal to that of the solvent.

Evidently, the hydrated density is equal to the hydrated weight of the biopolymer divided by its hydrated volume. The hydrated weight is equal to $(M_2 + W)$ where M_2 is the molecular weight of the biopolymer (anhydrous) and W is the weight of the solvent associated with one mole of the biopolymer. The hydrated volume is $(M_2\bar{V}_2 + W\bar{V}_1)$ where $\bar{V}_2$ is the partial specific volume of the biopolymer in solution and $\bar{V}_1$ is the partial specific volume of the solvent associated with the biopolymer. Setting the density of the solution, at zero rate of sedimentation, equal to ratio of the hydrated weight to the hydrated volume and solving for W gives

$$W = \frac{(1 - \rho_1\bar{V}_2)M_2}{(\rho_1\bar{V}_1 - 1)} \qquad 18$$

where ρ_1 is the density of the solvent at zero sedimentation rate. Per gram of biopolymer, eq. 18 becomes

$$W' = \frac{(1 - \rho_1\bar{V}_2)}{(\rho_1\bar{V}_1 - 1)} \qquad 19$$

where W' is the grams of solvent associated with one gram of anhydrous biopolymer.

Figure 4 shows a plot of ρ against $S\eta$ for ribonuclease at pH 7 in solutions of $NaCl$, $(NH_4)_2SO_4$ and KCl. Evidently a good straight line is obtained which extrapolates to a hydrated density of 1.284 which leads to a calculated hydration of about 0.35 grams of solvent associated with one gram of ribonuclease. Without exploring the problem further, it can be said in summary that, whereas the method is attractive, there still remain several unsatisfactory aspects.

Equilibrium Method. Whereas in rate sedimentation and for molecular weights below about 250,000 the centrifuge is operated at about 60,000 r.p.m., in the equilibrium method, and for molecular weights of between 10,000 and 200,000,

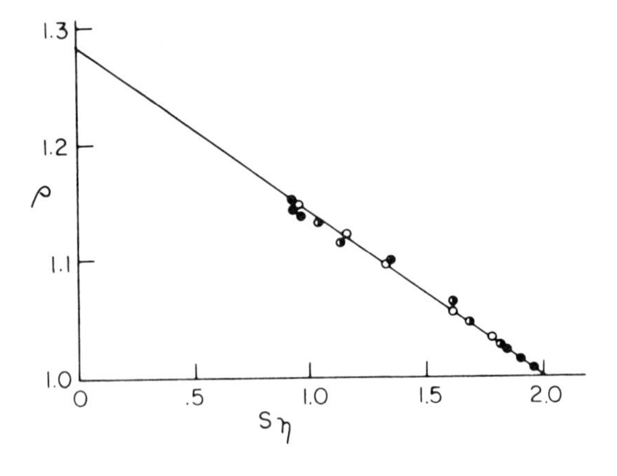

Fig. 4. Dependence of $S\eta$ on ρ for ribonuclease, pH 7, 20-25°. The salts used to increase the density were NaCl, ○; $(NH_4)_2SO_4$, ◑; KCl, ●. (D. J. Cox and V. N. Schumaker: J. Amer. Chem. Soc. *83:* 2433, 1961.)

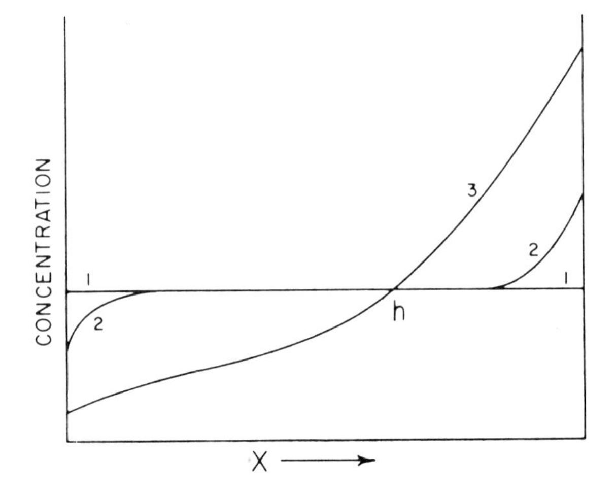

Fig. 5. Concentration in an ultracentrifuge cell as a function of the distance in the cell from the axis of rotation. Curve 1, Before start of centrifuge. Curve 2, After centrifugation for relatively short interval. Curve 3, After equilibrium has been attained. h is hinge point.

speeds between 15,000 and 3,000 are employed. Prolonged rate sedimentation simply results in the accumulating of the solute in the bottom of the centrifuge cell in the form of a pellet. In the equilibrium method the rate of sedimentation is balanced against the rate of back diffusion. Starting with a uniform concentration in the centrifuge cell, the concentration of the solute is gradually decreased at the meniscus and increased at the bottom of the cell, the center portion remaining unchanged. As equilibrium between sedimentation and diffusion is approached, the plateau region in the cell disappears and there will be only one position in the cell at which the concentration is the same as that initially present; this position is known as the hinge point. Shown in Figure 5 are diagrammatic plots of the concentration against the distance along the cell before centrifugation is started, after a short interval (suitable for the calculation of the molecular weight by the Archibald method) and finally at equilibrium.

The contrast between the concentration gradients existing in the centrifugal cell as a function of the distance from the axis of rotation for rate sedimentation and for equilibrium centrifugation is shown in Figure 6A and B.

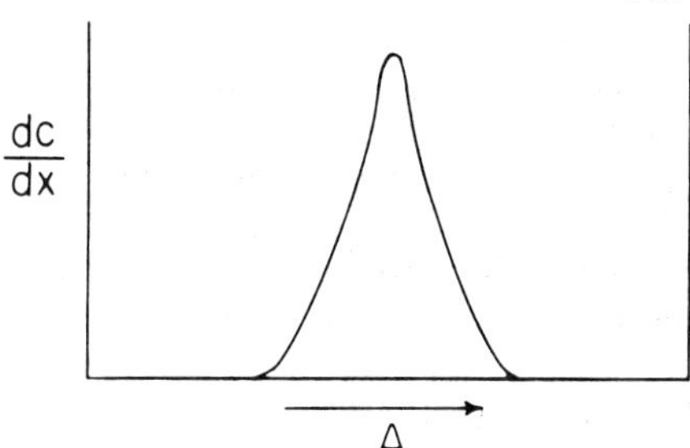

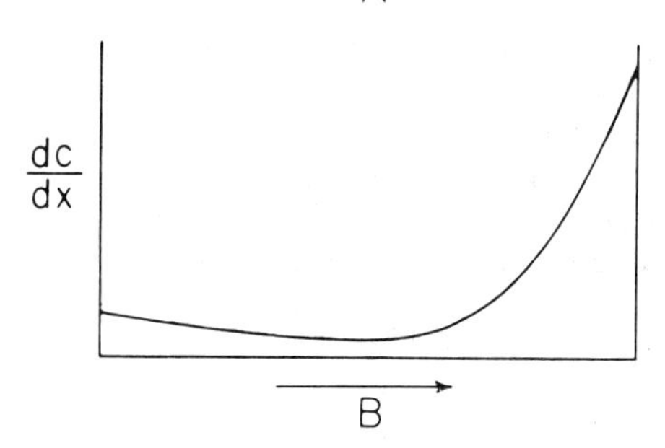

Fig. 6. Concentration gradients in centrifuge cell plotted against distance from axis of rotation. Arrows indicated direction of centrifugation. *A*, Rate sedimentation. *B*, Equilibrium centrifugation.

As a result of centrifugation, the rate flow of the mass of the solute, m, outward is

$$\frac{dm}{dt} = \frac{\omega^2 x(1 - \overline{V}_2\rho_1)MAC}{f} \qquad 20$$

where A is the cross sectional area of the centrifuge cell. The rate of flow backwards as a result of diffusion is (Fick's first law of diffusion)

$$\frac{dm}{dt} = -DA\frac{dC}{dx} \qquad 21$$

At equilibrium these two rates are equal and opposite, and combining eqs. 20 and 21 and rearranging gives

$$\frac{dC}{dx} = \frac{\omega^2 x(1 - \overline{V}_2\rho_1)MC}{Df} \qquad 22$$

Since D is equal to RT/f, eq. 22 can be written

$$\frac{dC}{dx} = \frac{\omega^2 x(1 - \overline{V}_2\rho_1)MC}{RT} \qquad 23$$

It is evident from eq. 23 that a plot of log C against x^2 should be linear with a slope of $\omega^2(1 - \overline{V}_2\rho_1)M/4.606\ RT$ from which the molecular weight of the sedimentating species can be calculated. A departure of the plot from linearity indicates interaction or heterogeneity or both.

The integration and rearrangement of eq. 23 gives

$$M = \frac{2\ RT \ln C_2/C_1}{\omega^2(1 - \overline{V}_2\rho_1)(x_2^2 - x_1^2)} \qquad 24$$

where C_1 and C_2 are the concentrations of the solute at distances x_1 and x_2 from the axis of rotation respectively. Equation 24 can be written in the form

$$C_2 = C_1\, e^{\frac{M\psi}{2RT}} \qquad 25$$

where ψ is the gravitational potential and is equal to $(1 - \bar{V}_2\rho_1)\,\omega^2\,(x_2^2 - x_1^2)$.

As formerly used, equilibrium centrifugation had several serious disadvantages. The attainment of equilibrium usually required several days of continuous centrifugation at a uniform speed during which time damage could occur to the protein or other biopolymers of interest. Furthermore, schlieren optics is poorly adapted for the determination of the concentration at different positions in the cell. For these and other reasons the equilibrium method had been largely abandoned. Two developments produced a revival of interest in the method. One of these was the introduction of short column cells which employ depths of solution of one to two millimeters. The time required for attainment of equilibrium is proportional to the square of the column height through which centrifugation occurs; accordingly, the short column cells have very greatly reduced the time required for equilibrium to be reached.

An associated development, which actually made possible the use of short column cells, was the inclusion of Rayleigh interference optics (see Chapter 11, Diffusion) in the optical system of the ultracentrifuge. The Rayleigh interference optics measures directly the concentration (refractive index) rather than the concentration gradient (refractive index gradient) as does schlieren optics. Rayleigh optics only requires the number of fringes that the zero order fringe has been displaced; the number of fringes is directly related to the concentration.

An advance which has made the equilibrium method still more attractive is the realization by Yphantis that it is best to use higher speeds of rotation to attain equilibrium. At higher speeds the meniscus region is cleared of sedimentating material and can thus be used as a reference for the Rayleigh interference fringe method. Also, since the sedimentating material is accumulating towards the bottom of the cell, more dilute initial solutions can be used without undue sacrifice of accuracy in the estimation of concentration. Another feature of the Yphantis method is that polymeric and higher molecular weight impurities tend to accumulate towards the bottom of the cell, leaving pure monomer in the region closer to the meniscus; the presence of impurities and polymers can be detected.

Archibald Method. The rate sedimentation is not suitable for the determination of the molecular weights below about 10,000. One difficulty encountered is that the front of the sedimenting boundary reaches the bottom of the cell before the rear of the boundary is completely away from the original meniscus; the boundary becomes very diffuse due to the rapid diffusion of the smaller molecules. Several methods have been suggested to avoid this difficulty and to extend centrifugation to smaller molecular weights. The most successful of these appears to be the Archibald method, the report of which was published some years before its importance was recognized; it might be called an approach to equilibrium method.

Equation 23 can be rearranged to give

$$\frac{1}{xC_m}\,\frac{dC}{dx} = \frac{M\omega^2(1 - \bar{V}_2\rho_1)}{RT} \qquad 26$$

After equilibrium between the rate of sedimentation and back diffusion has been attained, eq. 26 is valid at any place in the cell. Before equilibrium, there are only two places at which it is valid, exactly at the meniscus and exactly at the bottom of the cell. Further, at these two places eq. 26 is valid any time after centrifugation has been started. The problem is, therefore, the extrapolation of the term $dC/xCdx$ to the meniscus or to the bottom of the cell; from the value of this extrapolation and a knowledge of ω and $\bar{V}_2$, the molecular weight of the solute can be calculated out of hand. The essential difficulty associated with the extrapolation is the estimation of the concentration, C_m, at the meniscus or C_b at the bottom of the cell. The equation for

C_m is

$$C_m = C_0 - \frac{1}{x_m^2} \int_{x_m}^{X} x^2 \left(\frac{dC}{dx}\right) dx \qquad 27$$

where x refers to a position in the plateau region where the concentration is independent of the position in the cell and the concentration is equal to C_0. With the Archibald method it is possible to determine the molecular weights of very large molecules such as myosin as well as of very small molecules with molecular weights of less than a thousand; the time required is short.

Partial Specific Volume. At various points in this book the terms partial specific volume or partial molar volume have been used. The partial specific volume of a solute may be defined as that volume increase in a very large volume of the solution resulting from the addition of one gram of solute. The partial molar volume is the volume increase produced by the addition of one mole of the solute. A more useful way of expressing the partial specific volume is

$$\overline{V} = \frac{dV}{dw} \qquad 28$$

where dV is the infinitesimal increase in the volume of solution due to the addition of an infinitesimal weight of the solute to a finite volume of solution.

The partial specific volume may be determined experimentally by plotting the weights of the contents of a pycnometer for several concentrations against the weight concentration of the solute expressed as weight fraction; the weight fraction is 0.01 times the weight percent concentration. The best smooth curve is drawn through the points of the plot and the slope of the line determined. This gives dm/dw_2, where m is the weight of the contents of the pycnometer and w_2 is the weight fraction of the solute. The partial specific volume can then be calculated by the equation

$$\overline{V}_2 = \frac{1}{\rho} - \frac{(1 - w_2)}{\rho m} \frac{dm}{dw_2} \qquad 29$$

where ρ is the density of the solution.

It will be found for many dilute solutions, particularly those of proteins, that the partial specific volume is independent of concentration. In this event, the apparent partial specific volume may be used in place of the true partial specific volume. The apparent partial specific volume of the solute is defined as

$$\overline{V}_a = \frac{1 - w_1 v_1}{w_2} \qquad 30$$

where w_1 and w_2 are the weights of solvent and solute, respectively, in one ml. of solution and v_1 is the specific volume of the solvent (water) all at the experimental temperature. The apparent partial specific volume of the solute can be determined directly from density measurements by means of pycnometers. Density gradient columns have also been used and magnetic densitometers of great precision have been reported. A density balance can be employed. Specific volumes can be calculated approximately from the amino acid composition of the protein. It is not easy to measure the partial specific volumes with precision, and density measurements accurate to about one part in 10^5 are necessary to yield a precision of one percent in $\overline{V}_a$. Since the measurement of the molecular weight by the ultracentrifuge involves the quantity $(1 - \overline{V}_2 \rho_1)$, and $\overline{V}_2$ for a protein is about .75 and ρ_1 is about unity, it is clear that an error of one percent in $\overline{V}_2$ produces an error of about 3 percent in the molecular weight.

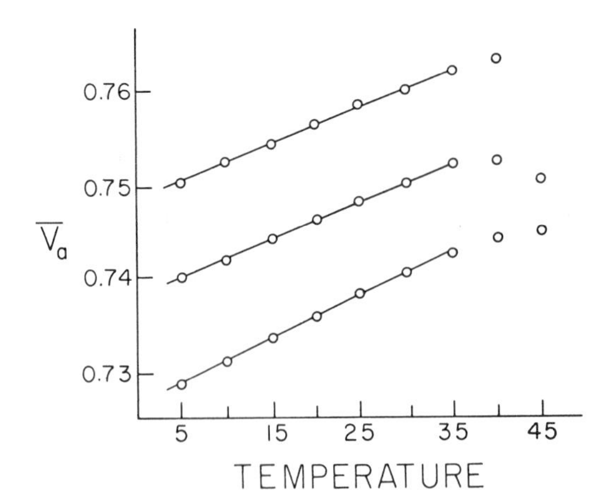

Fig. 7. Plots of the apparent partial specific volumes of proteins against temperature. Curve 1, Bovine methemoglobin; curve 2, egg albumin; curve 3, bovine serum albumin. (H. B. Bull and K. Breese: J. Phys. Chem. *72:* 1817, 1968.)

The apparent partial specific volume, $\overline{V}_a$, is related to the partial specific volume, $\overline{V}_2$, as a ratio of a finite increment to the corresponding differential coefficient. Since, as noted above, the partial specific volumes of proteins are, over a broad range of concentration, independent of protein concentration, the partial and apparent volumes are identical.

The partial specific volume is a measure of the volume of solvent displaced by the solute. It does not represent and is not directly related to the actual volume occupied by the anhydrous or hydrated solute molecule in solution. However, it is of interest that the partial specific volumes of proteins are not greatly different from the specific volume of dry protein. For example, the partial specific volume of egg albumin in aqueous solution at 25° is 0.7477, whereas the specific volume of the dry protein at 25° is 0.781. Incidentally, the apparent partial specific volume of the water of hydration associated with egg albumin at 25° is 0.80.

Figure 7 shows the partial specific volumes of bovine methemoglobin, of egg albumin and of bovine serum albumin as functions of the temperature, the volumes being measured with a density balance. It appears likely that the positive thermal coefficient of the partial specific volumes reflects the thermal expansion of the protein as well as the release of water of hydration with increasing temperature.

Molecular Weight Methods. A molecule has been defined as any particle whose energy of translation in solution is $3kT/2$ (see Chapter 2, Energetics). The molecular weight of the particle is the weight of the particle in grams multiplied by Avogadro's number. Attention has been called to several methods by which the molecular weight of a substance can be evaluated. It is perhaps useful to recapitulate some of this material.

If the substance is not homogeneous in respect to its molecular weight, there are various ways of averaging the molecular weight and each may be appropriate to what is needed. This is a perfectly general problem and arises in many connections. For example, the arithmetic mean of a series of values is frequently employed; however, in the calculation of errors it is found best to average the square of the errors.

The number average molecular weight, M_n, is simply the total weight of the substance divided by the total number of moles present and

$$M_n = \frac{\Sigma M_i N_i}{\Sigma N_i} \tag{31}$$

On the other hand, the weight average molecular weight, M_w, is

$$M_w = \frac{\Sigma N_i M_i^2}{\Sigma N_i M_i} = \frac{\Sigma C_i M_i}{\Sigma C_i} \qquad 32$$

and the Z-average molecular weight, M_z, is

$$M_z = \frac{\Sigma N_i M_i^3}{\Sigma N_i M_i^2} = \frac{\Sigma C_i M_i^2}{\Sigma C_i M_i} \qquad 33$$

and other molecular weight averages are possible.

It is only if the material is completely homogeneous in respect to molecular weight that all of the averages would be exactly equal to each other. If the material is not homogeneous

$$M_z > M_w > M_n \qquad 34$$

Indeed, the ratios of the various averages to each other gives some idea of the degree of heterogeneity.

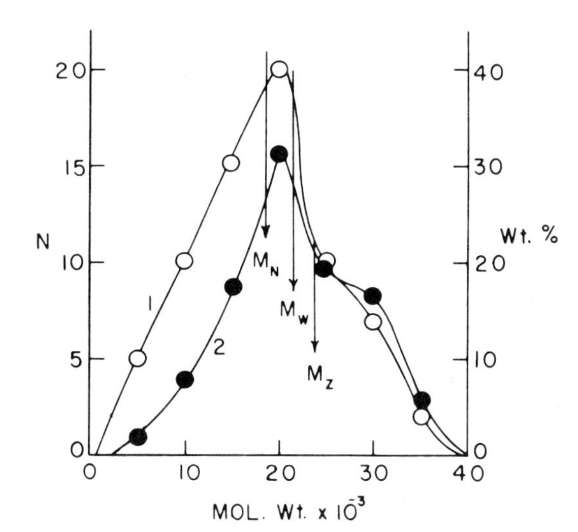

Fig. 8. Hypothetical mixture of a polymeric material. Curve 1, Number of moles corresponding to the indicated molecular weight. Curve 2, Weight percent of material corresponding to the indicated molecular weight.

TABLE. 1. SUMMARY OF MOLECULAR WEIGHT METHODS

Method	*Mol. Wt. Average*	*Range and Remarks*
Thermodynamic; entropy of mixing, $\Delta S = R \ln N_2$ as $C \to 0$		
Freezing point depression	M_n	Mol. wt < 1,000
Vapor pressure lowering	M_n	Mol. wt < 10,000
Osmotic pressure	M_n	Mol. wt 10,000-100,000
Spread monolayers	M_n	Mol. wt 1,500-90,000 peptides and proteins
Partly thermodynamic		
Light scatter	M_w	Depends on technique
Hydrodynamic (two acting forces opposing)		
Equilibrium centrifugation	M_z	Depends on technique
Archibald Method	M_w	Wide range
Hydrodynamic (frictional force equal acting force)		
Rate sedimentation and diffusion	M_w	Mol. wt > 10,000
Viscosity and diffusion	M_w	Undefined and approximate
Structural Methods		
End group titration	M_n	Requires suitable chemical method
Electron microscope	M_n	Mol. wt > 1,000,000
X-Ray diffraction and density	M_n	Requires number molecules per unit cell

Figure 8 shows the plot of the number of moles as well as the weight percent composition against the molecular weight of a hypothetical mixture of a polymeric material. The calculated values of M_n, M_w, and M_z are indicated.

Table 1 gives a partial summary of the various methods for molecular weights.

This chapter is a very brief exposition of the problems related to sedimentation. For those who actually intend to use the ultracentrifuge it will be necessary to seek more detailed sources of information: the Spinco Division of Beckman Instruments will be glad to oblige; the book by Schachman will also prove of great value.

GENERAL REFERENCES

McMeekin, T. L. and K. Marshall: Specific volumes of proteins and the relationship of their amino acid contents. Science *116:* 142, 1952.
Orr, C., Jr. and J. M. Dallavalle: Fine Particle Measurement. The Macmillan Company, New York, 1959.
Schachman, H. K.: Ultracentrifugation in Biochemistry. Academic Press, New York, 1959.
Svedberg, T. and K. O. Pedersen: The Ultracentrifuge. Oxford University Press, London, 1940.
Yphantis, D. A.: Equilibrium ultracentrifugation of dilute solutions. Biochem. *3:* 297, 1964.
Yphantis, D. A.: Advances in ultracentrifugal analysis. Ann. N.Y. Acad. Sci. *164:* 1, 1969.

PROBLEMS

1. A centrifuge rotates 22,000 r.p.m. at 25°C. The menisci of the aqueous solutions in the centrifuge tubes are 5 cm. from the center of rotation and the column of liquid in the tubes is 5 cm. in height. The tubes swing normal to the axis of rotation.
   (a) How long would it take to completely clear a suspension of spherical particles whose density is 1.340 and whose diameter is 0.1 micron?
   (b) What would be the ratio of concentrations one centimeter from the meniscus to that one centimeter from the bottom of the tube after equilibrium has been established for a solution of egg albumin? The mol. wt. of egg albumin is 45,000 and its partial specific volume is 0.749.

   Ans: (a) 573 secs. (b) 7.59×10^{23}

2. Describe how you could determine (*a*) the partial specific volume of a protein; (*b*) the partial specific free energy of a protein.

   Ans: See text.

3. (a) What are the nature of the acting forces and of the resisting forces in (1) electrophoresis, (2) diffusion, (3) centrifugation, and (4) molecular rotation?
   (b) What is a molecule?

   Ans. See text.

4. A suspension of subcellular particles in 0.001 M potassium phosphate buffer containing 0.0005 M $MgCl_2$ at pH 7.4 was centrifuged at 27,690 r.p.m. at 20.6°. The positions of the schlieren peak as a function of the time in minutes were as follows.

Time	*Positions of Peak in cm. from Axis of Rotation*
0	6.157
4	6.258
8	6.360
12	6.460
16	6.566

   Calculate S_{20},

   Ans: 79.8 *S*
   This problem was submitted by Dr. John P. Hummel at the request of the author.

5. A 0.50 percent solution of a protein with a partial specific volume of 0.700 was centrifuged at 20,410 r.p.m. at 24° for 24 hours at which time it was judged that equilibrium had been attained. The density of the solvent was 1.005. Rayleigh interference optics were employed. There was a shift of 13.7 fringes between the original protein solution and the buffer. The positions of the fringes in centimeters from the axis of rotation at the end of centrifugation were: 7.039, 7.059, 7.080, 7.095, 7.110, 7.123, 7.135, 7.147, 7.157, 7.168, 7.176, 7.185, 7.193, 7.201, 7.208. The hinge point was 7.147. Calculate the molecular weight of the protein.

   Ans: 17,060.
   This problem was submitted by Dr. John P. Hummel at the request of the author.

6. The kinetic particle in a protein solution is the hydrated molecule. The ultracentrifuge, however, yields an anhydrous molecular weight. Explain how this is possible.

   Ans: See text.

CHAPTER 15

Kinetics and Enzyme Activation

The biochemist is continually measuring rates of reactions; his experimental systems are in a state of flux. In many respects chemical kinetics is of more importance to the biologist than thermodynamics. Biochemical processes can, in general, occur in a variety of ways, i.e., the free energy relations permit a number of possible reactions arising from a given situation. The relative speeds of the various reactions determine which reaction will predominate.

The rates of biochemical reactions are sensitive to numerous factors such as concentrations (activities) of the reacting species, temperature, enzymes, pH, specific and non-specific ion effects, surface effects, etc., and it is the purpose of this chapter to explore some of these factors.

The velocity or rate of a chemical reaction is measured and expressed by the rate of disappearance or accumulation of a molecular species so that in the reaction

$$n_1a + n_2b + n_3c + \rightarrow m_1x + m_2y + m_3z + \qquad 1$$

the velocity at any time, t, is $-da/dt$, $-db/dt$, $-dc/dt$ or dx/dt, dy/dt, dz/dt, etc. Thus, there is a velocity in respect to each component and, whereas these several velocities are obviously related, they are not necessarily equal to each other.

If the concentration of component a is plotted against time, t, the slope of the line at any point is the reaction velocity providing the volume of the reaction mixture is maintained constant (Fig. 1).

Unfortunately, the velocity as measured by the slope of the line after a given time (Fig. 1) varies with the time and concentration and, accordingly, is awkward to use in comparing the influence of such factors as pH, temperature, etc., on reaction velocity. What is clearly needed is a standard reference state in respect to concentrations of the reactants.

There are a number of appropriate reference states. For example, the specific rate constants for the reaction are entirely suitable for this purpose. Before the specific rate constants can be obtained, however, it is necessary to formulate the kinetics of a reaction and to assign an order to the reaction. The order of a reaction is the number of molecules participating in the reaction whose concentrations

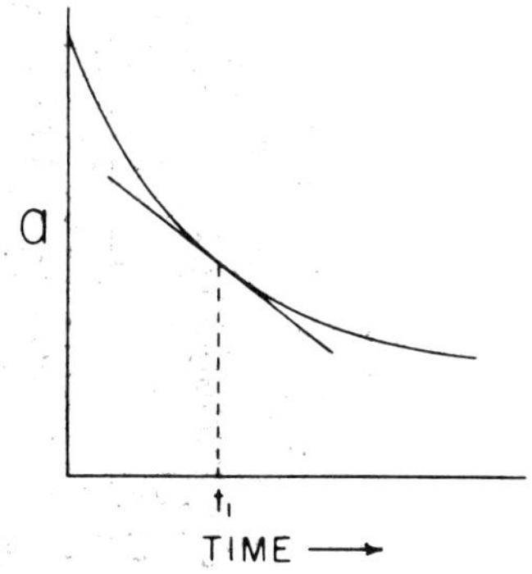

Fig. 1. Hypothetical plot of the concentration a against time. The velocity of the reaction after a given time, t, is equal to the slope of the curve.

influence its velocity. The reaction order is to be distinguished from the molecularity of the reaction. Molecularity refers to the total number of molecules undergoing reaction irrespective of the influence of their concentrations upon the reaction rate.

Zero Order. In a zero order reaction, the velocity is independent of the concentrations of any and all reactants and the reaction velocity can be expressed by

$$da/dt = -k \tag{2}$$

The integration of eq. 2 yields at any time t

$$a = -kt + a_0 \tag{3}$$

and if a is plotted as a function of time, t, there is obtained a graph such as is shown in Figure 2.

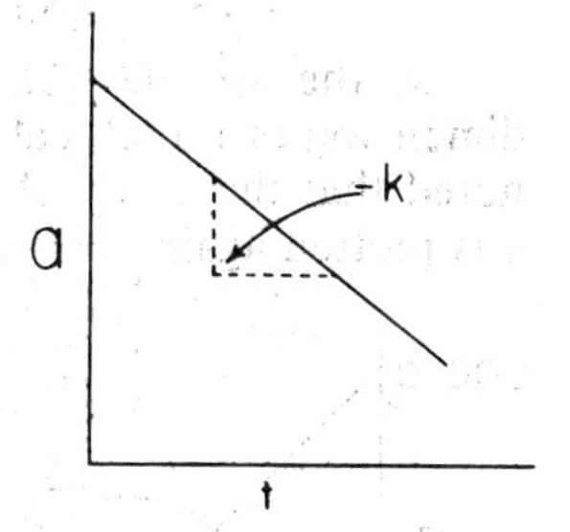

Fig. 2. Plot of a zero order reaction. The slope of the line is equal to $-k$.

k, the velocity constant of a zero order reaction, has the dimensions of $L^{-3}t^{-1}$ and can be expressed in moles per liter-seconds. Zero order reactions are characteristic of surface reactions at substrate concentrations in excess of that required to saturate the surface. Obviously, this is true since the reaction velocity is proportional to the number of molecules at the surface and the surface is saturated. For example, if a protein solution is shaken surface coagulation results and the amount of insoluble protein can be measured. Shown in Figure 3 is the percent surface coagulation of a solution of heat-denatured egg albumin produced by shaking at a constant rate as a function of time of shaking.

First Order Reactions. Those reactions whose velocity is directly proportional to the concentration of only one component are first order reactions and their velocity is expressed as

$$\frac{da}{dt} = -ka \tag{4}$$

which on integration and conversion to log to the base 10 gives

$$\log a = -\frac{kt}{2.303} + \log a_0 \tag{5}$$

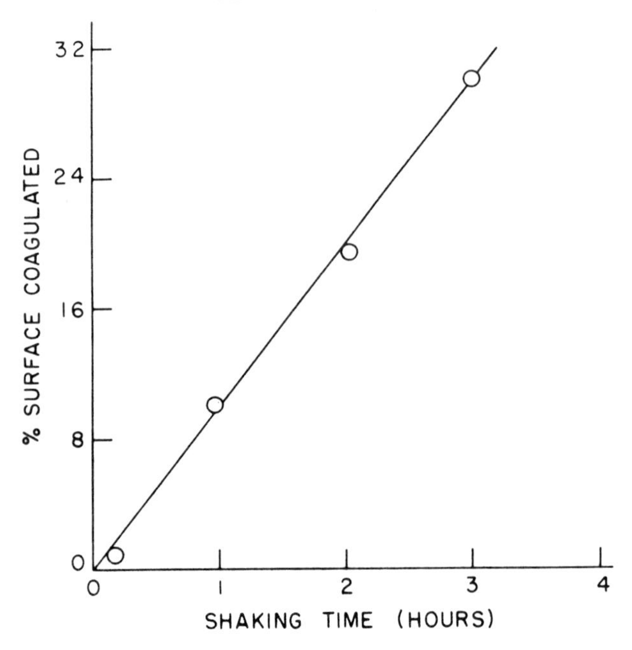

Fig. 3. Surface coagulation produced by shaking a clear, 0.495 percent solution of heat denatured egg albumin at pH 3.0, 25°. Showing zero order kinetics. (H. B. Bull: J. Biol. Chem. *125:* 585, 1938.)

k, the specific reaction velocity constant of a first order reaction, has the dimension of reciprocal time and can be expressed in reciprocal seconds; it is to be noted that the value of k is independent of the concentration units. Clearly, if log a is plotted against t, a linear relation will be obtained (Fig. 4).

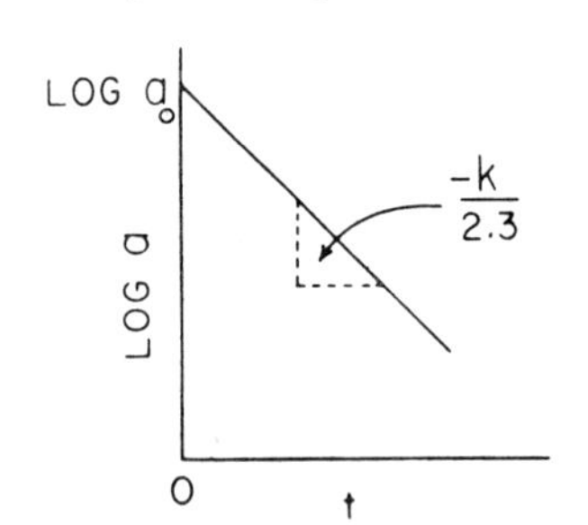

Fig. 4. Hypothetical plot of a first order reaction. The slope of the line is equal to $-k/2.303$.

Equation 5 can be written

$$\frac{a}{a_0} = \frac{1}{e^{kt}} \tag{6}$$

It is, therefore, evident that k, the first order reaction constant, is equal to the reciprocal of the time required for $1/e^{th}$ part of the reactant to react.

The condition for a first order reaction is that, in a given time and independent of the concentration, a constant fraction of the substance has decomposed or accumulated; this is a statement of the compound interest law. Biochemical reactions characteristically involve the reaction of two or more molecular species; however, many of these reactions exhibit first order kinetics because the rate is determined by the concentration of one molecular species; the other reactants are in excess concentration.

Radioactive isotopes have found wide use in biochemistry, particularly in the study of intermediary metabolism. Radioactive isotopes decay at a rate which obeys first order kinetics and the number of disintegrations per unit time is proportional to the number of radioactive atoms present and

$$kt = 2.303 \log \frac{n_0}{n} \tag{7}$$

where n_0 is the number of atoms at time zero and n is the number of atoms after time t. The half-life for the decay process is the time after which the concentration of the decomposing substance is reduced to one-half of its initial value and, accordingly, n is equal to $n_0/2$. Then

$$t_{1/2} = \frac{2.303}{k} \log 2 = \frac{0.693}{k} \tag{8}$$

and k is the decay constant.

A bacterial colony exhibits, after a certain lag phase, a logarithmic growth phase. For example, if a nutrient is inoculated with n_0 cells and each colony divides, there will be two n_0 cells at the end of one generation time. The rate of increase of the bacteria is proportional to the number of bacteria present and

$$\frac{dn}{dt} = kn \tag{9}$$

and

$$\log \frac{n}{n_0} = \frac{kt}{2.303} \tag{10}$$

The mean generation time (t_m) is that time at which n is equal to 2 n_0 and, accordingly,

$$\log 2 = \frac{kt_m}{2.303} \tag{11}$$

and

$$t_m = \frac{0.693}{k} \tag{12}$$

The diffusion of a substance across a plane surface follows first order kinetics as can be shown by the appropriate substitutions in Fick's first law of diffusion.

Second Order Reactions. If two molecules react, whether they are two molecules of the same compound or molecules of two different compounds, the reaction is known as a bimolecular reaction. If the velocity of the reaction is sensitive to the concentrations of both reacting species, the reaction is said to follow second order kinetics and the reaction velocity is

$$\frac{da}{dt} = -kab \tag{13}$$

where a and b are the concentrations of the two reactants. The integration of eq. 13 leads to

$$t = \frac{2.303}{k(a_0 - b_0)} \log \frac{b_0(a_0 - x)}{a_0(b_0 - x)} \tag{14}$$

where a_0 and b_0 are the initial concentrations and x is the amount of substance decomposed in time t. A plot of log $b_0\,(a_0 - x)\,/a_0\,(b_0 - x)$ against t should yield a straight line whose slope is $k(a_0 - b_0)/2.303$.

A special case arises if a_0 is equal to b_0 or if two molecules of a substance are reacting with each other. Under these circumstances the velocity of the reaction is

$$\frac{da}{dt} = -ka^2 \qquad 15$$

which on integration is

$$\frac{1}{a} = kt + \frac{1}{a_0} \qquad 16$$

and a plot of $1/a$ against time will give a straight line whose slope is equal to k (see Fig. 5). k for a second order reaction has the dimensions L^3t^{-1} and can be expressed in liters per mole per second. It is to be noted that the value of k will depend, unlike k for a first order reaction, on the units in which concentration is expressed.

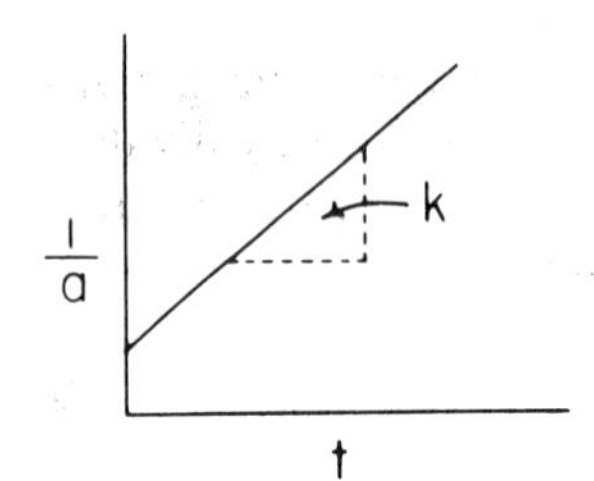

Fig. 5. Second order reaction plot for the condition that the initial concentrations of a and b are equal or that two molecules of A are reacting together. Slope of the line is equal to k, the reaction constant.

Sometimes it is difficult to reach a decision as to the order of a reaction on the basis of plots suggested above; for example, the order of the reaction may change during the course of the reaction and complications can lead to fractional orders. Reactions of very high orders can occur such as the denaturation of proteins by urea where the rate of the reaction can be proportional to the 10th and higher powers of the urea concentration.

The general expression for the velocity of a chemical reaction can be written

$$v = -ka^{n_1} \cdot b^{n_2} \cdot c^{n_3} \cdot \cdots \qquad 17$$

Taking the logarithm of both sides of eq. 17 gives

$$\log v = \log(-k) + n_1 \log a + n_2 \log b + \qquad 18$$

A plot of log v against log a should give a linear relation and the order of the reaction in respect to A is given from the slope of the line. To obtain a linear plot of log v against log a it is necessary to maintain the concentration of all the other reactants effectively constant. This can be done by determining the initial reaction velocities for several different concentrations of A or by having B, C, D, etc., in excess so that their relative concentrations remain essentially unchanged.

Casey and Laidler studied the rate of inactivation (denaturation) of pepsin at pH 4.83 and at 57.4° and when log v was plotted against the log of the activity of pepsin remaining the graph shown in Figure 6 was obtained.

A possible mechanism for the inactivation of pepsin and which would be in accord with the results of Casey and Laidler is that the transformation of the native pepsin to the denatured form is monomolecular and reversible, that is

$$\text{Native pepsin} \rightleftharpoons \text{Denatured pepsin} \qquad 19$$

Five denatured pepsin molecules then associate to produce a complex which stabilizes the denatured form; this last reaction is irreversible

$$5 \text{ Denatured pepsins} \rightarrow (\text{Denatured pepsin})_5 \qquad 20$$

At low pepsin concentration, reaction 20 is the rate determining step whereas at

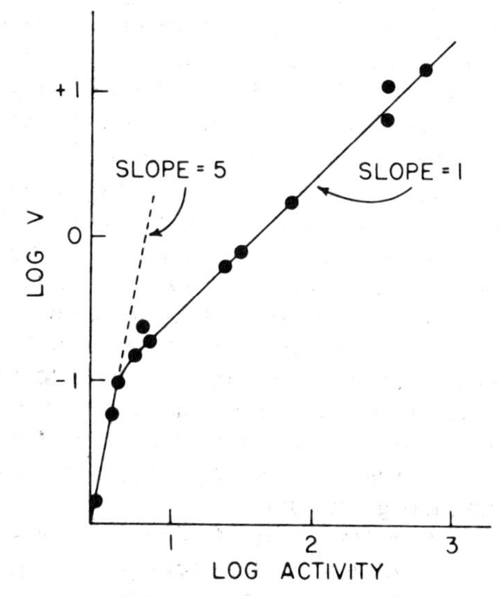

Fig. 6. Log V vs. log activity pepsin at pH 4.83 and 57.4° to determine order of reaction. (E. J. Casey and K. J. Laidler: J. Amer. Chem. Soc. *73:* 1455, 1951.)

higher concentrations reaction 20 becomes much faster and reaction 19 is rate determining. Thus it is easily possible for the order of a reaction to change during the course of the reaction. Later, enzymatic reactions are to be discussed; the order of an enzymatic reaction usually changes from first to zero order as the concentration of the substrate is increased. The results of Casey and Laidler illustrate another point. It is frequently possible to formulate a reaction mechanism which is consistent with the kinetics, but the kinetics do not establish the particular mechanism as true; there may be other mechanisms which will equally well satisfy the kinetics.

The standard states of the reactants in each of the situations discussed above are quite clear. The standard state for zero order is a saturated reacting surface; for a first order, unit concentration; for a second order, the product of the concentration terms is unity. In addition to the specific rate constants there are other expressions for reaction rates.

Change in a Given Time. In this method the change in the concentration of a reactant is measured over a fixed time interval. That is, the reaction is stopped after say 30 minutes and the concentration of the substance compared with its initial concentration. Due to its experimental convenience, this is probably the most popular mode of expressing the velocities of biochemical reactions. It is, however, subject to considerable ambiguity. For example, suppose there is a reaction which follows the course of curve 1 in Figure 7. Now suppose the conditions of the reaction

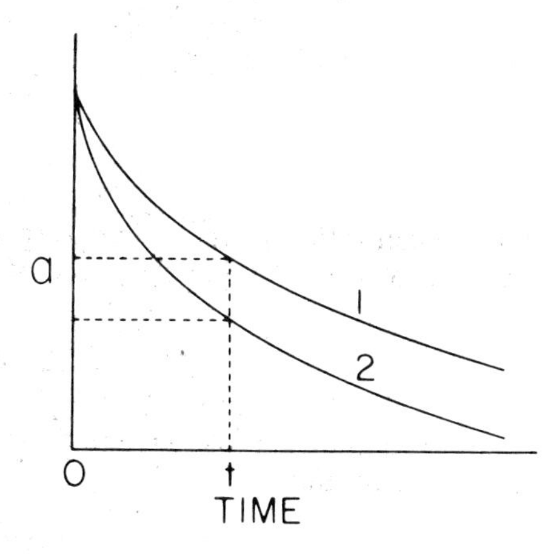

Fig. 7. Comparison of reaction velocities on the basis of the amount of reactants changed after an arbitrary time, t.

mixture are changed in respect to some factor such as pH or temperature and curve 2 is obtained and it is desired to compare the reaction velocities under the two conditions; this is done by determining the amounts of the reactants changed after a given time. It is evident from Figure 7 that the ratio of the amounts of the reactants changed after a given time and which has been selected as a measure of comparison of the rate or reaction will depend on the arbitrary selection of the time chosen. The only exception to this difficulty arises with zero order reactions in which the ratio of the amounts changed will be independent of the time interval chosen. If the time interval after the start of the reaction is made sufficiently small so that the relation between the concentration of the reactant and time is effectively linear, the method loses its arbitrary character and comparison of the amount of change after a given time is a suitable measure of the reaction rate. The slope of the tangent drawn to the kinetic curve at zero time gives the initial velocity of the reaction and the elapsed time has been reduced to an infinitesimal. The initial reaction velocity is frequently used to express the velocity of enzymatic reactions and has great utility.

Time for a Given Change. Another means for the expression of reaction rates is to determine the time required for a given change in the reactants. An examination of the integrated kinetic equations reveals that the product of the specific reaction rate constant and time is proportional to a concentration term; if the concentration is held constant as the conditions of the reaction are changed, the time is inversely proportional to the specific reaction rate constant. Thus the reciprocal of the time required for a given change in the reactants is a direct and suitable measure of the reaction rate. However, if the order of the reaction changes during the experiment, this method of expressing reaction velocities is no longer valid because the concentration term has changed its form.

A special case of this method is represented by the half-life of a reaction. As noted previously, the half-life of a first order reaction is equal to $0.693/k$ where k is the specific rate constant. Sometimes it is more useful with slower reactions to use shorter time intervals, perhaps the tenth-life of a reaction; this would be the time required for the reactants to decrease to 90 percent of the original value. Except for first order reactions, fractional lives of a reaction are meaningful only if the initial concentration is specified. Thus for a zero order reaction $t_{1/2}$ is equal to $a_0/2k$ and for a second order reaction it is equal to $1/ka_0$ (for two molecules of a substance reacting with each other).

The reactions discussed above are simple in the sense that a single reaction resulted from the reactants and furthermore this reaction was irreversible. In practice, the situation can become much more complex; some of these complications will be considered briefly.

Reversible Reactions. When a process is significantly reversible under the conditions of the experiment, the rate of the reverse reaction must be considered. For a reaction of the type

$$\mathrm{A} \underset{k_2}{\overset{k_1}{\rightleftharpoons}} \mathrm{B} \qquad 21$$

the concentration of A at time zero is a_0; after a time t its concentration is $a_0 - x$ and that of B is x and the net rate of the reaction is

$$\frac{dx}{dt} = -k_1(a_0 - x) + k_2 x \qquad 22$$

At equilibrium, the rates of the forward and reverse reactions are equal and dx/dt is zero and

$$k_1(a_0 - x_e) = k_2 x_e \qquad 23$$

Note that at equilibrium k_1/k_2 is equal to $x_e/(a_0 - x_e)$ and is equal to the equilibrium constant of the reaction. Solving eq. 23 for k_2 and substituting its value in eq. 22, there results

$$\frac{dx}{dt} = -k_1(a_0 - x) + k_1 \frac{(a_0 - x_e)x}{x_e} \qquad 24$$

$$\frac{dx}{dt} = -\frac{k_1 a_0}{x_e}(x_e - x) \qquad 25$$

Integration of eq. 25 gives

$$\ln \frac{x_e}{x_e - x} = \frac{k_1 a_0}{x_e} t \qquad 26$$

Evidently, by the appropriate plot of the kinetic data of reversible reactions it is possible to obtain both k_1 and k_2 as well as the equilibrium constant. An illustration of a reversible reaction which has been much studied is the mutarotation of glucose. Crystalline D-glucose dissolves in water to give α-D-glucose. The α-D-glucose is partially converted into β-D-glucose through an intermediate with a ruptured pyranose ring. The closure of the ring leads either to α-D-glucose or to β-D-glucose so that an equilibrium mixture of α- and β-glucose finally results. Each of these forms has characteristic specific optical rotatory values and so the mutarotation can be followed with a polarimeter.

Simultaneous Reactions. It is possible for a reactant to decompose or react in more than one way. For example, substance A can decompose into B and into C (see Fig. 8).

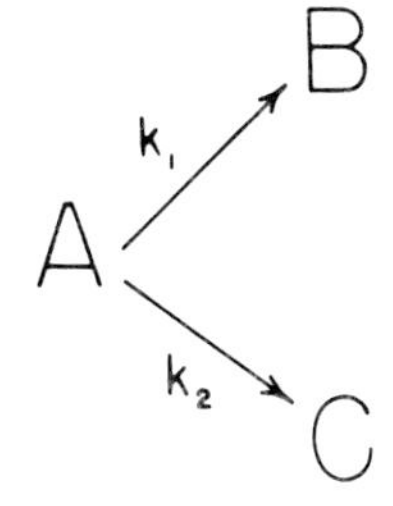

Fig. 8. Illustrating the occurrence of simultaneous reactions.

The rate of decomposition of A (Fig. 8) is

$$-\frac{da}{dt} = (k_1 + k_2)a \qquad 27$$

The integration of eq. 27 gives

$$\ln \frac{a_0}{a} = (k_1 + k_2)\, t \qquad 28$$

Solving eq. 28 for a gives

$$a = a_0 e^{-(k_1 + k_2)t} \qquad 29$$

The rate of formation of B is

$$\frac{db}{dt} = k_1 a = k_1 a_0 e^{-(k_1 + k_2)t} \qquad 30$$

The integration of eq. 30 results in

$$b = -\frac{k_1 a_0}{k_1 + k_2} e^{-(k_1 + k_2)t} + \text{constant} \qquad 31$$

and when t is zero eq. 31 becomes

$$\text{constant} = +\frac{k_1 a_0}{k_1 + k_2} \qquad 32$$

Substituting the value of the constant in eq. 31 gives

$$b = \frac{k_1 a_0}{k_1 + k_2}(1 - e^{-(k_1 + k_2)t}) \qquad 33$$

and similarly for C

$$C = \frac{k_2 a_0}{k_1 + k_2}(1 - e^{-(k_1 + k_2)t}) \qquad 34$$

Dividing eq. 33 by eq. 34 yields

$$\frac{b}{c} = \frac{k_1}{k_2} \qquad 35$$

Thus, the ratio of the amount of the substances transformed for two simultaneous first order reactions is equal to the ratio of the respective rate constants.

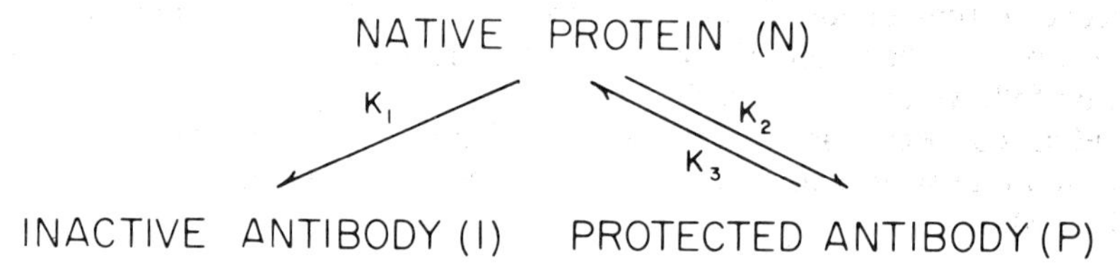

Fig. 9. Simultaneous reactions resulting from the action of urea on diphtheria antitoxin. (G. Wright and V. Schomaker: J. Amer. Chem. Soc. *70:* 356, 1948.)

Wright and Schomaker studied the inactivation of diphtheria antitoxin by urea and found that their results could be best represented by the scheme shown in Figure 9. One of the simultaneous reactions is reversible and the other is irreversible. At the beginning of the reaction, the transformation of the native into the inactive and into the protected antibody represents the important reactions and, whereas the native protein disappears according to first order kinetics, the total antibody activity does not. Instead it decays at a rapidly decreasing specific rate, corresponding to an incomplete limiting degree of inactivation. Gradually, however, the reverse reaction, the conversion of the protected antibody back to the native form, becomes important and, finally, by providing a continuing source of native protein such that ultimately a constant fraction of the native material is present

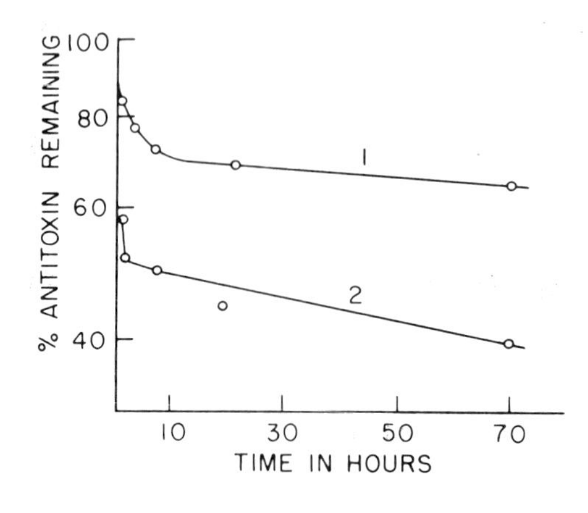

Fig. 10. Semi-log plot of the percent diphtheria antitoxin remaining as a function of time in 8 M urea at 25°. Curve 1, pH 6.25; curve 2, pH 5.38. (G. Wright and V. Schomaker: J. Amer. Chem. Soc. *70:* 356, 1948.)

in the native form in steady-state relation with the protected antibody, complete inactivation of the antibody according to first order kinetics eventuates. Some of the results of Wright and Schomaker are shown plotted in Figure 10.

Consecutive Reactions. Many if not all chemical reactions consist of several steps each of which produces an intermediate leading to the final reaction product. Frequently, the velocities of the intermediate steps vary greatly and only one step is rate determining; the kinetics of the entire series of consecutive reactions reflects only the kinetics of the slowest reaction. Consider, however, the following reaction in which the two kinetic constants k_1 and k_2 have similar magnitudes.

$$A \xrightarrow{k_1} B \xrightarrow{k_2} C \qquad 36$$

The rate equations for reaction 36 are

$$\frac{da}{dt} = -k_1 a \qquad 37$$

$$\frac{db}{dt} = k_1 a - k_2 b \qquad 38$$

$$\frac{dc}{dt} = k_2 b \qquad 39$$

Integration of eq. 37 gives

$$a = a_0 e^{-k_1 t} \qquad 40$$

Substitution of eq. 40 into eq. 38 yields

$$\frac{db}{dt} = k_1 a_0 e^{-k_1 t - k_2 b} \qquad 41$$

The integration of eq. 41 gives

$$b = \frac{k_1 a_0}{k_2 - k_1} e^{-k_1 t} - e^{-k_2 t} \qquad 42$$

It is to be noted that eq. 42 is in the form of a catenary.

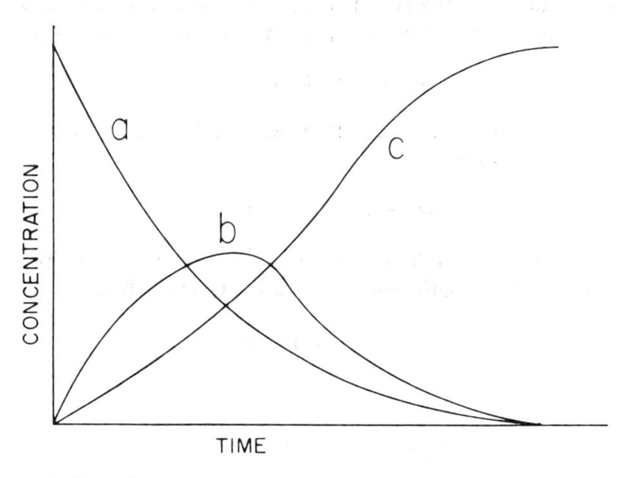

Fig. 11. Hypothetical plots of the concentrations of the reactants in consecutive reactions (see reaction 36).

Since at all times

$$a + b + c = a_0 \tag{43}$$

the value of C becomes, with the use of eqs. 40, 42 and 43,

$$C = a_0 - a_0e^{-k_1t} - \frac{k_1a_0}{k_2 - k_1}\,(e^{-k_1t} - e^{-k_2t}) \tag{44}$$

Equation 44 can be rearranged to give

$$C = \frac{a_0}{k_2 - k_1}\,[k_2\,(1 - e^{-k_1t}) - k_1\,(1 - e^{-k_2t})] \tag{45}$$

It is clear that, even for the simplest case, the mathematical expressions for consecutive reactions can become quite complex. Pictured graphically and depending on the magnitudes of k_1 and k_2 the concentrations of the three species A, B and C as a function of time might resemble the plots shown in Figure 11.

Synthesis and Degradation of Polymers. The synthesis of starches, glycogens, proteins and nucleic acids is understood in varying degrees and is described in modern biochemistry texts. The principle of the biological synthesis appears to be the end-wise addition of activated monomer units to a growing polymer chain. Unfortunately, there is little quantitative information available about such syntheses. On the other hand, many kinetic investigations of the synthesis of high polymers of industrial interest have been done and attention has been paid to the molecular weight distribution of such polymers.

If the polymer can only grow by the addition of monomers to other monomers or to chains already formed and it is not possible for chains to react with each other, linear polymers will show a very sharp molecular weight distribution. An example is the Leuch's synthesis of polypeptides through the interaction of the N-carboxy-amino anhydride. On the other hand, if the linear chains can react to produce larger molecules and all the groups have the same reactivity, the molecular weight distribution will be broad and become more widely distributed as the synthesis proceeds to completion. The distribution of molecular sizes is clearly a problem in statistics and is subject to the restrictions imposed by the conditions of the synthesis.

Consider a linear polymer composed of bifunctional units, each group having the same reactivity and chains combining with monomers as well as with other chains. The probability that any group has reacted is p, and x is the number of units in a chain. Then the probability that a given chain contains x units is p^{x-1}. The probability that the xth group is unreacted, thus limiting the chain to x units, is $1 - p$. Therefore, the probability that the chain is composed of exactly x units is

$$p_x = p^{x-1}\,(1 - p) \tag{46}$$

The probability p_x that any molecule selected at random contains x units is equal to its mole fraction and

$$p_x = N(1 - p)p^{x-1} \tag{47}$$

where N is the total number of molecules of all sizes. If N_0 is the total number of units in the reaction mixture both reacted and unreacted then

$$N = N_0(1 - p) \tag{48}$$

and

$$p_x = N_0(1 - p)^2\,p^{x-1} \tag{49}$$

The molecular weight of each species is directly proportional to x and the weight fraction for a given chain length is

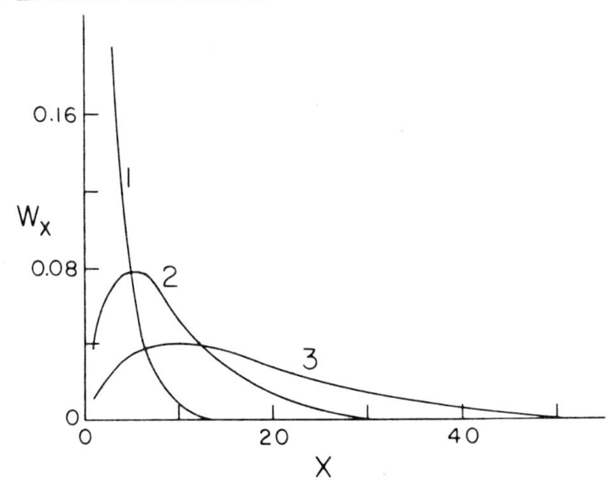

Fig. 12. Weight fraction distribution for moderate extent of reaction, *P*. Curve 1, *P* equals 0.5. Curve 2, *P* equals 0.8. Curve 3, *P* equals 0.9. (P. J. Flory: J. Amer. Chem. Soc. *58:* 1877, 1936.)

$$W_x = \frac{xp_x}{N_0} \qquad 50$$

Substitution of eq. 49 into eq. 50 gives

$$W_x = x(1-p)^2 p^{x-1} \qquad 51$$

Figure 12 shows the weight fraction distribution calculated according to eq. 51.

The random degradation of randomly condensed polymers should lead to a reversal of the condensation distribution shown in Figure 12 and the low molecular weights should predominate as the hydrolysis proceeds.

A pure protein may have only one molecular species present and if it were subject to random hydrolysis by some agent, such as hot mineral acid, the molecular weight distribution obviously cannot be calculated by the use of eq. 51, but it is a statistical problem which can be treated by another approach. Actually, all available evidence indicates that the susceptibility of peptide bonds to acid hydrolysis varies widely depending upon the nature of the amino acid residues adjacent to the bond; the attack is far from random. For example, aspartic acid residues are released during the early part of the hydrolysis reaction. It is also known that peptide bonds adjacent to hydrophobic residues such as those of valine and leucine are hydrolyzed with difficulty; the residue provides an umbrella for the peptide bond protecting it from the attack of a water molecule. The nature of the residue on the carbonyl side of the bond is important, indicating that the initial attachment of the water molecule is on the carbonyl group. Shown in Figures 13A and 13B are the molecular weight distributions of the resulting peptides when egg albumin is exposed to 7.95 N HCl at 30° and to 3.7 N $Ba(OH)_2$ at 60° respectively and after the removal of the isoelectric heat coagulatible material.

The above brief discussion is to serve as an introduction to the degradation of biopolymers by enzymes, a topic which can be described more conveniently under enzyme activation.

Metabolic Reactions. Organ or organism kinetics ordinarily deals with open systems in more or less steady state condition and the income is substantially equal to the outgo. In the living tissue there exist vast reaction networks and these reactions bear various and complex relations to each other. It is proper to ask if the kinetics of such a situation really can be reduced to meaningful quantitative terms and lead to fruitful interpretations.

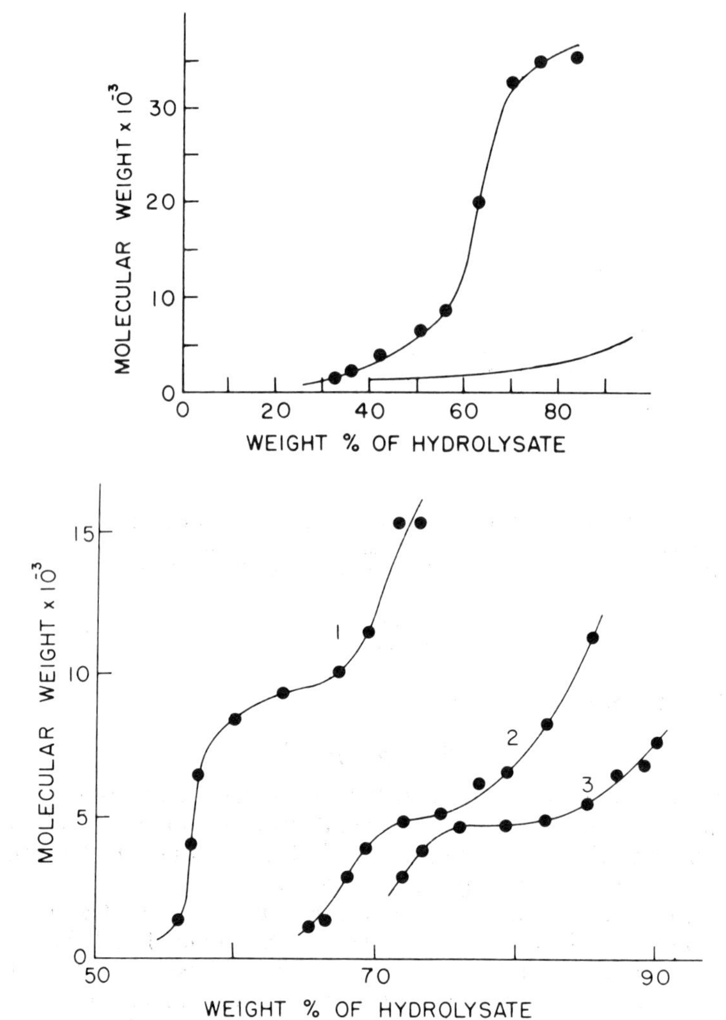

Fig. 13A (Upper figure). Molecular weight distribution of peptides produced by the action of 7.95 N HCl at 30° for two hours; 12.7 percent peptide bonds hydrolyzed. Solid line is the distribution to be expected on the basis of random attack. (H. B. Bull and J. W. Hahn: J. Amer. Chem. Soc. *70:* 2123, 1948.)

Fig. 13B (Lower figure). Molecular weight distribution of peptides resulting from the action of 3.7 N $Ba(OH)_2$ on egg albumin at 60°. Curve 1, 30 minutes hydrolysis; curve 2, 60 minutes hydrolysis; curve 3, 90 minutes hydrolysis. (H. B. Bull: Cold Spring Harbor Symp. Quant. Biol. *14:* 1, 1950.)

The study of whole body kinetics is an old one and respiration rates, urea clearances, etc., have long been measured and interpreted on an empirical basis and at times with great profit. More recently, metabolic reactions have been followed by the use of isotopes in which a labelled compound is given to an animal, the rate of disappearance of the labelled substance measured, and the appearance of the isotope in a suspected metabolic intermediate detected. By this approach, much valuable and useful information has been obtained.

Figures 14A, B and C illustrate three typical situations encountered in the use of isotopic labels in whole organisms.

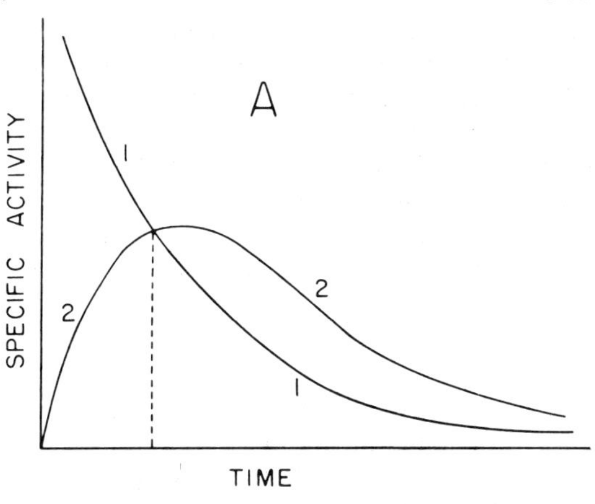

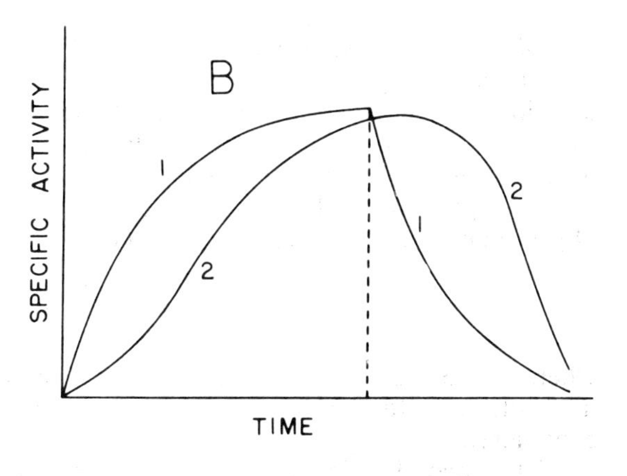

Fig. 14. Specific activity changes resulting from the feeding of labelled precursor and the production of isotopically labelled product. Curves indicated by 1, precursor; curves by 2, product. *A*, Organism given one initial dose at zero time. *B*, Labelled food fed until organism is maximally labelled. *C*, Labelled food withdrawn before the organism becomes maximally labelled. (J. M. Reiner: Arch. Biochem. Biophys. *46:* 53, 80, 1953.)

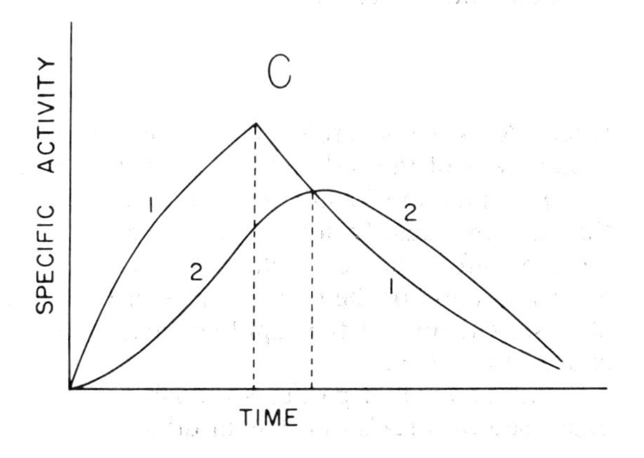

It is customary to express the rate of metabolism of a given labelled substance in terms of the half-life or as the turnover rate of the substance. The half-life is simply the time taken for half of the isotopically labelled compound to disappear from the animal body and, as has been noted, for a first order reaction is equal to $0.693/k$ where k is the first order reaction constant. Providing the compound disappears by first order kinetics, the half-life has a clear and unambiguous meaning. Suppose, however, that the disappearance follows second order kinetics and the rate is responsive to the concentrations of a second metabolic species. The time required for half of the labelled compound to disappear will then depend on the concentration of the second species as well as the initial concentration of the labelled compound; a wide range of half-lives for a given compound can be found depending upon the experimental conditions. In general, then, the reaction mechanism of metabolic reactions must be known before a half-life measurement can be interpreted with confidence.

The turnover rate is conveniently defined as the ratio of the amount of substance disappearing in unit time to the total amount of substance present and

$$\frac{dn/dt}{n} = \nu \qquad 52$$

where n is the amount of the substance present and ν is the turnover rate. It can be seen that ν has the properties of a first order rate constant and for a first order reaction is equal to $0.693/t_{1/2}$. As a general measure of the rate of biological reactions the turnover rate suffers from the same limitations as the half-life and its meaning will certainly depend upon the particular reaction mechanism followed.

Molecular Collisions and Reaction. Before two molecules can react to produce a chemical change they must first come in contact; there has to be a molecular collision. Under most circumstances the rates of chemical reactions are not limited by the collision rate but by factors leading to activation; the collision rate at ordinary concentrations is very high.

In a gas, the molecular collision rate can be derived from the kinetic theory of gases and for like molecules is given by the expression

$$Z = 8\left(\frac{\pi kT}{m}\right)^{1/2} r^2 n^2 \qquad 53$$

where k is Boltzmann's constant, T the absolute temperature, m is the mass of the molecule, r is the molecular radius and n is the number of molecules per cubic centimeter.

The second order rate constant for a gas in which every collision results in a reaction would, therefore, be given by

$$k = \frac{dC/dt}{C^2} = \frac{NZ}{(n)^2 \times 10^3} \qquad 54$$

where N is Avogadro's number. For gas molecules of average size and cross section, k is of the order of 10^{10} to 10^{12} liters per mole per second.

The situation in solution (where our main interest lies) is somewhat otherwise than it is in a gas. In solution the reacting molecules must diffuse through a viscous medium before they can achieve contact with each other. The point of departure for a consideration of the collision rate of molecules in solution is provided by Smoluchowski's theory of the rapid coagulation of suspensions of particles; the problem is one of diffusion.

The number of particles diffusing to and colliding with a given particle can be expressed by Fick's first law of diffusion

$$J = 4\pi r_1^2 D_1 \frac{dn}{dr_1} \qquad 55$$

where r_1 is the radius of the spherical surface around the particle, n is the number of particles per cubic centimeter and D_1 is the diffusion coefficient of the particles. The solution of eq. 55 is

$$n = n_0 - \frac{J}{D_1 4\pi r_1} \qquad 56$$

where n_0 is the number of particles per cubic centimeter at a great distance from the central particle, i.e., the average particle concentration. At the surface of the particle n is zero. r_1 is equal to the distance of closest approach which is twice the radius (r) of particles for a suspension of uniform size and from eq. 56

$$J = 8\pi D_1 r n_0 \qquad 57$$

The central particle can move in response to Brownian motion as well as can the other particles. Since the motion of the particles is completely independent of each other, the diffusion coefficient to be used is the sum of two diffusion coefficients and, if the particles are of equal size, eq. 57 becomes

$$J = 16\pi D r n_0 \qquad 58$$

where D is now the diffusion coefficient of the suspension of uniform particles measured in the usual way and J is the number of particles colliding with one individual particle. The rate of disappearance of the primary particles is

$$\frac{dn}{dt} = -\ 16\pi D r n^2 \qquad 59$$

At the beginning of the coagulation process, a primary particle disappears by collision with another primary particle; this is a second order process and

$$\frac{dC}{dt} = -\ kC^2 \qquad 60$$

C, the concentration in moles per liter, is 1000 n/N where N is Avogadro's number. Then

$$\frac{dn}{dt} = -\ \frac{1{,}000}{N}\, kn^2 \qquad 61$$

and comparison with eq. 59 shows that the second order constant k is

$$k = 16 \times 10^{-3}\pi D r N \qquad 62$$

For larger spherical particles

$$D = \frac{RT}{6\pi r \eta N} \qquad 63$$

where η is the viscosity coefficient of the solvent.

Substituting eq. 63 into eq. 62, there results

$$k = \frac{8 \times 10^{-3} RT}{3\eta} \qquad 64$$

which in water at 30° is 8.39×10^9 liters per mole per second and is independent of the particle size.

The Smoluchowski expression is closely obeyed in the rapid coagulation of a suspension of particles where every collision results in a reaction. However, if the collision efficiency is less than unity and not all of the collisions result in a reaction, the concentration of particles at a distance $2r$ from a given particle will no longer be zero and eq. 56 cannot be transformed into eq. 57. It is customary to insert a retardation or stability factor, W, in the denominator of the right side of

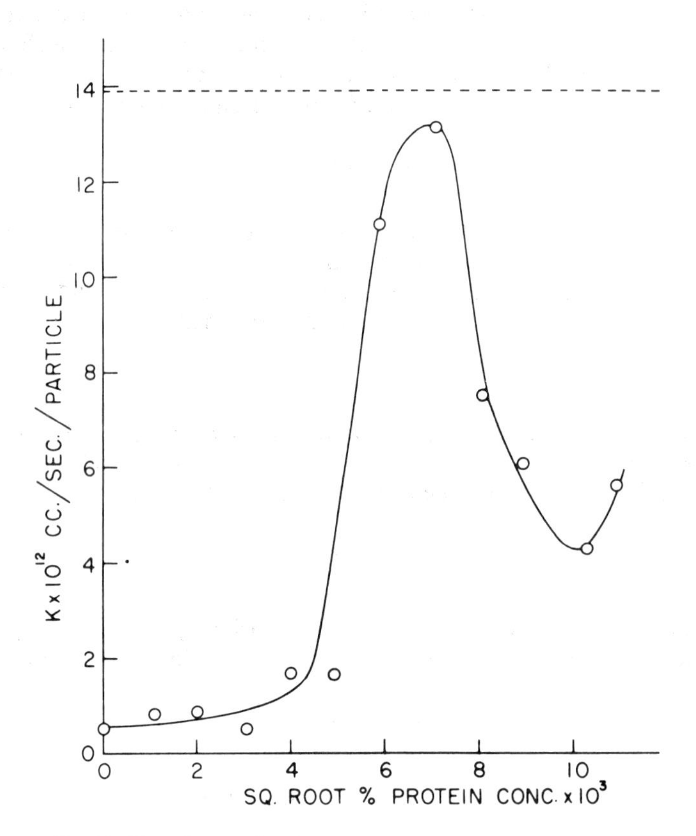

Fig. 15. Plot of second order rate constants for the disappearance of emulsion particles of *n*-octadecane as a function of the square root of the percent bovine serum albumin concentration, pH 4.88; ionic strength, 0.02; 30°. Dotted line is Smoluchowski's limiting rate. (S. Bandyopadhyay: Private communication.)

eq. 64. W is related in a complex manner to the electrical potential (the ζ-potential) of the particle. W can vary from unity, in which case every collision is fruitful, to infinity, when the suspension is completely stable and does not coagulate.

The gold number of cerebrospinal fluid was introduced into medicine a number of years ago as a diagnostic aid and is still widely used. The gold number is defined as that weight of a substance that is just insufficient to prevent a change in color from red to violet when one ml. of a 10 percent sodium chloride solution is added to 10 ml. of a red gold sol to which the hydrophilic substance has been added. Actually, more valuable information is obtained by plotting the change in color of the gold sol against the dilution of the spinal fluid in the presence of 0.4 percent sodium chloride. The gold curve is highly characteristic of the protein or split products employed in the test and reflects in a most intriguing way obscure properties of the protein. The problem has not been investigated in depth and characterized with modern scientific tools. Evidently, the gold number is an example of slow coagulation of a suspension.

It is possible to study the rate coagulation of a hydrocarbon emulsion by means of the Coulter counter as a function of protein concentration; the protein is adsorbed on the surface of the emulsion. Figure 15 shows a plot of the second order rate constants for the disappearance of emulsion particles as a function of the bovine serum albumin concentration.

Influence of Temperature. Arrhenius found that the velocity of a chemical reaction could be related to the temperature by

$$\ln k = -\frac{E}{RT} + B \tag{65}$$

Evidently, the plot of log k against $1/T$ should yield a straight line with a slope of $-E/2.3R$ from which E can be estimated. The constant E is known as the energy of activation and represents the energy that a mole of the reactant must have before a reaction is possible.

A consideration of the energy of activation is of aid in understanding the remarkable observation that a ten degree rise in temperature can double and sometimes more than double the rate of a reaction. The number of molecules with an energy content greater than the energy of activation can be calculated from the integrated Maxwell-Boltzmann equation which is

$$\frac{n_1}{n_0} = e^{-E/RT} \tag{66}$$

where n_1 is the number of molecules with an energy in excess of E, and n_0 is the total number of reactant molecules present. Suppose E equals 12,000 calories and the temperature is 300° (26.85°C). Then, from eq. 66, n_1/n_0 is equal to 2.06×10^{-9} and at 310° n_1/n_0 is 3.93×10^{-9} and the number of molecules capable of reacting has been almost doubled; the average kinetic energy has been increased by a little over 3 percent. The situation is illustrated in schematic manner in Figure 16 where the number of molecules having a given energy is plotted against the energy per mole. The distribution shown in curve 1 is contrasted with that of curve 2 which is at a higher temperature. Note the considerable increase in area under the curve for energies greater than E, the activation energy as the temperature is increased.

The velocity of a reaction will evidently be equal to the number of molecules reacting per second and for a bimolecular reaction this rate equals the number of molecules colliding per second multiplied by the chance that the colliding particles have sufficient energy to react. Equation 65 can be written

$$k = Ze^{-E/RT} \tag{67}$$

where $\ln Z$ is equal to the integration constant B. From what has been said, $e^{-E/RT}$ is evidently the chance that the colliding particle will have sufficient energy for a reaction to occur and Z is then the number of molecules colliding per second in unit volume. In a monomolecular reaction, Z is to be regarded as the frequency of vibration of the activated bonds in the reacting molecules and is approximately the same for all monomolecular reactions (of the order of 10^{13}). The plot of k against the absolute temperature is given schematically in Figure 17.

Not every activated collision can be effective in producing a reaction because the entire molecular surface is not reactive and the colliding molecules must be properly oriented in respect to each other. Equation 67 should then be written

$$k = PZe^{-E/RT} \tag{68}$$

where P is the so-called steric factor which may have any value from zero to unity.

It is generally true that the rate of reactions increases with increasing temperature and further a plot of log k vs $1/T$ yields a linear relation over a significant range in accordance with the Arrhenius equation; however, such behavior is not always observed; an Arrhenius plot can show curvature and sometimes distinct breaks probably reflecting changes in reaction mechanism. Beyond a certain temperature the speeds of enzymatically activated reactions always decrease with increasing temperature due to thermal denaturation of the enzyme.

The variation of many biological activities with temperature has been examined. For example, the logarithm of the velocity of the creeping of ants has been plotted

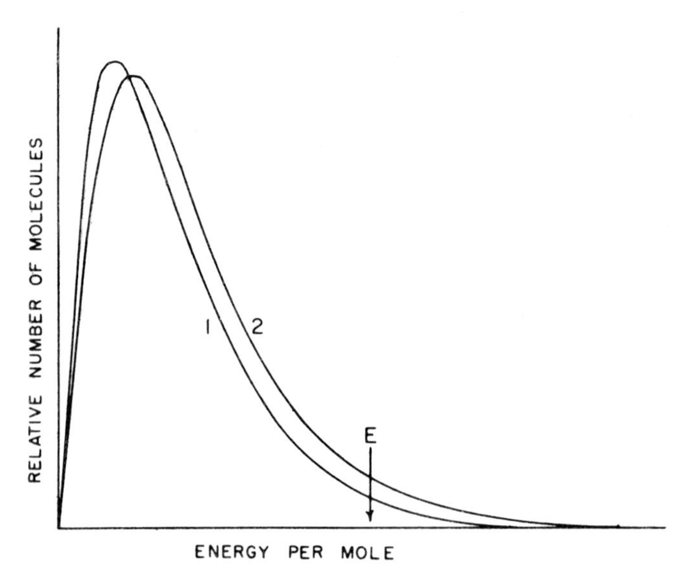

Fig. 16. Plot of the number of molecules with a given energy against the energy per mole. E indicates the activation energy. Curve 1 is for a lower temperature than curve 2.

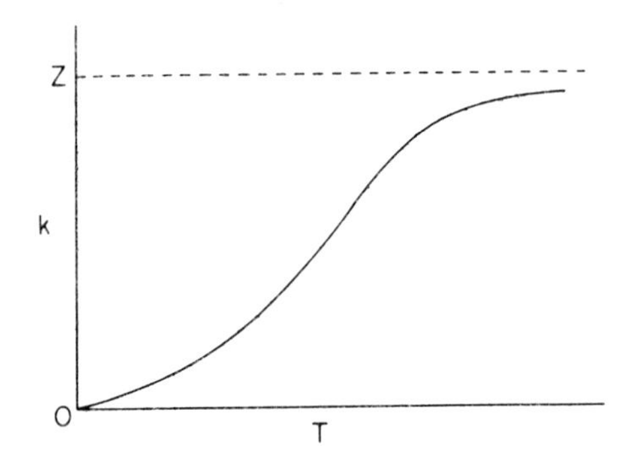

Fig. 17. Variation of the specific rate constant of a reaction with temperature according to Arrhenius (eq. 65).

against the reciprocal of the absolute temperature and a linear relation found with a distinct break in slope. At lower temperatures the energy of activation calculated was 25,900 calories and at the higher temperature 12,220 calories. Such phenomena are far too complex to be interpreted in a simple way.

Transition State Theory. The energy of a reacting system can be described by potential-energy contour diagrams, and for a reaction to occur there must be a saddle point in the contour which leads to a lower energy state. To pass over the saddle point the molecules must exist in an activated state which has an energy greater than that of the normal reactants and the average of this energy increment is the activation energy. Figure 18 shows a diagrammatic representation of the cross-section of such a pass, B^*, leading from one energy valley, A, which represents the normal state of the reactants, into a second valley, C, which is the energy level of the products of the reaction. In Figure 18, E_1^* represents the energy of activation of the forward reaction and E_2^* that of the opposing reaction.

The molecules pass from the normal state, A, to the activated state, B^*, and back again to the normal state, A. Therefore, between the normal reactants and the activated complex an equilibrium is established, which, however, is being con-

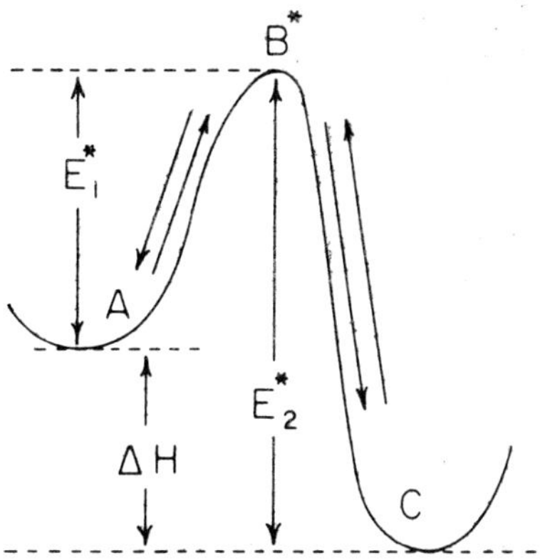

Fig. 18. Energy pass of a reaction.

tinually disturbed by the activated complex "spilling" over into C. The problem resolves itself into the calculation of the concentration of the activated complex and the rate at which this complex passes over the energy hump; the mean velocity of crossing the barrier is related to the energy of the activated molecules. From quantum mechanics, it is found that the specific reaction velocity is

$$k' = \frac{\gamma kT}{h} K^* \tag{69}$$

where K^* is the equilibrium constant between the normal reactants and the activated complex; kT/h is a frequency; k is the Boltzmann constant and is equal to 1.38×10^{-16} erg per degree per molecule; h is Planck's constant, equal to 6.62×10^{-27} erg-second; T is the absolute temperature. γ is the transmission coefficient and represents the probability of the system in state B^* either returning to the normal state A or progressing to state C. The transmission coefficient is a complex function but for many reactions approaches the value of unity. K^*, the equilibrium constant, at constant pressure and temperature is given by

$$RT \ln K^* = T\Delta S^* - \Delta H^* = -\Delta F^* \tag{70}$$

and from eq. 69 (assuming γ to be unity)

$$RT \ln \frac{k'h}{kT} = T\Delta S^* - \Delta H^* = -\Delta F^* \tag{71}$$

where ΔS^*, ΔH^*, and ΔF^* are the standard entropy, heat and free energy changes for the reaction in which the activated complex is formed from the reactants.

At constant temperature but variable pressure, P, the equation corresponding to eq. 70 is

$$\Delta F^* = \Delta E^* + P\Delta V^* - T\Delta S^* \tag{72}$$

from which is obtained

$$\Delta V^* = -RT\left(\frac{d \ln k'}{dP}\right)_T \tag{73}$$

and, if the logarithm of the velocity constant is plotted against the pressure exerted on a system at constant temperature, the volume change, ΔV^*, experienced in the formation of the activated complex can be calculated.

The importance of the transition state theory lies largely in its conceptional advantage. The actual interpretation of the calculated values of ΔF^*, ΔH^* and ΔS^* usually presents serious difficulties. The free energy of activation of a given reaction must lie between fairly narrow limits to be subject to experimental measurement. If it is too small, the reaction is too rapid, and if too large the rate is too slow and most reactions whose rate is capable of measurement will be in the neighborhood of 20,000 calories per mole. The heat of activation is interpretated in terms of the number of chemical bonds activated in the formation of the activation complex. Unfortunately, in a large molecule (proteins and nucleic acids) no one knows what energies to assign to the individual bonds; however, for simple molecules the length of the bonds are increased about 10 percent of their normal values and the energy of activation is approximately one fourth of the sum of the bond energies. The entropy of activation is related to the degree of randomness of the activated molecule as compared with that of the molecule in its normal state.

E, the energy of activation, is equal to ΔH^* plus RT, a fact which can be demonstrated by differentiating the logarithm of the specific rate constant in respect to temperature using eqs. 65 and 71 and comparing the results. At lower temperatures and appreciable energies of activation, ΔH^* and E are practically equal and eq. 71 can be written

$$k' = \frac{kT}{h} e^{\Delta S^*/R} e^{-E/RT} \tag{74}$$

Comparing eqs. 68 and 74, it is seen that

$$\frac{kT}{h} e^{\Delta S^*/R} = PZ \tag{75}$$

Since Z and kT/h have the dimensions of frequency and $e^{\Delta S^*/R}$ and P are dimensionless

$$\frac{kT}{h} \sim Z \tag{76}$$

and

$$P \sim e^{\Delta S^*/R} \tag{77}$$

The so-called steric factor of collision is, therefore, related to the entropy of activation.

Writing eq. 71 for the forward reaction as well as for the reverse reaction and subtracting the reverse from the forward expression and recalling that k_1'/k_2' is equal to the equilibrium constant for the reaction, it is evident that ΔF is equal to $\Delta F_1^* - \Delta F_2^*$, ΔH is equal to $\Delta H_1^* - \Delta H_2^*$ and ΔS is equal to $\Delta S_1^* - \Delta S_2^*$ where the subscript 1 indicates the forward reaction and the subscript 2 the reverse reaction.

Protein Denaturation. When proteins are treated with certain chemical agents such as urea, alkyl sulfates, alcohol, etc., or subject to some physical condition such as elevated temperatures, surface forces, high pressures, etc., they characteristically undergo changes which are grouped under the heading of denaturation. Among these changes are a loss in solubility at or near the isoelectric point or in the presence of a higher concentration of neutral salts, exposure of some chemical groups such as the sulfhydryl, increase in the intrinsic viscosity, changes in the optical rotation, loss of enzymatic properties if the protein is an enzyme, etc.

The kinetics of protein denaturation have been studied on numerous occasions and various criteria for denaturation have been employed. The earliest studies dealt with heat denaturation as judged by the amount of insoluble protein produced. Heat denaturation has a very large temperature coefficient and the energies of activation as obtained from Arrhenius plots are large and, for many proteins at or near their isoelectric points, in excess of a hundred kilo calories per mole. It was also noticed that the rates of denaturation as well as the energies of activation were marked functions of pH. Steinhardt showed that the rate of inactivation of pepsin was proportional to the fifth power of the hydrogen ion activity. He was led to postulate five successive ionizations of hydrogen ions in pepsin, the last one of which was crucial for denaturation. He concluded that a large part of the energy of activation for the heat denaturation of pepsin could be explained in terms of the heats of ionization of the protons.

Levy and Benaglia extended Steinhardt's approach to the denaturation of solutions of crystalline ricin, a protein from castor beans, as functions of temperature and pH. Their criterion of denaturation was insolubility of the ricin; denaturation followed first order kinetics. Plots of the logarithm of the first order velocity constants as functions of the pH at several temperatures are shown in Figure 19.

By assuming six ionizable groups to be closely associated with the denaturation process, Levy and Benaglia were able to accommodate all of their kinetic data (there are altogether about 50 prototropic groups in the pH range considered). The general picture of the process which they present is shown in Figure 20 in which each ionic form of the protein has a specific rate constant for conversion into the denatured protein. There is considerable overlap between several of the ionization constants and the actual evaluation of the constants has to be done by successive

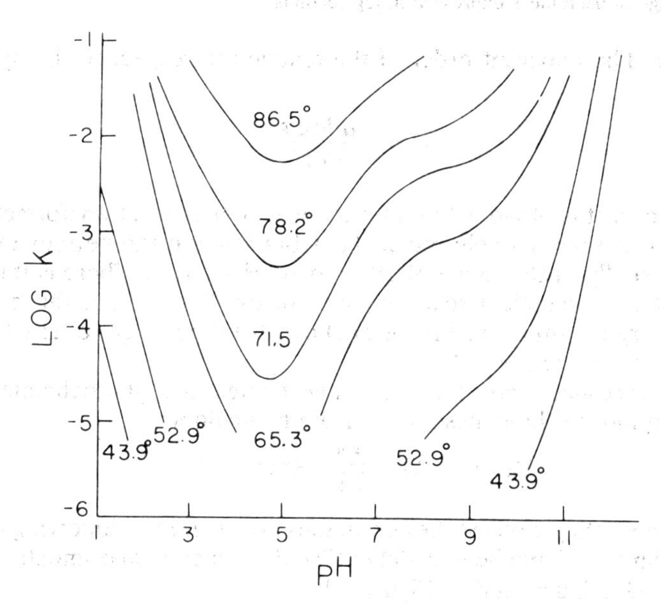

Fig. 19. Log k values for the denaturation of ricin as functions of the pH at the several indicated temperatures. Ionic strength, 0.4. (M. Levy and A. E. Benaglia: J. Bioi. Chem. *186:* 829, 1950.)

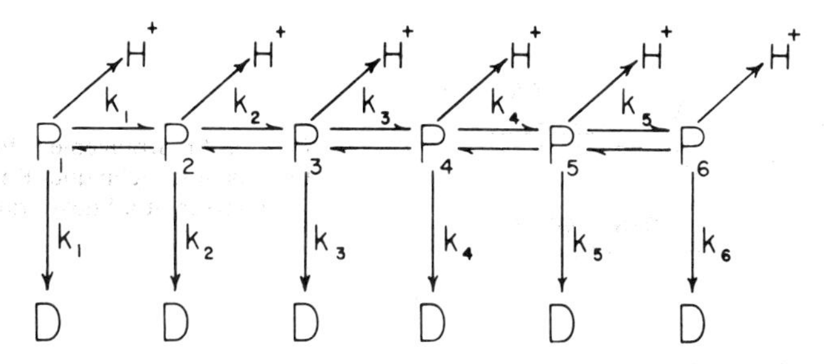

Fig. 20. Ionization of a protein and the conversion of each ionic form into the denatured protein (D).

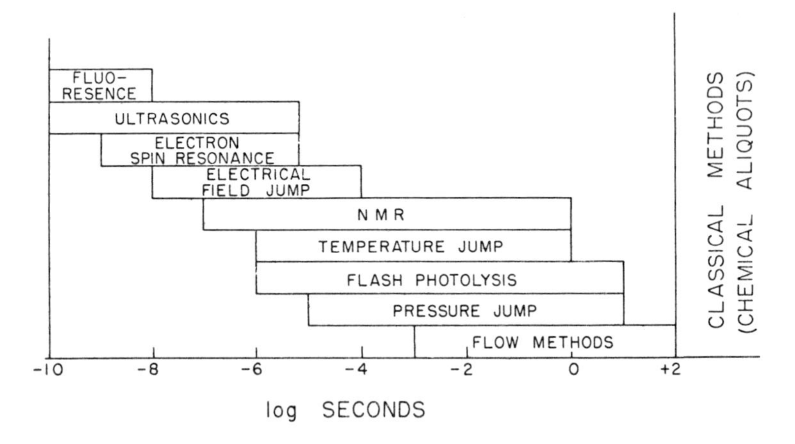

Fig. 21. Methods for the study of the rates of chemical reactions with the range of usefulness expressed in log seconds.

approximations. The apparent order of the reaction in respect to the hydrogen ion is given by

$$n = -\frac{d \log k}{dpH} \qquad 78$$

Fast Reactions. It is obvious that intramolecular chemical transformation cannot occur in times less than a single molecular vibration and the periods of molecular vibrations are usually in the range of 10^{-12} to 10^{-13} seconds. There is thus an upper limit to the rates of chemical reactions but, below this ceiling, the half times of kinetic events vary enormously. Figure 21 shows the time ranges over which various kinetic methods are effective.

In 1923, Hartridge and Roughton investigated the photochemical reaction whereby carbon monoxide is dissociated from hemoglobin

$$HbCO + O_2 \underset{\text{dark}}{\overset{\text{light}}{\rightleftharpoons}} HbO_2 + CO \qquad 79$$

and for this purpose they devised the continuous flow method. Following this pioneer study, flow techniques have been widely utilized in kinetic experiments. The principle of the method is illustrated in Figure 22.

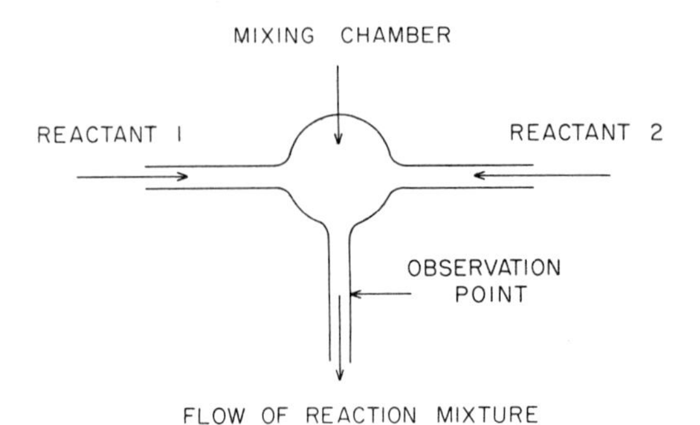

Fig. 22. An arrangement for rapid, continuous flow technique. Reactants 1 and 2 are mixed at a known rate.

Two solutions containing the reactants are mixed in the mixing chamber and flow out through a transparent tube. The reaction starts upon mixing and continues as the reaction mixture flows through the reaction tube which is scanned at different distances from the mixing chamber. Time resolution depends on the rate of flow and the distance from the mixing chamber and half times down to about 0.01 second can be measured. Providing the intermediates of the reaction have suitable absorption spectra, much information concerning the nature of the intermediates and their concentrations can be had. There are several variations of the experimental arrangement for continuous flow and light absorption is not the only means of detecting chemical change. Conductivity, pH or oxidation-reduction potentials and heat evolution can be used. The technique has been employed to good advantage by Chance in his studies on oxidation-reduction enzymes and the enzyme substrate complexes. These complexes have distinctive and characteristic absorption spectra and are especially suited to rapid-flow studies. Another application of the rapid flow technique is the fast, acid titration of hemoglobin by Steinhardt, the results of which were given in Chapter 5. A serious objection to continuous flow methods is the relatively large amounts of material required.

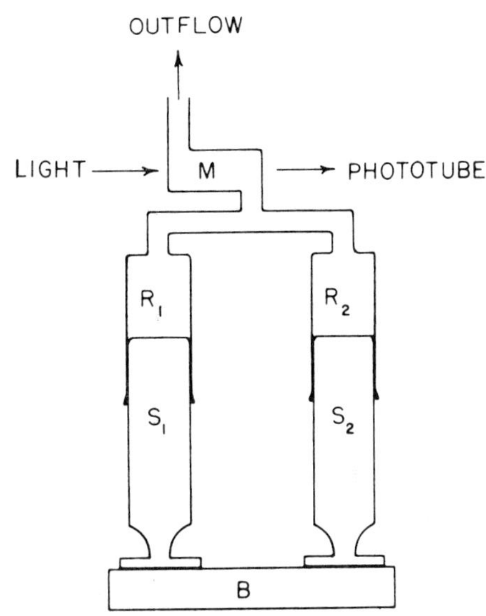

Fig. 23. Stopped flow apparatus for measurement of rates of fast reactions. S_1 and S_2 are syringe plungers connected to a common base, *B*. R_1 and R_2 are solutions of two reactants. *M* is the mixing chamber scanned by a beam of light of suitable wavelength. The syringe plungers are pushed upward and observation of the reaction begun when their movement is stopped. (Adapted from T. Spencer and J. M. Sturtevant: J. Amer. Chem. Soc. *81:* 1874, 1959.)

Fast reactions in the millisecond range can also be measured by means of a stopped flow apparatus (see Fig. 23) in which the reactants are mixed and the start of the reaction is taken to be the instant the flow of liquid is stopped; the mixing chamber is monitored by some suitable measuring device. So far, observations on enzymatic reactions have been made using only optical density and fluorescence changes.

A reaction of interest in respiration is the rate of hydration of carbon dioxide in the presence and absence of the enzyme carbonic anhydrase. The stopped flow apparatus has been employed by C. Ho and J. M. Sturtevant as well as by B. H. Gibbons and J. T. Edsall in the study of the rate of hydration and dehydration of carbon dioxide in the absence of the enzyme. The reactions involved can be written

$$\begin{array}{ccc} H^+ + HCO_3^- & \overset{k_{12}}{\underset{k_{21}}{\rightleftharpoons}} & H_2CO_3 \\ {\scriptstyle k_{13}}\downarrow\uparrow{\scriptstyle k_{31}} & & {\scriptstyle k_{23}}\swarrow\nearrow{\scriptstyle k_{32}} \\ & H_2O + CO_2 & \end{array} \quad (80)$$

The rate of ionization of carbonic acid is very large, k_{12} is about 5×10^{10} liters per second and k_{21} is about $1 \times 10^7 \text{sec}^{-1}$ and, since the concentration of H_2CO_3 is very small, the significant rate constants are those concerned with the conversion of CO_2 to bicarbonate and reverse, i.e., k_{13} and k_{31}. The overall rate equation for the conversion of dissolved CO_2 to bicarbonate is

$$-\frac{dCO_2}{dt} = k_{31}CO_2 - k_{13}H^+ \times HCO_3^- \qquad 81$$

Since in the hydration of CO_2 hydrogen ions are released and in the dehydration hydrogen ions disappear, the rate of reaction can be measured with an indicator (P-nitrophenol) in a dilute buffer system. The buffer-indicator solution is titrated with standard HCl or NaOH to determine the equivalency between the absorbance of the indicator and the release or uptake of hydrogen ions. A solution of carbon dioxide in equilibrium with water is placed in one syringe (see Fig. 23) and the buffer indicator in the other. The two are mixed by pushing the plungers upward until stopped by a predetermined setting of the apparatus. At this time the scanning of the mixing chamber by the light beam is begun and the rate of release or uptake of hydrogen ions measured by the change in the color value of the indicator. The rate of hydration of carbon dioxide is independent of pH and, at 25°, the rate constant is 0.0375 sec^{-1}. The rate of dehydration is equal to $k_{13} \times H^+ \times HCO_3^-$ and, of course, is pH sensitive. The value of the rate constant, k_{13}, at 25° is 5.5×10^4 liters per second. The product $k_{13}K_{H_2CO_3}$ is 13.7 reciprocal seconds. A complication is introduced at elevated pH values by the combination of the unhydrated but dissolved carbon dioxide directly with hydroxyl ions.

Rapid flow studies are limited by the rate of mixing of the reactants and to investigate the kinetics of ultra fast reactions, such as usually occur between ions, it is necessary to approach the problem in an entirely different manner; for this purpose the relaxation time of the reaction is determined by techniques introduced principally by M. Eigen.

If a chemical reaction in equilibrium is displaced from this condition by a small amount, it will return to the equilibrium state in a process characterized by a relaxation time τ which is the time required to restore $1/e^{\text{th}}$ part of the displacement. Consider, for example, the ionization of a weak acid

$$HA \underset{k_2}{\overset{k_1}{\rightleftharpoons}} H^+ + A^- \qquad 82$$

The equilibrium is disturbed by some sudden and momentary change in the physical enviroment such as a shock wave (pressure change), temperature, etc., by the amount ΔA^- such that ΔA^- is equal to $A^- - A_e^-$ where A_e^- is the concentration of the anion at equilibrium. The reaction will proceed towards equilibrium according to

$$\frac{dA^-}{dt} = k_1(HA_0 - A^-) - k_2(A^-)^2 \qquad 83$$

where HA_0 is the total concentration of acid. Also available is the statement that

$$k_1(HA_0 - A_e^-) = k_2(A_e^-)^2 \qquad 84$$

Combining eqs. 83 and 84, and with the knowledge that A^- is equal to $\Delta A^- + A_e$, permits both HA_0 and A^- to be eliminated from eq. 83. If $(\Delta A^-)^2$ is neglected because it is a second order term, eq. 83 can be written

$$\frac{dA^-}{dt} = \frac{d\Delta A^-}{dt} = -(k_1 + 2k_2A_e^-) \qquad 85$$

the integration of which is

$$\Delta A^- = \Delta A_0 e^{-(k_1 + 2k_2 A_e^-)t} \qquad 86$$

when ΔA^- is $1/e^{th}$ part of ΔA_0^-, the relaxation time τ is

$$\frac{1}{\tau} = k_1 + 2k_2 A_e^- \qquad 87$$

From the above argument, it is clear that the deviation of the concentration from equilibrium for a single chemical reaction after a step-wise perturbation is $\Delta C_0 e^{-t/\tau}$ where ΔC_0 is the concentration deviation at time t equal to zero.

If τ and the equilibrium constant are known, both k_1 and k_2 can be calculated. For the ionization of acetic acid at 25°, k_1 is 8×10^5 sec^{-1} and k_2 is 4.5×10^{10} liters per mole per second. The rate of the reaction

$$H^+ + OH^- \rightarrow H_2O \qquad 88$$

at 25° is 1.4×10^{11} liters per mole per second.

For many acids and bases, the association reaction between the proton and the anion or the hydroxyl ion and cation are diffusion controlled; reaction occurs on every collision and the second order reaction constant is about 10^{10} liters per mole per second. For such acids and bases, the acidity or basicity is determined by the rate of dissociation.

As indicated in Figure 21, the relaxation time of a reaction can be measured in a variety of ways. A shift in the temperature will produce a change in the equilibrium constant according to the relation (see Chapter 2)

$$\frac{d \ln K}{d T} = \frac{\Delta H^\circ}{RT^2} \qquad 89$$

where K is the equilibrium constant, ΔH° is the enthalpy of the reaction and T is the absolute temperature. Likewise, a change in pressure induces a shift in the equilibrium constant given by the relation

$$\frac{d \ln K}{d P} = \frac{\Delta V^\circ}{RT} \qquad 90$$

where ΔV° is the net molar change in the volume of the reactants and products of the reaction.

The temperature jump is produced by the discharge of a condenser charged to 10,000 volts through the reaction mixture. The energy available for heating is equal to $\frac{1}{2}CV^2$ where C is the capacity of the condenser and V is the voltage. Usually, about 45 joules are delivered in one microsecond, producing a temperature rise of 5° to 10° in the reaction mixture.

Convective mixing obscures the concentration changes after about one second, thus limiting the use of the method to faster reactions.

High pressures of 10 to 100 atmospheres are exerted on the reaction mixture. The pressure is suddenly released by the rupture of a membrane. The time constant for the reaction is limited by the speed of sound. Related to this technique is the attenuation of sound. A reaction subject to sonic waves (pressure changes) absorbs part of the energy, and the relaxation time of the reaction is related to the frequency of the sound at which maximum absorption occurs. Often there are several relaxation times of a reaction corresponding to several intermediate steps.

Of interest are relaxation studies on the transformation of a helical polypeptide chain to a disordered coil which requires about 10^{-7} second. For example, the relaxation time for the transition of poly-L-ornithine in water-methanol is about

1.7×10^{-7} second. The unwinding of two strands of DNA takes place in 10^{-4} second to several seconds. The formation of a hydrogen bond (in a non-aqueous medium) requires about 5×10^{-9} second.

Enzyme Activation. Almost all reactions occurring in living systems are controlled by enzymes. Some 700 enzymes have been described and about 100 have been crystallized. In a number of cases, the entire amino acid sequence is known and the three dimensional structure of a few enzymes has been obtained from X-ray crystallographic studies. Enzymes activate a wide variety of reactions including hydrolysis, phosphorolysis, cleavage of C-C bonds as in decarboxylation, addition of water without hydrolysis, oxidation-reduction and others. So far as is known all enzymes are proteins which may or may not have an attached prosthetic group; the presence of prosthetic groups appears necessary for the fulfillment of an oxidation-reduction function. Enzymes have had a long history extending back to at least 1833; the word enzyme was suggested by Kuhne in 1878 and was derived from the Greek words *en zyme,* meaning in yeast.

In common with all true catalysts, enzymes do not and cannot influence the equilibrium point of a reaction; they serve only to decrease the time required to attain equilibrium and to "lubricate" the reaction. Since an enzyme increases the speed of reaction in the forward direction without contributing energy to the system, it follows that the speed of the reverse reaction must also be increased. This conclusion is clear from the simple relation

$$\frac{k_f}{k_r} = K_e \qquad 91$$

where k_f and k_r are the velocity constants in the forward and reverse direction respectively and K_e is the equilibrium constant of the reaction which is unchanged by the enzyme. This concept, however, needs qualification. It is true that an enzyme does not change the total energy of a system, but no one in his right scientific mind believes that pepsin, placed in a peptic digest of a protein along with a dehydrating agent to shift the free energy change in the direction of synthesis, would ever succeed in synthesizing the original protein. Pepsin would no doubt facilitate the synthesis of peptide bonds but these bonds would not necessarily be the same as those originally present in the protein. In simpler systems, such as the hydration of fumarate to form malate under the influence of fumarase, catalysis of the forward and reverse reaction has been demonstrated.

The velocities of enzymic reactions are studied by the same methods employed for non-enzymic reactions. Light absorption can be used if the reactant or product has characteristic absorption spectra in a convenient wavelength region. If it is possible to use absorption technique, the experimenter is indeed fortunate. Other reactions require specific chemical methods for the determination of the concentrations of the reactants or the products. The pH-stat has been of help in investigation on reactions in which acidic or basic groups are produced such, for example, as during the hydrolysis of peptide or ester bonds. The pH-stat is electronically operated and maintains the pH of the reaction mixture constant by the addition of standard acid or alkali, the extent of the reaction being measured by the amount of acid or base required. Knowledge of the strength of the acidic and basic groups released must be known in order to calculate the ratio of the groups actually released to those titrated.

Manometric systems have found wide use in following the rates of enzymic reactions and are especially convenient when a gas is released, as is CO_2 in a decarboxylation reaction, or is consumed, as is oxygen in an oxidase system. It is sometimes possible to modify enzyme systems to provide for the production of a gas. For example, a bicarbonate buffer can be used and, if acid is released during the enzymic reaction, CO_2 will be given off.

Enzyme Kinetics. The point of departure for a consideration of enzyme kinetics is the enzyme-substrate complex. It was an investigation on the inversion of sucrose by invertase which led Michaelis and Menten in 1913 to propose their now well-known formulation of enzyme kinetics based on the creation of the enzyme-substrate complex. The formation and decomposition of the enzyme-substrate complex can be expressed as

$$E + S \underset{k_2}{\overset{k_1}{\rightleftharpoons}} ES \overset{k_3}{\rightarrow} E + P \quad 92$$

in which E is the enzyme, S the substrate, ES is the enzyme-substrate complex and P is the reaction product. From the simple reaction given in eq. 92, the rate equations for the change of the substrate, the enzyme-substrate complex and the product can be immediately set down as follows:

$$\frac{dS}{dt} = - k_1(E \times S) + k_2(ES) \quad 93$$

$$\frac{dES}{dt} = k_1(E \times S) - (k_2 + k_3)ES \quad 94$$

$$\frac{dP}{dt} = k_3(ES) \quad 95$$

There are no simple and unique solutions to eqs. 93, 94 and 95. It is clear that if the rate constants are suitable and if the substrate, enzyme-substrate complex

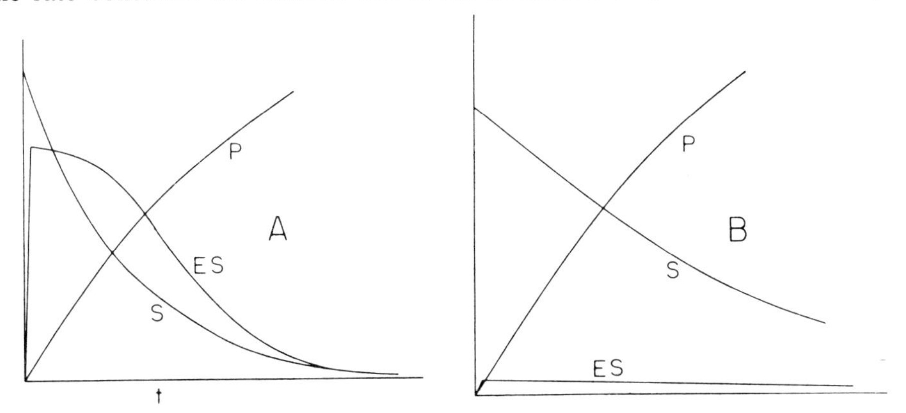

Fig. 24. *A*, Transient state kinetics. *B*, Steady state kinetics.

and the products possess distinctive absorption spectra, the sequence of the concentration changes can be determined by a rapid flow technique, the distance along the reaction tube being proportional to the time of the reaction.

It is evident from a consideration of eqs. 93, 94 and 95 that the concentrations of S, ES and P at any time will depend upon the magnitudes of E, S, k_1, k_2 and k_3; however, two extreme cases can be considered, the results of which are diagramed in Figure 24.

When the enzyme and substrate exist in a reaction mixture at comparable concentrations and k_3 is not very much smaller than k_1/k_2, the concentrations of each of the components of the reaction mixture change significantly with time and the reaction is said to follow transient state kinetics (Fig. 24A).

Typically, a series of fast intermediate steps occurs prior to the final stage of the reaction. As noted above, rapid flow and stop flow techniques are especially useful when applied to relatively concentrated enzyme solutions in which the molar concentration of the enzyme approaches that of the substrate.

Relaxation techniques can be and have been employed in the investigation of the several transient steps in enzymic reactions. If only enzymic reactions at equilibrium could be studied by relaxation methods, the approach would be severely limited because few enzymic reactions reach a state of true equilibrium. Equilibrium relaxation techniques are not usually applicable to essentially irreversible reactions.

Fortunately, a combination of rapid mixing and relaxation methods can circumvent this difficulty; a temperature jump can be applied immediately after mixing. The reactants are thereby perturbed from their steady state concentrations and allow measurements on the fast intermediate steps which occur before the rate determining step. The enzymes studied exhibit a series of relaxation times corresponding to the several intermediate steps. The identification of the observed relaxation times with discrete chemical reactions is a formidable task; the solution of this problem amounts to the formulation of the reaction mechanism, and there is ample room for error.

The formation of the enzyme-substrate complex is a necessary step in all enzymic processes and the second order rate constant approaches that of a diffusion controlled reaction (K is about 10^9 liters per mole per second). The formation of the enzyme-substrate complex is immediately followed by an isomerization reaction which has a time constant of about 1×10^4 sec $^{-1}$. These reactions may be accompanied by or followed by relaxations related to ionization, to hydration or to dehydration.

Steady State Kinetics. Suppose the concentration of the substrate is very much larger than that of the enzyme and/or k_3 is very much smaller than k_1/k_2, then the concentration of ES will build up to a constant level early in the reaction and will remain substantially constant for a significant interval. This is the situation for steady state kinetics (Fig. 24B) and is the kinetic situation considered in the Michaelis-Menten formulation in which dES/dt is set to zero and eq. 94 becomes

$$k_1(E \times S) = (k_2 + k_3)ES \qquad 96$$

Rearrangement of eq. 96 gives

$$\frac{k_2 + k_3}{k_1} = \frac{E \times S}{ES} = K_m \qquad 97$$

K_m, the Michaelis-Menten dissociation constant, is not a true dissociation constant because the concentration of the enzyme complex is being continually disturbed by the reaction leading to the formation of the products of the reaction. The dissociation constant of the complex (K_s) is equal to k_2/k_1 and, accordingly,

$$K_m = K_s + k_3/k_1 \qquad 98$$

If k_3/k_1 is small relative to K_s, K_m is in effect the true dissociation constant of the complex; further comment on this will be made later.

It is evident that the total enzyme concentration, E_0, is equal to the concentration of the free enzyme, E, plus the concentration of the enzyme-substrate complex or

$$E_0 = E + ES \qquad 99$$

According to eq. 97, E is equal to K_mES/S. Substituting this information into eq. 99 and rearranging gives

$$E_0 = \left(\frac{K_m}{S} + 1\right)ES \qquad 100$$

The rate of disappearance of S in the steady state is

$$\frac{-dS}{dt} = v = k_3ES \qquad 101$$

At sufficiently high concentration of the substrate, the active center on the enzyme surface is saturated and the maximum velocity of the reaction is

$$V_m = k_3 E_0 \tag{102}$$

Substituting the values of v (eq. 101) and V_m (eq. 102) into eq. 100 and rearranging, there results

$$v = \frac{V_m S}{K_m + S} \tag{103}$$

According to eq. 103 the plot of v, the velocity of the enzymatic reaction against S, the substrate concentration, should yield a plot of the kind shown in Figure 25.

From eq. 103, the value of S, corresponding to v equal to $V_m/2$, is equal to K_m. The estimation of both V_m and K_m with a plot of the kind shown in Figure 25 is awkward since v approaches V_m asymptotically as S is increased. It is customary to linearize eq. 103 in various ways to facilitate the determination of V_m and K_m and to exhibit the kinetic data in a more meaningful form and, by rearrangement, eq. 103 can be stated as follows:

$$\frac{1}{v} = \frac{K_m}{V_m}\frac{1}{S} + \frac{1}{V_m} \tag{104}$$

$$\frac{S}{v} = \frac{1}{V_m} S + \frac{K_m}{V_m} \tag{105}$$

$$v + K_m \frac{v}{S} = V_m \tag{106}$$

Figures 26, 27 and 28 show hypothetical plots of eqs. 104, 105 and 106, respectively.

It is indeed astonishing how generally applicable the Michaelis-Menten formulation turns out to be, and it is a tribute to the simplicity and genius of its derivation.

An accurate and unambiguous evaluation of the Michaelis constants can be a demanding and tedious experimental task. The initial velocities of the enzymatic reaction at a series of substrate concentrations must be determined. This means that the substrate or the reaction product concentrations as a function of time have to be measured for each initial substrate concentration. These results are plotted as functions of time and the slopes of the lines at zero time determined; these are the initial reaction rates and are the rates to be used to evaluate the Michaelis constants.

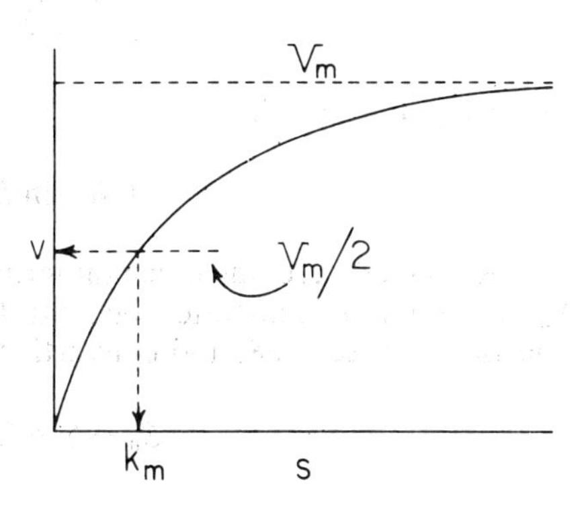

Fig. 25. Velocity of an enzymatic reaction as a function of substrate concentrate.

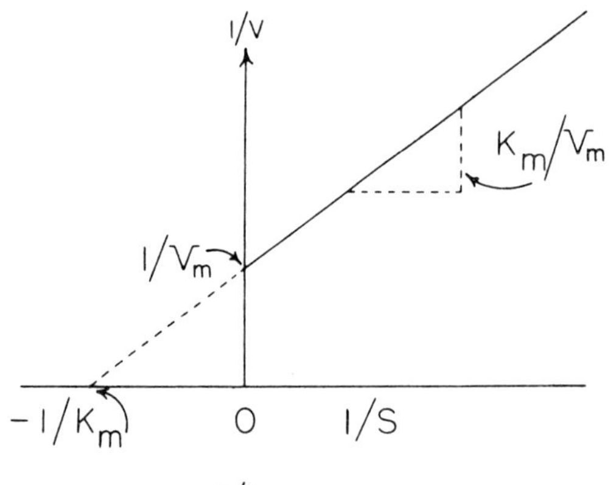

Fig. 26. Hypothetical plot of eq. 104.

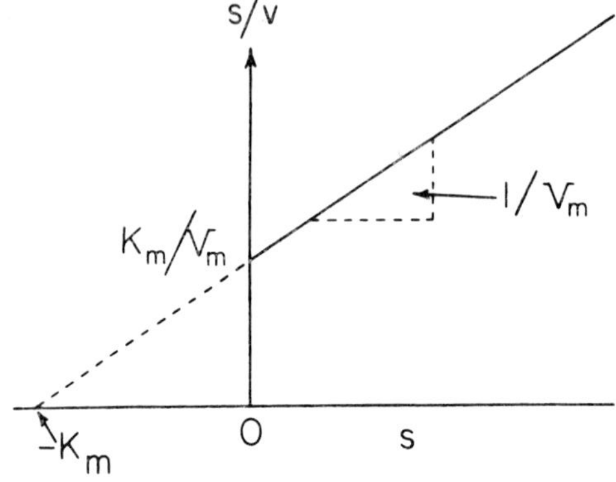

Fig. 27. Hypothetical plot of eq. 105.

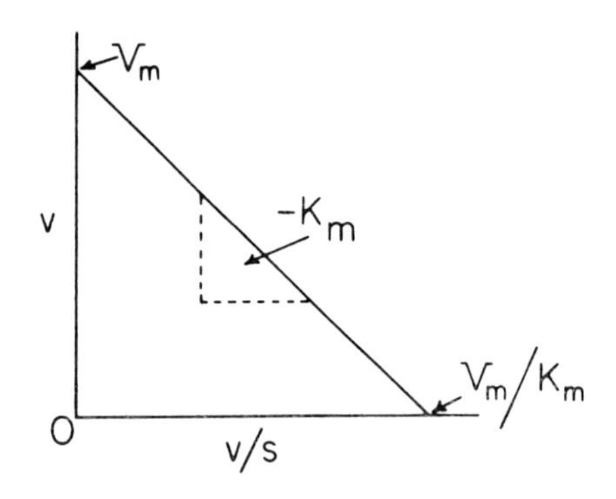

Fig. 28. Hypothetical plot of eq. 106.

Grave error can be introduced by considering the amount of substrate changed after a given time as representing the velocities of the enzymatic reactions.

Equation 103 can be written.

$$\frac{K_m dS}{S} + dS = - V_m dt \qquad 107$$

the integration of which is

$$S + K_m \ln S = - V_m t + \text{constant} \qquad 108$$

At t equal zero, the constant term in eq. 108 is seen equal to $S_0 + K_m \ln S_0$ where S_0 is the initial substrate concentration. Substituting the value of the integration constant into eq. 108, and converting to logarithm to the base 10, gives

$$S_0 - S + 2.3\ K_m \log \frac{S_0}{S} = V_m t \qquad 109$$

From eq. 109, it is clear that the kinetics of an enzyme reaction in the steady state involves a mixed zero and first order reaction. For small values of S_0, first order kinetics are followed and

$$\log \frac{S_0}{S} = \frac{V_m}{2.3\ K_m} t \qquad 110$$

and for large values of S_0

$$S_0 - S = V_m t \qquad 111$$

It is to be noted from eqs. 110 and 111 that the kinetic constant changes from $V_m/2.3\ K_m$ to V_m as the reaction changes from first to zero order kinetics; it is also to be recalled that k_3 is equal to V_m/E_0.

It would appear that the Michaelis-Menten constants could be determined from a single kinetic experiment through the use of eq. 109; however, such a procedure suffers from two serious defects. In the first place, in order to derive eq. 109, it was necessary to assume steady state kinetics throughout the time of the reaction, i.e., dES/dt is zero, and this is certainly not true over an appreciable time interval. Secondly, inhibition by the reaction products as they accumulate becomes more and more likely as the reaction proceeds.

The velocities of enzymatic reactions have been expressed in several ways. The initial reaction velocities in moles per second per mole of enzyme are suitable for most purposes. This method of expression is almost the same as the conventional "turnover number" which is the number of substrate molecules activated per molecule of enzyme per minute. If the enzyme is saturated with substrate, then the maximum velocity V_m is equal to k_3E_0 where k_3 is the specific rate of decomposition of the enzyme substrate complex to form the products and E_0 is the molar concentration of the enzyme; k_3 is then equal to or directly proportional to the turnover number, depending on the units used to express the velocity. However, if the turnover number has been determined at a concentration less than that corresponding to saturation of the enzyme by the substrate, the term loses its precise meaning and is then simply an expression of the rate of enzymatic activation under arbitrary conditions.

On occasion, the velocities of enzymatic reactions have been expressed in terms of first order reaction constants; for lower concentrations of substrate, first order kinetics will be approximately obeyed. It will be noticed, however, from eq. 110 that the first order constant is equal to V_m/K_m and has no simple meaning.

Two Substrates. The number of possible combinations of substrates, enzymes, activators, inhibitors and products is very large indeed. Once the knack of producing steady state kinetic equations for enzymic reactions has been acquired, it is easy to handle any given situation. To illustrate still further the derivation of kinetic equations, consider an enzymic reaction involving two substrates. Most enzymic reactions actually occur between two substrate molecules, but it often happens that one of the substrates is in such large excess that the reaction rate does not reflect its presence. For example, water is one of the substrates in all hydrolysis reactions.

In the two substrate situations, the substrates can bind either with the enzyme in a random sequence or in a compulsory sequence. The reactions leading to the respective sequences are:

$$\begin{aligned} E + S_a &\rightleftharpoons E\ S_a \\ E + S_b &\rightleftharpoons E\ S_b \\ E\ S_b + S_a &\rightleftharpoons E\ S_aS_b \\ E\ S_a + S_b &\rightleftharpoons E\ S_aS_b \\ E\ S_aS_b &\rightarrow E + \text{products} \end{aligned}$$

Random Sequence

$$\begin{aligned} E + S_a &\rightleftharpoons E\ S_a \\ E\ S_a + S_b &\rightleftharpoons E\ S_aS_b \\ E\ S_aS_b &\rightarrow E + \text{products} \end{aligned}$$

Compulsory Sequence

In the compulsory sequence pictured above, S_a must bind to the enzyme before S_b can be bound. The equilibria corresponding to the random sequence are as follows:

$$K_1 = \frac{E \times S_a}{E\ S_a} \qquad K_2 = \frac{E \times S_b}{E\ S_b} \tag{112}$$

$$K_3 = \frac{E\ S_b \times S_a}{E\ S_a S_b} \qquad K_4 = \frac{E\ S_a \times S_b}{E\ S_a S_b} \tag{113}$$

The total concentration of enzyme, E_0, is

$$E_0 = E + E\ S_a + E\ S_b + E\ S_a S_b \tag{114}$$

Substituting the appropriate equilibria expressions in eq. 114 gives

$$E_0 = \left[\frac{K_1 K_4}{S_a \times S_b} + \frac{K_4}{S_b} + \frac{K_3}{S_a}\right] E\ S_a S_b \tag{115}$$

Since v, the velocity of the reaction, is equal to $k_3\, E\ S_a S_b$ and the maximum velocity, V_m, is equal to $k_3 E_0$, eq. 115 can be written

$$V_m = \left[\left(\frac{K_1 K_4}{S_b} + K_3\right)\frac{1}{S_a} + \frac{K_4}{S_b} + 1\right] v \tag{116}$$

from which is obtained

$$\frac{1}{v} = \left(\frac{K_1 K_4}{S_b} + K_3\right)\frac{1}{V_m S_a} + \left(\frac{K_4}{S_b} + 1\right)\frac{1}{V_m} \tag{117}$$

It is evident from eq. 117 that, if S_b is held constant and $1/v$ is plotted against $1/S_a$, a straight line should be obtained the slope of which is $(K_1 K_4 + K_3 S_b)\ /S_b V_m$ with an intercept on the $1/v$ axis of $(K_4 + S_b)\ /S_b V_m$. Diagrammatic plots of $1/v$ against $1/S_a$ at three different concentrations of S_b are shown in Figure 29.

Setting the $1/v$ values equal for the various concentrations of S_b and solving the simultaneous equations resulting after substitution in eq. 117, it turns out that at the intersection point of the straight lines (see Fig. 29), $1/S_a$ is equal to $-1/K_1$. From the

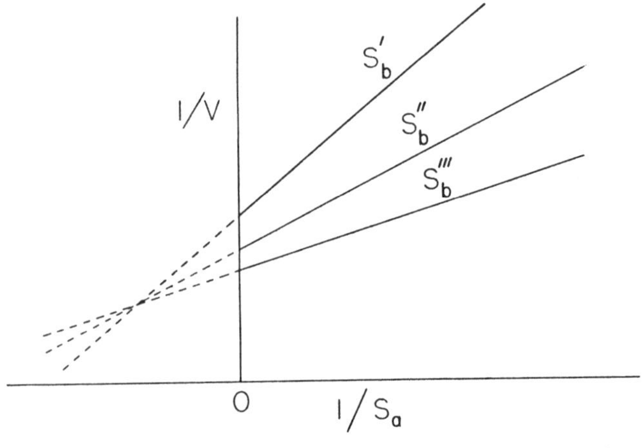

Fig. 29. Plot of $1/v$ vs $1/S_a$ for a two substrate enzymic reaction holding component S_a constant at three different concentrations. (Adapted from J. R. Florini and C. S. Vestling: Biochem. Biophys. Acta *25:* 575, 1957.)

slopes, intercepts and the concentration of S_a at the intersection point, it is possible to solve for K_1, K_3 and K_4. Since eq. 117 is symmetrical in respect to S_a and S_b, a plot of $1/v$ against $1/S_b$ at constant S_a permits K_2 to be evaluated.

It is clear that if S_a is in great excess, eq. 117 reduces to

$$\frac{1}{v} = \frac{k_3}{V_m S_a} + \frac{1}{V_m} \qquad 118$$

which is the Michaelis-Menten expression for a one substrate situation.

The compulsory sequence rate equation corresponding to eq. 117 for a random sequence is

$$\frac{1}{v} = \left[\frac{K_1}{S_a} + 1\right] \frac{K_2}{V_m S_b} + \frac{1}{V_m} \qquad 119$$

It is thus possible by the proper plots to distinguish between a random and a compulsory binding sequence.

Inhibitors. Obviously, any agent which destroys the specific protein structure of the enzyme will act as an enzymatic inhibitor; this destruction may be reversible or irreversible. Suppose, however, that a substance inhibits by combining with the same site on the enzyme surface as does the substrate; it competes with the substrate for the active site and the inhibition can be reversed by excess substrate. Such a substance is known as a competitive inhibitor and the reactions which describe this type of inhibition are

$$\mathrm{E} + S \overset{K_m}{\rightleftharpoons} \mathrm{ES} \rightarrow \mathrm{P} + \mathrm{E} \qquad 120$$

$$\mathrm{E} + \mathrm{I} \overset{k_i'}{\rightleftharpoons} \mathrm{EI} \qquad 121$$

where EI is the enzyme-inhibitor complex.

Suppose, however, the inhibitor combines with the enzyme in such a manner that there is no competition with the substrate and the inhibition cannot be abolished by increasing the substrate concentration. The inhibitor is, however, reversibly bound to the enzyme and the inhibition can be removed by decreasing the concentration of the inhibitor.

An inhibitor of this type is called a non-competitive inhibitor and in some way must change the structure of the enzyme to render it catalytically inactive even though the enzyme might still bind substrate at the formerly active site. Actually, the alteration in structure of the active center would be expected to prevent or greatly decrease substrate binding by the enzyme. There are no doubt many variations on this theme. For purposes of formulating the kinetic equation, the pertinent reactions are

$$E + S \overset{K_m}{\rightleftharpoons} ES \rightarrow P + E \qquad 122$$

$$E + I \overset{K_i'}{\rightleftharpoons} EI \qquad 123$$

$$ES + I \overset{K_i''}{\rightleftharpoons} EIS \qquad 124$$

The equilibrium dissociation constants for the complexes in reactions 122, 123 and 124 are

$$K_m = \frac{E \times S}{ES} \qquad 125$$

$$K_i' = \frac{E \times I}{EI} \qquad 126$$

$$K_i'' = \frac{ES \times I}{EIS} \qquad 127$$

The total enzyme concentration E_0 is

$$E_0 = E + ES + EI + EIS \qquad 128$$

Eliminating E, EI and EIS from eq. 128 by the use of eqs. 125, 126 and 127 gives

$$E_0 = \left[\frac{K_m}{S} + 1 + \frac{K_m I}{K_i' S} + \frac{I}{K_i''}\right] ES \qquad 129$$

Remembering that v is equal to $k_3 ES$ (eq. 101) and V_m is equal to $k_3 E_0$ (eq. 102), eq. 129 becomes

$$v = \frac{K_i' K_i'' V_m S}{K_i' K_i'' K_m + K_i' K_i'' S + K_m K_i' I + K_i'' IS} \qquad 130$$

Rearrangement of eq. 130 gives

$$\frac{1}{v} = \left(1 + \frac{I}{K_i'}\right)\frac{K_m}{V_m S} + \left(1 + \frac{I}{K_i''}\right)\frac{1}{V_m} \qquad 131$$

Comparing eq. 131 with eq. 104, it is clear that a non-competitive inhibitor will increase both the intercept and the slope when $1/v$ is plotted against $1/S$. If the inhibitor has the same affinity for the enzyme alone as it has for the enzyme substrate complex, K_i' and K_i'' are equal to each other. If K_i'', for a competitive inhibitor, is infinitely large (see eqs. 127 and 130) eq. 131 reduces to

$$\frac{1}{v} = \left(1 + \frac{I}{K_i'}\right)\frac{K_m}{V_m S} + \frac{1}{V_m} \qquad 132$$

The plot of $\frac{1}{V}$ vs $\frac{1}{S}$ for a competitive inhibitor gives an increased slope but the intercept remains unchanged. These conditions are diagramed in Figure 30.

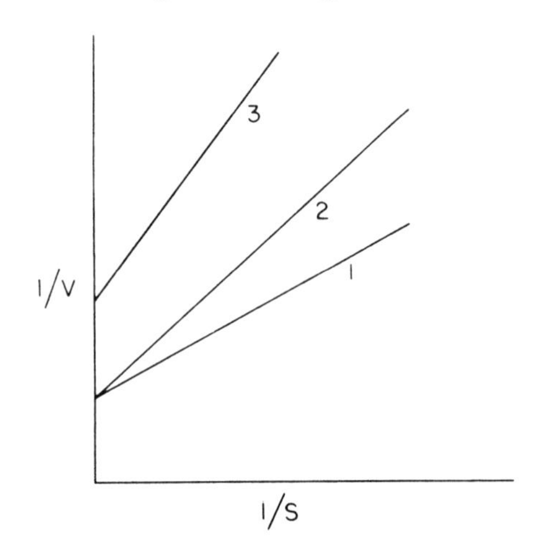

Fig. 30. Plots to reveal nature of inhibitor of an enzymatic reaction. 1, Unhibited; 2, competitive inhibition; 3, non-competitive inhibition.

A competitive inhibitor decreases the affinity of the enzyme for the substrate, whereas a non-competitor decreases the catalytic rate of the enzyme. A competitive inhibitor could be called an affinity inhibitor and a non-competitive inhibitor a rate inhibitor.

Dixon has suggested that competitive inhibition can be distinguished from non-competitive inhibition and the inhibition constants evaluated by plotting $1/v$ against the concentration of inhibitor, keeping the substrate concentration constant. Straight lines should be obtained and for two different substrate concentrations these lines will intersect at a point to the left of the $1/v$ axis (see Fig. 31 A and B).

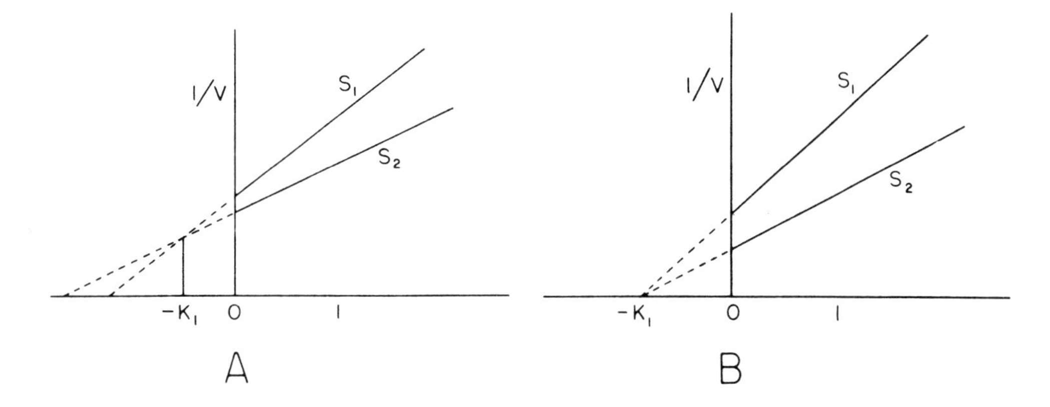

Fig. 31. Plots for the determination of the type of inhibition and the evaluation of the inhibition constants. *A*, Competitive inhibition; *B*, non-competitive inhibition.

Since an enzyme increases the rate of the forward reaction without altering the equilibrium, it follows that the rate of the reverse reaction must also be increased. Clearly, the enzyme must be able to bind with the products of the forward reaction. Thus, the products of the forward reaction should act as inhibitors to the forward reaction. It is frequently useful to study the effect of the addition of the products of the reaction on the initial rate of the forward reaction; such measurements can be helpful in the formulation of the mechanism of the enzyme reaction.

Activators. Many enzymes can be prepared in inactive forms and are activated in various ways. For example, the digestive enzymes pepsin, chymotrypsin and trypsin are secreted as inactive zymogens and are activated by the removal of naturally occurring peptides. Other enzymes such as papain need to be activated by a reducing agent and still other enzymes require the presence of metal ions. Hydrogen ions are especially important in enzymatic reactions. It is possible that the enzyme combines with the substrate only over a limited pH range; the hydrogen ions are acting as an affinity activator and/or as an affinity inhibitor. It is also true that the hydrogen ions can act to render the catalytic site active or inactive, i.e., as rate activator or inhibitor. Suppose an enzyme shows the following reactions in its active center.

$$E + S \overset{K_m}{\rightleftharpoons} ES \qquad 133$$

$$ES + H^+ \overset{K_a}{\rightleftharpoons} ESH^+ \overset{k_3}{\rightarrow} P + E + H^+ \qquad 134$$

$$ESH^+ + H^+ \overset{K_i}{\rightleftharpoons} ESH_2^2 \qquad 135$$

The equilibrium dissociation constants for eqs. 133, 134, and 135 are

$$K_m = \frac{E \times S}{ES} \tag{136}$$

$$K_a = \frac{H^+ \times ES}{ESH^+} \tag{137}$$

$$K_i = \frac{ESH^+ \times H^+}{ESH_2^{2+}} \tag{138}$$

The total enzyme, E_0, is

$$E_0 = E + ES + ESH^+ + ESH_2^{2+} \tag{139}$$

Eliminating E, ES, and ESH_2 from eq. 139 by the use of eqs. 136, 137, and 138 and remembering that v is equal to k_3ESH (eq. 101) and V_m is equal to k_3E_0 (eq. 102), eq. 139 becomes by rearrangement

$$v = \frac{K_iV_mH^+S}{K_mK_aK_i + K_aK_iS + K_iH^+S + (H^+)^2S} \tag{140}$$

At high substrate concentration and at all pH values such that the system exhibits zero order kinetics in respect to the substrate, the velocity, V_s, becomes

$$V_s = \frac{K_iV_mH^+}{K_aK_i + K_iH^+ + (H^+)^2} \tag{141}$$

It is implicity assumed that the rate of ionization of the various ionic forms greatly exceeds the rate of formation and breakdown of the enzyme-substrate complex.

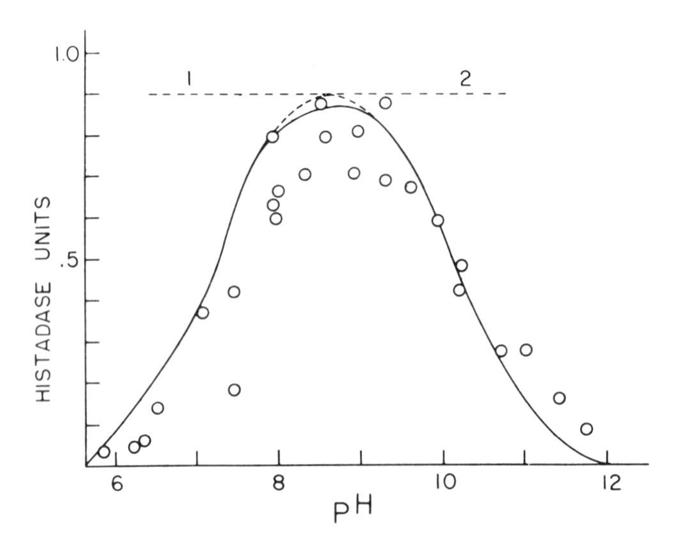

Fig. 32. Action of histidase on histidine expressed in histidase units as a function of pH. Circles are the experimental data. Solid line is a plot of eq. 141. (Data of A. C. Walker and C. L. A. Schmidt: Arch. Biochem. *5:* 445, 1944.)

Examination of eq. 141 shows that it provides for a maximum in V_s as the pH is changed; bell-shaped enzymatic activity-pH curves are of fairly common occurrence and shown in Figure 32 is a plot of the data of Walker and Schmidt on the action of histidase on the amino acid histidine as a function of pH. The enzymatic velocity is expressed arbitrarily in "histidase units."

It can be seen that eq. 141 is able to represent the data of Walker and Schmidt in a fairly satisfactory manner. The result shown in Figure 32 can be interpreted in the following manner. In the alkaline range, histidase is activated by the addition of hydrogen ions (eq. 142), and as the hydrogen ions are increased still further they result in inhibition (eq. 143). The dotted curve 1 in Figure 32 represents the hypothetical enzymatic activity if the hydrogen ions had no inhibitory effect, i.e., K_i of eq. 141 is infinitely large. Dotted curve 2 shows the course of the reaction if K_a of eq. 141 were very small, i.e., the first hydrogen did not dissociate as the pH was raised. The maximum in the rate hydrolysis for a hydrolyzing enzyme is a necessity if the enzyme is to act in this capacity. The active area of such an enzyme must have in it a proton donating and a proton accepting group to fulfill the Bronsted requirement for acid-base catalyst; K_a of eq. 141 refers to the dissociation constant of the proton donating group and K_i to the proton accepting group.

Shown in Figure 33 is the rate of attack of pepsin on egg albumin with excess substrate as a function of pH. Inspection of eq. 141 reveals that if an excess of hydrogen ions does not lead to inhibition (K_i is very large), this equation reduces to

$$V_s = \frac{V_m H^+}{K_a + H^+} \qquad 142$$

Equation 142 is plotted as a solid line in Figure 33. From eq. 142, if H^+ becomes large in respect to K_a, V_s is equal to V_m and, when H^+ is equal to K_a, V_s is equal to $V_m/2$; this provides a method for the estimation of K_a in Figure 32 as well as in Figure 33. K_a calculated for the egg albumin-pepsin system is 1.4×10^2 which is

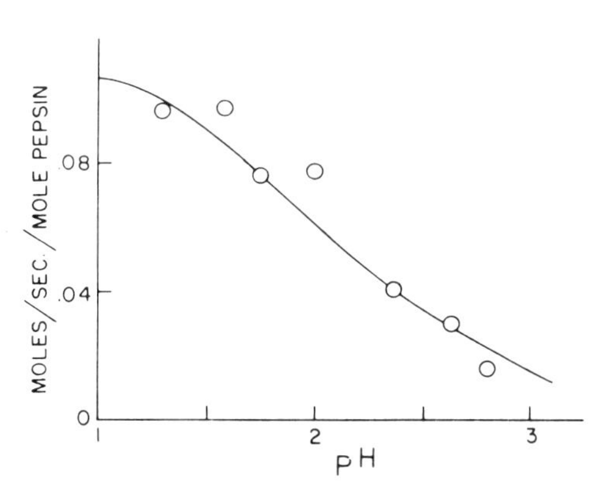

Fig. 33. Maximum velocity of digestion of egg albumin by pepsin against pH at 30°. The solid line was calculated from eq. 142. (Data from H. B. Bull and B. T. Currie: J. Amer. Chem. Soc. *71:* 2758, 1949.)

approximately the value to be expected of a carboxyl group. By conducting the kinetic experiments at two different temperatures, it is possible to calculate the heats of formation of the enzyme-substrate complex, of the ionization of the proton and the energy of activation for the formation of the products by methods which have already been discussed. Such a study was made of the egg albumin-pepsin system and the energy profile of the reaction is shown in Figure 34. The calculated heat of ionization associated with K_a is about 11,000 calories per mole which is far too large for the heat of ionization of a carboxyl group.

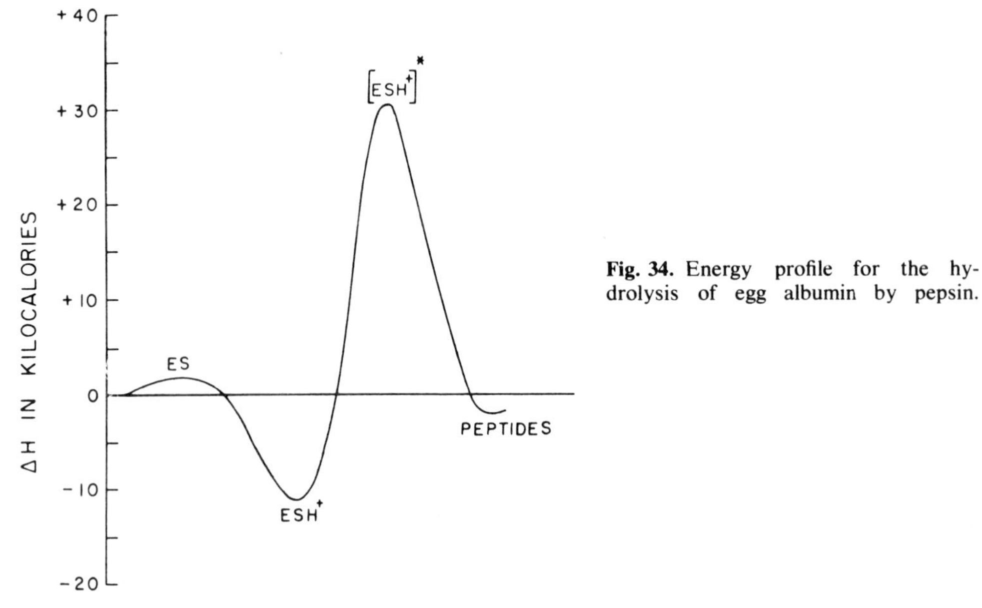

Fig. 34. Energy profile for the hydrolysis of egg albumin by pepsin.

The Michaelis-Menten complex does not correspond to the activated complex; the concentration of the activated complex is far smaller than is that of the Michaelis-Menten complex.

Gutfreund has determined the pH dependence of the catalytic activity of trypsin using benzoyl-L-arginine ethyl ester as the substrate. The rate of hydrolysis was measured by observing the time required for the pH to regain a certain value from which it has been slightly displaced by the addition of a small known amount of one molar sodium hydroxide. Sufficient substrate was used to ensure zero kinetics throughout the course of the experiments. Gutfreund's results at 25° are shown in Figure 35, the rates being expressed in arbitrary units.

The course of the hydrolysis as a function of pH in Figure 35 means that, as

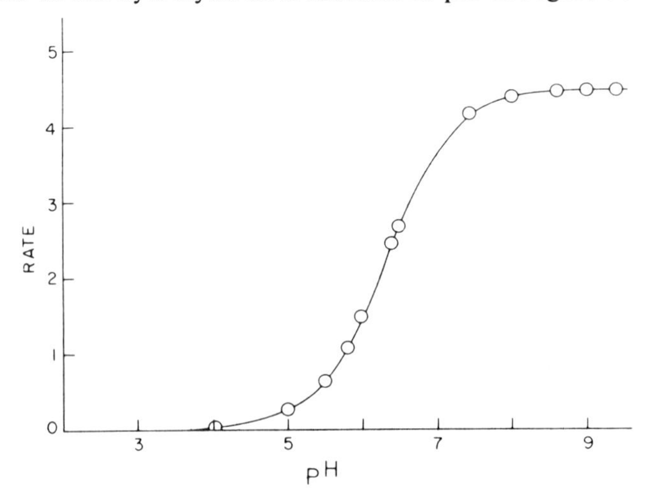

Fig. 35. Rate in arbitrary units of the hydrolysis of benzoyl-L-arginine ethyl ester by trypsin as function of pH, 25°. (H. Gutfreund: Trans. Faraday Soc. *51*: 441, 1955.)

the hydrogen ion concentration increases, trypsin is inhibited and the proton accepting group is being titrated. The fact that the rate curve does not show a maximum indicates that K_a is very small. Under this condition, eq. 141 reduces to

$$V_s = \frac{K_i V_m}{K_i + H^+} \qquad 143$$

From eq. 143, it is clear that, when H^+ is much smaller than K_i, V_s is equal to V_m and also, when K_i is equal to H^+, V_s is equal to $V_m/2$ which permits an estimation of K_i to be made. Gutfreund concluded that the curve shown in Figure 35 corresponds to the ionization of a group whose pK value is 6.25 and from this value, and likewise from the heat of ionization, identified the group as the imidazole in a histidine residue in trypsin. The magnitudes and heats of ionization of kinetically determined dissociation constants have to be interpreted with some caution since they may not be related in a simple manner to the values to be expected of a particular group in an uncomplicated environment. For example, the ionization of a proton may be linked to substrate binding in which event the ionization constant as kinetically determined may be very greatly shifted from its normal value.

It is also possible for the substrate to ionize and to influence the formation of the enzyme-substrate complex. The production of an unfavorable ionic form of the substrate by a shift in pH has the same effect as decreasing the concentration of the substrate. Thus, if the influence of the ionization of the substrate is only on the affinity between the substrate and enzyme, the maximum velocity over a broad range of pH should be independent of the ionization of the substrate.

Enzyme-Substrate Affinity. In the reaction

$$E + S \underset{k_2}{\overset{k_1}{\rightleftarrows}} ES \overset{k_3}{\rightleftarrows} P + E \qquad 144$$

K_m is defined as

$$K_m = \frac{k_2 + k_3}{k_1} = K_s + k_3/k_1 \qquad 145$$

Thus K_m is not a true measure of the affinity of the enzyme for the substrate except under the condition that $k_2 >> k_3$.

The magnitude of k_1 should be a function of the state of ionization of both E and S, whereas k_2 and k_3 should be independent of the ionization of these two components. In the simple situation pictured above, a variation of V_m without a variation in K_m in response to the addition of activators or inhibitors would indicate that k_3 is small in comparison with k_2 and, accordingly, K_m would be a true dissociation constant. However, in any real situation K_m would be expected to be a complex function of the state of ionization of the components of the enzyme system and, in general, no safe conclusion can be drawn as to the relative magnitudes of k_2 and k_3.

Whereas the existence of an enzyme-substrate complex is necessary for understanding the description of the enzyme kinetics, efforts to isolate this complex have not often been successful. If the enzyme-substrate complex has a characteristic absorption spectrum with a significant extinction coefficient, the concentration of the enzyme-substrate complex can be determined directly and, accordingly, the affinity of the enzyme for the substrate estimated. The results of Chance are very impressive because he actually demonstrated the separate existence of the enzyme-substrate complex. He concerned himself, along with other enzymes, with the peroxidase from horseradish. This enzyme has the property of combining with hydrogen peroxide, forming a very unstable complex which is green. In this complex, the iron of the peroxidase probably forms ionic bonds. This complex passes rapidly over into a red complex. During this transition the ionic bonds of the iron probably revert to covalent bonds. The complex then combines with the appropriate oxygen acceptor such as ascorbic acid and liberates the free enzyme. By using a rapid flow apparatus, it was possible to study the transit color changes as the colors proceeded from the brown of the free enzyme to the green complex, to the red complex, and finally to the reaction products with the regeneration of the brown complex. The reversible decomposition of the enzyme-substrate complex does not appear to play an important role in the reaction kinetics; transit state rather than steady state kinetics are followed.

It is possible to precipitate pepsin with egg albumin at pH 4.0. The egg albumin and pepsin remain native and the molar ratio at maximum precipitation is one to one. The complex is soluble in dilute salt solutions and by means of light scatter the dissociation constant between the egg albumin and the pepsin can be estimated. The light-scatter association constant is about 1.46×10^7 ml. per mole whereas the constant from kinetic measurement is 0.14×10^7 ml. per mole.

Hayes and Velick studied the binding of DPNH and of DPN by yeast dehydrogenase in the ultracentrifuge. They sedimented the protein in the presence of DPNH and measured the number of DPNH molecules bound per mole of protein and found 3.6. They also found 3.6 sites per mole of enzyme for DPN. The dissociation constant for the DPNH-enzyme complex was about 1.5×10^{-5} and that for DPN was about 2.6×10^{-4}. The K from kinetic data was about 2.2×10^{-5} for DPN.

Doherty and Vaslow using equilibrium dialysis investigated the binding of N-acetyl 3,5-dibromo-L-tyrosine by chymotrypsin. The enzyme-catalyzed reaction consisted of the exchange of the oxygen of the carboxyl group with water enriched with 0^{18}. The binding of the substrate by the enzyme should be unaffected by the exchange of the oxygen isotopes and, accordingly, a good measure of the affinity between enzyme and substrate should be obtained. It is evident, however, that the binding constant which is calculated is pH-dependent and must involve the ionization of the substrate and possibly also the enzyme.

Chymotrypsin treated with p-nitrophenylacetate at pH 5 binds one mole of substrate per mole of enzyme. The p-nitrophenol is split off and the enzyme is acylated. The acylated enzyme hydrolyzes to the original enzyme and acetate at pH 7. Since there is a marked color change in the substrate during these transformations, the reaction and the formation of the intermediate can be conveniently followed by stopped flow technique. The acylated intermediate can be isolated at pH 5. The reactions are

$$E + NPA \underset{k_2}{\overset{k_1}{\rightleftharpoons}} E \cdot NPA \qquad 146$$

$$E \cdot NPA \underset{k_4}{\overset{k_3}{\rightleftharpoons}} E \cdot A + NP \qquad 147$$

$$E \cdot A \rightleftharpoons E + A- \qquad 148$$

where NPA represents p-nitrophenylacetate, NP, p-nitrophenol and $A-$ acetate. At pH 7, chymotrypsin rapidly hydrolyzes p-nitrophenyl acetate and the intermediates are missed.

Inhibition by Excess Substrate. Inhibition by excess substrate is not an unusual occurrence and this situation can be described by the following reactions

$$E + S \overset{K_m}{\rightleftharpoons} ES \rightarrow P + E \qquad 149$$

$$ES + S \overset{k_1}{\rightleftharpoons} ES_2 \qquad 150$$

These two reactions lead directly to the following kinetic expression

$$v = \frac{k_1 V_m S}{K_m k_1 + k_1 S + S^2} \qquad 151$$

Shown plotted in Figure 36 are the data of Laidler and Hoare on the decomposition of urea by urease as a function of urea concentration.

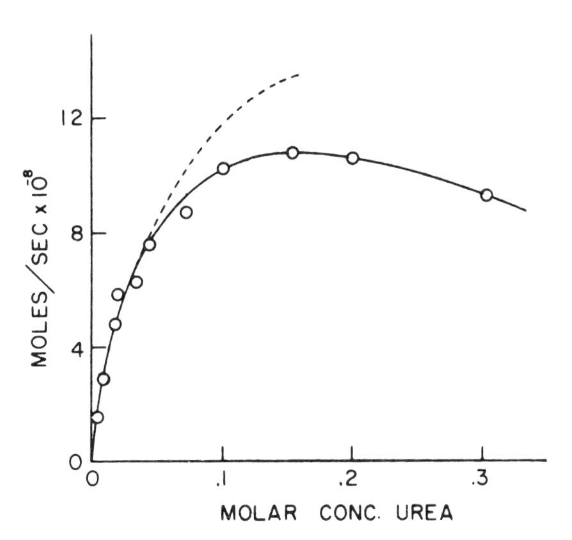

Fig. 36. Decomposition of urea by urease, 1.5×10^{-8} mg. urease per 100 ml., as a function of the urea concentration; pH 6.6; 30°. (Data of K. J. Laidler and J. P. Hoare: J. Amer. Chem. Soc. *71:* 2699, 1949.)

The solid line shown in Figure 36 is a plot of eq. 151, whereas the dotted line is a plot under the condition that k_1 is infinitely large, i.e., no substrate inhibition. It has been suggested that the decrease in activity at higher urea concentrations is due to the existence of neighboring sites in the active center of the urease molecule one of which binds urea and the other water, and at high urea concentration urea is adsorbed on the water site. Actually, urease appears to have two active enzymatic centers per urease molecule, each with its own individual Michaelis-Menten dissociation constant.

A more general interpretation of excess substrate inhibition involves the conception of three points of attachment of the substrate to the active center which permits the enzyme to discriminate between optical isomers of the substrate. At high substrate concentration two or more substrate molecules compete for the active site; each is bound to one of the points of attachment but none is successful in achieving the appropriate orientation on the active site.

The kinetics of enzymic reactions involving two substrates have been discussed. It is also possible to have one, two or more products produced. Evidently, the complete kinetic description of such reactions can become complex. However complex they may be, it is possible to derive the appropriate kinetic expressions using the simple methods outlined above. The difficulty is that the expressions become inordinately cumbersome. Cleland has introduced shorthand methods of dealing with such equations (see references at the end of the chapter). In addition, Cleland uses simple graphical representations of enzymic reactions which are helpful in visualizing the course of the reaction. He employes such terms as Uni Bi, Bi Bi, etc. to indicate that there is one reactant and two products or two reactants and two products respectively. Also if the enzyme oscillates between two stable forms during the course of the reaction, he designates this as a ping pong mechanism.

A compulsory Bi Bi reaction can be represented by

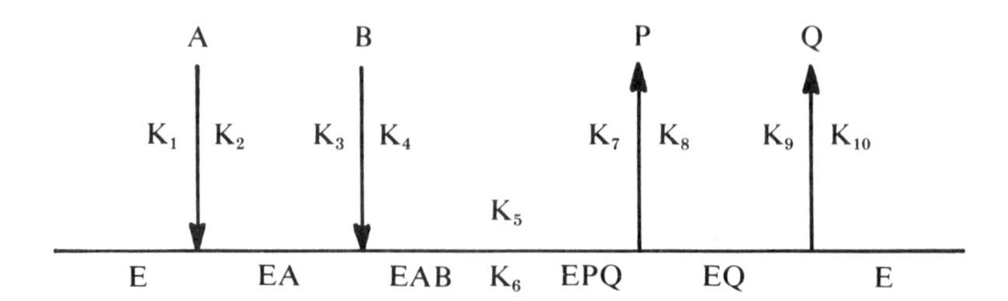

where A and B are the reactants and P and Q are the products and E is the enzyme. A random sequence for a Bi Bi reaction can be written

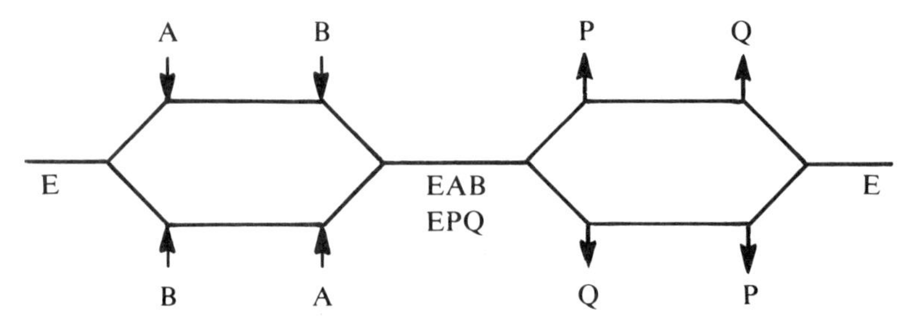

A ping pong mechanism in which the enzyme oscillates between two stable forms with two reactants and two products (a ping pong compulsory Bi Bi) can be expressed by

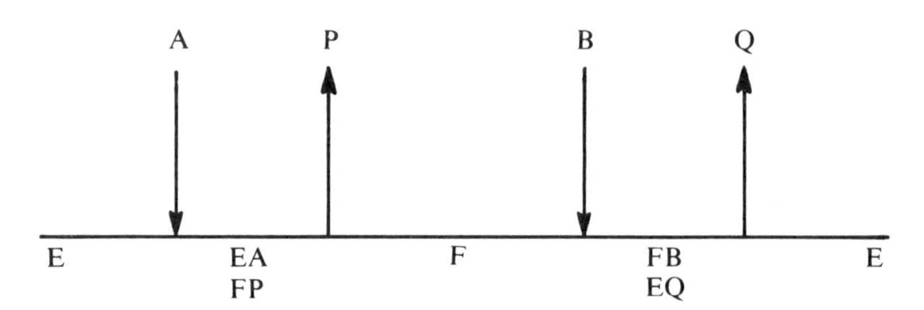

where E and F represent the two stable forms of the enzyme.

Allosteric Enzymes. Usually enzymes follow Michaelis-Menten kinetics, i.e., plots of the initial velocities of the enzyme reaction against the substrate concentrations yield hyperbolic curves and the corresponding reciprocal plots are linear. However, there are enzymes which give complex kinetics, and linear reciprocal plots are not obtained. There are various reasons why linear reciprocal plots do not result. For example, if more than one substrate molecule is being bound by the enzyme and the binding of the first molecule of substrate per molecule of enzyme changes the affinity of the enzyme for the second substrate molecule, complex kinetics will result. This is the classical situation of the binding of oxygen by hemoglobin.

The term *allosteric* has been proposed to encompass the more complex enzymes, and several theories of allosteric effects have been proposed. The two which have received the most attention are that of Monod and others (see reference at end of chapter) and that of Koshland and Neet (Ann. Rev. Biochemistry *37*:359, 1968).

Monod considers that allosteric enzyme molecules are composed of identical subunits. A change in the conformation of one subunit induces a conformational change in the other subunits leading to a catalytic change in the enzyme. Thus, the conformational change in the enzyme occurs through an isomerization between two states of the enzyme; the addition of a small molecule which is specifically bound to the enzyme shifts the equilibrium between the states of the enzyme.

Koshland's approach to allosteric properties is through the notion of an induced fit produced in the enzyme by the addition of the specific small molecule. It is assumed that there exist at least two substrate binding sites, a catalytic site and a regulation site. The binding at the regulatory site induces a conformational change in the enzyme that results in a change of affinity of the catalytic site for the substrate.

There are allosteric activators as well as allosteric inhibitors. The structure of the activator or the inhibitor need not resemble that of the substrate molecule. In fact, if an added molecule influences the enzymic rate and its structure is significantly different from that of the substrate, an allosteric situation can be suspected. Allosteric inhibitors are of especial importance in that they are frequent metabolic products of an enzymic reaction chain. The accumulation of the products can inhibit the early steps of the reaction chain; they can regulate their own synthesis. Thus, they constitute a metabolic feed back mechanism.

It is unfortunate that the conception and definition of the term allosteric is not as clear cut as could be desired. Such ambiguity is certain to lead to confusion. For example, some workers use the term to designate an enzyme which yields sigmoid curves when the reaction velocity is plotted against the substrate concentration. Others consider the word to apply to enzymes occurring at a branch point in a metabolic pathway and still others to describe enzymes composed of identical subunits, the change of structure of a single subunit induces changes in the other subunits with a resulting change in the catalytic behavior of the enzyme. The property which an enzyme must have to be considered allosteric is that the molecule must have at least two binding sites one of which is regulatory and the other is catalytic. Further, binding at the regulatory site induces a conformational change in the enzyme.

Reversible Reactions. Not many enzymatic reactions lend themselves to kinetic studies on both sides of the equilibrium point. The hydration of fumarate to malate by the enzyme fumarase is such an enzyme system and has been studied with great profit by Alberty and co-workers. The reaction is

$$\begin{matrix} {}^{-}OOC{-}CH \\ \| \\ H{-}C{-}COO^{-} \end{matrix} + H_2O \rightleftharpoons \begin{matrix} HOCH\ COO^{-} \\ | \\ CH_2COO^{-} \end{matrix} \qquad 152$$

FUMARATE MALATE

The free energy change for this reaction at 25° as written is −825 calories per mole. The kinetics to be considered are

$$F + E \underset{k_2}{\overset{k_1}{\rightleftharpoons}} EX \underset{k_4}{\overset{k_3}{\rightleftharpoons}} M + E \qquad 153$$

and

$$F + E \underset{k_2}{\overset{k_1}{\rightleftharpoons}} EX \underset{k_4}{\overset{k_3}{\rightleftharpoons}} EY \underset{k_6}{\overset{k_5}{\rightleftharpoons}} M + E \qquad 154$$

Both of the above mechanisms give rise to the same steady state equation which is

$$v = -\frac{dF}{dt} = \frac{dM}{dt} = \frac{(V_F/K_F)F - (V_m/K_m)M}{1 + F/K_F + M/K_m} \qquad 155$$

and, accordingly, it is impossible to distinguish between the reaction schemes on the basis of kinetic studies alone. Shown in Figure 37 is a reciprocal plot of the initial velocities vs substrate concentration. The Michaelis-Menten constant for L-malate

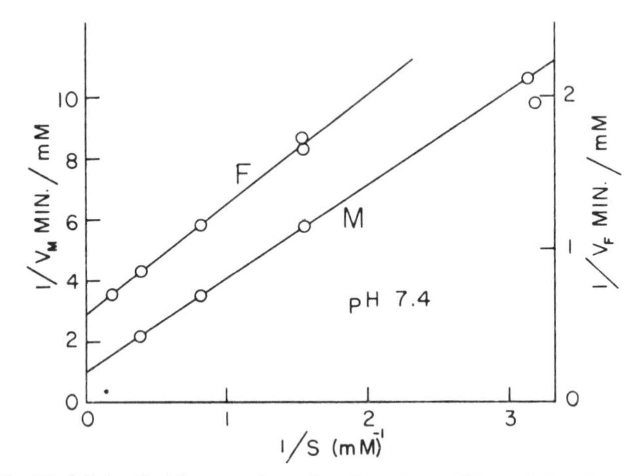

Fig. 37. Michaelis-Menten plots for L-malate (M) and for fumarate (F), pH 7.4, 0.05 phosphate buffer. (R. M. Bock and R. A. Alberty: J. Amer. Chem. Soc. *75:* 1921, 1953.)

at pH 7.4 in 0.05 M sodium phosphate buffer at 25° was estimated to be 4.81 ± 0.28 mM and the constant for fumarate, under the same conditions, 1.37 ± 0.14 mM. The kinetic constants of an enzymatic reaction are related to the equilibrium constant for the overall reaction by

$$K = \frac{V_F K_m}{V_m K_F} \qquad 156$$

where K_m and K_F are the corresponding Michaelis-Menten constants. Substitution of the kinetic data in the above gave K equal to 4.2 in agreement with the directly determined value of 4.45.

Depolymerizing Enzymes. There are a variety of enzymes capable of splitting biopolymers into monomers or into short segments. Among these may be mentioned ribonuclease acting on ribonucleic acid, amylase hydrolyzing starch and glycogen and various proteolytic enzymes attacking proteins. There are a few special problems associated with the degradation of biopolymers by enzymes which will be briefly reviewed at this point.

The essential difference between enzyme action on biopolymers and on ordinary substrates is that the product itself becomes a substrate. It can happen that the product is more readily hydrolyzed than is the original substrate. In this event, an all-or-none situation develops and the biopolymer once attacked appears to explode into small fragments. Some enzymes can act only on one end of a biopolymer removing one residue at a time, or the enzyme may attack specifically or randomly along the chain splitting the biopolymer into segments; evidently, the molecular weight distribution arising from these two modes of hydrolysis will differ significantly.

A possibility associated with end-wise attack is that of multiple end-splits before the biopolymer is released from the enzyme. This type of splitting would yield a different molecular weight distribution than would be expected from single attacks on different chains. Carboxypeptidase progressively hydrolyzes amino acid residues starting at the carboxyl end of the peptide chain. There are residues, however, such as proline, which prevent further hydrolysis. Nevertheless, carboxy-

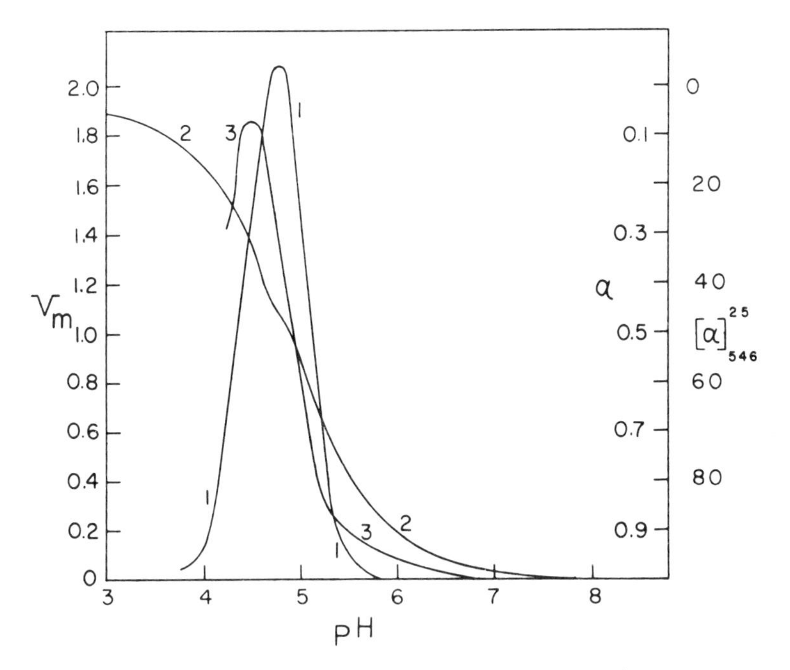

Fig. 38. Rate hydrolysis of poly α-L-glutamic acid by papain (curve 1, V_m) as a function of pH. Curve 2, Degree of ionization, α. Curve 3, Optical rotation, $[\alpha]^{25}_{546}$. [α] is a measure of the helix-coil transformation. (W. G. Miller: J. Amer. Chem. Soc. *83:* 259, 1961.)

peptidase has found use in the determination of the carboxyl terminal sequence of the amino acid residues; the sequence of release corresponds to the sequence of position. Evidently, a multiple attack on one chain at a time would lead to confusion as to the sequence of the residues in the chain. The action of β-amylase on amylose which is the linear form of starch would be more apt to fulfill conditions for a multiple attack than would carboxypeptidase since in starch the first attack does not alter the nature of the chain but merely changes its length. The character of the attack of an enzyme on a biopolymer can change with pH. For example, the average number of peptide bonds split by pepsin acting on egg albumin below pH 2.2 is 44 per mole of protein attacked, whereas above pH 2.5 the number is 84 and is independent of the time of the reaction.

The conformation of the biopolymers is important. For example, neither trypsin or chymotrypsin will digest a native globular protein and it is only after the protein has been denatured that hydrolysis proceeds. An interesting example of influence of the conformation of a peptide chain is afforded by the work of Miller who studied the hydrolysis of poly-α-L-glutamic acid by papain as a function of pH. Plots of V_m, of [α] (the specific rotation) and of α (the degree of ionization) as functions of pH are shown in Figure 38. It will be recalled that the variation in [α] indicates a helix-coil transformation. The poly-α-L-glutamic acid exists as a helix in the acid region and is transformed into a disordered coil as the pH is raised. It will be noticed from Figure 38 that the value of V_m rises sharply to a maximum at the pH at which the helical conformation is changing into the coiled form. No such sharp maximum is found with small molecular weight substrates. It is concluded that the papain attacks only the disordered coil with the groups adjacent to the peptide bond in an un-ionized condition.

Enzyme Concentration. The effect of substrate concentration on the velocity of enzymatic reactions has been detailed to some degree. Naturally, increasing the enzyme concentration, while maintaining the substrate concentration and other conditions constant, leads to increased reaction rate. In general, at lower enzyme concentrations the rate of reaction is directly proportional to the concentration of the enzyme; this is the region of zero order kinetics. As the enzyme concentration is still further increased the rate of increase of the reaction rate with increasing enzyme concentration diminishes. The steady state kinetics changes into transient state kinetics.

Mechanism of Enzyme Action. In principle, enzymatically activated reactions do not differ from non-enzymatic activation; the difference is a matter of degree. There are many examples of the rates of reaction being increased by the addition of ions or of small molecules. The catalytic efficiency of enzymic reactions is apt to be high. For example, a fumarase molecule saturated with fumaric acid hydrates the substrate to malic acid with a first order rate constant of 2×10^3 sec $^{-1}$ at 25°, whereas the corresponding rate constant in one molar acid is about 2×10^{-8} sec $^{-1}$; the enzymic reaction is about 10^{11} times more efficient than is acid catalysis. Also, enzymes tend to show a fairly narrow specificity. Some enzymes, for example, urease, show absolute specificity, whereas others such as pepsin exhibit hardly greater specificity than does hydrochloric acid acting on a protein at room temperature in respect to the peptide bonds hydrolyzed. Experimental work reveals three features common to enzymatic reactions: (1) Acid-base catalysis appears to function in most enzyme systems. (2) The enzymic reaction is divided into a considerable number of separate and individual steps; the activation energy for each step is characteristically low. (3) The enzyme undergoes cooperative conformational changes during the enzymic reaction, so-called allosteric transformations. In the end, the mechanism of enzyme activation will, no doubt, turn out to be a special chapter in the general subject of the mechanism of organic reactions.

There are certain characteristic and more or less obvious features which the enzymatically active center of the enzyme molecule must possess. The active spot must be on the surface of the protein molecule to permit ready access by the substrate molecules. Evidence indicates that the active spot is small relative to the bulk of the enzyme molecule and probably involves not more than 4 or 5 amino acid residues in the enzyme (the total number of amino acid residues may be several hundred). The catalytic power of the enzyme is closely related to the tertiary structure of the enzyme; the activity of the enzyme is destroyed by denaturation and is much more sensitive to denaturation than are the immunological properties. This means that the enzymic activity does not reside in a single sequence of a few amino acid residues in a single peptide chain but rather requires the presence of at least one fold in the peptide chain of the enzyme such that residues far removed in sequence are brought in juxtaposition.

Enzymes generally can distinguish between D and L isomers of a substrate and this would require at least a three point attachment of the substrate to the enzyme. The active spot must provide a complemental fit for the substrate and, since most enzymic reactions are bi-substrate, there have to be two such fits; further, the two binding sites must be immediately adjacent to bring the substrate molecules into a position favorable for a reaction to occur. It appears likely that whereas the configuration of the active spot must be accurately defined it must also have a measure of flexibility. Incidental to the activation and reaction of the substrate molecules, a significant volume increase of the substrate molecules is to be expected and the active spot would have to accommodate the volume change.

Molecular forces of various kinds would be expected to participate in enzymic reactions. Electrostatic forces are certainly important; the rates of many enzymic reactions are sensitive to the ionic strength of the medium. The so-called hydrophobic bond is no doubt operating in the reactions of hydrophobic substrates. Hydrogen bonding could also provide a measure of stability for the enzyme-substrate complex. The most important forces, however, are the quantum forces leading to covalent bonding of the substrate to the enzyme.

The complete amino acid sequence of ribonuclease is known without, however, defining the active spot. The amino acid sequence in at least part of the active spot of several enzymes has been explored. The nerve gas diisopropyl-fluorophosphate combines stoichiometrically with a number of enzymes and deactivates the enzyme. The enzyme is subjected to partial hydrolysis and the peptide containing the diisopropylphosphate isolated. The isopropylphosphate is always attached to a serine residue and the amino acid residues in the vicinity of the serine have been identified. Table 1 shows the amino acid residues present on either side of the reactive serine residue for those enzymes which have been investigated.

Apparently, serine plays a key role in the catalytic activity of the enzymes listed in Table 1. There has been considerable speculation as to the function it performs; it is possible that the alcohol group of serine provides a point of attachment of the co-substrate, a water molecule.

TABLE 1. SERINE ENZYMES. RESIDUES ADJACENT TO REACTIVE SERINE ARE SHOWN.

Enzyme	*Amino Acid Residue Sequence*
Pseudocholinesterase	Glu Ser Ala
Liver aliesterase	Glu Ser Ala
Alkaline phosphatase	Glu Ser Ala
Thrombin	Asp Ser Gly
Trypsin	Asp Ser Gly
Chymotrypsin	Asp Ser Gly
Elastase	Asp Ser Gly
Subtilisin	Thr Ser Met
Phosphoglucomatase	Ala Ser His

GENERAL REFERENCES

Amdur, I. and G. G. Hammes: Chemical Kinetics. McGraw-Hill Book Co., New York, 1966.

Cleland, W. W.: The kinetics of enzyme-catalyzed reactions with two or more substrates or products. Biochem. Biophys. Acta *67:* 104, 173, 188, 1963. (A general discussion of enzyme kinetics by Cleland is to appear in The Enzymes, Vol. 1, 3rd ed., to be published by Academic Press, New York, in the fall of 1970.)

Dixon, M. and E. C. Webb: Enzymes. Academic Press, New York, 1958.

Frost, A. A. and R. G. Pearson. Kinetics and Mechanisms, 2nd ed. John Wiley and Sons, New York, 1961.

Jencks, W. P.: Catalysis in Chemistry and Enzymology. McGraw-Hill Book Co., New York, 1969.

Monod, J., J. Wyman and J. Changeux: On the nature of allosteric transitions: a plausible model. J. Mol. Biol. *12:* 88, 1965.

PROBLEMS

1. What are the various ways by which the rates of chemical reactions can be expressed? Evaluate each method in terms of its theoretical and experimental convenience.

   Ans: See text.

2. The half-life of carbon 11 is 20.35 minutes. How long would it take for 95 percent of a sample of this isotope to disintegrate?

   Ans: 87.8 minutes.

3. The concentration of compound A in a reaction mixture at pH 7.0 and at 25°C was studied as a function of time with the following results:

Time in Hours	*Concentration in Moles per Liter*
0	0.100
1.26	0.089
2.84	0.077
4.50	0.066
6.28	0.056
8.65	0.045
9.89	0.040
10.48	0.038
11.49	0.035
12.69	0.032
14.13	0.029
16.91	0.025
19.03	0.022

Formulate a possible mechanism for this reaction and determine the orders and velocity constants.

Ans: First to second order, $K_1 = 2.56 \times 10^{-5}$ sec^{-1}, $K_2 = 6.20 \times 10^{-4}$ liter per mole per second.

4. (a) Explain why a 10 degree rise in temperature will frequently double the velocity of a chemical reaction whereas the kinetic energy increases only a few percent.
   (b) Draw a diagram showing the relation between stable states of reactants and products and the activated complex in terms of energy.
   (c) What is the relation between the specific velocity constant of a reaction and the equilibrium constant between the reactants and the activated complex?
   (d) Outline an experiment which would enable one to determine the entropy of activation of a chemical reaction.

Ans: See text.

5. Egg albumin solutions were heated at 65°C and at 70.2°C. The weight of soluble protein per cubic centimeter remaining at various time intervals were measured with the following results:

Hours	*mg. Soluble Protein at* 65°	*Minutes*	*mg. Soluble Protein at* 70.2°
0	7.77	0	6.75
5	7.37	30	6.13
10	7.00	80	5.25
20	6.32	120	4.68
30	5.72	200	3.72
40	5.22	300	2.76

Calculate (a) the velocity constant of the heat denaturation of egg albumin, (b) the free energy, (c) the entropy and (d) the heat of activation per mole of egg albumin at 65°C. The molecular weight of egg albumin is 45,000.

Ans: (a) 2.8×10^{-6} sec.$^{-1}$ (b) ΔF^* 28,600 cals
(c) ΔS^* 294 EU (d) ΔH^* 128,000 cals

6. The initial rates of decomposition of a substrate by an enzyme at concentrations sufficient to saturate the enzyme with substrate were determined as a function of pH with the following results:

pH	*v in Moles per Minute per Mole of Enzyme*
11	0.20×10^{-2}
10	0.45×10^{-2}
9	0.80×10^{-2}
8.5	0.88×10^{-2}
8.0	0.72×10^{-2}
7.0	0.28×10^{-2}
6.0	0.02×10^{-2}

Calculate the first and second ionization constants of the two acid groups in the active spot on the enzyme. What is the maximum velocity at complete activation of the enzyme by hydrogen ions but in the absence of inhibition by hydrogen ions?

Ans: pK_1, 7.40; pK_2 9.97; V 0.0099

7. Derive the expression for the velocity of an enzymatically catalyzed reaction in the presence of a non-competitive inhibitor. Show how it is possible to determine the Michaelis-Menten dissociation constant, the dissociation constant of the inhibitor–enzyme complex and the maximum velocity by the appropriate plots.

   Ans: See text.

8. One milligram of crystalline pepsin was dissolved in 100 cc. of each of three solutions containing 0.486, 0.953 and 2.97 grams of egg albumin at 25°C and at pH 2.0. The initial velocities of digestion of these three solutions were 0.117, 0.165, and 0.255 grams of egg albumin per hour respectively. The molecular weight of pepsin is 35,000 and that of egg albumin is 45,000. Calculate the maximum velocity of digestion in moles of egg albumin per second per mole of pepsin, as well as the Michaelis-Menten dissociation constant in molar quantities. What is the turnover number of pepsin at saturation?

   Ans: V_m 0.0627 K_m 1.61×10^{-4}

9. Doherty and Vaslow (J. Amer. Chem. Soc. *74:* 913, 1952) report that the enzyme chymotrypsin will catalyze the carboxyl oxygen exchange between water enriched with oxygen 18 and acetyl 3,5-dibromo-L-tyrosine without producing any other changes in the substrate. Outline an experimental procedure by which you would determine the number of active enzymatic spots on the chymotrypsin molecule. How would you calculate the dissociation constant of the enzyme-substrate complex? Would this dissociation constant be smaller or larger than that determined by Michaelis-Menten plots? Explain your answer.

   Ans: Read reference and text.

10. (a) The inhibition of an enzyme reaction by the reaction products is a fairly common finding. Using the simple Michaelis-Menten approach, derive an equation for the velocity of such a reaction; assume the enzyme is inhibited by a single reaction product.
    (b) What would a graph of $1/v$ against $1/S$ yield in case (a) above?

    Ans: (a) See text. (b) Straight line.

11. The initial rates of decomposition of urea by 1.5×10^{-8} mg. of urease in 100 cc. of reaction at 30°C were (mol. wt. urease = 483,000):

Urea Conc. Moles/Liter	*Initial Rates in Moles per Second*
.01	3.6×10^{-8}
.02	4.8×10^{-8}
.025	5.9×10^{-8}
.03	6.5×10^{-8}
.05	7.7×10^{-8}
.07	8.9×10^{-8}
.10	10.9×10^{-8}
.15	11.2×10^{-8}
.20	11.0×10^{-8}
.30	9.5×10^{-8}

Calculate max. vel. without substrate inhibition, the first and second association constants between urea and urease (molar basis).

Ans: V_m 7.35×10^9, K_1 0.074, K_2 0.278.

12. It has been found that cell free extracts of Escherichia coli catalyze the reaction

$$\text{Citrate} \xrightarrow[\text{Mg}^{++}]{} \text{Oxaloacetate} + \text{Acetate}$$

The following data* were obtained by incubating the reaction mixture for 30 seconds at 37°C in phosphate buffer at pH 7.0 with the enzyme concentration constant.

mM Citrate	*mM Mg^{++}*	*μM/Liter Keto Acid*
0.143	1.63	58.8
0.312	1.63	80.0
0.333	1.63	83.3
0.500	1.63	86.9
0.500	1.63	95.2
1.000	1.63	100.0
1.000	1.63	102.0
1.658	1.63	104.1
2.0	0.357	38.9
2.0	0.769	66.6
2.0	1.00	90.8
2.0	1.33	117.5
2.0	2.00	142.9
2.0	4.00	200

CITRATE CONCENTRATION CONSTANT

mM Mg^{++}	*ml. 0.025 M CaCl$_2$ Added to 3 ml. Reaction Mixture*	*μM/Liter Keto Acid*
8.35	0	392
	0.1	321.5
	0.2	241.5
	0.3	205
	0.4	180.0
2.88	0	255.0
	0.1	169.0
	0.2	120.0
	0.3	92.5
	0.4	79.0
0.96	0	144.0
	0.1	70.0
	0.2	47.6
	0.3	37.6
	0.4	27.6

Calculate (a) the maximum velocity and (b) the Michaelis constant for the enzyme-substrate complex in the presence of 1.63 mM Mg^{++}. (c) Does the Mg^{++} activate by altering the Michaelis constant or the maximum velocity or both? (d) Determine the type of inhibition produced by Ca^{++} and calculate (e) the dissociation constant of the enzyme-inhibitor complex.

Ans: (a) V_m 3.64×10^{-6}; (b) K_m 1.6×10^{-4}; (c) rate activator; (d) competitive inhibitor; (e) K_i 5.0×10^{-4}.

*Data of S. Dagley and E. A. Dawes: Biochem. Biophys. Acta *17:* 177, 1955; Quantitative Problems in Biochemistry by E. A. Dawes, E. and S. Livingstone Ltd., Edinburgh, 1956.

CHAPTER 16

Elasticity and Structure

Molecular structure is the basis for molecular behavior and biological structure is the basis for biological function. Little information is available concerning the molecular forces which direct the formation of specific biological configurations; the exploration of this area is for the future. The tools available at the present time for the investigation of the organization of living cells at a molecular level are extremely limited.

This chapter will first outline the separation of phases since the beginning of the formation of structure is the establishment of separate phases. The properties of gels with special emphasis on their rigidity and elasticity will be reviewed followed by a section on the elasticity of protein fibers. Methods and ways of observing biological structures at or approaching molecular dimensions will be summarized.

Phase Separations. There is a natural tendency for two liquid phases or a liquid and a solid phase to mix when brought into contact just as two gases tend to mix; the entropy of the system is increased by the mixing process. The entropy of mixing of n_1 solvent molecules with n_2 solute molecules is

$$\Delta S = -k[n_1 \ln V_1 + n_2 \ln V_2] \qquad 1$$

where k is Boltzmann's constant and V_1 and V_2 are the volume fractions of the components, solvent and solute respectively. However, if the interaction between molecules in one or both phases is sufficiently large as compared with the interaction energy between the two kinds of molecules in the two separate phases, there will be no mixing.

A most intriguing phase separation occurs when certain high polymers are dissolved together in water. For example, two phases separate in approximately equal volumes when 0.36 percent methylcellulose and 1.1 percent dextran are dissolved in water. The upper phase contains 0.39 percent dextran, 0.65 percent methylcellulose and 98.96 percent water. The lower phase contains 1.58 percent dextran, 0.15 percent methylcellulose and 98.27 percent water. The two aqueous phases are in equilibrium with each other with a very low surface tension between them. It is possible to separate large molecules such as those of a virus between the two phases. Other high polymers besides dextran and methylcellulose can be used to bring about aqueous phase separation and in the appropriate system it is even possible to distribute male and female bacteria between two phases.

In the distribution of particles and molecules between two phases, the larger the particle size of the partitioned substance, the more one sided the distribution. The approximate formula to describe the distribution is

$$K = e^{\frac{M\lambda}{kT}} \tag{2}$$

where K is the partition coefficient, M the molecular weight, k is the Boltzmann constant, T the absolute temperature and λ is a constant characteristic of the substance and the phase system. Counter current distribution has long been used to separate and purify materials of interest. Immiscible liquid phases are employed in a large number of connecting tubes; in effect, the system consists of a series of separatory funnels.

Coacervation is another example of phase separation. For example, if gelatin is mixed with gum arabic in solution at pH 3.5, complex coacervation occurs and viscous droplets separate out. At this pH, gelatin is positively charged and gum arabic is negatively charged. If the coacervate is placed in an electric field, the gelatin moves towards the negative pole and the gum arabic towards the positive pole; the two materials retain their individuality. In order for a coacervate to form, the electrostatic free energy tending to produce association has to exceed the contribution due to the entropy of mixing which acts in the opposite sense. However, there is more to the phenomenon than the operation of these two factors; a coagulum is not formed. Both the gelatin and the gum arabic are firmly hydrated (they are both so-called hydrophilic colloids) and the particles of opposite sign cannot come in direct contact with each other.

As noted above, in general, the entropy of mixing tends to produce solution; however, there are situations where the entropy of a system is increased by a separation of a solid phase from a solution. Consider, for example, a solution of rod-like molecules such as these of tobacco mosaic virus protein. If the solution is so concentrated that the molecules cannot execute random rotation freely, there will be a gain in entropy of the system if some of the solute molecules precipitate in an ordered arrangement. There is a dilution of the solution thereby and the molecules in solution can acquire full random rotary movement.

Possibly the above effect is responsible for tactoid formation. Tactoids are large boat-like or spindle-shaped particles which have been observed to form in concentrated solutions of iron oxide and of vanadium pentoxide, and also in solutions of several dyes. Tactoids are optically anisotropic as revealed by the polarizing microscope. This indicates that they are composed of oriented particles and the direction of orientation coincides with the long axis of the tactoid; tactoids are one dimensional crystals. Tobacco mosaic virus protein exhibits well-defined tactoids. The larger the virus tactoids, the more nearly they approach a spherical form; the smaller they are, the more cylindrical they become.

Gels. Gels are characterized by rigidity; biological structures aside from the bones and teeth are gels of varying degree of rigidity. Gels result when there is sufficient degree of cross bonding between molecules to form a three dimensional network throughout the system; it is characteristic that water and other small molecules are dispersed in the spaces between the network. It is evident that the units forming the gel must contain tri- or higher functional units in addition to such bifunctional units as may be present otherwise; it is impossible for a three dimensional structure to form; bifunctional units can form extremely viscous solutions but not gels. Thus polymerizations which are propagated by the reactions of bivalent or bifunctional molecules lead to soluble, diffusible products whereas the incorporation of units of higher functionality permits formation of gelled or insoluble products. Consider a substance containing groups A and B. A—A and B—B represent bifunc-

A

tional units, whereas A (branched to A and A) represents the trifunctional unit. AB or BA is the prod-

A

uct of the condensation of two functional groups. Polymerization of bi- and trifunctional units is shown in Figure 1.

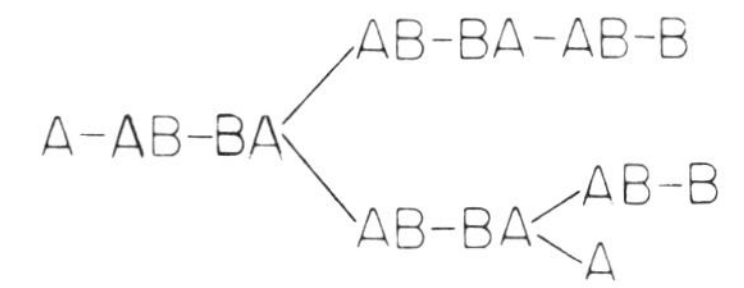

Fig. 1. Polymerization of bi- and trifunctional units.

Let α be the probability that a chain leading from a trifunctional unit eventually leads to another branched unit. If $\alpha < 1/2$, the infinite extension of a network is impossible since a given chain has less than an even chance of reproducing two new chains. On the other hand, if $\alpha > 1/2$, branching of successive chains may continue the network indefinitely. The condition of α equals 1/2 is, therefore, critical for the formation of infinite structures or for the formation of a gel. The equation for the general case is

$$\alpha_c = \frac{1}{f-1} \qquad 3$$

where f is the functionality of the branching units interpolymerized at random with bifunctional units. α_c is the critical probability above which a gel may form and below which no gel can form. As proteins, in general, contain many reactive groups, the possibility of gel formation is large.

Ordinarily, when a solution changes to a gel there is little change in gross physical and chemical properties of the system other than the obvious appearance of mechanical structure. The electrical conductance through a gel in the presence of electrolytes is substantially equal to that in the liquid condition. It has even been shown that the electrophoretic mobility of quartz particles is the same through a gelatin gel as through a gelatin solution.

On the basis of volume changes, it is possible to group gels into three classes. Thixotropic gels show no volume change when gelation or solution occurs at a fixed temperature. The second type, exemplified by gelatin or agar, displays a small decrease of volume when setting; heat is evolved in the process. The third type, which is represented by aqueous solutions of methylcellulose, shows a small increase in volume and an absorption of heat during gelation. Thus, it is found that gels of the second type set upon cooling, whereas those of the third type liquefy as the temperature falls. A mixture of gelatin and methylcellulose dissolved in water separates into two phases. The lower phase is largely gelatin, and the upper largely methylcellulose. Upon cooling such a separated system, the lower phase sets to a gel, while the upper phase containing the methylcellulose remains liquid. Heating above 50° produces the opposite effect; the upper phase sets, and the lower phase liquefies.

Thixotropy. Some gels when mechanically agitated undergo a reversible, isothermal gel-sol transformation; the gel re-forms when the agitation is stopped. The word thixotropy is derived from the Greek "thixis" meaning a touch and "trope" meaning a turn, to change.

Almost any gel will show a certain degree of thixotropy, but some gels show this property to a particularly striking degree. The clay bentonite is very convenient for the demonstration of thixotropy. The particles are very asymmetric, being long, thin plates. Bentonite is derived from volcanic dust, and its main constituent is the mineral montmorillonite; it is one of the few inorganic substances that swells in water. To produce a thixotropic bentonite gel, all that is needed is to mix with the right amount of water and shake in a test tube; depending on the concentration of the clay, the time of gelatin can be varied at will.

Swelling of Gels. A characteristic feature of dry proteins is their tendency to imbibe water and to swell. Dry isoelectric gelatin when placed in water imbibes water and swells to a definite limit, whereas other proteins, such as egg albumin, swell indefinitely and eventually disperse completely. Proteins swell as a result of two factors: (1) The polar groups on the protein molecules are hydrated, and water is bound to the proteins by hydrogen bonds. (2) The entropy of mixing of the water and protein is positive; the greater degree of randomness of the mixture makes the mixing a spontaneous process.

The cross-links between gelatin molecules prevent the solution of the gelatin and confine the swelling to a definite limit. If the temperature is increased sufficiently, the weak intermolecular bonds are broken and the gelatin disperses completely; however, the bonds re-form when the temperature is lowered. The solution of a gel at higher temperature is roughly analogous to the melting of a crystal, although the solution temperature is far from sharp and the solution temperature is always somewhat higher than the gelation temperature. The melting point of a gelatin gel increases considerably as the concentration is increased.

The swelling pressure of a gel can be calculated from the lowering of the vapor pressure of the liquid associated with the gel. Consider again the relation between the vapor pressure lowering and the osmotic pressure as discussed in Chapter 7, Osmotic Pressure and Related Topics. On the basis of the maximum work principle, it is found that

$$P_S\overline{V}_1 = RT \ln \frac{P_0}{P} \qquad 4$$

where P_S is the swelling pressure, P_0 is the vapor pressure of pure solvent and P is the vapor pressure of the solvent in equilibrium with the gel. $\overline{V}_1$ is the partial molar volume of the solvent. The swelling pressures exerted by proteins at low moisture content are impressive. For example, isoelectric salt-free gelatin shows swelling pressures of 3,900, of 2,200 and 600 atomspheres per square centimeter for 5, for 10 and for 20 grams of water per 100 grams of gelatin respectively at 25°.

It will be recognized that the term $P_S\overline{V}$, in eq. 4, is equal to the free energy change involved in the hydration of a gel. The free energy change can be resolved into a heat term and into the entropy of swelling in analogy with the resolution of the osmotic pressure discussed in Chapter 7. It is then possible from the entropy contribution to formulate a relation between the maximum extent of swelling and the average molecular weight of the gel molecules between the cross-links. The maximum extent of swelling when the gel is in equilibrium with the pure solvent is

$$qV_2^2 + \ln (1 - V_2) + V_2 = -\rho_2 V_1 V_2^{1/3} M_c \qquad 5$$

where V_2 is the volume fraction of polymer at equilibrium, V_1 is the molal volume of the solvent, p_2 is the density of pure polymer and q a coefficient characterizing the degree of interaction between the solvent and the polymer. M_c is the average molecular weight between the cross-links in the polymer.

If the pH of a completely swollen isoelectric gelatin gel is shifted either to the acid or basic side of the isoelectric point, additional swelling is observed. The reason for the increased swelling is undoubtedly due to the electrical charges on the gelatin and to the development of a Donnan equilibrium. As pointed out in Chapter 7, a

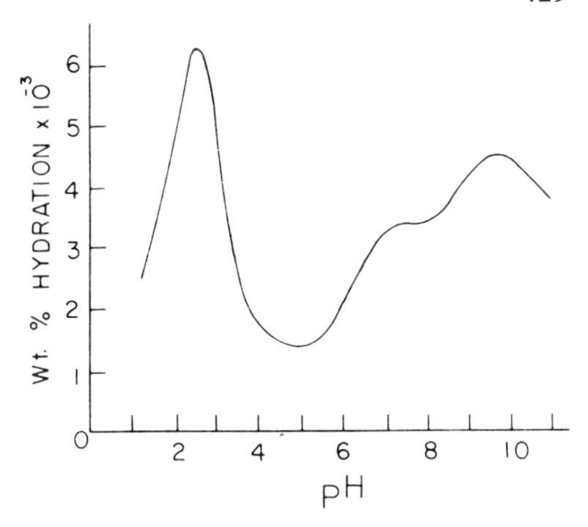

Fig. 2. Weight percent uptake of water by dry gelatin as a function of pH. HCl or NaOH added to produce the desired pH. No other electrolyte present. (Adapted from D. Jordan Lloyd and W. B. Pleass: Biochem. J. *21:* 1352, 1927.)

Donnan equilibrium leads to an increased osmotic pressure because of the accumulation of electrolytes inside the gel, this accumulation being brought about by the electrically charged groups on the protein. The gel then swells in response to the increased osmotic pressure. Figure 2 shows the water uptake by dry gelatin as a function of the pH of the bathing solution. The addition of neutral salt greatly decreases the water uptake which results from Donnan swelling.

Rigidity of Gels. When a physician thumps the chest of a patient he is, in his own way, obtaining a measure of tissue rigidity. More precisely, the behavior of gels is described in terms of their moduli of rigidity and elasticity. Figure 3 shows a stress applied to a block of gel resulting in a given deformation without change of volume. The modulus of rigidity, G, is given by the relation (see Fig. 3)

$$G = \frac{\text{stress}}{\text{strain}} = \frac{F}{A \tan \alpha} \qquad 6$$

where A is the area of the top face of the block of gel, F is the stress in dynes acting and α is the angle of deformation.

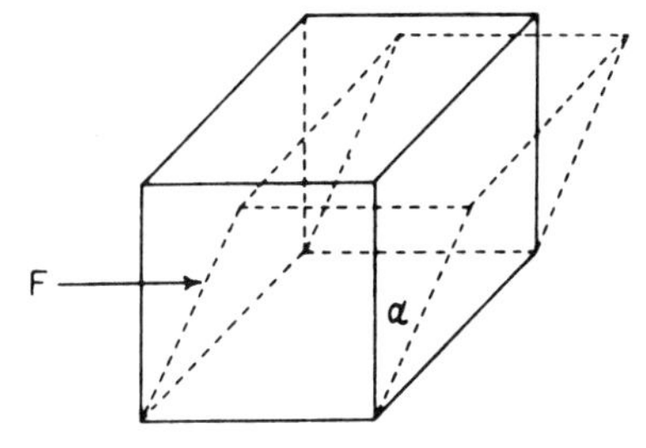

Fig. 3. Deformation of a gel resulting from a stress, F. This figure is the basis for the definition of the modules of rigidity.

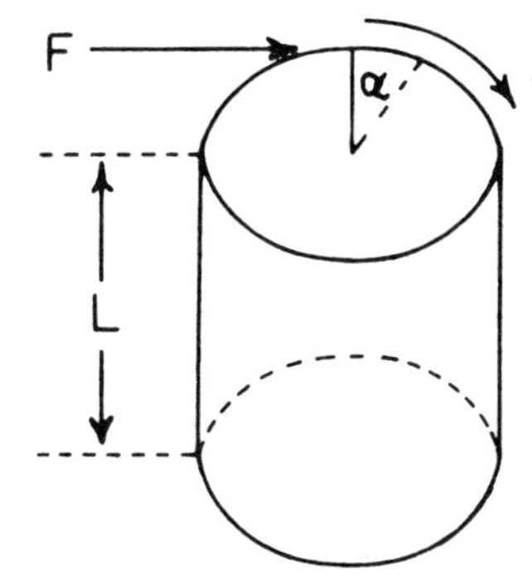

Fig. 4. Bar of gel twisted through an angle α.

The modulus of rigidity can also be defined in terms of the torque required to twist a bar of gel as shown in Figure 4 and the modulus of rigidity is given by

$$G = \frac{\text{stress}}{\text{strain}} = \frac{2FxL}{\pi r^4 \theta} \qquad 7$$

where x is the distance in centimeters from the center of the bar of gel to the periphery and is equal to r, the radius of the bar. Fx is the torque applied which produces a twist of θ radians in a bar of length L. The dimensions of the modulus of rigidity are force per unit area or pressure. In Figure 5 is shown a diagrammatic sketch of an apparatus capable of measuring the modulus of rigidity of a gel.

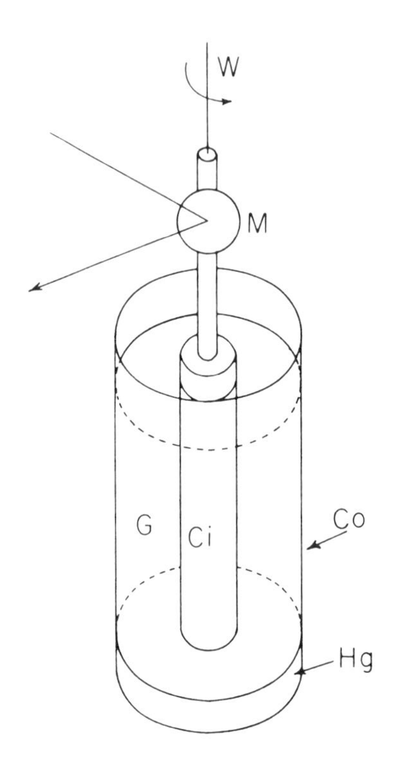

Fig. 5. Diagrammatic sketch of apparatus for the measurement of rigidity of gels. W, Steel wire; M, mirror for optical lever; C_i, inner cylinder suspended by steel wire, W; C_0, outer stationary cylinder containing gel, G; Hg, mercury at bottom of outer cylinder into which inner cylinder dips to reduce end effects.

The apparatus diagramed in Figure 5 resembles the arrangement for a Couette viscometer. The modulus of rigidity is calculated from the experimental data by the relation

$$G = \frac{\text{stress}}{\text{strain}} = \frac{T}{4\pi L} \left(\frac{1}{r_1^2} - \frac{1}{r_0^2}\right) \frac{\phi - \omega}{\omega} \qquad 8$$

where T is the torque per unit twist of the wire, L is the height of the gel, r_i and r_0 are the radii of the inner and outer cylinders respectively, ϕ is the angle through which the torsion head has been turned and ω is the angle through which the inner cylinder turns.

The viscosity as well as the relaxation time of a gel must be considered. The rotation of a bar of pure liquid without structure would require the exertion of a force, but as soon as the displacement had been achieved the stress would fall to zero and there would be no tendency for the bar of liquid to return to its former position. The stress on a rigid body required to maintain a given strain decays with time; the rigid structure relaxes. The time of application of the stress must, therefore, be short compared with the relaxation time if the initial modulus of rigidity is to be measured.

Provided the relaxation time, defined as the time required for the stress at a given strain to decay to $1/e^{th}$ part of the initial stress, is long enough, the modulus of rigidity as well as the rate of relaxation of the gel can be measured with the apparatus shown in Figure 5. Frequently, however, the relaxation time of a gel is short, a matter of a fraction of a second, and some kind of dynamic loading of the system has to be resorted to. For example, the inner cylinder (Fig. 5) is suspended from a wire of known stiffness and the outer cylinder is oscillated with a known amplitude and frequency. It will be observed that the strain lags behind the stress. The stress can be resolved into two components, one in phase with the strain and the other 90° out of phase. The ratios of these stresses to the observed strain give two moduli of rigidity. The one in phase is the real component and the 90° out of phase is the imaginary component and the complex rigidity modulus is given by

$$G^* = G' + iG'' \qquad 9$$

The phase angle, γ, between the stress and strain is related to the real, G', and the imaginary, G'', moduli by

$$\tan \gamma = \frac{G''}{G'} \qquad 10$$

The modulus of rigidity can also be calculated from the velocity of transverse waves through the medium, the velocity of such shear waves being related to the modulus of rigidity by

$$V = \left(\frac{G}{\rho}\right)^{1/2} \qquad 11$$

where ρ is the density of the gel. By a suitable optical arrangement using polarized light, shear waves through an optically clear gel can be visualized; shear induces birefringence in the gel. The velocity, amplitude, and attenuation of the waves can

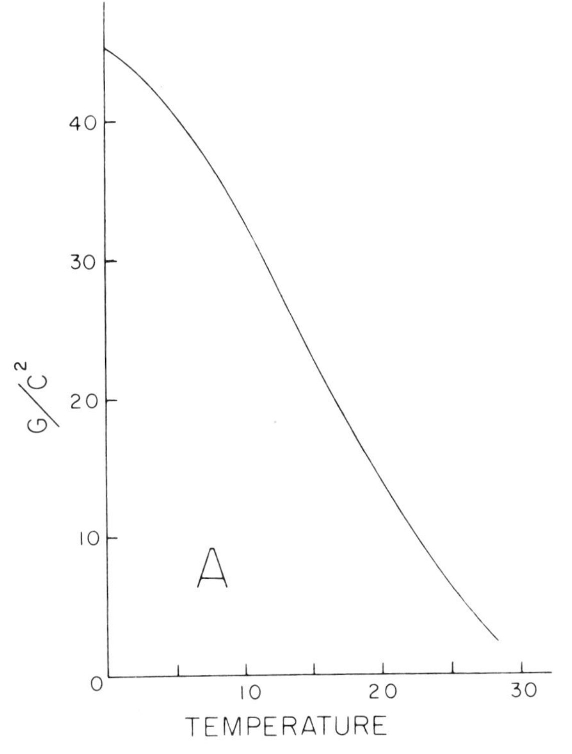

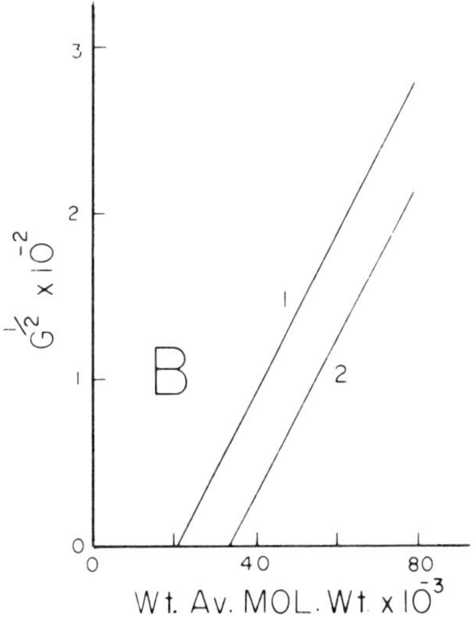

Fig. 6. Modulus of rigidity, G, in dynes per square centimeter of gelatin gels in water. *A*, The ratio G/C^2 as a function of the temperature. C, the concentration of the gelatin, is expressed in grams per liter. *B*, $G^{1/2}$ as a function of the weight average molecular weight of the gelatin. Curve 1, 5°. Curve 2, 15°. (Adapted from J. D. Ferry: J. Amer. Chem. Soc. *70:* 2244, 1948.)

be directly measured and related to the modulus of rigidity. What is badly needed in this difficult area is a micromethod suitable for living cells and parts of cells. Figure 6A shows the modulus of rigidity of gelatin gels as a function of temperature and Figure 6B as a function of the molecular weight of gelatin in water.

When thrombin is added to a solution of fibrinogen a clot forms. The nature of this clot (opaque or transparent) is dependent on a number of factors such as pH and electrolyte concentration. The time of clotting is also a function of several factors. Shown in Figure 7 is a plot of the increase in opacity and rigidity of fibrinogen undergoing clotting.

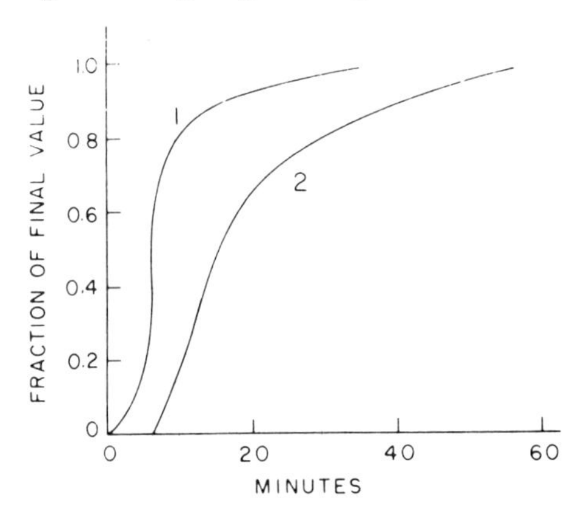

Fig. 7. Opacity (curve 1) and rigidity (curve 2) plotted against time, during clotting of solution of fibrinogen. 16 grams fibrinogen per liter, 1.0 unit thrombin per milliliter, ionic strength 0.3, pH 6.33. (J. D. Ferry and P. R. Morrison: J. Amer. Chem. Soc. *69:* 388, 1947.)

Elasticity of Gels. The modulus of stretch or of elasticity is defined as

$$E = \frac{\text{stress}}{\text{strain}} = \frac{FL}{A\Delta L} \tag{12}$$

where F is the stress in dynes, A is the cross-sectional area and L the length of the fiber. ΔL is the increase in length of the fiber produced by the stress. The modulus of elasticity is also known as Young's modulus of stretching and has the dimensions of pressure. In the region where the modulus is independent of the strain, Hook's law applies; Hook's law states that the strain produced in a body is directly proportional to the stress applied. Provided Hook's law is obeyed, the modulus of elasticity is the force required to extend a bar of unit cross-sectional area by 100 percent. Curiously, the physical definition of elasticity is contrary in sense to the popular meaning. To express the concept of stretchability, the term compliance is used; the modulus of compliance is equal to the reciprocal of the modulus of elasticity.

Many fibers do not obey Hook's law over an appreciable stretch and a more satisfactory expression for the elastic modulus at a given degree of stretch, L, is

$$E = \frac{L}{A}\frac{dF}{dL} \tag{13}$$

The stress at a given stretch, in general, decays with time and for a single elastic element

$$E = E_0 e^{-t/\tau} \tag{14}$$

where τ is the relaxation time of the stress and E_0 is the modulus at time zero. For two or more elastic elements in parallel

$$E = E_1 e^{-t/\tau_1} + E_2 e^{-t/\tau_2} + \tag{15}$$

where E_1, E_2, etc., are the moduli at time zero for the individual elastic elements. If the stress is maintained constant, the fiber stretches as time passes; this behavior is known as creep.

The modulus of elasticity is closely related to the velocity of longitudinal waves (compression waves) and V, the velocity of such waves, is given by

$$V = \left(\frac{a + 2G}{\rho}\right)^{1/2} \quad 16$$

where a is equal to $E/(1+\mu)\ (1-2\mu)$. μ is Poisson's ratio and is the ratio of the lateral contraction to the longitudinal extension; for most materials the ratio is about 0.3 to 0.5. Unlike shear in which there is no volume change, compression and extension result in a change in the total volume.

The viscosity of a series of elastic elements is related to the elastic moduli and to the respective relaxations by the equation

$$\eta = \frac{1}{2(1+\mu)}\left[E_1\tau_1 + E_2\tau_2 + E_3\tau_3 + \cdots\right] \quad 17$$

Often it is of help in dealing with elastic problems to resolve them into springs and viscous jackpots such as is done in Maxwell and Voigt elements as shown in Figure 8.

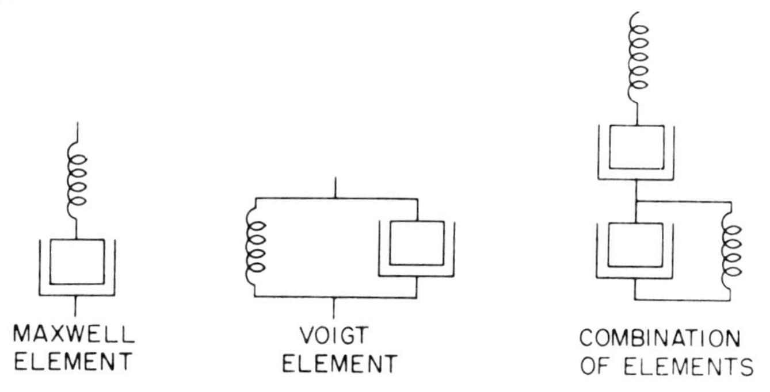

Fig. 8. Combination of springs and viscous jackpots to represent elastic problems.

The approximate elastic moduli of some common materials are shown in Table 1.

TABLE 1. ELASTIC MODULI OF SOME COMMON SUBSTANCES

Substance	*Temperature*	E_0
Copper wire	20°	11×10^{11}
Steel wire	20°	20×10^{11}
Cellophane	20°	4.2×10^{11}
Silk	20°	1.0×10^{11}
Rubber		
E_1	20°	1×10^{11}
E_2	20°	2×10^{6}
Hair (human)	25°	3.6×10^{9}
Ice	0°	0.98×10^{11}
H_2O	20°	1×10^{11}

The elastic modulus of glass changes only about 2 percent in the temperature interval from 0° to 500° whereas τ, the relaxation time, decreases sharply with increasing temperature. At very low temperatures, the elastic moduli of most substances are of the order of 1×10^{11} to 1×10^{12}.

If the relaxation time is very short, a body can exhibit a high modulus of elasticity which decays rapidly; this behavior is shown by "bouncing putty" which can be deformed by slow application of stress as can ordinary putty (a plastic body); yet if the bouncing putty is dropped on the floor it will bounce like a rubber ball. One can attain any quality of elastic behavior by variation in the modulus, the relaxation time, and the viscosity.

Molecular Basis of Elasticity. The random coil is the most probable form of a completely flexible molecule, the average distance between the ends of a random coil being (Chapter 6, Biopolymers)

$$\bar{d} = \ln^{1/2} \tag{18}$$

where l is the length of the links between the flexible joints in the chain. The maximum length of a long stretched fiber molecule is

$$d_{\max} = nl \sin \theta/2 \tag{19}$$

where θ is the average valence angle. The ratio between the root mean square distance for a random coil and the maximum distance the molecule can stretch is

$$\frac{\bar{d}}{d_{\max}} = \frac{n^{1/2}}{n \sin \theta/2} \tag{20}$$

or the ratio $\bar{d}/d_{\max}$ is approximately equal to $l/n^{1/2}$.

A random coil owes its existence to the thermal motion of the flexible molecules and the force, f, required to extend the coil to its maximum stretched length is

$$f = 3kTn\left(\frac{\bar{d}}{d_{\max}^2}\right) \tag{21}$$

where k is Boltzmann's constant and T is the absolute temperature.

There exist cross-links between the isoprene chains in vulcanized rubber and the cross-links hold the neighboring chains together and prevent slippage of the chains relative to each other as the rubber is stretched. In between the cross-links, the isoprene chains are very nearly randomly arranged and, accordingly, the force exerted by a piece of stretched rubber is due to the heat motion of the chains.

From a molecular point of view there are two kinds of elasticity: (1) rubber-like noted above and (2) normal body elasticity such as shown by a steel wire. In the first instance, the heat motion of the molecules is responsible for the elastic behavior just as the heat motion of gaseous molecules is responsible for the elastic behavior of gases. However, heat motion tends to expand a gas whereas heat motion tends to contract a fiber. Normal body elasticity results from the distortion of chemical bonds. Rubber-like elasticity is characterized by: (1) generation of heat upon stretching; (2) contraction when heated at a given stress; (3) an increase in the stress exerted by stretched rubber as the temperature is increased.

The stress at constant length (isometric) can be resolved into two components: (1) the force due to the heat motion of the molecules and (2) that due to the deformation of valence bonds. The stretching of a fiber results in the ordering of the molecules with a decrease in entropy. The stretch also leads to an increase in the internal energy of the fiber because of the distortion of the chemical bonds. It is then possible to write

$$F = F_E + F_s \tag{22}$$

where F is the total stress exerted by the fiber and F_E and F_s are the parts of the stress resulting from the increase in internal energy and from the decrease in entropy respectively. For a reversible stretch, the force multiplied by the change in length is equal to the free energy change of the process. The Helmholtz free energy change, ΔA, is given by (contrast with the Gibbs free energy, Chapter 2, Energetics)

$$\Delta A = \Delta E - T\Delta S = F\Delta L \tag{23}$$

The change in the internal energy, ΔE, is equal to $\Delta H - P\Delta V$ where ΔH is the enthalpy, P is the pressure on the system and ΔV is the change in volume of the system. F is the stress on the fiber, ΔL is the change in fiber length and ΔS is the change in entropy. Rearranging eq. 23 gives

$$F = \frac{\Delta E}{\Delta L} - T\left(\frac{\Delta S}{\Delta L}\right) \qquad 24$$

If $\Delta E/\Delta L$ and $\Delta S/\Delta L$ are independent of temperature, the differential of the stress in respect to temperature gives

$$\frac{dF}{dT} = \frac{\Delta S}{\Delta L} \qquad 25$$

Substitution of eq. 25 into eq. 24 gives

$$F = \left(\frac{\Delta E}{\Delta L}\right)_T - T\left(\frac{dF}{dT}\right)_L \qquad 26$$

This expression is known as the Wiegand-Snyder equation. The slope of the plot of the stress at constant stretch against the temperature multiplied by the absolute temperature gives the stress, F_s, due to the entropy change: F_E is obtained by difference between the observed stress, F, and F_s. Experiments by which F_E and F_s can be estimated are difficult to perform; the observed stress must be an equilibrium value. The resolution of stress of biological fibers in terms of internal energy and entropy is generally associated with the name of E. Wohlisch who was the first to undertake work of this character.

From eq. 21 it is clear that there should exist a relation between stress due to the entropy change and the length of the stretched fiber. By methods of statistical mechanics it has been shown that for randomly linked cross-linked chains

$$F = \alpha\left(L - \frac{1}{L^2}\right) \qquad 27$$

where F is the stress per unit area of the unstretched fibers, α is a coefficient related to the molecular weight of the flexible segment and for such kinked chains is equal to RTW/M where W is the weight of the elastic material per cubic centimeter, R is the gas constant and T is the absolute temperature. For randomly arranged rigid

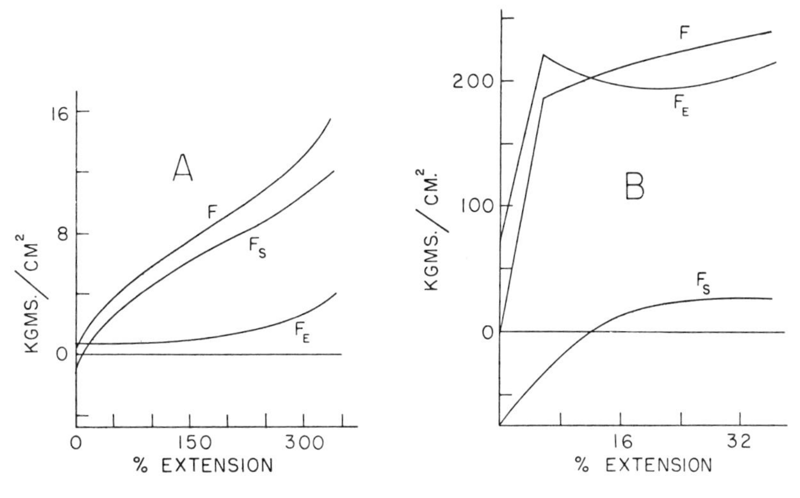

Fig. 9. Stress vs percent extension at 25°. *A*, Rubber; *B*, human hair saturated with water.

rods connected end to end by flexible links and imbedded in an isotropic fluid matrix, α is equal to $RTW/5M$.

Shown in Figure 9 is the resolution of the stress of vulcanized rubber and of human hair into F, F_E and F_S.

Elasticity of Protein Fibers. The animal body makes use of a number of proteins for structural purposes; these are all fibrous proteins and with the exception of myosin and actin from muscle are extracellular and, again with these exceptions, are insoluble in aqueous solutions unless subjected to special solubilizing techniques. By suitable treatment, practically any globular protein can be spun into fibers. Artificial protein fibers are not without interest but, for present purposes, discussion is to be limited to silk fibroin, keratin, resilin, elastin and collagen.

Table 2 shows the stress of keratin, collagen, elastin and resilin resolved into that due to internal energy, F_E and into the entropic part, F_S.

TABLE 2. STRESS RESOLUTION FOR FOUR PROTEIN FIBERS (L IS THE RATIO OF THE STRETCH LENGTH TO THE ORIGINAL LENGTH. THE STRESSES ARE EXPRESSED IN KILOGRAMS PER SQUARE CENTIMETER AT ROOM TEMPERATURE.)

Protein	L	F	F_E	F_S
Keratin	1.00	0.0	75	−75
(human hair)	1.08	197	215	−18
	1.25	228	200	+28
Collagen	1.032	1.0	5.0	−4.0
	1.042	6.4	21.7	−15.3
	1.051	15.1	28.1	−13.0
Elastin	1.00	0.0	−19.7	+19.7
	1.25	3.4	−30.2	+33.5
	1.52	9.3	−17.5	+26.9
Resilin	1.1	1.8	2.8	−1.0
	1.4	5.5	2.9	+2.6
	2.0	15.0	2.0	+13.0

A brief review of some of the properties of polypeptide chains (see Chapter 6, Biopolymers, and examine Figs. 3 and 4) will serve as an introduction to a consideration of the fibrous proteins. A stretched polypeptide chain is shown diagrammatically in Figure 10.

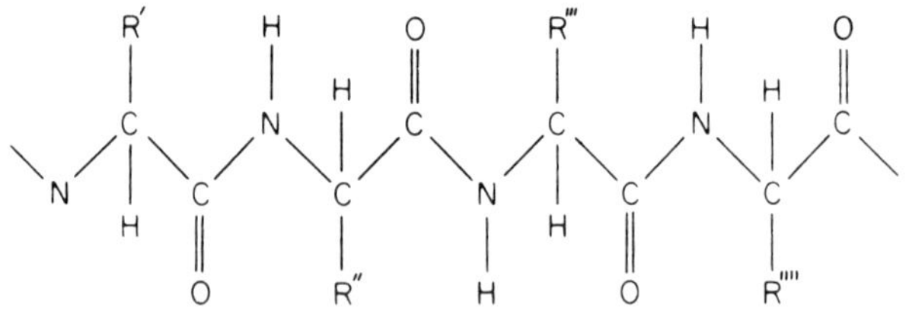

Fig. 10. Diagrammatic representation of a stretched polypeptide chain. *R* indicates amino acid side chains.

The peptide bond —C(=O)—N(H)— has considerable double bond character and, accordingly, is not free to rotate and, furthermore, the groups attached to it are forced into a planar configuration. Rotation of the —N(H)—C(R)— and the —C(R)—C(=O)— bonds is permitted, but if the R-groups are large severe limitations on their movement can be imposed. Peptide chains in the solid state do not exist alone and attention has to be paid to the relation of peptide chains to each other.

Pauling and Corey proposed two pleated sheet structures for an assembly of stretched peptide chains. In these structures the neighboring chains run in the same direction (parallel chains) or in opposite directions (anti-parallel). Interchain hydrogen bonds hold the chains together in sheets. The bond angles of the peptide chains lead to puckered configurations and hence the name pleated sheet. In Figure 11 is a diagrammatic representation of a pleated sheet configuration.

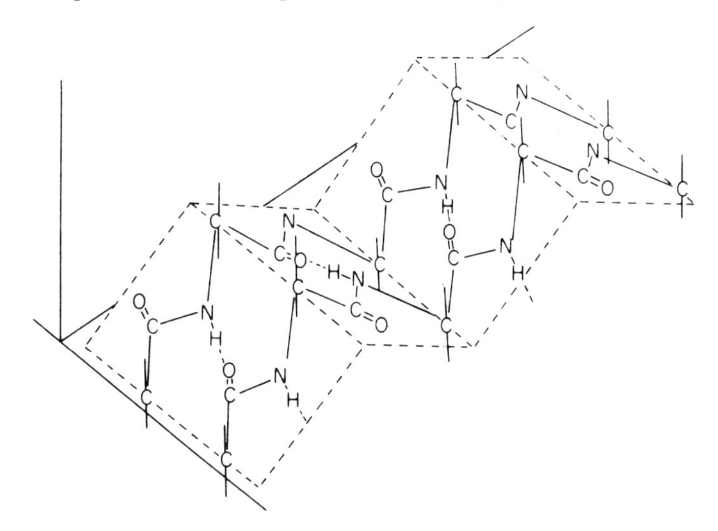

Fig. 11. Pleated sheet configuration of stretched parallel peptide chains.

Whereas the peptide chains in a single sheet are held together primarily by hydrogen bonds between $\rangle$CO and $\rangle$NH groups of neighboring chains, the intersheet forces are of the Van der Waal type. It is evident that long range elastic stretch of peptide chains in an already stretched condition is not to be expected.

Peptide chains can also exist in the form of helices and a brief discussion of the α-helix was given in Chapter 6. This structure contains 3.6 amino acid residues per turn of the helix and is stabilized by the formation of hydrogen bonds between neighboring carboxyl and amino groups adjacent to the peptide bonds as the chain coils back on itself. The conversion of an α-helix to a completely stretched peptide chain would provide for about 117 percent stretch and could thus serve as a basis for long range elastic behavior. There is a strong tendency for polypeptide chains of a fibrous protein in the α-helical configuration to associate together to form super helices containing two or more chains with a slow twist around a common axis.

In addition to the well-recognized α-helix, there exists a helix of the form shown by poly-L-proline which occurs in collagen and in gelatin. Collagen, as will be noted presently, contains considerable quantities of proline and hydroxyproline. Proline is an imino acid containing a five membered ring. The introduction of a proline residue into a peptide chain interrupts an α-helix, and if the peptide chain contains appreciable quantities of proline a helix of a different kind results. Two kinds of helices are possible, one with the peptide bond in a trans configuration and the other with the peptide bond in a cis configuration. Shown diagrammatically in Figure 12 are these two possibilities. A peptide chain which contained substantial quantities of proline or of hydroxyproline, due to the difficulty of rotating the proline residues, would be expected to be fairly stiff and would become flexible only at higher temperatures.

In addition to the stretched peptide chains, α-helices and the poly-L-proline

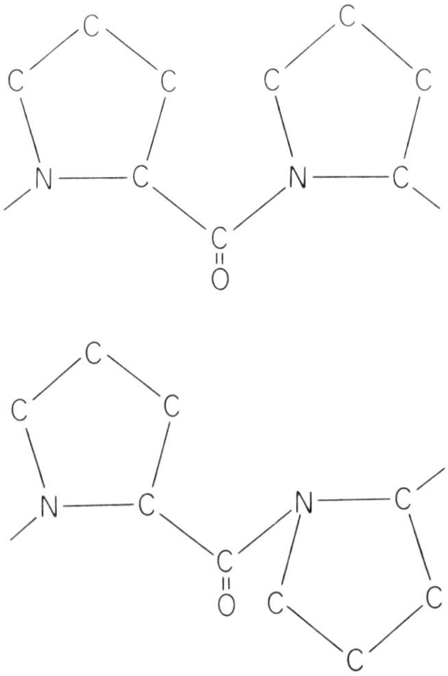

Fig. 12. Showing cis and trans configurations about the peptide bond as in poly-L-proline.

helices which can lead to a high degree of crystallinity in the peptide fiber, there are, however, amorphous forms in which there is no ordered arrangement of the chains; the polypeptide chains due to steric interference do not fit well together.

Silk Fibroin. Historically, silk played an important role in the development of ideas about the structure of fibrous proteins as well as of proteins in general; it is ideal as a model because of its comparatively simple amino acid composition (see Table 3). The earliest X-ray diffraction studies were done on silk fibroin.

Raw silk can be separated into two fractions by the action of hot water. The soluble fraction is called sericin (silk gelatin) and the insoluble portion is the fiber silk fibroin which makes up about 80 percent by weight of the raw silk. Actually, there are several varieties of silks, the silk spinning moths producing those which are commercially important. Silk produced by spiders usually does not contain sericin.

The cocoon thread of Bombycidae mori consists of two fibroin filaments cemented together by sericin. In the silk gland, the soluble viscous fibroin is distinct and separate from the sericin and the fiber is extruded from a pair of spinnerets, coated and glued together with sericin. The cross-sections of the filaments are approximately an equilateral triangle with rounded corners; the average diameter of the fiber of two filaments is from 5 μ to 25 μ.

Silk fibroin dissolves easily in cupriethylene diamine solution as well as in concentrated solutions of lithium bromide or thiocyanate. The salts can be dialyzed out to yield soluble silk fibroin with the peptide chains in a folded configuration. The molecular weight of the soluble silk has been variously reported as being from 30,000 to 300,000.

Silk fibroin is characterized by high content of glycine and alanine as well as a considerable amount of serine, some tyrosine, and small amounts of a number of other residues. The amino acid content of two varieties of silks is shown in Table 3. The various silks differ somewhat in composition mainly in the relative amounts of glycine and alanine. The high content of the smaller glycine and alanine residues permits a close and comfortable fit of neighboring peptide chains and provides the possibility of a solid structure consisting of stretched polypeptide chains with a relatively high degree of crystallinity (about 40 percent). In the crystalline regions the fiber consists of antiparallel pleated sheets. The relative absence of extensibility of the fiber under high stress can thus be understood.

TABLE 3. AMINO ACID CONTENT IN RESIDUES PER 10^5 GRAMS SILK

Amino Acid	*Bombyx*	*Tussah*
Glycine	581	318
Alanine	334	529
Serine	154	140
Tyrosine	70	58
Aspartic acid	21	57
Valine	31	7
Phenylalanine	20	6
Arginine	6	31

Keratin. Keratin has a far more complex composition and structure than does silk fibroin. The outstanding feature of keratin is its high sulfur content as well as a large assortment of amino acid residues many of which are bulky. Two types of keratin can be distinguished on the basis of its physical character, i.e., soft and hard. The outer layer of the epidermis is an example of soft keratin and horn, wool, hair and feathers contain hard keratin. Soft keratin has its sulfur mainly in the form of sulfhydryl groups whereas in hard keratin the SH groups have been oxidized to form disulfide cross-links between peptide chains.

Human hair has many attractive features as a material for elastic investigations. It is, however, a complex biological structure. It is composed of two main parts, the cuticle and the cortex; the cuticle forms a sheath around the cortex, the cortex is the shank of the hair fiber. The cuticle makes up 10 to 20 percent of the hair fiber and is composed of shingle-like flat cells which show no birefringence and play no significant role in the elastic properties of hair.

The cortex of the hair fiber has spindle-shaped cells arranged with their principal axis parallel to the fiber axis. The cells are about 120 μ in length and a few micra in diameter and contain fibrils which are parallel to the fiber axis. The fibrils have indistinct longitudinal strictions. It has been shown that, when hair is stretched, the cortical cells also stretch and by nearly the same proportion as the fiber; the elastic properties of hair reside in the cortex.

The disulfide bonds cross-linking the peptide chains (about every eighth amino acid residue in keratin is cystine) can be ruptured by oxidation with per acetic acid and about 70 percent of the fiber becomes water soluble; the molecular weight of the soluble fraction is about 70,000.

Both hair and wool in a wet condition show long-range extensibility with little or no plastic flow; the disulfide bonds of cystine serve to "vulcanize" the fiber and the fiber returns to its original length after release of the stress. Figure 13A shows a stress-strain curve of a wet human hair; note the hysteresis loop between extension and contraction.

Astbury and Street showed many years ago from X-ray diffraction studies on wool that a definite relation exists between the molecular structure of keratin and the extent of stretch of the protein fiber. Ordinary native hair in an unstretched condition exists as α-keratin which has a low degree of crystallinity. It appears likely that it consists principally of grossly distorted α-helices; the high content of disulfide cross-links as well as the presence of various sized amino acid side chain residues would make the fit and folds between neighboring peptide chains something less than ordered. As the fiber is stretched, it is progressively converted to a β-keratin form (more or less fully extended peptide chains). The conversion of α to β keratin is complete after the fiber has been stretched about 70 percent of its original length. The tight cross-linking between the peptide chains in keratin would be expected to produce a highly viscous fiber with considerable hysteresis between stretch and contraction (see Fig. 13A).

In addition to the α-β transformation noted above, keratin fiber stretched

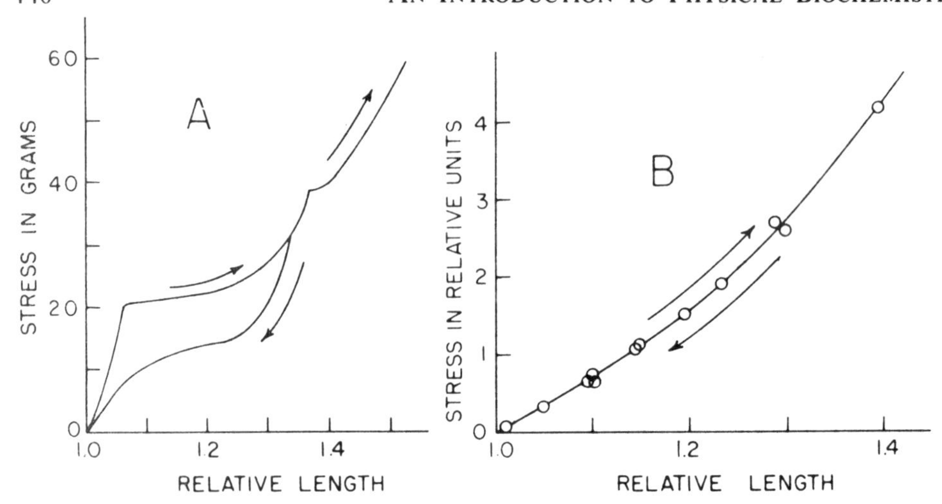

Fig. 13. *A*, Extension and contraction curve of human hair; *B*, extension curve of ligamentum nuchae which is completely reversible. (G. C. Wood: Biochem. Biophys. Acta *15:* 311, 1954.)

rapidly in steam and held stretched for about two minutes and then released while still in steam will contract to about two thirds of its original unstretched length. Apparently, the peptide chains have folded in an amorphous condition which is known as the supercontracted form of keratin.

Thus, the elastic properties of keratin can be understood at least in broad outline in terms of the chemical composition of the fiber containing, as it does, frequent cross-linking (disulfide bonds) along with amino acid residues of various sizes which tend to produce a poor fit between neighboring peptide chains. Hair owes its elastic properties mostly to change in internal energy rather than to entropic changes (see Table 2 and Fig. 9).

Collagen. Collagen is the most abundant protein in the animal body and is present in cartilage, tendons, ligaments, the matrix of bone, underlayers of the skin and blood vessels and provides the intracellular binding substance in muscle and in other organs; collagen is the predominating fiber protein of white connective tissue. Collagen can be solubilized and converted into gelatin. Due to the industrial importance of collagen and gelatin and to medical interest in connective tissue diseases, intensive research has been conducted on collagen.

Collagen has a characteristic amino acid composition. About one third of the total residues are glycine, about one eighth are proline and about one in ten is hydroxyproline. Table 4 gives the amino acid content of pigskin collagen.

TABLE 4. PARTIAL AMINO ACID ANALYSIS OF PIGSKIN COLLAGEN EXPRESSED IN NUMBERS OF RESIDUES PER 10^5 GRAMS PROTEIN

Amino Acid	*Residues*
Glycine	357.0
Proline	142.8
Alanine	121.4
Hydroxyproline	104.6
Glutamic acid	78.9
Arginine	52.8
Aspartic acid	51.3
Lysine	28.7
Leucine	26.0

Collagen treated in the cold with weak organic acids, for example, citrate buffer, pH 3.0-4.5, dissolves to form tropocollagen (also called procollagen). The tropocollagen molecule is rod-like with a length of about 2,800 A-units and a diameter of about 15 A-units. The molecule consists of three peptide chains, two α-chains and one β. The molecular weights of the α and β chains differ from each other and probably their amino acid compositions are also different. The molecular weight of the tropocollagen molecule is about 320,000. The question of the exact molecular weight is difficult to settle because of the tendency of collagen to degrade under relatively mild conditions and, under slightly different conditions, to associate.

Tropocollagen molecules in a suitable ionic environment will reassemble themselves into a fiber structure apparently identical with the structure of the original fiber. If heated, tropocollagen denatures to form gelatin. If collagen is autoclaved or treated with hot dilute acids or bases, the collagen is solubilized and gelatin produced directly. Figure 14 illustrates the sequence of reactions.

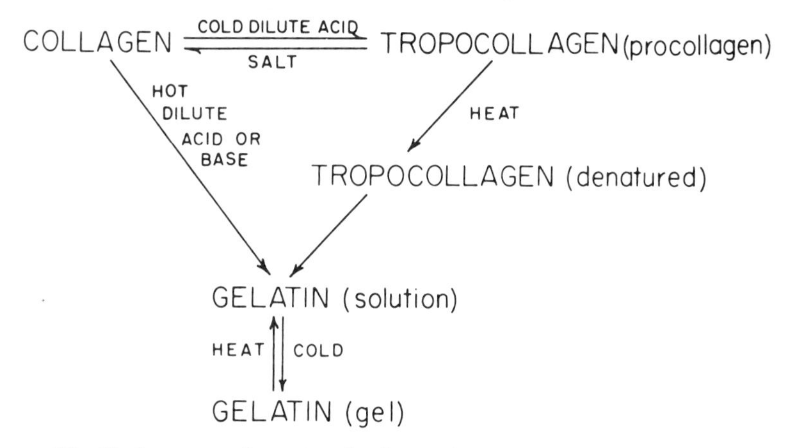

Fig. 14. Sequence of reactions leading to the production of tropocollagen and of gelatin.

Collagen fibers exhibit a relatively high degree of crystallinity and a number of distinct X-ray diffraction spacings are observed, the most characteristic of which is the 2.86 A-unit spacing along the fiber axis. The spacing corresponding to the distance between the peptide chains, 11-15 A-units, increases with the water content of the fiber. Small angle X-ray diffraction reveals a strong reflection along the fiber axis of 640 A-units.

Electron micrographs of collagen stained with phosphotungstic acid show a pronounced cross banding of the fiber. The length of the bands averages 644 A-units composed of a less dense band of about 200 A-units and a more dense band of 444 A-units in length. The cross banding of a collagen fiber is diagramed in Figure 15. There has been considerable speculation concerning the origin of the cross banding of collagen. It was suggested by R. S. Bear that the longer, denser, portions of the bands represent less tightly packed and more amorphous material which swells easily and into which the phosphotungstic acid penetrates and hence becomes

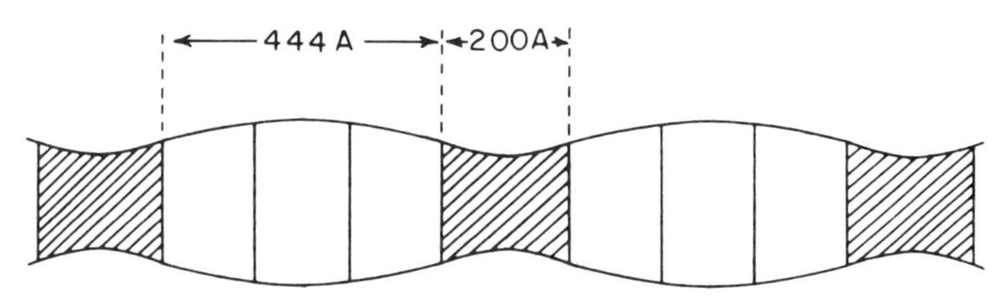

Fig. 15. Diagrammatic representation of a collagen fibril based on electron-micrograph after staining with phosphotungstic acid (negative print).

opaque to electrons. The less dense (after staining) is much more compact and crystalline. The alignment of the collagen molecules to form the fiber and to give rise to the bands is due to a regular and suitable periodic arrangement of the amino acid residue sequence in the peptide chains in the collagen.

The molecule contains alternating regions dominated restively by polar and non-polar amino acid residues.

It has been established that the collagen molecule consists of three helices of the poly-L-proline type with three residues per turn of the helix wound around a common axis giving a 2.86 A-unit translation between a residue on one chain and the corresponding residue on another chain. Tropocollagen in its native banded structure is mostly arranged so that one molecule in respect to a second one is displaced laterally by one quarter of its length with molecular "heads" all pointed in the same direction. Tropocollagen is a complex molecule. The single peptide chains are $\alpha 1$ and $\alpha 2$ which differ in amino acid content. Mild treatment with hydroxylamine (hydroxylamine ruptures ester bonds) splits $\alpha 1$ into five and $\alpha 2$ into seven sub-units. Two α-chains of about 100,000 molecular weight each combine to give the β-component; the β-component may be either $(\alpha 1)_2$, $\alpha 1\ \alpha 2$ or $(\alpha 2)_2$. The β-component then combines with another α-chain to produce the γ-component which is the tropocollagen molecule. When gelatin is formed from the collagen, the helices are separated from each other and become disordered coils in solution at higher temperatures. Upon cooling below a temperature which is dependent on the type of collagen (for many collagens the temperature is between 35° and 40°) the helices reform and the gelatin sets to a gel. However, the helices now have a random arrangement in respect to each other. There must, however, be fairly strong interaction between the helices to account for the rather rigid gels which result at concentrations above 0.5 percent. The helices, no doubt, become entangled with short segments of neighboring helices with a certain amount of interwinding of the helices. When gelatin sets to a gel the optical rotation undergoes a large negative increase.

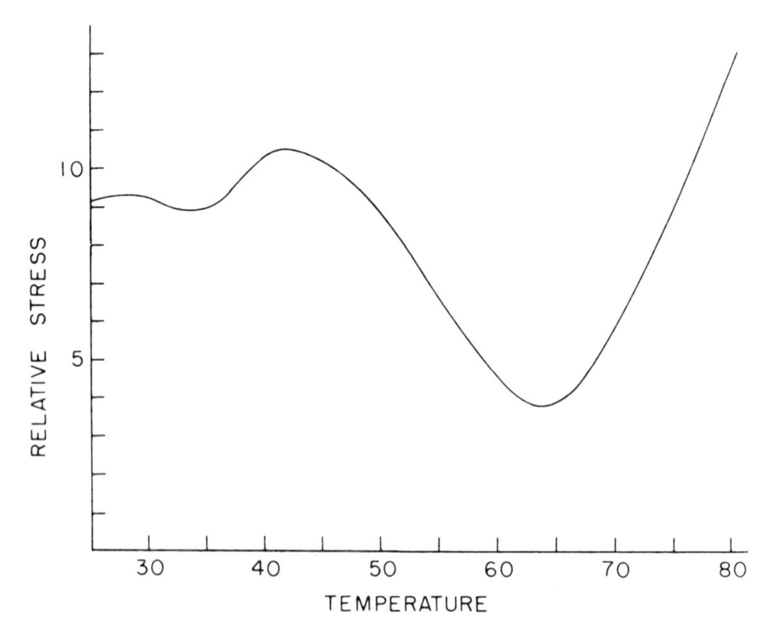

Fig. 16. Stress as a function of temperature. Collagen fibril from 7 year old human male. Achilles tendon. Initial stress of 3×10^5 grams per square centimeter at 25°, held at constant length, temperature raised one degree per minute; pH 7.77; ionic strength, 0.10. (Adapted from C. D. Nordschow: Ph.D. Thesis, University of Iowa, 1964.)

Native collagen shows limited extensibility at room and body temperatures. From the stress resolution (Table 2) it appears that the internal energy is responsible for the elastic modulus with the entropy term favoring extension. Behavior of this kind is to be expected of a fiber showing the high degree of crystallinity which collagen exhibits. As the temperature of a collagen fiber under stress is raised, the stress at constant length undergoes a fairly complex course and, at about 60°, the collagen melts; if stress is removed, the fiber shortens to about 1/3 of its native length. The temperature of melting depends somewhat on the rate of heating and also on the type of collagen, the pH and ionic environment. The melted fiber is distinctly rubber-like in its elastic behavior. It has been mentioned that gelatin undergoes a transition at about 40°. The melting of collagen is converted into disordered coils. Figure 16 shows the variation of stress of a collagen fibril as a function of the temperature.

Figure 17 shows the melting temperature of unstressed collagen as a function of pH of the solution in which the fiber was immersed.

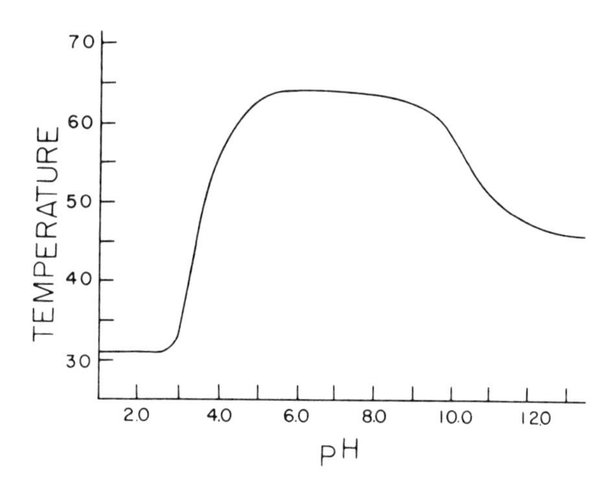

Fig. 17. Melting temperature of unstressed collagen as a function of pH. Ionic strength, 0.10; 12 year old human female; Achilles tendon. (Adapted from C. D. Nordschow: Ph. D. Thesis, University of Iowa, 1964.)

Elastin. Elastin predominates in yellow connective tissue although most connective tissue contains a mixture of elastin and collagen. Thus, the collagen content of the Achilles tendon is about 20 times that of the elastin, whereas in the ligamentum nuchae the elastin content is about 5 times that of collagen. The fenestrated membrane of the great aorta has a high content of elastin. Ligamentum nuchae can be extended in excess of 100 percent of its initial length and the stress-strain curve is reversible and independent of time up to a 50 percent extension (see Fig. 13B). Elastin is an exceedingly inert protein and is separated from other tissue proteins by mild alkaline or acid hydrolysis (the tissue is boiled with one percent acetic acid for about 16 hours). The amino acid content of elastin resembles somewhat that of collagen; both are high in glycine and in proline, although elastin contains little or no hydroxyproline and the content of hydrophobic residues in elastin is much larger than in collagen. X-ray diffraction studies show the elastin fiber to be completely amorphous with no order in the arrangement of the peptide chains. The entropic part of the stress is large (see Table 2); elastin is rubber-like in its elastic behavior.

The peptide chains are cross-linked with pyridinium derivatives containing amino acid side chains. The cross-links prevent the peptide chains from slipping past each other when subject to stress.

Resilin. Resilin is the so-called insect rubber occurring in the prealar arm and wing-hinge ligament of locust, the elastic tendon of dragon flies and no doubt in many other insects and other organs. Resilin has few cross-links between the peptide chains and the chain segments are about 60 amino acid residues long. The peptide chains of resilin are cross-linked with tyrosine residues. X-ray diffraction shows the fiber to be completely amorphous. In neutral buffer, the dry protein swells easily and the volume fraction of the protein in the wet tissue is about 0.45. Thus, the resilin fiber, in its behavior, resembles somewhat that which would be expected of a concentrated gelatin gel equipped with permanent covalent cross-links. Resilin, however, contains no hydroxyproline. As noted in Table 2, the entropic part of the stress is positive at greater extensions but it is not nearly so rubber-like as is elastin.

Whereas the relation between elasticity and structure of fiber proteins is not very clear-cut, certain generalizations can be made. One certain requirement for long-range rubber-like elasticity is disorder of the molecular chains. Related to this requirement is that the number of cross-links between the molecular chains must be minimal. Some cross bonding is, however, necessary; otherwise slippage between chains occurs on stretching. Hydrophobicity of the molecular chains per se does not lead to long-range reversible extensibility. In order to achieve disorder there should be a poor structural fit between the chains. For example, rubber and gutta percha are both isoprene chains. Rubber is the cis isomer and gutta percha is the trans isomer. Gutta percha is a relatively hard non-extensible solid; the trans chains fit well together. If a rubber band is stretched sufficiently it undergoes crystallization, and the isoprene chains align themselves in paralleled array as shown by X-ray diffraction studies. Thus, a state of parallel order can arise without the aid of significant intermolecular forces. It is hard to overemphasize the importance of water in the elastic behavior of protein fibers. The water acts to lubricate the process of extension. A completely dry protein fiber will rupture before it will extend to any considerable extent.

The final relation between protein structure and the elasticity of tissue is very apt to be most complex and not easily predictable. The microscopic structure of the tissue can certainly modify the elastic properties of the various elastic elements in a profound manner; consider a nylon stocking.

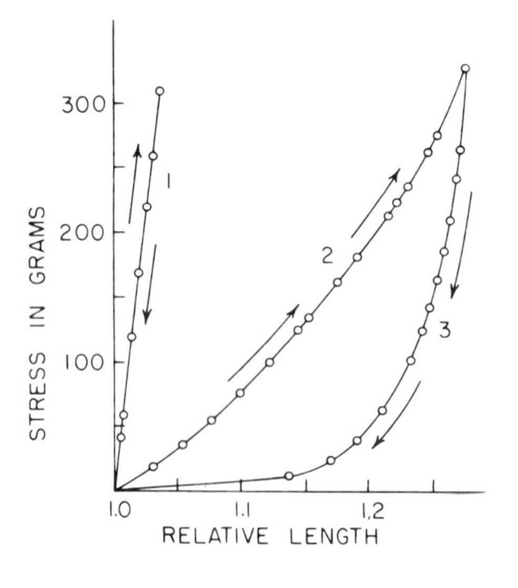

Fig. 18. Stress-stain curve for nylon thread and stocking. Curve 1, Thread. Curve 2, Extension of nylon stocking. Curve 3, Contraction of nylon stocking.

Straight nylon thread at room temperature is not significantly extensible and in this respect resembles collagen fibers. When, however, nylon thread is woven into a lady's stocking, the stocking acquires a two-way stretch which is impressive.

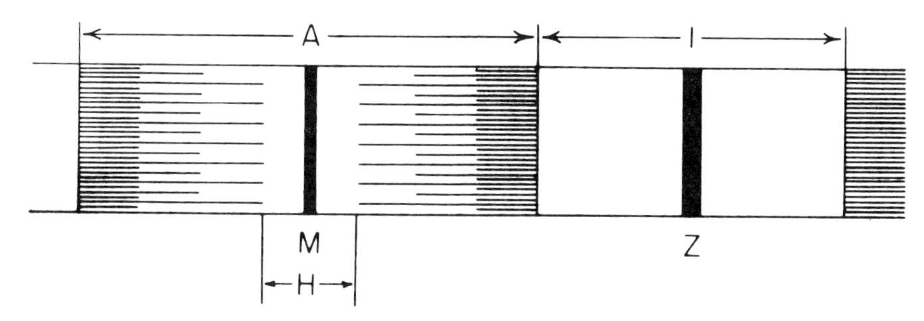

Fig. 19. Skeletal muscle sarcomere in resting unstretched state.

A circular section was cut from a nylon stocking perpendicular to the direction of the leg. The section was subject to stress in the direction of the cut; the stress-strain curve for the application and removal of stress is shown in Figure 18. As can be seen, the stretching of a nylon stocking exhibits a hysteresis loop and the curve resembles that which can be obtained by a peripheral stretch of the aorta of a young person.

Muscle Fibers. Muscle is a biological machine of remarkable character. Under favorable conditions it attains a mechanical efficiency up to 60 percent. It requires little attention and is subject to relatively few diseased states. The heart muscles of a man may continue working steadily and without mishap for 100 years. It is not suprising that muscle investigation has attracted numerous researchers from many areas and the literature has grown to voluminous and overwhelming proportions. The following section will attempt to express the problem in the simplest possible terms.

There are three kinds of vertebrate muscle: (1) smooth muscle located in the gastrointestinal tract, uterus, etc., (2) cardiac muscle of the heart which resembles skeletal muscle except that the fibers are branched and (3) striated or skeletal (voluntary) muscle. The discussion will be confined to striated muscle.

A skeletal muscle properly stained and viewed with a microscope is seen to be cross striated with fine dark bands about 2 microns in length. Upon enlargement and especially with the use of the electron microscope the fiber has the appearance shown in Figure 19. Distinct are the birefringent A bands and the isotropic I bands. Also clearly visible are the narrower M and Z bands along with the H regions.

Muscles are composed of thousands of fibrils. Individual fibrils show a threshold of stimulation on the all-or-none principle. The body achieves graded response by activating, at any instant, a certain fraction of the total number of fibrils in a muscle. A nerve controls 100 to 150 fibrils in striated muscle.

Upon stimulation of a muscle, there is a short latent period of the order of a millisecond followed by brief relaxation (the muscle lengthens). The tension then develops and the muscle shortens if allowed to do so. At constant length (isometric contraction), the tension reaches an early maximum with a slower decay to zero. The heat release exhibits a complex course involving two maxima as a function of time, and the birefringence of the muscle decreases as the twitch develops. Shown in Figure 20 are the relative magnitudes of heat release, the total birefringence and the tension exerted for a single isometric twitch as a function of time. There is also a complicated series of electrolyte changes. Potassium ions are released inside the fibril and exchange with sodium ions outside. There is a rapid and small decrease of pH followed by an increase and then a decrease. The increase of pH is due to liberation of creatine and the final acidification to lactic acid.

A relaxed muscle is soft and plastic and is easily stretched which makes it difficult to measure the length of a resting muscle. The resistance to stretch is about

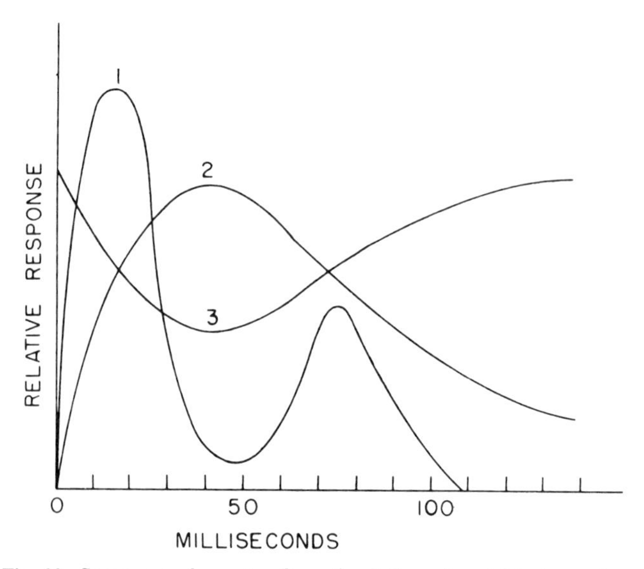

Fig. 20. Sequence of events after stimulation of a striated muscle. Isometric contraction. Curve 1, Heat liberated. Curve 2, Tension. Curve 3, Total birefringence.

100 grams per square centimeter. During contraction, the resistance to stretch increases ten and fifty fold and is several kilograms per square centimeter. The magnitude of the tension developed during isometric contraction depends on the length of the muscle when stimulated. Figure 21 shows the percent of the maximum isometric tension exerted by single muscle fibers from the semitendinous muscle of the frog (Rana pipiesus) as a function of the length at stimulation.

The contraction of a muscle is dependent on protein interactions. Muscle contains about 20 percent protein and muscle extracted with a dilute salt solution yields a considerable number of individual proteins. However, only two of these proteins, myosin and actin, have been implicated in the immediate process of contraction.

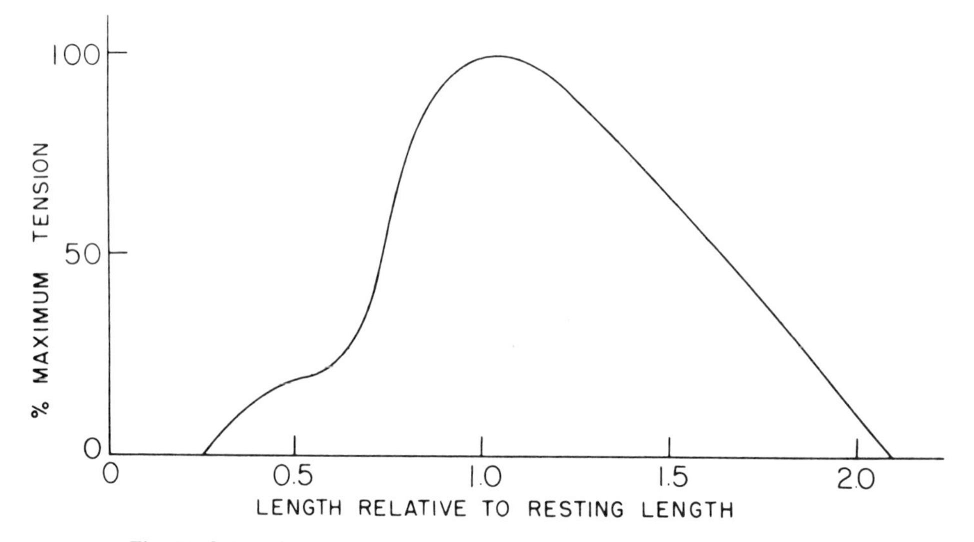

Fig. 21. Isometric tension as function of length of stimulated fiber. 11°, semitendinous muscle of frog. (Adapted from R. W. Ramsey and S. F. Street: J. Cell. Comp. Physiol. *15:* 11, 1940.)

Striated muscle contains two sets of filaments, the thicker consisting of myosin and the thinner of actin. The thicker filaments center in the A-band whereas the thinner extend out from the I-band. The two sets of filaments interdigitate. The thicker filaments in the A-band are spindle shaped about 170 A-units in diameter, with an overall length of 1.5 μ containing 200 to 400 myosin molecules parallel to each other and in a staggered configuration along the filament. The myosin molecule is rod-like about 1,600 A-units long of non-uniform diameter with a terminal globule. The diameter of the rod part is about 20 A-units and that of the globular part 40 to 50 A-units, extending 150 to 200 A-units along the molecule. The molecular weight of myosin is about 500,000. Trypsin acts on myosin to produce a water insoluble fibrous protein called light meromyosin and a more soluble globular protein known as heavy meromyosin. The heavy meromyosin retains the adenosinetriphosphate activity of the original myosin, whereas the light meromyosin is devoid of enzymic activity. The light fraction has a high content of α-helices whereas the heavy part is helical to only a moderate extent.

The monomer of actin is known as G-actin and is nearly spherical, with a diameter of about 55 A-units and a molecular weight of about 57,000. Upon addition of 0.1 M KCl and traces of Mg^{++}, G-actin polymerizes into F-actin (fibrous form) which in turn can be depolymerized back into G-actin by treatment with excess ATP. In muscle, each filament of actin (the thin filaments) consists of two interwined strands of F-actin with the polymerized spherical beads of G-actin clearly visible with the electron microscope. When F-actin is added to a myosin solution, immediate combination between the two proteins occurs; the result is known as actomyosin, which forms a highly viscous solution. The addition of excess ATP leads to a dissociation of the actomyosin with a significant drop in the viscosity of the solution; the actomyosin is depolymerized.

Whatever else a muscle does, it does work and, accordingly, needs a fuel supply. A muscle does not require oxygen to contract nor does it need oxygen to recover from contraction. However, if muscle is repeatedly stimulated in the absence of oxygen, it builds up an oxygen debt which must be paid eventually if the muscle is to continue its activity. To furnish the energy necessary for muscle work, muscle glycogen is broken down by anaerobic glycolysis to generate high energy phosphate bonds which are stored in creatin phosphate and released to ADP to produce ATP as needed. The concentration of ATP in muscle is held at a low level. The glycogen under anaerobic conditions ends up as lactic acid which diffuses out of the muscle fiber and into the blood fairly rapidly. It is believed that the hydrolysis

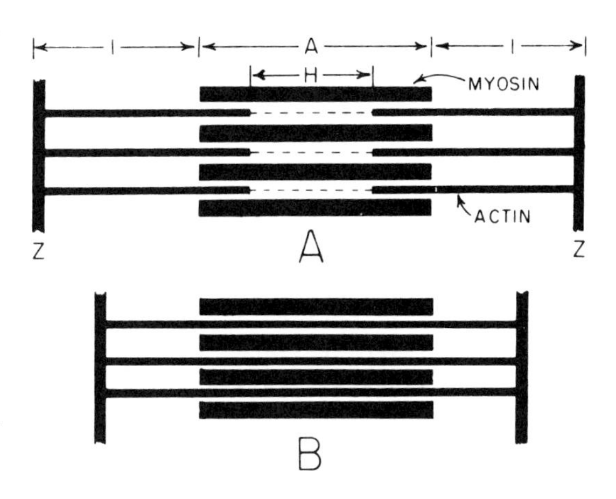

Fig. 22. Diagrammatic sketch of striated muscle fibril at the sarcomere level as shown by the electron microscope. *A*, Resting muscle. Filaments connecting actin fibers are shown as dashed lines. *B*, Contracted muscle. (Adapted from H. E. Huxley and J. Hanson: Ann. N. Y. Acad. Sci. *81:* 403, 1959.)

of ATP to ADP is coupled more or less directly with muscle contraction or recovery but how this takes place is unknown.

The electron microscopic studies of Huxley and Hanson of the structure of muscle fibers have had a profound influence on modern thinking regarding the nature of muscle contraction. Figure 22A is a diagrammatic sketch of a resting muscle fibril at the sarcomere level as revealed by the electron microscope and Figure 22B is a sketch of a contracted fibril (compare with Fig. 19). Figure 23 is a cross-section through a striated muscle through the denser regions of the A band on either side of the H-zone.

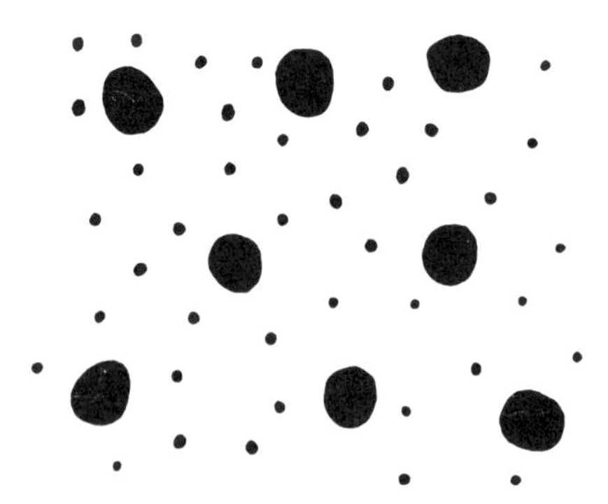

Fig. 23. Cross-section through denser region of A-band of striated muscle. Note the larger myosin fiber between which are the smaller actin fibers. (Adapted from H. E. Huxley and J. Hanson: Ann. N. Y. Acad. Sci. *81:* 403, 1959.)

On the basis of their electron microscopic studies, Huxley and Hanson propose that changes in length of the muscle, whether resulting from a passive stretch or from a contraction, require the myosin and actin filaments to slide past each other and the extent of overlap of the filaments is thereby altered. More recently, attention has centered on cross-bridges which form during contraction between the myosin and actin filaments through a distance of about 130 A-units which separates their surfaces. Projections on the myosin filaments can be seen in the resting condition and are located on the heavy meromyosin part of the myosin molecule. It appears that these projections carry the ATPase activity of myosin as well as the binding sites for actin. The cause of the active sliding of myosin and actin relative to each other during muscle contraction is, however, unknown, i.e., we still don't know what makes a muscle contract.

It has not yet been established whether the contraction of a muscle is spontaneous and occurs because of an entropy increase or whether contraction primarily requires a change in internal energy. A muscle contracts with great rapidity and, accordingly, it is likely that the rate limiting step is ionic. When the soft flexible muscle contracts, it becomes rigid and stiff; the process is essentially a sol-gel transformation. There have been many theories of muscle contraction but these will not be explored.

Microscopy. Electromagnetic radiation has been used to scan biological structures since Anton Leeuwenhoek looked at living cells with his microscope. The resolving power of a microscope is limited by the wavelength of light used as well as by the numerical aperture of the viewing system. The smallest particle which can be resolved in the visible range is about 2,000 A-units in diameter. Some extension of this resolution can be achieved by ultraviolet light microscopes.

Delineation of microscopic specimens can be improved by phase contrast where observation is based primarily on differences in refractive indices rather than on differences in light absorption. It is also sometimes advantageous to use a dark-field microscope with a condenser system which illuminates the specimen against a dark background. Polarizing microscopes also have great utility. They are equipped with a polarizer and an analyzer and are able to detect and to measure the degree of birefringence in the specimen.

The light microscope, however, is unable to resolve biological structures at

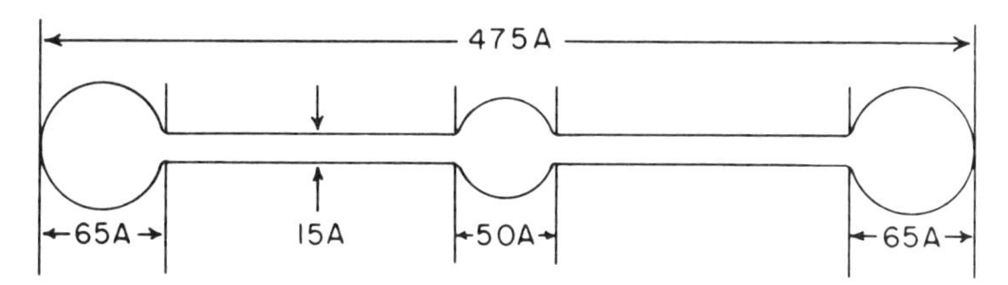

Fig. 24. Fibrinogen molecule from electron micrograph. (Adapted from C. E. Hall and H. S. Slayter: J. Biophys. Biochem. Cytol. *5:* 11, 1959.)

molecular levels and for this purpose the electron microscope is used. The electron microscope is based on the same optical principles as is an ordinary microscope except that the electron beam is focused with magnetic fields instead of with glass lenses. The electron beam originating from a heated tungsten filament passes through the specimen and is brought to focus on a photographic plate or on a fluorescent screen. The wavelength of the electron beam depends on the potential through which the electrons are accelerated; the larger the potential the shorter the wavelength. Most electron microscopes use between 50 and 100 kilovolts corresponding to wavelengths of about 0.05 A-units. For various technical reasons, resolutions of this order of magnitude are not possible but under favorable conditions present-day electron microscopes do have a resolution of about 5 A-units.

The specimen shows differential scatter of electrons; electrons which are not scattered form the image of the specimen. Biological material is apt to show more or less uniform scatter and, accordingly, unstained specimens lack contrast. Further, the supporting medium scatters and there is confusion between the object and the background.

Methods have been developed to increase contrast and to see structures at molecular levels. These include staining with phosphotungstic acid or osmium oxide thereby greatly increasing the electron scatter (see discussion of collagen and Fig. 15). It is also possible to stain a specimen negatively; the background is stained but not the object. Shadowing and replication are also of help in bringing out details of structure. In shadowing, a beam of metal atoms evaporated from a heated filament in a high vacuum is directed obliquely at the specimen. The metal covers the surface of the supporting medium as well as the specimen except for those areas protected by the vertical projection of the specimen, and these produce shadows. It is possible to estimate the thickness of the specimen by the length of the shadow as well as to visualize the structure itself. In replication, a plastic or

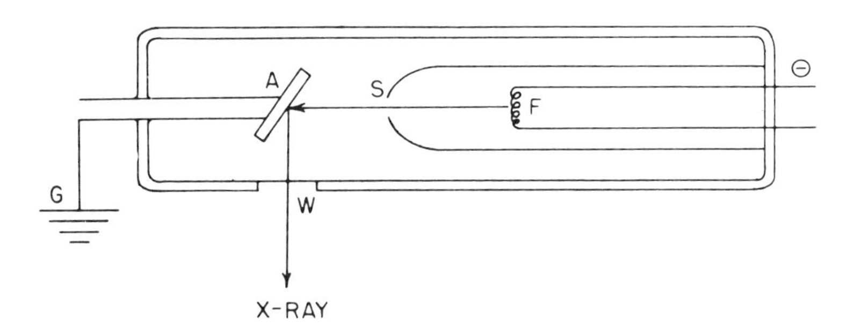

Fig. 25. X-ray tube. *A* is the metal anode. *F* is a heated tungsten filament made negative by several kilovolts relative to the ground *G*. *S* is metal focusing shield. *W* is a window of thin aluminum or beryllium foil which is transparent to X-rays.

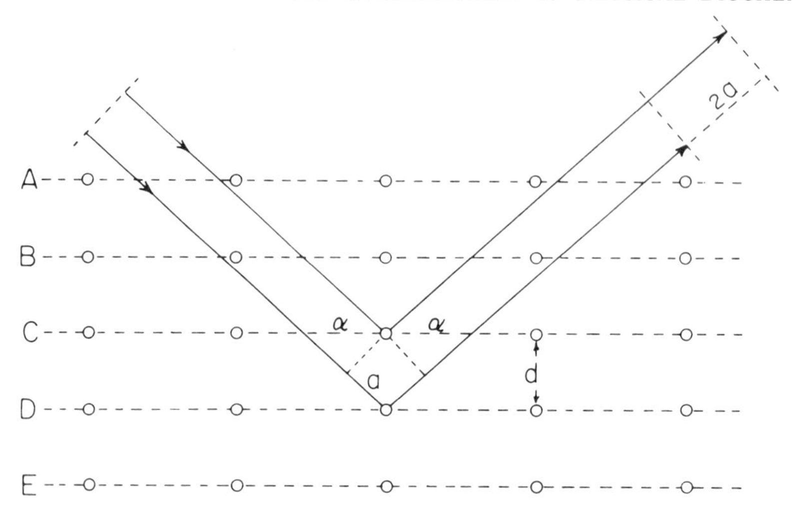

Fig. 26. X-ray diffraction from atomic planes d distance apart.

carbon film is spread over the surface of an object which is then removed and a cast of the surface is left on the plastic film. The replica is then shadowed.

Since air and water vapor scatter electrons, it is necessary that the electron beam travel through a high vacuum which in turn means that the specimen must be completely dry. The drying of a biological structure can produce serious distortions and introduce artifacts. In working with tissue, it is necessary to employ sections 200 to 300 A-units in thickness and artifacts can be produced in the process of sectioning. Changes in structure can also be introduced by staining. Caution, therefore, has to be exercised in the interpretation of electron micrographs.

Many handsome electron micrographs have been published of various virus preparations permitting molecular dimensions to be estimated. Still smaller molecules have been resolved including those of fibrinogen, which consist of a chain of three balls connected by thin filaments. A diagrammatic sketch of the fibrinogen molecule is shown in Figure 24.

X-ray Diffraction. Electrons from a heated filament are accelerated in a high vacuum by the application of several kilovolts potential and strike a metal target thereby producing X-rays. The general arrangement for an X-ray tube is shown in Figure 25.

The production of X-rays is exactly analogous to the production of visible emission spectra. The optical spectra are caused by transitions of the valence electrons and X-ray spectra arise from transitions of electrons in the inner electron shells. As a result of a collision with a high speed electron from the external beam, an electron in the metal atom may be driven completely from its orbit leaving a vacancy in the target atom. When electrons in the outer shells drop into the vacancy, X-rays having an energy $h\nu$ per quantum are emitted, where h is Planck's constant and ν is the frequency. Since a quantum of X-rays has a very high energy, ν is also very large and λ, the wavelength, is very small as compared with visible radiation.

The wavelength of X-rays is determined by the position of the activated electron in the metal atom. There are K, L, M rays corresponding to the K-, L-, M-shells of electrons around the nucleus of the atom. K-radiation has the shortest wavelength and is more penetrating. There are several electrons in each shell, and, accordingly, the rays arising from a given electron shell will be multiple. For example, K X-rays are designated as K_{α_1}, K_{α_2}, K_{β_1}, and K_{β_2}. The metal targets most frequently employed are copper, iron, and chromium. Table 5 gives the wavelengths of the K radiations of these metals.

TABLE 5. WAVELENGTHS OF K X-RAYS IN A-UNITS

Element	*Wavelengths*			
	K_{α_1}	K_{α_2}	K_{β_1}	K_{β_2}
Chromium	2.2889	2.2850	2.0806	–
Iron	1.9310	1.9321	1.7530	–
Copper	1.5412	1.5374	1.3894	1.3782

It is highly desirable to have monochromatic X-rays for diffraction analysis, and the usual method for obtaining such radiation is either to absorb out the weaker rays with metal foil or to pass the X-ray beam through a large crystal and isolate the desired wavelength by diffraction.

When an X-ray strikes an object that is composed of repeating units whose distances of separation are commensurate with the wavelength of the X-ray, the X-rays are diffracted. This diffraction is of the same kind as that observed when ordinary light is passed through a diffraction grating. Diffraction involves the reinforcement of rays whose waves are in phase and the cancellation of rays whose waves are out of phase. The situation is shown in Figure 26. A parallel beam of X-rays impinges on a crystal so that the glancing angle is α. Part of the beam will be reflected from plane A. Some of the beam will penetrate the crystal and be reflected from plane B, from plane C, etc. The difference in distance traveled by rays from planes C and D is $2a$. If this distance is equal to whole numbers of wavelengths, then the various reflections will reinforce each other and a strong diffracted beam will be observed at this particular angle of reflection. It is evident from an inspection of Figure 26 that sin α is equal to a/d and, since $n\lambda$ is equal to $2a$, Bragg's equation is obtained which gives the condition for a maximum reflection

$$n\lambda = 2d \sin \alpha \qquad 28$$

where d is the distance between the reflecting planes. As α is increased, a series of positions is found, corresponding to n equal 1, 2, 3, 4, etc., at which maximum reflection occurs. These diffraction maxima are called first, second, third, etc., orders according to the value of n.

In practice, a collimated beam of monochromatic X-rays is allowed to impinge on the specimen. At suitable distance from the specimen and in line with the undeviated beam is placed a photographic plate; the positions as well as the intensities of the diffracted rays are measured. Figure 27 is a diagrammatic representation of the arrangement employing a fiber specimen.

Aside from quantitative considerations, much can be learned from a study of the quality of X-ray pictures and the following aspects of such pictures can be listed.

1. Diffuse concentric rings indicate repeating units but no orientation is present. This type of picture is given by liquids and unoriented fibers.

2. Sharp well-defined rings mean no orientation present. However, there are

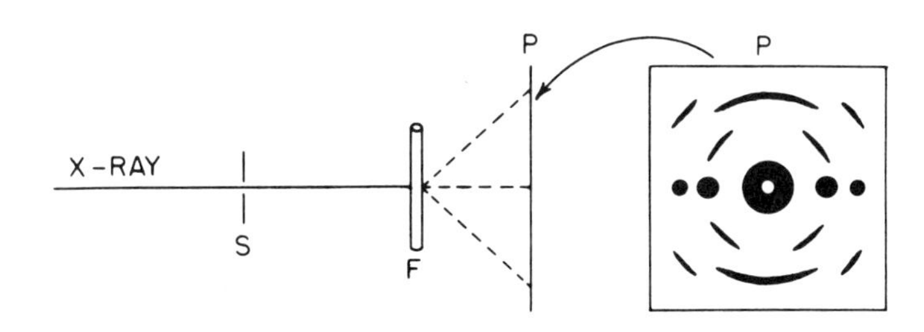

Fig. 27. Arrangement for obtaining an X-ray diffraction pattern of a fiber, F, on a photographic plate, P. S is a collimating pinhole.

precise repeating distances present. Such a picture is given by a large number of small crystals randomly oriented—so-called powder patterns.

3. Concentric rings are not complete but exist rather as arcs (see Fig. 27). This is the typical fiber photograph; the angle subtended by the arcs gives an idea of the goodness of orientation, and the sharpness of the arcs gives information on the constancy of the repeating distance. Order and spacings normal or parallel to the fiber axis can be distinguished.

4. Sharp and regular pattern of spots is characteristic of true crystals with order in three dimensions.

As can be seen from eq. 28, the distance of the spots or rings from the central undeviated beam is an inverse measure of the length of the spacing in the specimen.

From the discussion above, the impression might be had that X-ray diffraction is indeed very simple; such an impression would be completely erroneous. X-ray diffraction analysis in all of its complexities is strictly for the professional and it is probably the most difficult area of physical chemistry. The ultimate objective of X-ray diffraction study is to deduce the three dimensional molecular structure from the two dimensional diffraction pattern. Since the scatter of X-rays is proportional to the electron density, the intensity of the scatter at any point on the photographic plate must represent the electron density of some locality in the unit cell. The unit cell is the smallest volume of the crystal which contains the structural pattern of the crystal and may have in it one or more molecules. In Chapter 1, Mathematical Review, it was shown that any repetitive function can be represented by a sine or cosine series (Fourier series). The atomic positions in a three dimensional crystal are repeated in the sense of a three dimensional wallpaper pattern. It should, therefore, be possible to represent the intensities of electron scatter by a Fourier series in three dimensions; this is exactly what is done. Unfortunately, without a knowledge of the structure of the crystal it is impossible to determine the phase angle to substitute in the Fourier series. With simpler molecular structures, it is possible to make intelligent guesses as to the molecular structure and to calculate the intensities of their positions. If there is good correspondence between the calculated and the observed intensities and positions, it is likely that the correct molecular structure has been deduced; it is not proven, however, because other structures might accommodate the diffraction picture as well or better.

The other method of proceeding is to attach heavy metal atoms to the molecule under investigation. The metal atoms produce intense scatter and serve as land-marks on the diffraction picture. It is then possible to establish the regular and simpler spatial relations of the heavy metal atoms and obtain the phase relations and the Fourier analysis carried to a successful conclusion.

It is usual to employ two dimensional projections of the electron densities and to obtain contour maps of the electron densities at regularly spaced intervals through the third dimension. Comparison of a series of consecutive maps then gives the three dimensional electron density from which the positions of the atoms in the molecule are obtained.

In Chapter 6, Biopolymers, attention was called to the tertiary structures of myoglobin and of hemoglobin as deduced by Perutz, Kendrew and their co-workers. An important contribution to their success was the use of isomorphic replacements with organic compounds containing heavy metal atoms. These compounds were attached to the protein molecules without changing the crystal structure of the proteins.

Digital computers are widely used in the calculations of molecular structure from X-ray diffraction.

GENERAL REFERENCES

Dickerson, R. E.: X-Ray Analysis and Protein Structure. *In* The Proteins, Vol. 2. Academic Press, New York, 1964.
Ferry, John D.: Viscoelastic Properties of Polymers. John Wiley and Sons, New York, 1961.
Flory, P. J.: Principles of Polymer Chemistry. Cornell Univ. Press, Ithaca, 1953.
Huxley, H. E.: The mechanism of muscle contraction. Science *164:* 1356, 1969.
Reiner, Markus: Deformation Strain and Flow. H. K. Lewis and Co., London, 1960.
Seifter, S. and P. M. Gallop: The Structure Proteins. *In* The Proteins, Vol. 4. Academic Press, New York, 1966.

PROBLEMS

1. The relative water vapor pressure associated with a block of gelatin at 25° is 0.90. What pressure would have to be exerted on the gelatin to squeeze pure liquid water from the gelatin?

   Ans: 14.76×10^4 cm. H_2O.

2. Outline experiments by which you would resolve the stress applied to the interior of a vein into that due to heat motion and that due to distortion of chemical bonds.

   Ans: See text.

Frequently Used Constants

ϵ	Unit ionic charge	4.802×10^{-10} e.s.u.
C	Dielectric constant of water	78.54 at 25°C
$-\log K_w$	pK_w of water	13.996 at 25°C
σ	Surface tension of water	71.97 dynes/cm. at 25°C
η	Viscosity of water	0.008937 poise at 25°C
k	Boltzmann constant	1.38026×10^{-16} ergs/degree
R	Molar gas constant	8.3144×10^{7} ergs/degree
	$2.3026\ RT_{25°}$	1364.25 cals./mole
	$2.3026\ RT_{30°}$	1387.13 cals./mole
	$2.3026\ RT/F_{25°}$	59.16 m.v.
	$2.3026\ RT/F_{30°}$	60.15 m.v.
F	Faraday	96,493 coulombs/eq.
$1/\kappa_{25°}$	Thickness ionic double layer	$3.043 \times 10^{-8}/(\Gamma/2)^{1/2}$cm.
$1/\kappa_{30°}$	Thickness ionic double layer	$3.033 \times 10^{-8}/(\Gamma/2)^{1/2}$ cm.
N	Avogadro's constant	6.0238×10^{23}
h	Planck's constant	6.6238×10^{-27} ergs/sec.
C	Velocity of light in vacuum	2.99790×10^{10} cm./sec.
	Absolute zero temperature	−273.15°C
	One calorie = 4.185 joules	
	One joule = 1×10^{7} ergs	
g	Gravitation acceleration	980.616 cm./sec.^{2}; sea level; 45° lat.
π		3.14159
e	Natural log base	2.71828

$\log_{10}x = 0.43429 \log_e x$: $\log_e x = 2.30258 \log_{10} x$

Greek Alphabet

Α	α	Alpha	Ν	ν	Nu
Β	β	Beta	Ξ	ξ	Xi
Γ	γ	Gamma	Ο	o	Omicron
Δ	δ	Delta	Π	π	Pi
Ε	ϵ	Epsilon	Ρ	ρ	Rho
Ζ	ζ	Zeta	Σ	σ	Sigma
Η	η	Eta	Τ	τ	Tau
Θ	$\theta \vartheta$	Theta	Υ	υ	Upsilon
Ι	ι	Iota	Φ	ϕ	Phi
Κ	κ	Kappa	Χ	χ	Chi
Λ	λ	Lambda	Ψ	ψ	Psi
Μ	μ	Mu	Ω	ω	Omega

Index

ABSORBANCE, 216
Absorbancy. See *Optical density.*
Absorption
  amino acids, 219
  infra-red, 210-212
  nucleic acids, 221, 222
  photochemical equivalent, 223
  of radiation, 215-224
  ultraviolet, 212-223
  visible light, 185, 215-218
Acetic acid
  buffers of, 107, 112, 115
  ionization
    rate of, 399
    temperature coefficient, 110
    in urea solutions, 308
Acid(s). See individual names.
  amino. See *Amino acids.*
  definition, 104
  ionization
    constants (Table), 109
    constants from conductance, 307, 308
    constants from e.m.f. cells, 109, 110
    dicarboxylic, 118, 119
    at interfaces, 353, 354
    rate of, 399
    temperature effects, 110, 111, 114
  pH, calculation of, 108
  titration curves, 106, 107, 123, 124
Actin, 446, 447
Action currents, 317
Activated complex, 292-294
Activators, enzymes, 409-413
Activity
  chemical potential, relation to, 51, 52
Activity *(continued)*
  coefficient, 49
    ionic, 69, 74, 91
  halide solutions, 76
  protein solubility, relation to, 84
  two component systems, 160
Actomyosin, 447
Adenosine diphosphate
  ionization constants (Table), 60, 61
  metal complexes, 60, 61
  relation to muscle contraction, 447
Adenosine triphosphate
  biological role, 59
  hydrolysis, energy changes, 61, 62
  ionization constants (Table), 60, 61
  metal complexes, 60, 61
  molecular model, 60
  relation to muscle contraction, 447, 448
  synthesis, 103
Adhesion, 234-236
Adiabatic processes, 41
ADP. See *Adenosine diphosphate.*
Adsorption
  binding, relation to, 164, 165, 245, 246
  energy of, 162, 163, 244, 245
  fatty acids, 238
  Gibbs equation, 236, 237
  heat of, 165
  Langmuir equation, 242
  of proteins, 243, 244
  statistical effects, 117-119
Alanine, optical isomers, 191
Aldolase, isoelectric point, 350
Allosteric effects, 147, 416, 417, 420

Alpha helix (alpha keratin)
elastic properties, 437
hydration of, 273
structure, 141
Alternating currents, 317, 318
Amino acid(s), 122-125
diffusion, 296, 297
ionization constants (Table), 125
metal ion complexes (Table), 87
residues, ionization (Table), 127
spectra, 219, 220
Amperometric methods, 94-96
Ampholytes, 123
Amylopectin, 134
Amylose, 134, 142
Anomalous osmosis, 315, 316
Ants, temperature coefficient of creeping, 391
Archibald method, 368, 369
Arrhenius equation, 390-392
Aspartic acid, ionization, 124, 125
Associating systems
electrophoresis of, 340,341
light scatter by, 200, 201
osmotic pressure of, 173, 174
ultracentrifugation of, 365, 367
ATP. See *Adenosine triphosphate.*

BACTERIA
generation time, 377
separation of sexual forms, 426
Bacteriophage, molecular weight of RNA form, 204
Bases, definition, 104
Beer's law, 216
Beta lactoglobulin
isoelectric point, 350
solubility, 151
surface films, 249
Bicarbonate buffers, 115-117
Bifacial films, 255, 256
Binding
and adsorption, 164, 165, 245, 246
of ions to proteins, 117-121, 120, 121, 130-132, 340
Binomial theorem, 14, 22
Bioelectric potentials, 92, 317
Biological oxidations, 100-103
Bioluminescence, 224
Birefringence
electric, 189, 300
flow, 189, 299, 300
muscle, 445, 446
Black films, 255, 256
Blood
buffers, respiration and, 115-117
flow in artery, 265
ionic composition, 85
Blood plasma
electrophoresis, 12, 341
ultracentrifugation, 364, 365
Boltzmann
constant, 33
distribution, 40, 72, 347, 391
Bond angles
peptides, 140
water molecules, 77
Booth, electroviscous equation, 356
Boundaries in electrophoresis, 336-338
Bovine serum albumin
absorption, spectra, 219
adhesion to paraffin, 235
adsorption on glass, 243
binding of azosulfathiazole, 221
electrophoresis, 324, 350-352
isoelectric points, 350, 351
light scatter, 201
partial specific volume, 370
surface films, 253
surface tension, 237
surface viscosity, 242
titration curve, 127
Bragg's law, 451
Bronsted theory of acids, 104
Brownian motion, 134, 281, 309
Bubble pressure, 229, 230
Buffers
acetate, 107, 108, 112, 114
bicarbonate, 115-117
blood, 115, 116
capacity, 111, 112
dilution, 113
ionic strength, 115
practical, 114, 115
standard, 115
temperature effects, 114
tris, 115
Built-up monolayers. See *Deposited monolayers.*

CALCITE, double refraction, 187, 188
Calcium electrode, 94
Calculus
differential, 8-10
integral, 10-13
Calomel half cell, 89, 93
Calorie, definition, 32
Calorimeters
measurements with, 36, 37
Van't Hoff heats, 47
Capacity
of cells, 318
of conductors (Table), 71
Capillary flow, 262-267
Capillary rise, 230
Carbon dioxide
hydration, 116
rate hydration, 397, 398
Carbonic acid, ionization, 115-117
Carbonic anhydrase, 116
Carrier ion, 323, 324
Cataphoresis. See *Electrophoresis.*
Cell constant, conductance, 304, 305
Cells
capacities of membranes, 318
conductance, 308, 309, 317, 318
ion accumulation, 319-324
osmotic equilibrium, 161
penetration by non-electrolytes, 293, 294
resistance of membranes, 319
Cells (e.m.f.)
concentration, 94
equation for, 91
half-cells, 89, 90
liquid-junctions, 91, 94, 311-313
Centigrade scale, 34
Centrifuge, forces acting, 20, 21

Chara ceratophylla, diffusion into, 294
Charge
  density in ionic solutions, 72
  distribution at interface, 328
  electrostatic, unit, 70
Charging, work of, 70, 71, 74
Chelates, 86, 87
Chemical potential, 50-52
  relation to diffusion, 295
Chemical reactions. See *Reactions.*
C.G.S. units, 27, 28
Chloride electrode, 93
Chlorophyll, 224, 225
Cholesterol, surface films, 248
Chromatography, 246
Chymotrypsin
  activity at an interface, 244
  catalytic sequence, 414
  isoelectric point, 353
Chymotrypsinogen denaturation, 46, 47
Circular dichroism, 192, 193
Clathrate crystals, 83
Clausius-Clapeyron equation, 165
Coacervation, 426
Cohesion, 234
Coils, random and disordered, 134-136
Collagen
  amino acid composition, 440
  elasticity, 442
  fibers, 436, 437, 440-443
  melting temperature, 443
  solubilization, 441
  structure, 441
Colligative properties, 158
Collision rate
  in gases, 388
  in solution, 388-390
Colorimeters, 217
Colors, wave length of, 185
Complexes
  ionic, 60, 86
  iron and porphyrin, 145, 146
Compressibility, surface films, 249
Compression waves, velocity of, 433
Computers, 29
Concentration
  cells (e.m.f.), 94
  expressions for, 49
  gradients, measurement, 281-286
Conductance
  cells, 305
  equivalent, 304
  function of frequency, 317-319
  ionization constants from, 307, 308
  measurement, 304, 305
  potassium chloride, 304, 305
  specific, 303, 304
  suspensions, 308, 309
  urea solutions, 308
Constants, physical, list of, 454
Constituent mobilities, 340
Contact angle, 234-236
Cooperative structure, 151
Copolymers, 134
Correlation coefficient, 27
Cotton effect, 192, 193, 195, 196
Counter ions, 327, 328
Couette viscometer, 267
Coulomb's law, 69
Coulter counter, 309
Counter current distribution, 246
Coupled reactions, 58, 59
Co-volumes, 172, 252
Cross of isocline, 300
Cylindrical lens for concentration gradients, 283
Cysteine, ionization, 125
Cytochrome C
  ionization constants of, 129
  oxidation-reduction of, 47, 101
  perturbation spectra, 222, 223

DEBYE-HÜCKEL theory of electrolytes, 71-77
Decarboxylation, 100
Dehydrogenases
  lactate, equilibrium, 57, 58
  oxidative chain, 100
  yeast, substrate complex, 414
Density gradients in ultracentrifuge, 365
Deoxyribose nucleic acid (DNA)
  rate unwinding, 400
  structure, 153, 154
  titration, 131
  ultraviolet absorption, 221, 222
Depolymerizing enzymes, 418, 419
Deposited surface films, 257, 258
Dialysis, equilibrium, 132
Diamagnetism, 100
Dichroism, 189, 193
  circular, 192, 193
  infra-red, 212
  nucleic acids, 221, 222
  visible and ultraviolet, 221, 222
Dielectric constant, 67, 71
  dispersion, 300
  index of refraction, relation to, 186, 196
  table of values, 71
  water, 71
Difference spectra, 220
Differential equations, 13, 14
Differential thermal analyzer, 37
Diffusion, 276-302
  amino acids, 296, 297
  concentration gradients, 281-285
  Fick's first law, 276
  Fick's second law, 279-281
  to flat surface, 289, 290
  frictional forces, 296
  in gels, 286-288
  influence of hydration, 297
  ionic, 309-311
  membranes, through, 290-294
  phenomenological, 292
  potentials, 278, 279
  probability relations, 280, 281
  reaction rates, 388-390
  rotary, 299-302
  size and shape molecules, 296-298
  thermodynamics of, 295
  viscosity relation, 298
Diffusion coefficient, 279
  concentration dependence, 286
  dimension and units, 276
  glucose, 279
  influence of heterogeneity, 286

Diffusion coefficient *(continued)*
  hydrochloric acid, 311
  optical methods, 282-285
  porous dish method, 277, 278
  potassium chloride, 279, 311
  rotary, 266-269
  sodium chloride, 311
  temperature correction, 286
  urea, 279
Digestion, 56
Dilution
  entropy of, 42, 423
  free energy of, 52, 53
  reversible and irreversible, 42
Dimensional analysis, 27-29
Diphtheria antitoxin, denaturation of, 382, 383
Dipolar ions, 123
Distribution(s)
  Boltzmann, 40
  coefficient, 51, 293, 294
  energies, 33
  errors, 21-23
  Gaussian, 22
  Maxwell, 33
  between phases, 427
Disulfide bonds, 144, 437
Donnan equilibrium, 174-179, 428, 429
Dorn effect, 329
Double refraction, 187, 188
Dropweight method for surface tension, 230, 231
Drude equation, 192, 194, 195
du Nouy ring, method for surface tension, 232
Dyne, definition, 38

EGG albumin
  adsorption on glass, 244
  binding of water and solutes, 164, 165
  denaturation, 149, 150
  deposited films, 235
  electrophoresis, 333
  helical content, 195
  hydrolysate, molecular weight of, 385, 386
  partial specific volume, 370
  pepsin hydrolysis, 201, 202, 414, 419
  rate surface coagu ation, 376
  solubility, 151
  surface films, 249
  surface potentials, 241
  surface tension, 237
Einstein
  diffusion equation, 296, 297
  unit of, 223
  viscosity equation, 268
Elasticity
  collagen, 436, 442, 443
  decay of, 432
  elastin, 436
  elements, 433
  keratin (hair), 435, 436
  modulus of, 432, 433
  molecular basis for, 434, 435
  muscle, 446
  nylon, 444, 445
  protein fibers, 436-444
  resilin, 436, 443, 444
  rubber, 435
  types, 434
Elastin, 436, 443
Electrical
  conductance, 303, 304
  double layer, 328, 347, 348
  forces in peptide structure, 139, 140
  resistance, specific, 303
  units, 70
Electrochemical potential, 319, 320
Electrodes
  calomel, 89, 93
  glass, 92
  hydrogen, 89, 92
  membrane, 93, 94
  mercuric oxide, 96
  non-polarizable, 92
  normal half cells, 90
  potentials (Table), 91
  oxidation-reduction, 97
  oxygen, 94, 95
  silver-silver chloride, 93
Electrokinetics
  comparison of methods, 344
  potentials, 329, 330
Electrolytes, activity of, 69
Electromagnetic radiation, fate of, 206
Electromagnetic spectrum, 186
Electromotive force cells. See *Cells (e.m.f.)* and *Electrodes*.
Electronic absorption spectra, 212-215
Electronic configurations of molecules, 212-215
Electron microscope, 448, 449
Electron orbitals, 212-214
Electron spin resonance, 207, 208
Electron transfer, enzymic systems, 101
Electrons
  forced oscillations by light, 197
  in optically anisotropic crystals, 187
  potential energy curves (fluorescence and phosphorescence), 215
Electroosmosis
  equation for, 329
  measurement, 342
Electrophoresis
  acceleration of particles, 14
  adsorbed nucleic acids, 334, 335
  adsorbed protein, 333, 334, 351-353
  blood plasma, 12, 341
  bovine serum albumin, 334, 350, 352
  complex mixtures, 341, 342
  constituent mobilities, 340
  cylindrical lens system, 283
  egg albumin, 333
  equation for mobility, 327-339, 344-346
  heterogeneity test, 339, 340
  interacting systems, 340, 341
  ions, mobility, 306, 338
  Komagata's correction, 332
  Longsworth scanning, 283
  micromethod, 330-334
  mobilities, calculation, 336
    relation to charge, 347-352
  moving boundary, 334-341
    boundary conditions, 336-338
    comparison with micromethod, 352
    method, 334-341
  nucleic acids, 334, 335
  optical methods, 282-285

Electrophoresis *(continued)*
particle size and shape, 344-346
proteins
concentration determination, 335, 336
fractionation, 341, 342
as function of charge, 349
red blood cells, 334
relaxation effect, 345
surface conductance effects, 345
zone, 341, 342
Electrostatic charge
on bovine serum albumin, 352
ionic, 69, 70
osmotic pressure, effect on, 174-178
titrations, effect on, 120-122, 127, 128
influence of surface ionizations, 353, 354
on surfaces, 347-353
unit, 69, 70
viscosity, effect on, 355, 356
Electrostatic forces, 357, 358
Electrostatic free energy, effect on ionization, 70, 120-122
Electrostatics, elementary, 69-71
Electroviscous effect, 355, 356
Ellipsoids
diffusion of, 297, 298
viscosity of suspensions, 270
Ellipsometer, 257
Ellipticity, molar, 193, 194
Emulsions
particle size distribution, 205
rate coalescence, 390
Endergonic reactions, 44
Energy
activation, 390-394
adsorption, 162, 163, 244, 245
adsorption fatty acids, 238
biological transformations, 56, 57, 62-64
conservation of, 41
conversion factors, 38, 39
definition, 33
distribution, 33, 392
free, 43, 44
heat, 32
internal, 41
ionic hydration, 80-83
pass for reaction, 393
profile for enzyme reaction, 412
rotation about bonds, 135, 136
surface, 232-234
Enthalpy
activation, 393, 394
ATP hydrolysis, 62
calculation, 38
calorimetric and Van't Hoff, 47
definition, 37
glucose oxidation, 37
surface, 232, 233, 254
Entropy
activation, 393, 394
disorder, relation to, 43
elasticity of fibers, 434-445
free energy, relation to, 43
glucose oxidation, 44
ion hydration, 81, 82
mixing, 42, 171, 172, 255
surfaces, 233, 234, 254,
Entropy *(continued)*
swelling of gels, 428, 429
Enzyme(s)
allosteric transformations, 147, 416, 417, 420
mechanism of action, 420, 421
metal ion activation, 86
nature of, 400
oxidation-reduction, 100, 101
substrate complex, 401, 413, 414
Enzyme reactions
activators, 409-413
depolymerizing, 418, 419
enzyme concentration, effect of, 420
flow techniques, 396, 397, 402
graphical representations, 415, 416
histidinase activity, 410, 411
inhibitors, 407-409, 414, 415
kinetics, 401, 407
mechanism, 420, 421
papain activity, 273, 419
pepsin activity, 411, 412
pH, influence of, 410-413
reversible, 400, 417, 418
trypsin activity, 412, 413
two substrate kinetics, 405-407, 415, 416
velocity measurements, 400
Eötvös constant, 234
Equations
differential, 13, 14
solution of, 3-6
Equilibrium
biological, 62
Donnan, 174-178
centrifugation, 365-369
constant, 44, 45
definition, 45
between phases, 425, 426
temperature effects, 46, 47
Erg, definition of, 38
Errors, distribution of, 21-23
Ethylacetate, free energy hydrolysis, 53-56
Ethylenediamine tetraacetic acid (versene), 87
Exergonic reaction, 44
Exponents, 1, 8
Extinction coefficient, 216

FARADAY, unit of, 90
Fast reactions, 396
Fatty acids
hydrogen bond formation, 137
orientation at surfaces, 238, 246, 247
protein denaturation, 150
surface tension, 238
Ferredoxin, 225
Fibrin clots, opacity and rigidity, 432
Fibrinogen molecule, electron micrograph, 449
Fick's diffusion equations
first law, 276
second law, 279-281
Field strength, electric, 70, 74
Films. See *Surface films*.
Filtration coefficient of membranes, 292
Flame photometer, 85
Flow
birefringence, 300
of blood, 265
in a capillary, 262-267

Flow *(continued)*
  methods for reaction rates, 396
Fluidity of protein solutions, 268, 269
Fluorescence, 214, 215
Fluorescence polarization, 300-302
Flux
  in diffusion, 276, 295
  ratio of ions, 320
Forces acting on suspended particles, 360-362
Free energy
  of activation, 393, 394
  adsorption, 162, 163, 245, 246
  ATP hydrolysis, 60-62
  biological reactions, 63
  calculation of, 47, 48, 52, 53
  cells (e.m.f.), 47, 90
  concentration, function of, 44, 45
  definition, 43
  dilution, 52, 53
  elasticity of fibers, 434-436
  electrostatic potential, 70
  equilibrium constant, relation to, 44, 45
  ethylacetate hydrolysis, 53-56
  of formation (Table), 54
  Gibbs, 44
  glucose oxidation, 44
  glycolysis, 56
  Helmholtz, 44
  ion accumulation, 319
  ion hydration, 81
  lactate oxidation, 57, 58
  leucylglycine hydrolysis, 48, 49
  oxidation-reduction reactions, 96, 97
  partial molar, 50
  protein hydration, 120-122, 163
  surface ionization, 353, 354
  surfaces, 232
  symbols for, 55, 56
  temperature effects, 46
  water ionization, 53
  wetting, 235
Free radicals, 98-100, 207, 208
Freezing point
  depression, 166, 167
  of tissue, 167
Fricke's equation of conductance of suspensions, 308, 309
Frictional coefficient, molecular, 295
Fumarase, kinetics of, 417, 418
Fumaric acid, ionization of, 120

GAMMA globulin, heterogeneity test, 339, 340
Gas
  collision rate, 388
  constant, 32, 33, 168, 454
  pressure, 13, 32
  work of compression, 12, 30
Gaseous surface films, 251-253
Gaussian distribution, 22
  diffusion rate, 280
Gelatin
  gels, rigidity of, 429-432
  production from collagen, 441
  swelling of gels, 428, 429
Gels, 426-433
  classification, 427
  deformation, 429, 430
Gels *(continued)*
  diffusion in, 286-288
  elasticity, 432, 433
  formation, 427
  molecular weight determination with, 288, 289
  rigidity, 429-432
  swelling, 428, 429
  thixotropic, 427, 428
Gibbs
  adsorption equation, 236, 237
  free energy, 43, 44
Gibbs-Helmholtz equation, 46
Glass electrode, 92
Globin, 147
Glucose
  diffusion coefficient, 279
  energy of oxidation, 37, 44
  metabolism, 56
  phosphate, 58
Glutamine hydrolysis, 62
Glycerol, osmotic coefficient, 181
Glycine, ionization, 122-124
Glycogen
  metabolism, 56
  structure, 134
Glycolysis, 56
Gold number, 390
Gouy
  balance, 100, 207
  equation for surface charge, 347
  interference optics, 283, 284
Greek alphabet, 456
Gutta percha, 444

HAIR, elastic properties, 435, 436, 440
Half-cells (e.m.f.), 90, 91, 93
Half life
  isotopes, 377, 387, 388
  reactions, 380
Halides, activity of, 76
Hardy-Schulze rule, 357
Hartley-Roe equation, 353
Harmonic motion, 18-20
Heat
  of activation, 393, 394
  of adsorption, 165
  capacity, 35, 36, 49
  definition, 32
  measurement, 35-37
  mixing, 171
  of muscle contraction, 37, 445, 446
  of reactions, 37, 38
  of surface expansion, 233
  protein hydration, 163, 165
  vaporization, 165
  work equivalency, 38, 39
Helical structures
  alpha, 141, 437
  in amylose, 142
  in collagen, 43, 437, 438, 442
  in nucleic acids, 154
  optical rotation of, 192-195
  poly-L-glutamic acid, 142
  poly-L-lysine, 142
  poly-L-proline, 212, 213, 437, 438
  in proteins, 144
  rates of unwinding, 400

Heme, 145-148
Hemoglobin
  absorption spectra, 218
  carbon monoxide reaction, 396
  globin, 147, 148
  oxygen uptake, 146, 147
  structure, 145, 146
  solubility, 84
  sulfhydryl groups in, 96
  titration, 130
Henderson equation, liquid-junctions, 312
Henderson-Hasselbalch equation, 106, 107
Henry's law, 151
Heterogeneity of biopolymers
  diffusion, 286
  electrophoretic test, 339, 340
  relation to Mw/Mn, 370-372
  solubility test for, 152, 153
Hill method for vapor pressure, 165, 166
Hinge point in ultracentrifuge, 366
Histidinase, activity as function pH, 410, 411
Hofmeister (lyotropic) series, 83, 84
Hook's law, 432
Hyaluronic acid, 272
Hydration
  of carbon dioxide, 116, 397, 398
  ionic, 80-83
  of proteins, 161-165
  relation to diffusion, 297
  relation to sedimentation, 365, 366
  relation to viscosity, 273, 274
Hydrocarbons, forces between, 138, 139
Hydrochloric acid, diffusion of, 311
Hydrogen bonds
  detection with infra-red, 211, 212
  influence on acid ionization, 120
  length and strength, 137
  rate formation and rupture, 400
  in peptides, 136, 137
Hydrogen electrode, 89-92
Hydrogen ions
  determination, 89-92
  ionization from acids, 104-108
  oxidation-reduction, relation to, 97-99
  secretion in stomach, 320, 321
Hydrogen transport, 100-103
Hydronium ion, 81, 106
Hydrophobic groups, 137, 138, 248
Hypochromicity, 221, 222
Hysteresis in fibers, 439, 440

i, meaning of, 5, 21
Ice structure, 77-79
Impedance, 317
Infra-red, 210-212
Inhibitors, enzymic reactions, 407-409, 414, 415
Injury potentials, 317
Insulin
  specific heat of, 48
  surface films, 254
Integral calculus, 10-14
Interacting systems. See *Associating systems.*
Interfacial tension, 232
  influence of dodecyl sulfate on, 4
Internal energy, 41
  elasticity of fibers, 434-436
Intrinsic viscosity, 269, 270
Iodopsin, 226, 227
Ionization
  causes of, 67
  constants
    of acids (Table), 109
    of amino acids (Table), 125, 127
    from conductance, 307, 308
    from e.m.f. cells, 109, 110
  electrostatic effects, 120-122
  heats of, 110, 114, 127
  organic phosphates (Table), 61
  at surfaces, 354
Ions
  accumulation by cells, 319-324
  activity coefficients, 69, 91
  atmosphere, 72
  binding
    from electrophoresis, 340
    electrostatic effects, 120, 121
    light absorption, 220, 221
    methods of study, 130-132
    multiple, 117-121
  biological role, 84, 85
  in blood serum, 85
  carrier models, 323
  complexes, 86, 87
  conductances, 303-309
  in crystals, 67
  diffusion of, 309-311
  in double layer, 327, 328, 347-349
  enzymic activation by, 409
  flux ratios, 320
  Hofmeister (lyotropic) series, 83, 84
  hydration, 80-84
  interactions, 71-74
  ionic strength,
    of buffers, 115
    definition, 67, 68
    protein titration, influence of, 126, 127
  mobilities, 305, 306, 338
  origin of, 67
  potentials of, 74
  radii of, 68
  red-ox pump, 102, 323
  transport, active, 319-321
  transport number, 305, 338
  zwitter, 123
Irreversible thermodynamics, 292
Isoelectric focussing, 342
Isoelectric points
  amino acids, 123
  from electrophoresis, 349-351
  of proteins, 126, 350, 353
Isoelectrophoretic points, 351
Isoionic points, 126, 351
Isopiestic method for vapor pressure, 161, 164
Isopropyl phosphate, attachment to enzymes, 421
Isothermal systems, 41
Isotropically labeled compounds in animal body, 387

JOULE, definition of, 32, 39

KAPPA, reciprocal distance, 74, 348
Kelvin temperature scale, 34
Keratin
  alpha, 141

Keratin *(continued)*
  beta, 140, 434
  branched, 134
  elasticity of, 436, 439, 440
Kerr effect, 180, 300
Kinetic theory, 32, 33
Kinetics. See *Reactions.*
Kirkwood forces, 358
Km, 402
Komagata's correction, 332
Krebs tricarboxylic acid cycle, 56

LACTATE-PYRUVATE equilibrium, 57, 58
Lactoglobulin. See *Beta lactoglobulin.*
Lambert-Beer's law, 216
Laminar flow, 263, 264
Lamm scale displacement, 283
Langmuir
  adsorption equation, 243
  surface balance, 246, 247
Least squares, method of, 26, 27
Lecithin surface films, 248
Ligamentum nuchae, 440, 443
Light
  absorption measurement, 217, 218
  polarized, 187-189
  refraction, 186, 187
  scatter, 195-206
  theory, 184, 185
  velocity, 185
  wave lengths, visible, 185
Liquid junction potentials, 91, 311-313
Liquids, flow through capillaries, 262-264
Living cells. See *Cells.*
Logarithms, 2
London forces, 139, 357
Longsworth scanning method, 283
Luciferase, 224
Luminescence, 224
Lyotropic ion series, 83, 84
Lysozyme
  ionization constants, 129
  surface films, 251

MACLAURIN series, 14, 15
Macromolecules. See *Biopolymers.*
Magnesium ions
  anesthetizing action, 85
  ADP, ATP complexes, 60, 61
Maleic acid, ionization, 120
Maxwell
  Demon, 63
  elastic elements, 433
  energy distribution, 33
  suspension conduction, 308, 309
Melting curves of tissue, 167
Membranes
  anomalous osmosis, 315, 316
  cellular
    electrical potential, 317-319
    pore diameters, 294
    structure, 293
  diffusion through, 290-294
  electrodes, 93, 94
  ion distribution across, 174-176
  penetration, lipid solubility, 293, 294
  pore size measurement, 288
Membranes *(continued)*
  potentials, 178, 179, 313-317
Metabolic reactions, 46, 385-387
Methemoglobin
  absorption spectra, 218
  partial specific volume, 370
  porphyrin-iron complex, 146
Mhos, definition, 303
Mica, wave plates, 188
Michaelis-Menten kinetics, 6, 7, 402-405
Microelectrophoresis, 330-334
Microscopy, 448, 449
Microwave spectroscopy, 207
Mitochondria swelling, 180
Mobility. See *Electrophoresis.*
Mobility coefficient in membranes, 292
Mobility, ionic, 305, 306, 338
Moffit equation, 192, 194, 195
Molal solutions
  definition, 49, 51, 52
  hypothetical, 51, 52
Molar solutions, definition, 49
Mole fraction, 49
Molecular
  collisions, 388-391
  ellipticity, 193, 194
  frictional forces, 295, 296
  motion, 32, 33
  orbitals, 212-214
  orientation in flow, 273, 274, 299, 300
  orientation of surfaces, 246, 247
  polarity, 137, 138, 246, 247
  sieves, 288, 289
Molecular dimensions from
  diffusion, 296, 298, 299
  electron microscope, 449
  electrophoresis, 351
  light scatter, 202-204
  viscosity, 270-272
  X-ray diffraction, 451, 452
Molecular weights
  averages, 169, 201, 370-372
  freezing point, 166
  gel filtration, 288, 289
  light scatter, 200-206
  osmotic pressure, 168-170
  sedimentation, 364, 367-369
  surface films, 251, 252
  vapor pressure lowering, 160
  viscosity, 271, 272
  summary of methods, 370, 371
Molecularity of reactions, 375
Molecule, definition of, 33
Monolayers. See *Surface films.*
Motion, harmonic, 18, 19
Moving boundary
  boundary conditions, 336-339
  concentration of components, 335, 336
  electrophoresis method, 334
Muscle
  contraction and structure, 445-448
  heat of contraction, 37
Myoglobin, structure of, 144
Myosin, 446, 447

NARCOSIS, 293
Nerve conduction, 324-326

Newtonian flow, 261, 273, 274
Nicol prism, 188
Nicotinamide adenine dinucleotide (NAD), 57, 58
Nuclear magnetic resonance, 208, 209
Nucleic acids
  absorption spectra, 221, 222
  hydration, 163
  ionization constants (Table), 131
  protein synthesis, relation to, 156
  structure, 153-155
  titration curves, 130, 131
  viscosity of solutions, 272, 274
Nucleoproteins, 156, 157
Number systems, 1
Nylon, elastic properties, 444, 445

OCTADECYLAMINE, surface ionization, 354
Ohm's law, 303
Onsager
  equation for electrical conductance, 306, 307
  reciprocal relations, 292
Open systems in thermodynamics, 41
Opsins, 226, 227
Optical
  density, 216, 218
  isomers, 191
  methods for concentration gradients, 282-285
  rotation, 189-191, 192-195
Orbitals, electronic, 212-214
Orientation of molecules
  in electrical fields, 189, 267
  in flow gradients, 273, 274, 299, 300
  of surfaces, 246, 247
Osmosis, anomalous, 315, 316
Osmotic coefficients, 180, 181
Osmotic pressure
  departure from ideality, 171-174
  Donnan equilibrium, 175, 176
  ideal, 39, 40, 168, 169
  interaction constants, 172, 173
  light scatter, relation to, 200
  measurement, 169, 170
  protein solutions, 169-171
  temperature influence, 39, 40
Osmotic work, 39, 40
Ostwald's dilution law, 307
Ostwald viscometer, 265, 266
Ovalbumin. See *Egg albumin.*
Overton theory of membranes, 293
Oxidases, 100
Oxidation-reduction
  biological, 100-102
  concentration, function, 96, 97
  definition, 89
  free energy changes, 96, 97
  hydrogen ions, influence of, 97, 98
  potentials, 97, 103
  pump, 102, 103
  soluble systems, 96, 97
  step-wise, 98-100
Oxidative chain (respiratory), 100, 101
Oxygen electrode, 94, 95

PAPAIN, action on poly-L-glutamic acid, 419
Paraffins, water interaction with, 83, 138, 139
Paramagnetism, 100, 207
Partial molar quantities, 50
Partial specific volumes, 369, 370
Partial vapor pressures, 158, 159
Particle movement, forces acting to produce, 360-362
Particle size
  Coulter counter, 205, 309, 390
  electrophoresis, 344, 345, 351
  light scatter, 205
  sedimentation, 361, 362
Pendant drop method for surface tension, 231, 232
Penetration through cellular membranes, 292-294
Penetration of surface films, 253, 254
Pepsin
  activity as function of pH, 411, 412
  denaturation, rate of, 378, 379
  egg albumin complex, 201, 202, 414
  surface film, 253
  surface potential, 241
Peptide bond
  absorption spectra, 219
  angles and lengths, 140
  double bond character, 436, 437
Peptides
  circular dichroism, 193, 194
  elasticity, 436-444
  infra-red spectra, 212, 213
  molecular models, 142, 143
  optical rotation, 194, 195
  structure, 140-142, 437
Permeability coefficient, 291
Peroxidase, 100, 413, 414
P_{CO_2} of blood, 117
pH
  calculation of, 107, 108
  definition, 3, 91, 92
  determination with glass electrode, 92
  influence on protein solubility, 151
  scale, calibration of, 92
  at a surface, 354, 355
  of water, 105
Phase angles, electrical circuits, 317
Phase potentials, 327
Phase separations, 425
Phenomenological coefficients, 292
Phenylalanine, absorption spectra, 219
Phosphates, organic, 59
  ionization constants (Table), 61
Phosphatidic acid, surface films, 248
Phosphoglucomutase, cotton effect in, 196
Phosphorescence, 214
Phosphorylation, oxidative, 102
Photochemical changes, 223-227
Photomultiplier, 217, 218
Photon, 206
Photosynthesis, 224, 225
Photovoltaic cells, 217
Physical constants, list of, 454
Pi-orbitals, 213
pK
  of acids, 92, 108
  of proteins, 129
  values at surfaces, 353, 354
Planck's constant, 206
Planck's equation for liquid junctions, 312
Poise, 261
Poiseuille's equation, derivation
  from dimensional analysis, 29

Poiseuille's equation *(continued)*
from hydrodynamics, 262-264
Poisson's equation, 72
Poisson's ratio, 432
Polar group, 137, 138, 247
Polarized light, 188, 189
Polarograph, 94
Polaroids, 189
Poly-L-glutamic acid
action of papain on, 419
helical form as function of pH, 142
viscosity of, 273
Poly-L-lysine, helical structure, 142
Poly-L-ornithine, rate helical transformation, 399
Poly-L-proline
cis and trans isomers, 437, 438
infra-red spectra, 212, 213
Polymers
bi and tri functional units, 426
kinetics of degradation, 384, 385
structure, 134
Pore diameters of membranes, 294
Porous disk method for diffusion, 277
Porphyrin—iron complexes, 145, 146
Potassium chloride
diffusion coefficient, 279, 311
conductance, 304, 305
Potentials
bioelectric, 92, 317
electrostatic, 71, 72
of group transfer, 59
injury, 317
of an ion, 74
liquid junction, 311, 312
membrane, 178, 179, 313-315, 317
at a surface, 72, 241, 327, 347
relation to ionization, 353, 354
zeta, 346, 347
Potentiometer, 90
Pressure, effect on reaction rates, 393
Pressure jump, 399
Probability
curve, 21-25
entropy relation to, 43
Proteins. See also names of individual proteins.
absorption spectra, 219, 220, 223
buffer capacities, 128, 119
denaturation, 46, 47, 148-150, 394-396
diffusion, 298, 299
electrophoresis, 347-352
fiber elasticity, 436-444
fluidity of solutions, 268
heterogeneity test, 339, 340
hydrodynamic hydration, 274, 365, 366
ion binding, 117-121, 130-132, 220, 221, 340
ionization of, 128, 129
isoelectric points, 126, 350, 353
isoionic points, 126, 351
of muscle, 446, 447
optical rotation, 192-195
osmotic pressure, 169, 170
partial specific volumes, 369, 370
resistivities, 309
sedimentation, 364-369
solubilities, 151-153
structure, 143-145
surface films, 249-254
Proteins *(continued)*
surface potentials, 242
surface tension, 237
titration, 126, 130
viscosities, 268, 272, 298, 299
water binding, 162-165
Protons, hydration of, 106
Protoporphyrin, 145
Pyocyamine, oxidation-reduction, 99
Pyruvate, reduction, 57, 58

QUADRATIC formula, 5
Quantum of radiation, 206, 223
Quantum yield, 223

R (gas constant), 32, 33, 168, 452
Radiation, energy of, 186
Radioactive isotopes, rate of decay, 376, 377
Raman spectra, 214
Random coils, 135, 136
Raoult's law, 150, 151
Rapid flow techniques, 396-398
Rayleigh optics, 284, 285, 368
Rayleigh's ratio, 199
Reactance, 317
Reactions
biomolecular, 377
change in given time, 379, 380
complex, 380-384
consecutive, 383, 384
coupled, 58, 59
diffusion controlled, 388-396
enzymic. See *Enzyme(s)*.
enthalpy of, 37, 38
entropy of, 42, 44
fast, methods of study, 396-400
free energy of, 43, 44, 48, 52, 53
half life, 380, 388
metabolic, 56, 57, 385-388
order of, 374, 375
pressure influence, 393
rate of, 8, 374, 375, 378
reversibility, rate of, 380, 381
simultaneous, 381, 382
standard states, 379
steric factor, 391, 394
temperature, influence on rate, 390-394
transition state theory, 392-394
Red blood cells
electrical conductance, 319
electrophoretic mobilities, 334
ion flux ratios, 321
osmotic equilibrium, 180
Reduced mass of atoms, 211
Reflection coefficient, 292, 294
Refraction
of light, 186-189
specific increment, 187
Refractometric methods for diffusion, 282-285
Relaxation, elastic element, 432, 433
Relaxation effect in electrophoresis, 345
Relaxation methods for chemical reactions, 398-400
Resilin, 436, 443, 444
Resistance, specific, 303
Resonance, mechanical and electrical, 20
Respiration, 115-117

Respiratory chain (oxidative), 100, 101, 103
Retinals, 226
Reynolds number, 264
Rhodopsin, 226
Ribonuclease
  denaturation, 150
  hydration, 365, 366
  ionization constants, 129
  photochemical oxidation, 223, 224
  surface potentials, 241
Ribonucleic acid (RNA)
  light scatter, 204
  structure, 153
  titration curve, 131
Ricin, denaturation of, 393
Rigidity, modules of, 429, 430
Rotary diffusion coefficient, 299-302
Rotation, angular velocity of, 20, 21
Rotatory dispersion, 192-196
Rubber, 134, 136
  stress-strain curve, 435

SALMINE, surface potentials of solutions, 241
Salt accumulation by cells, 319-324
Salt bridges, 93
Salting out constant, 76, 83
Scatchard plot, 131
Schlieren optics, 282, 283
Sea urchin eggs, water penetration, 180
Secretion of hydrochloric acid by the stomach, 320, 321
Sedimentation. See also *Ultracentrifuge*.
Sedimentation
  coefficient, 364
  equilibrium, 365-369
  particle size distribution, 362
  potentials, 329, 367, 368
  rate, 363, 364
Series
  arithmetic and geometric, 1
  convergent and divergent, 16
  expansion, 14-16
  harmonic, 18
Serine peptides of enzymes, 421
Serum albumin. See *Bovine serum albumin*.
Sialic acid in red cell membranes, 333
Sigma electron orbitals, 212, 213
Sigmoid curves, 25, 26
Significance, test for, 24, 25
Silk, 438, 439
Silver-silver chloride electrode, 93
Simha viscosity equation, 270
Slopes of curves, measurement, 10
Smoluchowski
  collision rate equation, 388-390
  electrophoresis equation, 327, 328
  electroviscous equation, 355, 356
Sodium acetate, ionization, 107, 108
Sodium chloride
  conduction, 307
  diffusion, 311
  osmotic coefficients, 181
Sodium dodecyl sulfate
  gaseous spread films, 253
  influence on oil-water tensions, 4
Sodium ions, transport through frog skin, 322, 323
Sodium pump, 321
Solubility
  activity coefficient relation to, 83, 84
  test for purity, 152, 153
Solutions
  concentrations, 49
  entropy of dilution, 42
  osmotic pressure, 39, 40, 167-177
  surface tension, 236-239
  work of concentrating, 40
Solvent and solute binding by proteins, 164, 165, 245, 246
Solvent perturbation spectra, 222, 223
Sorensen pH scale, 92
Specific conductance, 303
Specific heats
  definition, 35, 36
  insulin, 48, 49
  water, 36
Spectra
  electronic, 212-217
  infra-red, 210-212
Spectrophotometers, 217, 218
Spectroscopy, 206, 207
Spread films (monolayers). See *Surface films*.
Squid, giant axon, 324-326
Standard deviation, 23, 24
Standard states of solutes and solvents, 51, 52
Starch, structure of, 134, 142
Starling concept of plasma-tissue water balance, 179
Stationary levels, microelectrophoretic cells, 331-333
Statistics, 21-27
Steady state kinetic, enzymic reactions, 402-407
Stearic acid, surface ionization, 354
Sterols, infra-red spectra, 211
Stokes frictional coefficient, 361
Stop-flow, 397, 398
Stream birefringence, 300
Streaming potential, 329, 330, 342, 343
Stress
  decay of, 432, 433
  thermodynamics of, 434-436
Sucrose
  inversion of, 2
  osmotic coefficient, 181
Sulfhydryl groups, amperometric titration, 96
Surface
  balance, 246, 247
  charge, 347-350
  conductance, 343
  energy, 232-234
  entropy, 233, 234
  forces acting on a liquid drop, 234
  pH, 355
  potentials and ionization, 353, 354
  wetting, 234, 236
Surface films
  bifacial, 255, 256
  cholesterol, 248
  compressibility, 249
  deposited, 256, 257
  equation of state, 238, 251, 252
  gaseous, 251-253
  high pressure, 248-250
  interfacial, 244, 254, 255
  lecithin, 248

Surface films *(continued)*
  molecular weights, 251, 252
  penetration, 253, 254
  phase changes, 247
  potentials, 239-241
  pressures, 246, 247
  protein, 249-253
  spread, 246-255
  viscosity, 240-242
Surface tension
  measurement, 229-232
  protein solutions, 237
  solutions, 15, 16, 236-239
  table values, 233
  temperature effects, 233
  time effects, 237
  water, 233
Suspensions
  coagulation, 390
  electrical conductance, 308, 309
Svedberg unit of sedimentation, 364
Swelling pressure of gels, 428
Szyszkowski equation, 15, 16

TACTOIDS, 426
Tangentmeter, 10
Temperature
  absolute (Kelvin), 34
  centigrade, 34
  definition, 32, 33
  fiber stress, 434, 435, 442
  free energy, 46
  ionization of acids, influence on, 110, 111
  jump, 399
  measurement, 34, 35
  protein denaturation, effect on, 46, 47, 150
  reaction rates, effect on, 390-392
Tetramethylsilane as NMR reference, 210
Thermistors, 34, 35
Thermocouples (Thermopile), 34
Thermodynamics
  biological systems, 63, 64
  defined systems, 41
  diffusion, 295, 296
  first law, 41
  membrane penetration, 292
  second law, 42
  statistical, 43
  survey of, 40
  third law, 48
Thixotropy, 427, 428
Tiselius electrophoretic method, 334
Titration curves, 106, 107, 123, 124
Tobacco mosaic virus, 156, 157, 426
Transducers in tissue, 57
Transition state theory, 392-394
Translational kinetic energy, 33
Transmission coefficient, 393
Transport numbers, 305, 338
Transverse waves, velocity of, 431
Tricarboxylic acid cycle, 56
Trigonometric functions, 16-21
Tris, 115
Tropocollagen, 441, 442
Trypsin
  activity as function of pH, 412, 413
  isoelectric point, 350
Trypsinogen, isoelectric point, 350
Tryptophane, absorption spectra, 219, 220
Turbidity, 197
Turbulent flow, 263, 264
Turn-over-number, enzymic reactions, 405
Turn-over-rate, isotopes in animals, 388
Tyndall cone, 195
Tyrosine, absorption spectra, 219

ULTRACENTRIFUGE, 362-369
  Archibald method, 368, 369
  blood plasma, 364, 365
  concentration gradients, 367
  diagrammatic representation, 363
  equilibrium method, 365-369
  partial specific volumes, 369, 370
  protein hydration, 365, 366
  rate sedimentation, 363-365
Ultra-violet spectra
  aromatic amino acids, 219, 220
  nucleic acids, 221, 222
  proteins, 219
Unit cell, 452
Urea
  diffusion coefficient, 279
  ionization of acetic acid in, 308
  osmotic coefficient, 181
  protein binding, 164
  urease action on, 415
Ussing's equation for ion flux ratios, 320, 321
Ussing's short-circuit of membranes, 321-323

VALONIA macrophysa, ion accumulation by, 321
Van der Waal forces, 139, 357
Van't Hoff equation, 46
Vapor pressure
  ideal, 158, 159
  measurement, 161, 162, 165, 166
Vaporization, heat of, 165
Versene, 87
Virial coefficient. See *Interaction constant.*
Viscosity, 261-275
  coefficient, 261
  DNA, 272, 274
  electroviscous effect, 355, 356
  elastic element, 433
  expressions for, 266, 269
  hydration of solute, 273
  kinematic, 261
  liquids (Table), 262
  measurement, 265-267
  molecular weight determination, 271, 272
  Newtonian, 261, 273, 274
  particle asymmetry, 270-272
  poly electrolytes, 355
  poly-L-glutamate, 271
  protein solutions, 268, 272, 298, 299
  silver bromide, 356
  structural, 275
  surface films, 242
  suspensions, 268-271
Visible spectrum, 185, 218-220
Vision, 225, 226
Voigt element, elasticity, 433

WATER
  binding by nucleic acid, 163

Water *(continued)*
  binding by proteins, 162-164
  biological importance, 84, 85
  in blood and tissue, 77-79
  clathrate crystals, 83
  dielectric constant, 71
  dipole moment, 77
  flow through membranes, 180
  hydrogen bonds in, 79, 80
  ice structure, 78, 79
  in ionic solutions, 82, 83
  paraffin
    adhesion to, 138, 139
    interaction with, 83
  penetration into cells, 180
  projected molecular area, 78
  protein denaturation, relation to, 149
  rate of ionization, 399
  specific heat, 36
  structure, 77-79, 83
  surface tension, 233
  vapor pressure of solutions, 161
  viscosity, 262
Wave
  length, 18, 185
  number, 185
  plate, 188, 189
  compression, velocity of, 433
  transverse, velocity of, 431
Wetting balance, 236
Wheatstone bridge, 304, 305
Wiegand-Snyder equation, 435
Wilhelmy slide, method for surface tension, 232
Work
  adhesion, 234
  biological, 56, 57
  definition, 38
  electrostatic charging, 70, 71, 74
  gaseous compression, 13
  heat equivalency, 38, 39
  osmotic, 39, 40

X-RAY diffraction, 450-452

YEAST dehydrogenase, binding by DPN and DPNH, 414

ZETA potential, 328-330
  role in coagulation, 346, 347
Zimm
  plot, 204
  viscometer, 267, 268
Zone electrophoresis, 341, 342
Zwitter ions, 123